BIOLOGY OF APPLES AND PEARS

Biology of Apples and Pears is a comprehensive reference book on all aspects of pomology at the organ, tree and orchard level. It provides detailed information on propagation, root and shoot growth, rootstock effects, canopy development in relation to orchard design, flowering, pollination, fruit set, fruit growth, fruit quality factors and quality retention in store. It also deals with mineral nutrition, water relations and irrigation, diseases and pests, and biotechnology. The book emphasizes the scientific basis of modern tree and orchard management, and fruit storage. It describes key cultivar differences and their physiology and genetics, and environmental effects and cultivar × environment interactions in tropical and subtropical as well as temperate zone conditions. It is written for fruit growers, extension workers, plant breeders, biotechnologists and storage and crop protection specialists as well as for researchers and students of pomology and horticulture.

JOHN JACKSON has a wealth of experience in the science of fruit growing. He held senior positions at East Malling Research Station and was chairman of the Orchard and Plantation Systems (High Density Planting) working group of the International Society for Horticultural Science for many years. Recently he was in charge of the Horticultural Research Centre, Marondera, Zimbabwe. He has visited fruit growing areas and lectured to fruit growers and scientists in 15 countries, spanning four continents. He has published extensively in the primary literature and has contributed to, edited or co-edited numerous volumes on various aspects of fruit trees and plantation systems.

BIOLOGY OF HORTICULTURAL CROPS

Existing texts in horticultural science tend to cover a wide range of topics at a relatively superficial level, while more specific information on individual crop species is dispersed widely in the literature. To address this imbalance, the *Biology of Horticultural Crops* presents a series of concise texts, each devoted to discussing the biology of an important horticultural crop species in detail. Key topics such as evolution, morphology, anatomy, physiology and genetics are considered for each crop species, with the aim of increasing understanding and providing a sound scientific basis for improvements in commercial crop production. Volumes to be published in the series will cover the grapevine, citrus fruit, bananas, apples and pears, and stone fruit.

The original concept for this series was the idea of Professor Michael Mullins, who identified the topics to be covered and acted as General Editor from 1983 until his untimely death in 1990.

Biology of the Grapevine, Michael G. Mullins, Alain Bouquet and Larry E. Williams

Biology of Citrus, Pinhas Spiegel-Roy and Eliezer E. Goldschmidt

BIOLOGY OF APPLES AND PEARS

John E. Jackson

CAMBRIDGE
UNIVERSITY PRESS

CAMBRIDGE UNIVERSITY PRESS
Cambridge, New York, Melbourne, Madrid, Cape Town, Singapore, São Paulo

Cambridge University Press
The Edinburgh Building, Cambridge CB2 2RU, UK

Published in the United States of America by Cambridge University Press, New York

www.cambridge.org
Information on this title: www.cambridge.org/9780521380188

First published 2003
This digitally printed first paperback version 2005

A catalogue record for this publication is available from the British Library

Library of Congress Cataloguing in Publication data

Jackson, John E., 1934–
Biology of apples and pears / John E. Jackson.
p. cm. – (Biology of horticultural crops)
Includes bibliographical references (p.).
ISBN 0 521 38018 9
1. Apples. 2. Pears. I. Title. II. Series.
SB363 .J33 2003 634′.11 – dc21 2002031549

ISBN-13 978-0-521-38018-8 hardback
ISBN-10 0-521-38018-9 hardback

ISBN-13 978-0-521-02105-0 paperback
ISBN-10 0-521-02105-7 paperback

Contents

Preface

Biology of Apples and Pears is written for undergraduates and postgraduate students of horticultural science, for scientists from other disciplines who need a core reference book on these crops, and for fruit growers and technical advisers.

There is already a vast amount of published work on apple and pear biology and production. However, much of this literature is difficult to interpret and apply in the context of rapid changes in areas of production – hence environmental constraints – and of cultivars with basic differences in physiology. This book addresses these issues directly by emphasising responses to environment, and cultural and storage technology at the cultivar level.

Biology of Apples and Pears deals with the biology of their eating quality and its retention as well as of their tree growth and cropping. It also emphasises the factors underlying the dramatic change from orchards of large trees to modern high-density orchards of dwarfed trees, and also those underlying modern techniques of fruit tree irrigation and nutrition.

Throughout the book the results of research on apple and pear biology are linked to the relevant current concepts in more basic seciences. The numerous references, therefore, cover both crop-specific and basic-science research papers and reviews.

Given the breadth of coverage, from anatomy and histology through physiology to biotechnology and disease, pest and environmental stress resistance there must inevitably be questions of relative emphasis. The guiding principle followed has been to concentrate on those aspects of the biology currently most important to the production of apple and pear fruits through to the point of consumption.

Acknowledgements

Special thanks are due to Horticultural Research International, East Malling for allowing access to their library after my retirement and to Mrs Sarah Loat, Librarian, for her help. C.J. Atkinson, E. Billings, P.S. Blake, D.L. Davies, N.A. Hipps, D.J. James, D.S. Johnson, J.D. Quinlan, J.R. Stow, K.R. Tobutt, R. Watkins, A.D. Webster and other previous colleagues gave generously of their time for discussion, but are not in any way responsible for any errors of weeknesses in their text. M. Blanke, R. Harrison-Murray, S.E. Horton, B.H. Howard, K. Hrotkó, J.S. Kamboj, A.N. Lakso, H.A. Quamme, S. Sansavini and J. N. Wünsche were particularly helpful with respect to sourcing text figures and tables. The co-operation of the Japanese Academic Association for Copyright Clearance is gratefully acknowledged. Mrs Molly Kurwakumire typed the initial text and Mrs Barbara Jackson subsequent modifications and correspondence.

The continued support of Dr Maria Murphy and colleagues at Cambridge University Press is gratefully acknowledged and the book is dedicated to the scientists upon whose work it is based.

Introduction

The special characteristics of apple and pear production: setting the scene for their scientific study

Apples and pears are among the oldest of the world's fruit crops, figuring in both the Bible and the tales of Homer. They are by far the most important of the deciduous tree fruits, are widely grown in temperate and, increasingly, in tropical regions, and figure prominently in world trade.

The fruits of apples and pears are primarily grown for the fresh fruit market, which is much more remunerative than that for processing. O'Rourke (1994) noted that in the United States a thousand tons of apples qualifying for fresh sale would, on average, generate more than three times the revenue of a thousand tons sold for juice and that even within the fresh fruit category the most desirable fruits may sell for three or four times the price of the least desirable fruits. Moreover, apples and pears for fresh consumption, and also even in some processed forms, are marketed by cultivar name to a much greater extent than has been traditional for other fresh fruits and vegetables, and the different cultivars command different prices. The culture of apples and pears is, therefore, directed towards the production of fruits of named cultivars and to the production of fruits of high perceived quality within each cultivar. In general the cultivars do not come true-to-type when grown from seed and the necessary uniformity is achieved by clonal propagation. Fruit flavour, texture, size and colour are all characteristics of the cultivar but can be influenced by tree nutrition and pruning and by fruit thinning.

Market demand extends throughout the year although any one cultivar in a particular region has only a limited harvesting season, excepting some relatively minor tropical production. To meet year-round demand the season of availability is extended in several ways.

1 Growing a range of cultivars with different harvesting seasons, from early to late, in the same orchard or locality. This approach is declining in importance in the major areas of commercial production as demand focuses increasingly on a fairly limited number of cultivars.
2 Growing the cultivars which are in greatest demand in areas with different seasons of production. Thus 'Golden Delicious' apple is available throughout the year on English markets by combining northern hemisphere sources, apples from southern France being available earlier than from northern France or the Netherlands, and southern hemisphere sources such as South Africa.
3 The use of long-term storage, which encompasses both the use of low temperature and controlled atmosphere stores to modify the ripening process and also the use of pre-harvest treatments to improve the intrinsic potential storage life of the fruits. Such treatments include those to modify chemical composition so as to maintain cell wall and membrane integrity, and also harvesting the fruits at an appropriate stage of maturity.

The perennial nature of apple and pear trees and their characteristic patterns of flowering and fruiting are also very important.

If seedling fruit trees are planted at a spacing appropriate to their natural size at maturity the yield per unit area of orchard would be low for many years after planting. The trees at maturity would also be so large as to require ladders to prune and pick, making these operations slow and expensive, and would be too shaded in their lower parts to produce fruits of acceptable commercial quality there. Moreover it is difficult and expensive to ensure thorough application of pesticides throughout the canopies of large trees and to carry out effective pest and disease monitoring. A major thrust in orchard system development has therefore been to use dwarfing rootstocks and pruning methods which control tree size at maturity. These enable the trees to be planted at close spacings (high-density planting) so as to achieve high levels of light interception and yield early in the life of the orchard without subsequent overcrowding (Jackson, 1980). Increasing the harvest index, i.e. the proportion of the total dry matter increment which is devoted to fruit, is one key objective of rootstock selection.

A second distinctive feature of perennial fruit trees is that factors in the year or season preceding that of cropping can have major effects on yield. Factors such as crop load and shade within the tree influence the formation of the flower buds that will give the next season's crop. Temperatures during the winter control emergence from dormancy and methods of inducing budbreak in tropical and subtropical climates with mild winters are a key element in successful apple and pear production there (Ruck, 1975; Saure, 1985). Even in temperate regions there are large effects of winter temperatures, prior to any leaf emergence, on fruit set and yield in the following season (Jackson *et al.*, 1983; Lakso, 1994).

A third feature is the generally 'spare capacity' of apple and pear trees. Their potential growth once mature usually exceeds requirement and the number of flower buds is greatly in excess of the number of high quality fruits that can be carried to maturity. Even the root system is highly mutable and can adapt to a very limited soil volume. Efficient tree management may therefore involve varying degrees of pruning or growth regulator use to control tree size at maturity, of pruning and fruit thinning to reduce fruit numbers, and irrigation methods in which water is applied to only part of the orchard surface.

The popularity of apples and pears as fruits, combined with their perishability, has led to two major trends in their production. The first is a concentration in regions of substantial climatic advantage combined with sophisticated technologies of production, storage and transportation. The second, in low-income countries, has been an expansion of production especially for local markets even in climates previously regarded as unsuited to temperate-zone fruits. In each case there has been rapid development based on an understanding of the scientific basis of fruit crop production and development of the relevant technologies. Furthermore, there is now very rapid transfer of fruit-growing technologies to the new areas of production.

Recommended reading

Childers, N.F. (1983). *Modern Fruit Science.* Gainesville, FL: Horticultural Publications.

Westwood, M.N. (1993). *Temperate Zone Pomology, Physiology and Culture.* Portland, OR: Timber Press.

References

Jackson, J.E. (1980). Light interception and utilization by orchard systems. *Horticultural Reviews* **2**, 208–67.

Jackson, J.E., Hamer, P.J.C. and Wickenden, M.F. (1983). Effects of early spring temperatures on the set of fruit of 'Cox's Orange Pippin' apple and year to year variation in its yields. *Acta Horticulturae* **139**, 75–82.

Lakso, A.N. (1994). Apple. In *Handbook of Environmental Physiology of Fruit Crops*, Vol. 1. *Temperate Crops*, ed. B. Schaffer and P.C. Anderson Chapter 2, pp. 3–42. Boca Raton, FL: CRC Press.

O'Rourke, A.D. (1994). *The World Apple Market.* Binghamton, NY: Food Products Press. 237 pp.

Ruck, H.C. (1975). *Deciduous Fruit Tree Cultivars for Tropical and Subtropical Regions.* Horticultural Review No. 3. East Malling, UK: Commonwealth Agricultural Bureaux.

Saure, M.C. (1985). Dormancy release in deciduous fruit trees. *Horticultural Reviews* **7**, 239–87.

1

The growing of apples and pears

The history of apple and pear growing

Apples and pears in the wild and in prehistory

The genus *Malus* has, according to most authorities, 25 to 30 species and several subspecies of so-called crab apples. These species are found in the wild almost continuously throughout temperate Eurasia and North America. The primary centre of diversity appears to be within a region stretching from Asia Minor to the western provinces of China (Janick *et al.*, 1996; Juniper *et al.*, 1999, 2001). Forests of wild apples are still found in this region (Roach, 1985), with fruits ranging from small and unattractive to ones similar to the traditional cultivated eating apples.

There is evidence that the fruits of apples were collected as food by prehistoric man. Carbonized fruits dating from 6500 BC were found at Çatal Hüyük in Anatolia and remains of both sour crab apples and a larger form, which may have been cultivated, were discovered in prehistoric lake dwellings in Switzerland. It seems likely that apples moved with human migration along the Old Silk Roads linking western China with the Near East and Danube valley even in Neolithic and Bronze Age times. These routes passed through Almaten (Alma Ata) in eastern Kazakhstan and the northern slopes of the Tien Shan Mountains, now thought to be the possible centre of origin of the domestic apple (Juniper *et al.*, 2001).

Cultivated pears appear to have arisen from three centres of diversity: a Chinese centre where forms of *Pyrus pyrifolia* and *P. ussuriensis* are grown, a centre in the Caucasus Mountains and Asia Minor where the domesticated forms of *P. communis* arose, and a Central Asian centre where *P. communis* and its hybrids occur (Vavilov, 1951; Bell *et al.*, 1996). Asian or Japanese pears are thought to have been domesticated in prehistoric times from wild *P. pyrifolia* and to have been cultivated in China for at least 3000 years (Lombard and Westwood, 1987).

Apples and pears in antiquity

Improved forms of apples and pears (*P. communis*) were spread through the civilizations of the Fertile Crescent, extending from the hills of Persia and south of the Caspian Sea to Turkey and through Palestine to Egypt. Apple trees apparently reached Palestine in about 2000 BC and feature in the Bible (Authorized King James Version) in the Song of Solomon. From Palestine they were taken to Egypt and apple plantations in the Nile Delta are mentioned in the Third Papyrus of Anastasi in the reign of Rameses II (1298–1235 BC). The Harris Papyrus of the time of Rameses III (1198–1166 BC) refers to 848 baskets of apples being delivered as offerings to the Temple of Ra in Heliopolis.

Both apples and pears were well known in the world of Ancient Greece. Homer, in the *Odyssey* written between 900 and 800 BC refers to a large orchard with both apples and pears; and Theophrastus spoke of the difficulty of propagating apples from cuttings, budding and grafting being the generally accepted methods.

Apple culture was well developed in the Roman empire. Columella described cleft- and rind-grafting and also a technique of propagation practically identical with modern patch-budding. Pliny described the apple as having the highest value among fruits. He noted that fruit cultivation was a very profitable enterprise, provided that the orchards were near to a town where the fruits could be sold, and that fruit cultivation in the villages near Rome was more profitable than any other form of farming. The Roman Varro (116–27 BC) described apple storage and the construction of an apple store so as to keep it cool and well ventilated. All the Roman writers included the names of a number of apple cultivars. Pears seem to have been favoured more for cooking and the Romans had many named cultivars, some already of considerable antiquity.

The earliest written record of cultivated Asian pear groves in Japan is in the manuscript of the Emperor Jito in AD 693 (Kajiura, 1994).

Apples and pears in medieval and pre-industrial times

Charlemagne, the Frankish king who was crowned Holy Roman Emperor in AD 800, introduced a law which laid down that crown lands in every city of the Empire should have a garden planted with herbs and fruits. The fruits included apples and pears. Over the ensuing centuries, both in England and throughout western Europe, the monasteries became major centres for apple and pear production. In England these monasteries were, from 1100, under the direction of Norman-French bishops and abbots and had the management not only of adjoining properties but also of lands allocated to them by the King. In Kent, in 1086, almost half of the entire county was owned by

Christ Church and St Augustine's priories at Canterbury. They grew apples and pears for eating and cooking and also apples for cider. The sale of cider was recorded at Battle Abbey in Sussex in 1275 and cider production was recorded in Yorkshire at around the same time. In the South Tyrol, also at around this time, apples were grown in the gardens of monasteries, castles and rural settlements to supply local markets, e.g. the Obstplatz (fruit square) in Bolzano (Oberhofer, 1981). In England two cultivars, the 'Pearmain' and the 'Costard' were grown extensively in the thirteenth century and there are records of apples and pears and their rootstocks being bought and sold. Pears were much planted in medieval England, with new cultivars brought over from the La Rochelle area of France. King Henry VIII ordered the importation of graftwood of the best available cultivars in 1533 and Richard Harris, Fruiterer to the King, imported many apple cultivars, especially pippins, from France and pear graftwood from the Low Countries (Netherlands). Subsequent to this Walloon refugees settled in Sandwich in Kent. Some of these Dutch settlers then moved to northwest Kent and Surrey to establish market gardens to supply London and later established apple and pear orchards for this purpose.

In the sixteenth century the use of dwarfing 'Paradise' rootstocks was described for the first time in Europe, by Ruellius in 1536. This 'French Paradise' probably had originated in Armenia as a form of *Malus pumila* or a *M. pumila* × *M. sylvestris* hybrid. 'Paradise' apple trees were grown and used as rootstocks to control the vigour of cultivars grafted on them in England in the late sixteenth and early seventeenth centuries. The first 'Paradise' rootstocks had been brought over from France but they were subsequently propagated in English nurseries.

The introduction of apple and pear culture in North and South America, South Africa, Australia and New Zealand accompanied European settlement. The first documented apple orchard in what is now the USA was planted near Boston in 1625, and this was almost certainly preceded by plantings in Latin America.

Apple and pear production in the modern era

The distinctively modern era, from the perspective of apple and pear production, dates from the development of cheap and rapid long-distance transportation by steamship, railway train and truck. It is characterized by the development of science-based technologies to improve the productivity of fruit trees in a wide range of environments and to enable apples and pears to be stored in good condition for many months.

Prior to these developments apple and pear growing for market was predominantly in areas close to large towns and cities, such as in villages near to Rome in the time of Pliny, and in Kent near to London in the Middle Ages and subsequently. There was some international trade, e.g. over the Brenner Pass from areas south of the Alps and over the Channel from France to England, but the distances involved and the bulks transported were limited.

APPLE PRODUCTION

In 1867 the railway from Bozen (Bolzano) in what is now Italy was opened across the Alps, connecting the South Tyrol with many populous cities and towns in central Europe (Oberhofer, 1981). Subsequently there was a major increase in apple exports from what is now the Alto Adige region of northern Italy and by far the greater proportion of fruit grown there is now exported, especially to Germany and Austria.

Apple growing in the United States followed the extension of settlement westwards. Climatic conditions differed markedly from those in Great Britain and there was much practical trial and error to select appropriate cultivars. The native crab apples were largely discounted even as a source of breeding material and large numbers of seedlings were imported, especially for the production of cider. As populations moved into areas with colder winters many northern European cultivars were introduced and by 1872 more than a thousand cultivars were listed. Production in the eastern and mid-western states of the USA centred on New York, Michigan and the Shenandoah Valley area of Pennsylvania, West Virginia and Virginia, all within easy reach of major markets. Apple planting in Washington State, remote from large centres of population, began early in the nineteenth century primarily to supply apples for the settlers themselves. Commercial orchards were planted extensively near Yakima by the 1890s (Marshall and Steigmeyer, 1995). Most of the plantings were along the banks of the Columbia River and its major tributaries, the Okanagan, Snake, Wenatchee and Yakima rivers. Steamboats on the Columbia were a major means of freight, rail service became available from Wenatchee town in 1893 and car-lot shipping started in 1901. Further extension of the railway and heavy promotion of apple planting led to a rapid increase in production and by 1920 Washington had become the leading state for apple production in the USA. Expansion continued and in the 1990s there were extensive new plantings on virgin land in the central Columbia River Basin. In 1994 Washington State produced 2.540 million tons of apples out of a total production in the USA of 4.909 million tons, i.e. more than 50% of the US total. It should be noted that this production is totally dependent on irrigation, average annual rainfall in Wenatchee and Yakima being only 252 and 202 mm respectively (Elfving, 1997). The industry is dependent on markets outside the state, international exports rising to 33% of total shipments by 1994. Two thirds of the production is of the cv. 'Red Delicious', followed by 'Golden Delicious', 'Granny Smith', 'Gala' and 'Fuji'.

The apple industries of Australia, New Zealand and South Africa underwent major expansion in the early twentieth century to take advantage of the

Table 1.1 *Annual apple production by country 1948/50–1996 (1000 MT)*

Rank		Year					
(1996)	Country	1948/50	1960	1970	1980	1990	1996
1	China	133	298	801	2383	4332	16009
2	USA	2521	2231	2902	4000	4381	4773
3	Former USSR	1033	1950	5046	5090	5800	4716
4	France	3615	5693	4423	2930	2346	2445
5	Turkey	109	208	748	1430	1900	2100
6	Iran	39	50	89	800	1501	2000
7	Italy	642	1834	2062	1937	2050	1940
8	Poland	141	627	691	844	812	1700
9	Germany	1191	3142	1975	2395	4147	1594
10	India	85	180	277	658	1093	1200
11	Argentina	155	431	445	958	980	1147
12	Japan	366	876	1021	960	1053	963
13	Chile	86	101	140	245	632	910
14	Spain	184	262	484	931	657	811
15	Korean DPR	69	87	115	460	645	660
16	Brazil	6	10	15	42	543	653
17	Mexico	46	64	146	262	376	645
18	Korean Rep.	46	104	212	410	629	630
19	Pakistan	3	3	33	107	240	600
20	Canada	325	304	406	553	540	560
21	Belg./Lux.	271	197	252	327	223	515
22	South Africa	41	150	222	426	530	500
23	Netherlands	253	384	450	470	431	490
24	New Zealand	43	72	140	211	361	480
25	Hungary	121	290	661	1017	945	475
26	Romania	57	111	176	414	683	457
27	Egypt	2	4	4	14	62	455
28	Austria	369	702	309	330	338	368
29	Morocco	2	3	14	27	300	360
30	UK	561	686	596	357	311	350
Total	1–30	12 515	21 054	24 855	30 988	38 841	50 506
Total	World	14 576	23 497	27 502	34 104	40 518	53 672

Data from FAO (1987, 1997).

opportunities offered by the English market. Indeed research in England on apple rootstock breeding and on fruit storage during this period was partly to support production in and shipment from these southern hemisphere countries.

Over the period 1948/50, when annual world production averaged 14 576 000 tons, France was the leading producer of apples, followed by, in order, the USA, Germany, the [then] USSR, Italy and the United Kingdom (Table 1.1). These six countries between them produced 65% of the total world output. In 1996, by comparison, China was the leading producer followed by the USA, the former USSR, France, Turkey, Iran and Italy, between them producing 59% of a world output of 53 672 000 metric tons (MT). In 2000 these were still the top six producing countries, but production in China had risen to 22.89 million MT out of a global total of 60.64 million MT (Belrose, 2001).

From 1948/50 to 1996 production in northwestern and central Europe (France, Germany, Belgium/Luxembourg, the Netherlands, Austria and the UK) was relatively constant, although that in the Netherlands increased and that in the UK declined. Output from Italy increased up to 1960 and then changed very little while production in Poland, Spain, Hungary and Romania increased substantially.

The modern era of apple production in Japan began in the 1870s, when more than 100 cultivars of apple were introduced from western countries and distributed by the government, mainly to prefectures in northern Japan (Fukuda, 1994). These replaced the poor crab-apple type of apples called 'Waringo' (*M. asiatica* Nakai) which had been grown in Japan from the fifteenth century. The cvs. 'Ralls Janet' and 'Jonathan' dominated until the 1960s when they were replaced by 'Fuji', which was bred in Japan, and by 'Delicious' strains, the latter then being largely replaced by newer, mainly locally-bred, cultivars.

Total production of apples in Japan remained relatively constant from 1960 onwards but elsewhere in Asia there were dramatic increases in production. In western Asia, production in Turkey increased steadily over the period, from 109 000 MT on average in 1948/50 to 2 100 000 MT in 1996, while that in Iran increased very rapidly from the 1970s onwards. In India expansion in production was fairly steady, while in Pakistan most of it came after 1990. In east Asia the increase in production in the two Koreas came well before that in China, where it was much greater in the 1990s than in any earlier period. Indeed, the increase in production in China was the dominant feature of world apple growing in the 1990s. The largest production was in Shandong province, across the Yellow Sea from South Korea and southern Japan, followed by Liaoning to the north, Hubei in central China, Hunan, Shaanxi and Shanxi (Shou-Chun, 1998). As has typically been the case where there have been large planned increases in production in 'new' areas, there was emphasis on production of new cultivars with high consumer demand. By the late 1990s approximately 50% of the apples grown in China were of the Japanese cv. 'Fuji', which returned twice the price per kilogram compared with the previously dominant 'Red Delicious'. Although the main market was within the country, apples were exported to Far East Russia, and to Hong Kong and Singapore from which they were re-exported to the rest of Asia.

In North America expansion of production was relatively moderate and steady in Canada and the USA, doubling between 1948/50 and 1996, whereas it increased 14-fold over the same period in Mexico with the greatest growth in absolute terms in the 1990s.

In South America most of the increase in production in Argentina was in the period up to 1980, whereas in Chile output continued to increase rapidly in the 1980s and 1990s and in Brazil the rapid increase in production was virtually confined to those two decades. Planting in Brazil was mainly in Santa Caterina province in the south, with emphasis on the relatively new cvs. 'Gala' and 'Fuji'.

In South Africa production increased 12-fold over the period 1948/50 to 1996, the rate of increase being fairly steady, while growth in apple production in Egypt and Morocco was much more rapid in the 1990s than previously. It is notable that in 1996 production in Egypt and Morocco together considerably exceeded that in South Africa.

Production in New Zealand expanded steadily, increasing 10-fold over the period considered, with a pattern of change very similar to those in the other southern hemisphere countries, Chile and South Africa, where apples are also grown mainly for export.

The figures given in Table 1.1 show dominance of a relatively few countries in apple production, but the concentration of output in relatively small favoured areas is even more striking. More than half of the total output of apples in the USA is from a small number of river valleys in Washington State, where the fruit trees are grown in irrigated semi-desert country in the rain shadow of the Cascade Mountains (Figure 1.1). Similarly, Italy is the world's seventh largest producer of apples and more than 40% of its production (Sansavini, 1990) and half of its exports (Oberhofer, 1981) are from a very small area of the Trentino and Alto Adige provinces. In Poland, another major producer, the area around Grojec, about 50 miles south of Warsaw, has one of the greatest apple orchard concentrations in Europe, accounting for 35% of Poland's fresh apple production in 1993 (Florkowski *et al.*, 1996).

PEAR PRODUCTION

Pear production, although on a much smaller scale than apple production, has followed a similar pattern of expansion in recent years. China was the leading country in 1996, having overtaken Italy by 1980. Italy, Spain, Argentina and Japan are all relatively higher-ranked for pear than for apple production although they each produce more apples than pears. Production in western and central mainland Europe (Italy, Spain, Germany, France, Belgium/Luxembourg, the Netherlands, Switzerland, Portugal and Austria) gave 46% of the world's output over the years 1961/65 but despite a slight rise in output by 1996 produced only 21% of the world output in that year. This relative decline was because of major increases in output in China, Argentina, Turkey, Chile, South Africa, Iran and others (Table 1.2).

There is a similar concentration of production within specific areas, as is found with apple. More than 70% of Italian output of pears comes from the

(*a*)

(*b*)

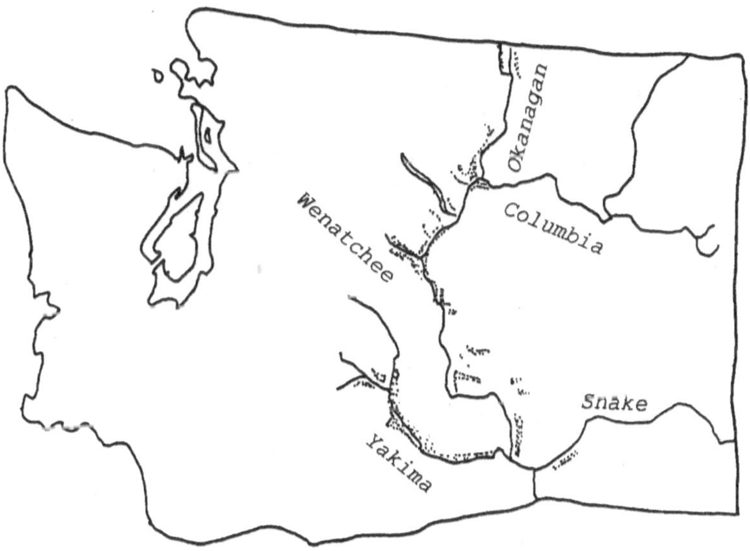

Figure 1.1 Concentration of apple production. More than half of all production in the USA is in Washington State (W on map(*a*)) and within that state is largely confined to the valleys of the Columbia, Okanagan, Snake, Wenatchee and Yakima rivers in the stippled areas on map (*b*). Map (*b*) redrawn from Marshall and Steigmeyer (1995).

Table 1.2 *Annual pear production of leading countries 1961/65–1996 (1000 MT)*

Rank (1996)	Country	Year 1961/65	1969/71	1980	1990	1996
1	China	826	1203	1645	2483	5615
2	Italy	934	1749	1318	968	937
3	USA	557	612	814	874	707
4	Spain	147	295	437	449	584
5	Argentina	96	90	155	210	513
6	Former USSR	N.A.	538	610	500	435
7	Japan	330	466	496	443	426
8	Turkey	141	172	330	413	410
9	Germany	499	536	452	380	370
10	France	410	551	445	331	350
11	Chile	20	31	39	140	250
12	South Africa	63	94	133	203	220
13	Iran	24	29	45	125	184
14	Korean Rep.	30	49	60	159	163
15	Australia	124	162	124	171	160
16	Belg./Lux.	55	73	76	59	158
17	Netherlands	111	120	115	90	130
18	India	40	52	64	105	130
19	Korean DPR	10	22	65	115	125
20	Switzerland	167	155	101	86	100
21	Egypt	12	17	51	75	95
22	Greece	96	112	111	51	90
23	Portugal	52	43	59	94	81
24	Austria	236	158	126	100	78
Total	1–24	4980	7330	7871	8624	12 311
Total	World	5655	8232	8726	9509	13 093

Data from FAO (1977, 1982, 1992, 1997).

lowlands of Emilio Romagna and Veneto (Sansavini, 1990). More than 35 % of the output of pears in the USA comes from the irrigated valleys of Washington State, which overtook California in pear production in the mid-1990s (USDA, 1995–96).

In Japan there were more than 1000 named cultivars of Nashi (Japanese or Asian pears) by 1860 and commercial production began around the capital. This was boosted in about 1895 with the introduction of two high-quality chance seedlings ('Nijisseiki' and 'Chojura'), and the development of the railway network enabled Nashi fruits to be transported to the big cities from distant agricultural regions such as the Tottori prefecture (Kajiura, 1994). 'Chojura' remained the dominant cultivar until 1971 when it was replaced by the newly-bred cvs. 'Kosui' and 'Hosui'. In 1992 'Kosui', 'Nijisseiki' and 'Hosui' were cultivated on 36.1 %, 21.8% and 21.2%, respectively, of the area growing Nashi

Table 1.3 *Apple exports. Exports in 1000 MT from the 12 leading apple exporters*

Country and rank in 1987	Exports 1987	1988	1989	Country and rank in 1995	Exports 1993	1994	1995
France (1)	762	707	681	France (1)	641	654	768
Hungary (2)	396	341	318	USA (2)	525	739	635
Italy (3)	347	340	280	Italy (3)	422	414	499
Chile (4)	331	347	326	Chile (4)	361	347	433
USA (5)	228	306	276	Netherlands (5)	366	401	412
Argentina (6)	202	208	215	Belg./Lux. (6)	276	271	368
Netherlands (7)	192	200	200	New Zealand (7)	225	201	302
South Africa (8)	190	221	225	Argentina (8)	145	147	243
New Zealand (9)	166	120	174	South Africa (9)	175	245	214
Belg./Lux. (10)	138	144	156	Iran (10)	216	190	190
Poland (11)	81	85	80	Poland (11)	175	115	139
China (12)	60	88	65	China (12)	119	107	109

Data from FAO (1990, 1996).

pears in Japan. 'Kosui' is an early-season pear and 'Nijisseiki' and 'Hosui' are mid-season pears.

Trade in fresh apples

Between 1980 and 1993 imports, which are a measure of between-country trade, averaged between 8% and 9% of production in most years (Belrose, 1996), i.e. most apples were consumed in the country of production. However, exports of apples are very important to a number of national economies and also, to at least some extent, provide an indication of the countries with comparative advantages in production. Exports are shown in Table 1.3.

Most exports are to nearby countries. In 1986 the major destinations of apples exported were as shown in Table 1.4. Exports from France, Italy, Poland, the USA, the Netherlands, Argentina and Belgium/Luxembourg were all to neighbouring countries; only those from Chile, New Zealand and South Africa were to distant countries. In these latter cases the need for long-term storage during transport is self-evident, though this is of course also true for apples sold within the country of production. The high level of exports from the Netherlands and Belgium-Luxembourg to some extent involves re-export of imported fruits, but in 1995 Dutch exports of 411 812 MT greatly exceeded imports of 284 851 MT and the corresponding figures for Belgium-Luxembourg were 368 337 MT of exports and 228 132 MT of imports (FAO, 1996). Imports into Poland in 1995 were only 18% of exports, no imports were recorded for

Table 1.4 *The major destinations of fresh apples from the top ten exporters in 1986*

Exporter	Rank	Major destinations
France	1	United Kingdom
Chile	2	Netherlands
Italy	3	West Germany
New Zealand	4	Belgium/Luxembourg
Poland	5	USSR
United States	6	Canada
South Africa	7	United Kingdom
Netherlands	8	West Germany
Argentina	9	Brazil
Belgium/Luxembourg	10	West Germany

From Vidyashankara and Wilson (1989). Credit Washington State University IMPACT Program.

Chile, Iran or South Africa, and those into New Zealand were only three hundredths of 1 % of exports.

Trade in pears

Whereas in 1987 France and Italy were ranked first and third, in terms of exports they were only ninth and seventh respectively in 1995, while Argentina, South Africa, Chile and the USA had improved their relative positions (Table 1.5).

Apple and pear products

Apples and pears have many uses: fresh fruits, fresh fruit juice, concentrated fruit juice, cider and perry, 'pop wine', and various canned and dried fruit products.

Fresh fruits are by far the most important in terms of total apple consumption (Table 1.6). They are even more so in terms of value: in the USA over the period 1962–72 the prices growers received for fresh apples averaged more than twice those received for apples for processing. Within the processing sector higher prices were paid for large apples suitable for peeling for canning and freezing; dried apples, from sound fruit, achieved about 80% of the canning and peeling price and apples for pressing for juice and cider averaged 60% of the canning and freezing price (Greig and Blakeslee, 1975). Prices of apples for

Table 1.5 *Pear exports. Exports in 1000 MT from the 12 leading pear exporters*

Country and rank in 1987	Exports 1987	1988	1989	Country and rank in 1995	Exports 1993	1994	1995
France (1)	117	70	94	Argentina (1)	142	142	222
Argentina (2)	97	116	142	South Africa (2)	115	99	150
Italy (3)	74	83	79	Belg./Lux. (3)	103	140	148
Netherlands (4)	70	90	87	Chile (4)	147	157	147
South Africa (5)	69	80	52	USA (5)	124	162	146
China (6)	53	127	59	Netherlands (6)	102	158	143
Spain (7)	49	32	43	Italy (7)	172	127	113
Chile (8)	45	63	76	China (8)	69	97	91
Belg./Lux. (9)	40	56	53	France (9)	50	82	80
USA (10)	39	54	83	Spain (10)	45	71	69
Australia (11)	36	40	21	Australia (11)	31	28	26
Japan (12)	13	12	9	Hong Kong (12)	18	21	23

Data from FAO (1990, 1992, 1996).

Table 1.6 *Major uses of apples in selected regions, 1989–90 (1000 MT)*

	Fresh consumption	Fresh exports	Processing	Withdrawal
EC	6414	1384	1521	276
USA	2229	340	1946	0
Turkey	1673	84	92	0
Argentina	223	225	620	0
South Africa	205	222	133	0

Data from O'Rourke (1994).

processing are declining and in the USA in 2000 were only about a quarter of the fresh apple price (Belrose, 2001).

Under many, perhaps most, circumstances the apples which are used in processing do not cover their full share of production costs. In Washington State in 1990 a typical orchard might produce 20 tons of apples per acre (50 t ha^{-1}) at a total cost of \$4000, i.e. \$200 per ton. The orchard would typically yield 15 tons of apples sold for fresh consumption at \$309.44 per ton, 2 tons of apples suitable for canning or peeling at \$108.47 per ton and 3 tons of apples suitable for juice only at \$69.94 per ton (Hinman *et al.*, 1992). Although the fruit sold for processing does not appear to meet its cost of production, as long as the price per ton exceeds the costs of harvesting and transporting to the processor this processing fruit makes a net contribution to the grower's income because it does not affect his pre-harvest costs.

Fresh apples and pears are, as mentioned earlier, sold in categories, classes or grades which reflect perceived quality. The quality criteria are specific to types (e.g. red, partially coloured or green apples) and even to individual cultivars. In general larger fruits are preferred to smaller ones, bright red colour is preferred in red cultivars, and downgrading results from any surface blemish whether this is the result of pest or disease attack or a physiological or environmental cause. This quality grading has very important consequences in terms of tree management. Large fruits develop when fruits and leaves are well exposed to sunlight and fruit-to-fruit competition for assimilates is reduced by fruitlet thinning. The anthocyanin pigment which gives fruit skin its red colour is only formed, in most cultivars, under the direct influence of exposure to sunlight. Downgrading can thus result from fruit development under shady conditions. When, however, light intensities are very high, inadequate shade can lead to sunburn, which can result in the fruit being unsuitable for the fresh market.

Apple juice is the second most important apple product. The apples are ground, pressed and filtered to remove skins and pulp. The juice is then pasteurized. It may be sold as such, in un-concentrated form, or concentrated to give a 6 to 1 concentrate, i.e. a concentrate which is reconstituted to apple juice by adding six parts of water to one of concentrate. The ability to concentrate apple juice has greatly reduced its transportation costs and made it possible to ship it economically from areas of production to distant markets. In 1985 Argentina processed more than 40% of its apple production into juice concentrate and exported 97% of this to the USA (O'Rourke, 1994). Apple juice is the cheapest of all fruit juices: as well as being sold as such it is therefore also widely used in fruit juice blends which are marketed under the name of the other ingredient. In England and in Canada apples are fermented to make an alcoholic drink called cider (most American cider is non-alcoholic juice). In England this is produced from special cultivars giving characteristic aroma and flavour. Conventional apple juice can also be used as the sugar source for producing 'pop wines' as an alternative to using the more expensive grape juice.

The second most important processed apple product is apple sauce. The apples are peeled, cored and trimmed, chopped and cooked with sugar. The cooked sauce is then filtered, and water and sugar automatically added to ensure consistency of product prior to canning. Other products include dried apple and apple chips.

Pears are primarily grown for the fresh market and for canning. One cultivar, 'Bartlett', dominates the canning market. In 1995, out of a total US production of 944 250 tons of pears, 493 000 tons were of 'Bartlett' grown in the states of California, Washington and Oregon of which 390 040 tons (79%) was processed (USDA, 1995–96). Over the ten years 1986–95 an average of

51 % of the utilized pear production was for the fresh market, the proportion being greater in the more recent years. 'Bartlett' used in processing, mainly grown under contract for canning, averaged about 67 % of the 'grower price' of 'Bartlett' for fresh consumption in the USA over the years 1985 and 1990–98. Processed pears of other cultivars tend to be used in juice and other low value products and achieved only about 23 % of their fresh fruit price (Belrose, 2000).

Climatic conditions in major centres of production

It is clearly impossible to give any narrow description of the climatic conditions required for successful apple and pear production. Apples and pears are grown for export, i.e. to an internationally competitive level, in climates as diverse as those of the Netherlands and South Africa. They can be grown with varying degrees of success throughout the temperate zone and, increasingly, in the subtropics and even the tropics. Successful production depends both on the climate, especially the local microclimate, and on effective adaptations to this in terms of cultivar selection and cultural practice. This is discussed in detail later.

Table 1.7 illustrates the range of temperatures in which apples and pears are successfully produced. The dominant climatic constraints range from winter-freeze damage in Poland to inadequate winter chilling for fruit bud development of most cultivars in Egypt, and summer heat stress and fruit sunburn in Washington State. It should be noted that whereas the means of the daily minima and maxima (daily means) are useful in defining growing conditions, the means of the monthly minima and maxima (monthly means) are much more informative as the levels of potentially limiting stress factors which occur on a regular basis.

At a more subtle level, temperature effects on fruit set, fruit size, fruit colour and fruit shape determine where the leading apple cultivar, 'Red Delicious', is best produced.

Table 1.8 shows the incoming solar irradiation in a number of apple and pear growing regions. At higher irradiation levels a greater depth of canopy can receive light at any given level so potential photosynthesis of the canopy is increased.

Irrigation is practised in most of the major regions of apple and pear production. Rainfall is probably adverse in most areas, leading to increased incidence of fungal diseases, especially apple scab, and bacterial diseases, especially fire blight, on pears. The high cloud cover associated with rainfall is also a major factor in reducing the available solar radiation in many fruit-growing areas.

Table 1.7 *Daily and monthly mean maximum and minimum temperatures (° C) in different apple growing areas*

	Jan	Feb	Mar	Apr	May	Jun	Jul	Aug	Sep	Oct	Nov	Dec
1. *Netherlands*												
Daily max	5.0	5.1	8.5	11.9	15.9	19.0	20.6	20.9	18.7	14.1	9.4	6.3
Daily min	1.0	0.6	2.5	5.4	8.7	12.0	14.1	14.4	12.6	8.9	5.2	2.4
Monthly max	10.2	10.6	15.3	19.4	24.1	26.8	27.6	27.5	25.0	19.1	13.9	11.4
Monthly min	−5.0	−5.7	−2.5	0.9	3.6	7.5	10.0	10.5	7.9	2.8	−0.3	−4.0
2. *Poland*												
Daily max	−0.4	0.3	5.6	11.8	19.6	22.6	24.1	22.9	19.0	12.9	5.6	1.9
Daily min	−5.5	−6.0	−2.0	2.9	9.0	12.3	14.6	13.5	9.5	5.1	0.7	−2.5
Monthly max	6.7	8.3	14.5	21.3	28.4	30.1	31.2	29.6	27.9	21.0	13.0	8.5
Monthly min	−17.2	−16.3	−8.7	−2.7	2.6	6.3	10.1	9.0	3.1	−1.1	−5.5	−11.6
3. *Italy (1)*												
Daily max	4.8	8.7	14.5	18.8	23.0	26.0	28.2	27.3	24.5	18.8	10.6	5.6
Daily min	−4.5	−1.4	3.0	7.4	10.8	14.2	15.8	14.9	12.3	6.6	1.4	−2.7
Monthly max	11.5	15.9	21.9	26.4	29.4	32.1	33.6	33.1	29.5	25.1	17.1	12.1
Monthly min	−10.0	−6.4	−2.3	1.9	4.8	8.8	11.0	9.8	6.6	−0.4	−4.6	−8.5
4. *Italy (2)*												
Daily max	5.1	8.4	13.5	18.4	23.2	27.2	30.1	29.9	26.0	19.6	18.8	7.3
Daily min	−0.6	1.5	5.4	9.5	13.7	17.5	19.7	19.3	16.5	11.7	6.3	1.9
Monthly max	11.7	15.9	20.6	24.7	29.5	33.3	35.3	35.0	30.7	25.5	18.3	14.6
Monthly min	−6.0	−3.5	−0.3	4.1	7.8	12.0	14.4	14.8	11.4	6.4	0.8	−4.6

5. *USA*												
Daily max	2.5	7.1	12.9	18.8	23.4	26.7	31.6	30.3	26.0	18.7	8.9	4.3
Daily min	−7.5	−4.9	−1.8	1.8	6.0	9.3	11.7	10.3	6.6	1.9	−2.9	−4.8
Monthly max	12.1	14.9	20.4	27.4	32.6	35.4	38.8	36.8	33.8	26.6	16.8	13.2
Monthly min	−15.5	−11.9	−6.6	−2.3	0.8	4.9	7.8	6.8	2.0	−2.9	−7.9	−10.9
6. *Egypt*												
Daily max	19.4	21.7	25.6	30.6	34.4	36.1	36.7	36.1	33.9	30.6	26.1	21.1
Daily min	4.4	6.1	8.9	12.2	17.2	18.9	20.0	20.6	18.3	15.6	11.7	6.1
Monthly max	26.1	30.0	34.4	39.4	43.3	42.2	41.7	41.1	39.4	36.1	32.2	27.2
Monthly min	1.1	1.7	4.4	7.8	11.7	15.6	17.2	17.8	15.5	11.7	7.2	2.2

Sites: 1. Vlissingen, Netherlands. 51° 28′ N, 3° 35′ E, 1 m altitude
2. Warsaw, Poland. 52° 13′ N, 21° 03′ E, 110 m
3. Bolzano, Italy. 46° 30′ N, 11° 21′ E, 271 m
4. Bologna, Italy. 44° 30′ N, 11° 21′ E, 60 m
5. Yakima, USA. 46° 34′ N, 120° 32′ W, 323 m
6. Cairo, Egypt. 29° 52′ N, 31° 20′ E, 116 m

Data from Meteorological Office (1980). The figures for Cairo are taken from the 1958 edition.

Table 1.8 *Incoming global radiation during a 5-month growing season at weather stations in major fruit-producing regions*

Location	Year	Radiation ($GJ\ m^{-2}$)
Davis, California	LT	4.13
Avignon, France	1971	3.23
Riwaka, New Zealand	LT	3.20
Ithaca, New York	1987	2.73
Wilhelminadorp, Netherlands	1986–89	2.50

LT, long-term average data.
From Wagenmakers (1995). Reproduced with permission.

Recommended reading

Janson, H.F. (1996). *Pomona's Harvest.* An illustrated chronicle of antiquarian fruit literature. Portland, OR: Timber Press.

References

Bell, R.L., Quamme, H.A., Layne, R.E.C. and Skirvin, R.M. (1996). Pears. In *Fruit Breeding*, Vol. 1. *Tree and Tropical Fruits*, ed. J. Janick and J.N. Moore, pp. 441–514. New York: John Wiley and Sons.

Belrose (1996). *World Apple Review 1996*. Pullman, WA: Belrose Inc.

Belrose (2000). *World Pear Review 2000*. Pullman, WA: Belrose Inc.

Belrose (2001). *World Apple Review 2001*. Pullman, WA: Belrose Inc.

Elfving, D.C. (1997). Trends in the Washington State tree-fruit industry. *Acta Horticulturae* **451**, 31–43.

FAO (1977). *FAO Production Yearbook*, Vol. 30. 1976. Rome: UN Food and Agriculture Organization.

FAO (1982). *FAO Production Yearbook*, Vol. 35. 1981. Rome: FAO.

FAO (1987). *World Crop and Livestock Statistics* 1948–1985. Rome: FAO.

FAO (1990). *FAO Trade Yearbook*, Vol. 43. 1989. Rome: FAO.

FAO (1992). *FAO Production Yearbook*, Vol. 45. 1991. Rome: FAO.

FAO (1996). *FAO Trade Yearbook*, Vol. 49. 1995. Rome: FAO.

FAO (1997). *FAO Production Yearbook*, Vol. 50. 1996. Rome: FAO.

Florkowski, W.J., Majewski, E. and Bagel, M. (1996). Opportunities in production and marketing of integrated fruit production (IFP) certified apples in Poland. *Journal of Tree Fruit Production* **1** (2), 15–31.

Fukuda, H. (1994). Apple. In *Horticulture in Japan*, ed. K. Konishi, S. Iwahori, H. Kitagawa and T. Yakuwa, pp. 23–7. Tokyo: Asakura Publishing Co. Ltd,.

Greig, W.S. and Blakeslee, L.L. (1975). *Potential for apple juice processing in the US with implications for Washington.* Pullman: Washington State University College of Agriculture Research Center Bulletin 808.

Hinman, H.R., Tvergyak, P., Peterson, B. and Clements, M. (1992). *1992 estimated cost of producing 'Red Delicious' apples in Central Washington.* Pullman,

WA: Washington State University Cooperative Extension, Farm Business Management Report, E.B. 1720.
Janick, J., Cummins, J.N., Brown, S.K. and Hemmat, M. (1996). Apples. In *Fruit Breeding*, Vol. 1. *Tree and Tropical Fruits*, ed. J. Janick and J.N. Moore, pp. 1–77. New York: John Wiley and Sons.
Juniper, B.E., Robinson, J., Harris, S.A. and Watkins, R. (2001). Origin of the apple (*Malus domestica* Borkh.). In *Encyclopedia of Genetics*, ed. E.C.R. Reeve, pp. 674–7. London: Fitzroy Dearborn.
Juniper, B.E., Watkins, R. and Harris, S.A. (1999). The origin of the apple. *Acta Horticulturae* **484**, 27–33.
Kajiura, I. (1994). Nashi (Japanese pear). In *Horticulture in Japan*, ed. K. Konishi, S. Iwahori, H. Kitagawa and T. Yakuwa, pp. 40–7. Tokyo: Asakura Publishing Co. Ltd.
Lombard, P.B. and Westwood, M.N. (1987). Pear rootstocks. In *Rootstocks for Fruit Crops*, ed. R.C. Rom and R.F. Carlson, pp. 145–83. New York: John Wiley and Sons.
Marshall, J. and Steigmeyer, R. (1995). *Washington Apple Country.* Portland, OR: Graphic Arts Center Publishing.
Meteorological Office (1980). *Tables of temperature, relative humidity, precipitation and sunshine for the world*, Parts I, III and IV. London: HMSO.
Oberhofer, H. (1981). Fruit growing in South Tyrol. *Acta Horticulturae* **114**, 23–42.
O'Rourke, A.D. (1994). *The World Apple Market.* New York: Food Products Press.
Roach, F.A. (1985). *Cultivated Fruits of Britain. Their Origin and History.* Oxford: Blackwell.
Sansavini, S. (1990). The fruit industry in Italy. In *Discovering Italy's Horticulture*, ed. S. Sansavini, pp. 10–13. XXIII International Horticultural Congress. Florence: Societa Orticola Italiana.
Shou-Chun, Q. (1998). Cited in: China pushes Fuji into overseas markets. *Asia fruit.* January/February 1998, p. 22.
USDA (1995–96). Agricultural Statistics 1995–96. Washington, DC: US Government Printing Office.
Vavilov, N.I. (1951). *The Origin, Variation, Immunity and Breeding of Cultivated Plants.* New York: Ronald Press and Waltham, MA: Chronica Botanica.
Vidyashankara, S. and Wilson, W.W. (1989). *World trade in apples 1962–1987.* Pullman, WA: Washington State University, IMPACT Center, Information Series No 31.
Wagenmakers, P.S. (1995). Light relations in orchard systems. Thesis, Wageningen.

2

Apples and pears and their relatives

Taxonomy

Apples and pears belong to the *Rosaceae*, subfamily *Pomoideae*, the pome fruits. Other members of this subfamily include quince and medlar. The flowers of the *Rosaceae* are actinomorphic with 5 sepals, 5 petals, numerous stamens and either one compound pistil or many simple pistils. The number of styles equals the number of carpels. The *Pomoideae* have mixed flower buds containing both leaf and flower initials, an epigynous ovary and 2–5 carpels. The *Pomoideae* have a basic chromosome number of 17 compared with 7–9 for the other subfamilies of *Rosaceae*.

The *Malus* (apple) inflorescence is determinate but the descriptive terminology is disputed (Pratt, 1988). It is variously described as a corymb, a corymbose raceme, a cyme and a false cyme. *Pyrus* (pear) inflorescences have been described as umbel-like simple corymbs (Clapham, Tutin and Warburg, 1952) and as racemes (Bell *et al.* 1996). *Cydonia* (quince) flowers are solitary.

The typical apple flower (see also Chapter 9) has 5 petals, varying from white to deep pink, 5 sepals, 20 stamens in three whorls (10 + 5 + 5) with yellow anthers, and a pistil which divides into five styles united at the base. The ovary has 5 locules, each usually containing 2 ovules giving a maximum seed content of 10 although some cultivars may have up to 30 (Janick *et al.*, 1996). *P. communis* flowers typically have 5, usually white, petals, 5 sepals and 20–30 stamens with red or purple anthers. The 2–5 carpels are completely united with each other and with the receptacle and there are 2 ovules per locule giving a maximum seed number of 10. The 2–5 styles are free although closely constricted at the base. *Cydonia* flowers have styles united at the base, 5 carpels and numerous ovules per locule.

The fruits of *Pomoideae* are pomes, commonly described as having a core with fleshy pith within it and a cortex of flesh outside the core line. There are two hypotheses as to the nature of these tissues, the receptacular and the appendicular (Pratt, 1988).

The apple or pear tree grown today is almost invariably a compound tree consisting of a fruiting scion grafted or budded on to a rootstock. In some cases it is made up of three components: the scion, the rootstock and an interstock (interstem). This latter may be used where the scion and the rootstock will not form a strong graft union with each other, i.e. are incompatible, but will each unite with the interstock. It is also used to achieve control of scion vigour by the influence of an interstock cultivar which is too difficult to propagate for it to be used as a rootstock.

The apple scion cultivars in commerce are classified as *Malus* × *domestica* Borkh. (Korban and Skirvin 1984). This species was considered to have originated as the result of interspecific hybridization, its main ancestor being *M. sieversii* with contributions from *M. orientalis*, *M. sylvestris*, *M. baccata*, *M. mandshurica* and *M. prunifolia*. Recent work, however, suggests that *M. sieversii* was the sole genetic ancestor of traditional European eating apples (Juniper *et al.*, 2001).

Resistance to apple mildew is being introduced from *M.* × *robusta*, *M.* × *zumi* and *M. hupahensis* and to apple scab from *M.* × *floribunda*. Resistance to apple scab and tolerance to very low winter temperatures are also being introduced from *M. baccata*.

Roach (1985) considered that the original 'Paradise' rootstocks, from which the 'Malling' ('M.') clonal rootstocks were selected, were forms of *M. pumila*. Seedlings of crab apple, *M. sylvestris*, were also commonly used as rootstocks in Europe. *Malus* × *robusta* has also been used because of its tolerance of cold winters. *M. sieboldii* (Regel) Rehder, *M. baccata* (L.) Borkh and *M. prunifolia* Borkh were generally used in Japan with a pendulous sport of the latter being predominant (Fukuda, 1994). Additionally *M. baccata* (L.) Borkh and *M.* × *micromalus* Makino were used in breeding programmes, each being crossed with 'M.9' (*M. pumila*) to produce rootstocks which are now commercially available (Fischer, 1997).

Most *Malus* species intercross and, given the length of time over which apple trees of different sorts have been selected, it is difficult to be certain of the authenticity of specific names. A modern list of sections, primary species and species hybrids (Janick *et al.*, 1996) is presented in Table 2.1. Fischer (1997) and R. Watkins (personal communication) consider *M. floribunda* and *M. micromalus* to be hybrids and therefore better designated as *M.* × *floribunda* and *M.* × *micromalus*.

The pear cultivars grown in Europe and much of the 'western' world are mainly of the species *Pyrus communis* L., possibly with a degree of hybridization with *P. korschinskyi* (which may be simply a synonym) and *P. heterophylla*. In Japan, China and other temperate Far Eastern countries the pear cultivars are derived from *P. pyrifolia* (Burm.) Nakai (*P. serotina* Rehder). In the English language these have been referred to as 'Sand pears', 'Asian pears' and 'Japanese pears', but

Table 2.1 *Malus sections and primary species*

Sections	$2n$	Apomixis	Fruit size (cm diam)	Calyx[a]	Carpel no.	Persistence of ripe fruit
SECTION I. Malus						
Subsection A. Pumilae						
Series a. Pumilae						
M. asiatica Nakai	34	No	>2	P	5	No
M. domestica Borkh.	34, 51, 68	No	>2	P	5	No
M. micromalus Makino	34	No	1	D	–	–
M. orientalis Uglitzk. ex Juz.	–	–	–	–	–	–
M. prunifolia (Willd.) Borkh.	34	No	2	P	5	Yes
M. pumila Miller	34	No	>2	P	5	No
M. sierversii (Lodeb.) M. Roemer	–	–	>2	P	5	No
M. spectabilis (Aiton) Borkh.	34, 68	No	2	P	5	Yes
M. sylvestris Miller	34	No	>2	P	5	No
Series b. Baccatae						
M. baccata (L.) Borkh.	34, 68	No	1	D	4, 5	Yes
M. floribunda (Siebold) ex van Houtte	34	No	1	D	4, 5	Yes
M. halliana Koehne	34	No	1	D	3, 4, 5	Yes
M. hupehensis (Pampan.) Rehder	51	Yes	1	D	3, 4	Yes
M. mandshurica (Maxim.) V. Komarov	34	No	1	D	5	Yes
M. sikkimensis (Wengzig) Koehne ex C. Schneider	51	Yes	1	D	4, 5	Yes

Taxon						
Subsection B. Sieboldianae						
M. sargentii Rehder	68	Yes	1	D	3, 4, 5	Yes
M. sieboldii (Regel) Rehder	34–85?	?	1	D	3, 4, 5	Yes
Subsection C. Kansuenses						
Series a. Kansuenses						
M. fusca (Raf.) C. Schneider	34	No	1	D	2, 3, 4	Yes
M. kansuensis (Batalin) C. Schneider	–	No	1	D	3, 4, 5	Yes
M. komarovii (Sarg.) Rehder	–	–	1	D	–	Yes
M. toringoides (Rehder) Hughes	51	Yes	1	D	4, 5	Yes
M. transitoria (Batalin) C. Schneider	–	–	1	D	5	Yes
Series b. Yunnanenses						
M. honanensis Rehder	–	–	1.5	P	4	Yes
M. ombrophila Hand.-Mazz	–	–	>2	P	5	No?
M. prattii (Hemsley) C. Schneider	34	No	1.5	P	5	Yes
M. yunnanensis (Franchet) C. Schneider	34	No	1.5	P	5	Yes
SECTION II. Sorbomalus						
M. florentina (Zuccagni) C. Schneider	34	No	1	P, D	3, 4, 5	Yes
SECTION III. Eriolobus						
M. trilobata (Poiret) C. Schneider	–	No	2	P	5	–
SECTION IV. Chloromeles						
M. angustifolia (Aiton) Michaux	34	No	>2	P	5	No
M. coronaria (L.) Miller	51(68)	Yes?	>2	P	5	No
M. ioensis (Alph. Wood) Britton	34	No	>2	P	5	No

(*cont.*)

Table 2.1 (*cont.*)

Sections	2n	Apomixis	Fruit size (cm diam)	Calyx[a]	Carpel no.	Persistence of ripe fruit
SECTION V. Docyniopsis						
M. doumeri (Boiss.) A. Chev.	–	No	>2	P	5	No
M. melliana (Hand.-Mazz) Rehder	–	–	>2	P	5	No
M. tschonoskii (Maxim.) C. Schneider	34	No	>2	P	5	No

Note. *Malus* species hybrids (secondary species) include

M. × *adstringens* Zabel (*baccata* × *pumila*)
M. × *arnoldiana* (Rehder) Sarg. (*baccata* × *floribunda*)
M. × *astracanica* Dum.-Cours. (*pumila* × *prunifolia*)
M. × *atrosanguinea* (Spaeth) Schneid. (*halliana* × *sieboldii*)
M. × *dawsonia* Rehder (*fusca* × *pumila*)
M. × *gloriosa* Lem (*pumila niedzwetskyana* × *scheideckeri*)
M. × *hartwiggii* Koehne (*halliana* × *baccata*)
M. × *heterophylla* Spach (*coronaria* × *pumila*)
M. × *magdeburgensis* Schoch. (*spectabilis* × *pumila*)
M. × *platycarpa* Rehder (*coronaria* × *domestica*)
M. × *purpurea* (Barbier) Rehder (*niedzwetzkyana* × *atrosanguinea*)
M. × *robusta* (Carr.) Rehder (*baccata* × *prunifolia*)
M. × *scheideckeri* (Spaeth) Zabel (*floribunda* × *prunifolia*)
M. × *soulardii* (Bailey) Britton (*ioensis* × *pumila*)
M. × *sublobata* (Dipp.) Rehder (*prunifolia* × *sieboldii*)
M. × *zumi* (Mats.) Rehder (*mandschurica* × *sieboldii*)

[a] P, persistent; D, deciduous.
Source: Adapted from Way *et al.* (1990).

with the increasing globalization of their production the name 'Nashi', which is Japanese for a pear grown in Japan, is now preferred (Kajiura, 1994). It is possible that the 'Nashi' may have some genetic influence from other wild pears, e.g. *P. hondoensis* Kik. & Nakai, and from the Chinese white pear, *P.* × *bretschneideri* Rehder (Kajiura, 1994). In northern China and Japan the much more cold-tolerant *P. ussuriensis* Maxim., hybrids of *P. ussuriensis* and *P. pyrifolia* and *P.* × *bretschneideri* Rehder are grown, and in warmer areas of southern China (Pieniazek, 1966) and northern India (Mukherjee *et al.*, 1969) selections of *P. pashia* D. Don are cultivated.

The main clonal rootstocks for *P. communis* are clonal selections of quince (*Cydonia oblonga* L.). Quince has been used as a rootstock for pears for centuries, different clones inducing differing degrees of dwarfing. Quince is easy to propagate by layering or by cuttings but is not winter-hardy in areas of very cold winters, is intolerant of calcareous soils and is incompatible with some important cultivars such as 'Bartlett' ('Williams'') although it can be used for these with an interstock of, for example, 'Beurré Hardy' or 'Old Home'. *Pyrus* clonal selections have been made but most are difficult to propagate vegetatively. Seedlings of the main scion cultivars of *P. communis* are very widely used as rootstocks, especially in North America.

The most commonly used rootstocks for Nashi pears in Japan are seedlings of *P. pyrifolia* with some use also of *P. dimorphophylla* and *P. betulifolia* (Kajiura, 1994). In south China *P. calleryana* is used and in northern China *P. betulifolia* and *P. ussuriensis* (Lombard and Westwood, 1987); *P. pashia* is used in northern India and southern China (Bell *et al.*, 1996). The distribution of *Pyrus* species is given in Table 2.2.

The place of cultivars in apple and pear production

Historically apples and pears have been high-value crops which can repay considerable expenditure on orchard and post-harvest management practices if these improve yield, quality and marketability. There has also been ready movement of germplasm, and development and transference of technologies of production, over the centuries as well as in recent times. In modern times there has been a concentration of production in climatically-favoured areas following the opportunities for nationwide and global marketing based on the development of transport and storage technologies. This latter trend has been accompanied by an enormous decline in the number of cultivars that are grown on a commercial scale. Way *et al.* (1990) reported that whereas more than 10 000 apple scion cultivars are documented, only a few dozen are now of major importance. This trend towards concentration on a relatively few

Table 2.2 *Pyrus species*

Species	Distribution	Synonyms
EUROPEAN		
P. communis L.	West to SE Europe, Turkey, Eurasia	*P. asiae-mediae* Popov *P. balansae* Decne. *P. boissieriana* Buhse *P. elata* Rubtzov *P. medvedevii* Rubtzov *P. korshinskyi* Litv. *P. pyraster* Burgsd.
P. caucasia Fed.	SE Europe, Greece, Turkey	*P. communis* L. (?)
P. nivalis Jacq.	West, Central, and Southern Europe	
P. cordata Desv.	SW England, W France, Spain, Portugal	
CIRCUM-MEDITERRANEAN		
P. amygdaliformis Vill.	Mediterranean Europe, Asia Minor	*P. sinaica* Dum.-Cours.
P. elaeagrifolia Pallas	SE Europe, Russia, Turkey	*P. kotschyana* Boiss. ex Decne.
P. syriaca Boiss.	Tunisia	
P. longipes Coss & Dur.	Algeria	*P. cossonnii* Rehder
P. gharbiana Trab.	Morocco, W Algeria	
P. mamorensis Trab.	Morocco	
MID-ASIAN		
P. glabra Boiss.	Iran	
P. salicifoia Pallas	NW Iran, NE Turkey, S Russia	
P. reqelii Rehder	South Central Asia (Afghanistan)	*P. heterophylla* Regel & Schmalh.
P. pashia D. Don.	Pakistan, India, Nepal	*P. kumaoni* Decne. *P. variolosa* Wall. ex G. Don *P. wilhelmii* C. Schneider
EAST ASIAN		
P. pyrifolia (Burm.) Nakai	China, Japan, Korea, Taiwan	*P. serotina* Rehder
P. pseudopashia Yu	NW China (Yunnan, Kweichow)	
P. ussuriensis Maxim.	Siberia, Manchuria, N China, Korea	*P. lindleyi* Rehder *P. ovoidea* Rehder *P. sinesis* Lindley
P. calleryana Decne.	Central & S China, Vietnam	
P. betulaefolia Bunge	Central & N China, S Manchuria	
P. fauriei Scheid	Korea	
P. hondoensis Kik. & Nakai	Japan	
P. dimorphophylla Makino	Japan	
P. kawakamii Hayata	Taiwan, SE China	*P. koehnei* Schneid.

(cont.)

Table 2.2 (*cont.*) *Interspecific hybrids*

Species	Parentage	Distribution
NATURALLY OCCURRING		
P. × *bretschneideri* Rehder	*P. ussuriensis* × *P. betulifolia*	Northern China
P. × *phaeocarpa* Rehder	*P. betulifolia* × *P. ussuriensis*	Northern China
P. × *serrulata* Rehder	*P. pyrifolia* × *P. calleryana*	Central China
P. × *compelexa* Rubtzov	Unknown	Caucasus
P. × *salvifolia* DC	*P. communis* × *P. nivalis*	Europe, Crimea
P. × *canescens* Sprach	*P. nivalis* × *P. salicifolia*	–
PROBABLE ARBOTEA OR ARTIFICIAL HYBRIDS		
P. × *lecontei* Rehder	*P. communis* × *P. pyrifolia*	–
P. × *michauxii* Bosc ex Poiret	*P. amygdaliformis* (?) × *P. nivalis*	–
P. × *uyematsuana* Makino	*P. dimorphophylla* × *P. hondoensis*	Korea

Source: Adapted from Rehder (1967), Westwood (1982), Wiersen (personal communication, 1985 and Zeven and Zhukovsky (1975).
From Bell *et al.* (1996). Reprinted by permission of John Wiley & Sons Inc.

cultivars of both scions and rootstocks may or may not continue: there are, indeed, some indications of demand for a greater choice. However, in considering the role of cultivars in apple and pear production the basic background question is to define why there are still different cultivars/cultivar groups rather than a universal cultivar with production methodologies tailored to different environments. Description of the main cultivars, where they are grown and their key differences in terms of environmental requirements and their biology, also assists in the interpretation of the results of horticultural and biological studies using different cultivars as the test plants.

Apple scion cultivars

The statistics of production by cultivar are incomplete and the rate of change is appreciable, but Table 2.3 gives an indication of the balance of production in the mid-1990s. Seven cultivars, 'Delicious' ('Red Delicious'), 'Golden Delicious', 'Fuji', 'Granny Smith', 'Jonagold', 'Gala' and 'Idared' accounted for more than 50% of the total production in those countries with relevant data, the first three alone accounting for more than 40%. The figures probably underestimate the importance of 'Fuji', given the scale of its planting in China, where Shou-Chun (1998) noted that it represented 50% of all apples. They also do not show the importance of cold-resistant cultivars grown in the former USSR, where Way *et al.* (1990) reported that half of the production consisted of 'Common Antonovka', 'Anis', 'Papirovka', 'Koricznoje Polosatoje' and 'Ossiennoje Polosatoje'.

Table 2.3 *Estimates of production of different apple cultivars*

Share of production (%) of the *c*. 80% of world production with available data 1992–93 to 1995–96.

Cultivar	% 1992/93	% 1994/95	% 1995/96
(Red) 'Delicious'	20.46	19.93	19.15
'Golden Delicious'	17.75	15.52	14.72
'Fuji'	4.50	6.46	7.00
'Granny Smith'	5.19	4.49	4.33
'Jonagold'	2.79	2.31	2.23
'Gala'/'Royal Gala'	1.37	2.03	1.96
'Idared'	1.95	1.97	1.76
'Jonathan'	2.74	1.89	1.58
'Rome Beauty'	2.11	1.78	1.60
'McIntosh'	2.07	1.57	1.64
'Elstar'	1.05	0.97	0.91
'Braeburn'	0.34	0.68	0.77
'Cox's Orange Pippin'[a]	1.12	0.66	0.47
Others	36.56	39.74	41.88

[a] In 1995/96 'Cortland' (0.56), 'Reinette' (0.53) and 'Newton' (0.48) all had higher output than 'Cox'.
Data from Belrose (1996, 1997). Reproduced with permission.

Of the cultivars listed in Table 2.3, 'Delicious', 'Golden Delicious', 'Granny Smith', 'Jonathan', 'Rome Beauty', 'McIntosh', 'Braeburn' and 'Cox's Orange Pippin' all originated as chance seedlings. 'Fuji' was bred by crossing 'Ralls Janet' with 'Delicious', 'Jonagold' is a 'Golden Delicious' × 'Jonathan' cross, 'Gala' is a 'Kidd's Orange Red' × 'Golden Delicious' cross, 'Idared' a 'Jonathan' × 'Wagener' cross, and 'Elstar' a 'Golden Delicious' × 'Ingrid Marie' cross. None of them comes 'true-to-type' from seed, so all of them are maintained and multiplied vegetatively before being grafted or budded on to rootstocks. This has not however, precluded selection for 'improved' strains within each of these cultivars. These strains have usually been naturally-occurring bud-sports. A bud-sport is a mutation arising in a cell from which a bud develops, which results in the production of a shoot that differs from the plant on which it was produced, often mainly in one character. The most commonly selected mutations are those which are highly visible. Increases in the amount of anthocyanin in the outer cell layers of the fruit skin give rise to 'red sports' which may have differences in both the surface area which is coloured red and the intensity of the pigmentation (Dayton, 1959). Other sports show a reduced tendency to russet. Yet others are of different growth habit: mutation fairly frequently giving rise to branches of very compact, heavily-spurred habit (i.e. with little lateral vegetative shoot production), which are clothed in fruiting spurs close together. These have given rise to many compact or 'spur-type' clones. In some cases the incidence of mutation has been increased by irradiation and desirable clones produced (Lacey and Campbell, 1987). Some 'sports' also differ from the parent cultivar in time of harvesting and in chilling requirement (Gonzalez-Cepeda, 1992; Hauagge and Cummins, 2000). The data given in Table 2.3 include 'sports' within each cultivar even though many of the 'sports' have been patented and given individual names.

There is little comparative data on changes over time in the global production of different cultivars (O'Rourke, 1994). However, the total production and market share of cultivars introduced after 1960 ('Fuji', 'Gala', 'Jonagold' and 'Elstar'), together with that of 'Braeburn' (discovered in 1952) has increased very rapidly, whereas that of 'Jonathan', 'Rome Beauty' and 'McIntosh' has declined in relative terms. The continued predominance of 'Red Delicious' in large part follows the steady introduction of new strains of this cultivar.

The key attributes of apple cultivars concern those characteristics which give them customer and marketing appeal and those which determine their adaptability to different environments.

Apple cultivar fruit quality

In common with some other tree fruits, but in sharp distinction from many other crops, apples are marketed to the end-purchaser, the consumer, by cultivar name. Surveys in the United States in 1974–75 (O'Rourke, 1994) showed 'Red Delicious', followed by 'Golden Delicious', to be the favourite cultivars but with 'McIntosh' in second place in Pennsylvania. Freedom from bruise or blemish, juiciness and crispness, and firmness were universally regarded as the most important quality factors; others were colour, shape and size. Fukuda (1994) noted that 'Ralls Janet' and 'Jonathan' had been the top two cultivars in Japan until the 'economic takeoff' in the 1960s when consumers shifted their choice to sweeter and larger apple cultivars. As a result these were replaced by 'Delicious' strains and 'Fuji'. 'Delicious' became the top cultivar in the 1970s but then lost popularity because, as grown in Japan, it quickly became mealy and also developed internal breakdown. By 1990 'Fuji' occupied 45% of the planted area and 'Delicious' only 7.4%. O'Rourke (1994) concluded that the relative sales of 'Red Delicious' and 'Granny Smith' indicated that more consumers prefer a red sweet apple than a green and tart one, but regarded these as separate markets.

Fruit flavour is dependent on the combination of acids, sugars, tannins and aromatic compounds, but consumer acceptability is basically determined by the balance between acidity and sweetness. The main sugars are fructose, sucrose and glucose; the acid in the mature fruit is almost all malic acid, although Hong *et al.* (1997) reported 10–20% of citric acid in 'Fuji'. Sweetness and acidity are inherited independently, so all combinations are possible. Table 2.4

Table 2.4 *Cultivar differences in days to maturity, sugar and acid content and fruit firmness in South Africa*

Days to maturity from blossoming to harvest, total soluble solids (TSS) as a refractometer measurement of the juice (%), titratable acid per 100 g of juice, and firmness, measured with an 11.1 mm diameter penetrometer, of apples harvested at the optimum maturity stage.

Cultivar	Days from full bloom to harvest	TSS (%)	Titratable acid (g/100 g)	Firmness (kg)
'Gala'/'Royal Gala'	121	12.1	0.41	7.90
'Golden Delicious'	131	12.9	0.54	7.63
'Starking'	136	11.7	0.31	8.04
'Topred'	135	11.6	0.31	7.90
'Starkrimson'	138	11.3	0.29	7.95
'Granny Smith'	171	11.7	0.71	7.50

Data from van der Merwe (1996a). Reproduced with permission.

shows the range of soluble solids (sugars) and acid in four important cultivars, harvested at the optimum stage of maturity for the maintenance of acceptable market quality under controlled atmosphere or normal refrigerated storage. The very high acidity of 'Granny Smith' and low acidity of 'Delicious' contrast with each other and with the higher sugar levels and intermediate acidity of 'Golden Delicious' and 'Gala'. Some consumers prefer a sweet apple while others prefer a more tart taste (Shewfelt, 1993). Flavour is discussed in detail in Chapter 10.

There are also differences in preference for apples of different sizes which are so great as to influence choice of cultivar. Different cultivars have fruits of different average size and weight, i.e. have different mean and modal values for their fruit size, with a normal distribution of smaller and larger fruits around the mean values. It is possible, indeed it is standard practice, to modify the size distribution by choice of rootstock, pruning and fruit thinning, but it is easier to obtain consistently large fruits of good storage quality from a naturally large-fruited cultivar than from an intrinsically smaller fruited cultivar. In general the production of small apples is not profitable but whereas the English prefer medium-sized apples the Dutch prefer large ones (Combrink and Von Mollendorff, 1996). Table 2.5 shows cultivar differences in fruit size, and sugars and acids, in Germany. Japanese preference for very large apples, weighing 300 g or more per apple, can best be met by cultivars like 'Fuji' which can be stored for more than 6 months in cold storage without developing flesh browning or mealiness even if over 300 g (Fukuda, 1994). In contrast the average fruit weight of 'Cox's Orange Pippin' is around 115 g with a range of 67–163 g (Gotz and Silbereisen, 1989).

Table 2.5 *Fruit quality parameters of apple cultivars grown in Germany*

Cultivar	Average fruit weight (g)	Average sugar (%)	Average acid ($g\ l^{-1}$)
'Golden Delicious'	145	13.5	5.6
'Jonagold'	189	13.2	6.1
'Idared'	154	11.5	6.1
'Jonathan'	125	12.5	7.7
'McIntosh'	125	10.9	5.5
'Elstar'	125	14.0	9.5
'Cox's Orange Pippin'	115	14.0	7.1

Data from Gotz and Silbereisen (1989).

Apple cultivar fruit storage potential and shelf life

Fruit storage life, in low temperature refrigerated stores or in controlled atmosphere (CA) stores, is very important because storage extends the marketing options for fruit growers and packers. A fresh apple left at ambient temperature after harvest continues to respire and steadily loses weight and shrivels. The respiration rate at 0 °C is only one sixth to one tenth as rapid as at 20–21 °C and controlled atmosphere storage, with low oxygen concentrations, reduces the respiration rate even more, so extending the storage life even further. Different cultivars differ in their basic rates of fruit respiration and these relative differences are maintained over a wide temperature range, from 0 °C to ambient. The rate of deterioration both in store and out is governed by equation (2.1) (Findlay and Combrink, 1996).

$$\phi = ke^{-0.13t} \qquad (2.1)$$

where ϕ is storage life in days, k is a constant for the specific cultivar and t is temperature in °C.

In general early-season (summer) apples have higher respiration rates and deteriorate more rapidly than late-season (autumn) apples; also, for example, 'Cox's Orange Pippin' has higher respiration rates and deteriorates more quickly than 'Golden Delicious' or 'Bramley's Seedling'. (Tables 10.3 and 10.4, pp. 352 and 353, give relevant data. The advances in storage technology have to some extent reduced the competitive advantage of some naturally long-storing cultivars. In the early 1950s most 'Red Delicious' had to be sold within four months of harvest, usually at a steadily declining price because of loss of firmness (O'Rourke, 1994). By the early 1980s 'Red Delicious' was being marketed in substantial quantities each month from October to August, and the market for 'Winesap', a naturally much longer storing apple but with less consumer appeal, was greatly reduced.

Storage technology is now cultivar-specific and is designed not just to extend the time for which the apples will remain firm and juicy but also to extend the period over which they retain their flavour and aroma. The characteristic aroma of 'Cox's Orange Pippin', for example, was reduced by ultra-low oxygen (1.25% O_2) storage even though this improved firmness. A brief raising of storage O_2 concentration to 2% was found to maintain aroma without loss of firmness (Smith, 1984). Similar development of 'dynamic' storage regimes has been done to extend the period of storage over which 'Gala' can retain its capability to emit its characteristic aroma volatiles (Mattheis and Fellman, 1997). Such developments are likely to add weight to the concept of flavour being a major discriminant between cultivars.

Apple cultivar yield

High yields, beginning in the early years after planting and continuing on a regular basis without excessive biennial bearing, are an important but not a sufficient criterion to establish the popularity of a cultivar. Many production costs, e.g. those of pruning, pesticide spraying, grass-mowing and herbicide use, are effectively on a per-hectare basis independent of yield as are most farm overhead costs. The cost of production per kilogram therefore rises markedly as yields fall. However, many costs, e.g. storage and transport, are post-harvest and per kg of fruit and, most importantly, the volume of apples which can be sold at any given price depends on customer demand for the particular cultivar, i.e. on customer perceptions of its quality (O'Rourke, 1994). It is notable that the very popular apple cv. 'Red Delicious' is classed as only fair for precocity and productivity (on a scale of poor, moderate, fair, good and very good) and shows a moderate degree of alternate bearing (Westwood, 1993). Much can be done to improve productivity by choice of rootstock, selection of improved clones and development of appropriate cultural practices. Large-fruited cultivars may give higher sustainable yields because the adverse effect of heavy cropping on yield in the following year is largely due to seed and hence fruit numbers, not yield *per se*. Climatic factors may determine which cultivars yield well in different areas (cf. pp. 36–46).

Apple cultivar tree vigour and growth habit

Orchards of large trees are slow to come into cropping and are expensive to manage at maturity. In general this problem is solved by the use of dwarfing rootstocks (see pp. 49–58) and no cultivar has attained importance in commercial fruit growing because of its compact habit or high ratio of yield to tree size. However, there has been very effective selection of compact or spur-type mutants within many of the important cultivars, usually giving trees about two thirds of the size of the original cultivar. Also, compact habit has been the main selection criterion for a particular group of apple cultivars for use in home gardens or even on patios. These are the Columnar apples which have a spur-type habit inherited from the mutant 'Wijcik McIntosh'. These trees bear fruits on spurs along the main central stem, have very few branches and need very little pruning; examples are 'Waltz', 'Polka', 'Bolero' and 'Flamenco'.

Independently of vigour, apple cultivars can be classified into four main morphological types (Lespinasse and Delort, 1986).

1 Spur-types, in which the scaffold branches are conical and tend to develop sub-branches on their lower surface (basitony). The central leader, the trunk, is not particularly dominant in all cultivars. Most fruits are on short spurs on branches two years old or older.
2 'Reine des Reinettes' types, with main branches making wide angles to the trunk, a relatively dominant centre leader but strong growth of lower branches.
3 'Golden Delicious' types, with a strongly dominant leader with wide-angled fruiting branches coming directly off this and numerous short shoots. These trees are well suited to vertical-axis and spindlebush types of tree management.
4 'Rome Beauty' types, which rarely develop lateral shoots in their lower regions but develop overarching and drooping branches in the upper part of the tree, i.e. with an acrotonic tendency. This results in the fruiting zone moving towards the upper and outer zones of the tree.

These types are shown in Figure 6.3 (p. 162).

Disease resistance of apple cultivars

Although there are cases where disease susceptibility has prevented a cultivar being grown in a particular environment, e.g. 'McIntosh' is too susceptible to apple canker to succeed in England, none of the cultivars listed in Table 2.3 is important primarily because of its disease resistance. The 'Delicious' group have a degree of resistance to powdery mildew and fire blight and are very resistant to cedar apple rust, but even they depend on fungicides for the production of blemish-free fruits.

There is, however, increasing interest in foodstuffs produced without the use of chemicals and a number of apple cultivars have been bred for resistance to the major diseases. Apple scab, caused by the fungus *Venturia inaequalis*, is the most serious disease of apples world-wide. Both polygenic and monogenic resistance are available and some new cultivars have shown virtual field immunity, to the extent that orchards of them have produced scab-free fruits for many years without being sprayed. Most resistant cultivars have the *Vf* resistance gene from *Malus* × *floribunda*, perhaps the most important at present being

'Judeline', an apple grown for juice production in France. Other sources of resistance genes are *M.* × *micromalus* (*Vm*), *M. pumila* R12740–7A (*Vr*), *M. baccata* jackii (*Vbj*), *Hansens baccata* H2 (*Vb*) and 'Antonovka' PI 172623 (*Va*). Dessert apples with scab resistance include 'Freedom', 'Liberty', 'Saturn', 'Enterprise', 'Goldrush', 'Topaz', 'Santana' and 'Florina'. Some of these may become important in the future if there is greater consumer demand for 'organically produced' apples.

It is also possible that with the spread of fire blight, caused by the bacterium *Erwinia amylovora*, the degree of resistance to this disease shown by 'Delicious' could increase its importance.

Apple cultivar requirements *re* length of growing season

The length of time which elapses between flowering and fruit maturation is a very important cultivar characteristic. It cannot be modified to any great extent by cultural practice, although it is influenced by rootstock and some plant growth regulators. It is one of the most important factors in determining where the world's major cultivars can be grown.

The actual length of the period from flowering to fruit maturation varies with temperature. Luton and Hamer (1983) showed that, over the years from 1959 to 1980, the length of season for 'Cox's Orange Pippin' in England varied from 127 to 153 days. Much of this variation was associated with variations in temperature, the harvest date being later by 4 days for each 1 °C decrease in mean temperature between 1 May and 3 September. The necessary length of growing season therefore varies from place to place (compare requirements in South Africa, Table 2.4, with those in northern Italy, Table 2.6, and with those mentioned for other areas in the text) but the rankings of different cultivars in terms of earliness are fairly consistent. 'Golden Delicious' can be grown over a particularly wide range of climates. 'McIntosh' fruits require 3–4 weeks less to mature than 'Golden Delicious' and it is still a leading cultivar in the colder, more northerly, apple producing regions such as Canada, the northern USA and Poland. 'Gala', 'Cox' and 'Elstar' mature in about three weeks less than 'Golden Delicious' and can be grown to high quality in England and the Netherlands. The 'Delicious' group can ripen earlier or later than 'Golden Delicious', but their tendency to small fruit size in cool summers limits their production in cool areas. In contrast, 'Jonagold', although harvested as late as or later than 'Delicious', is a large-fruited cultivar which is widely grown in northern Europe especially Belgium and Germany. The long-season group of cultivars, 'Braeburn', 'Fuji' and 'Granny Smith', cannot generally be grown successfully in northern apple areas and this applies even more strongly to ultra-late cultivars. such as 'Pink Lady' and 'Sundowner'. There are, however,

Table 2.6 *Cultivar requirements to reach maturity (days from full bloom to harvest) at Laimburg, Italy*

Cultivar	Date of full bloom (April)	Date of harvest (Aug–Oct)	Days to harvest
'Gala'	16.04	24.08	130
'Elstar'	15.04	26.08	133
'Jonathan'	15.04	26.08	133
'Jonagold'	15.04	11.09	149
'Golden Delicious'	16.04	15.09	152
'Red Delicious'	15.04	16.09	154
'Idared'	12.04	21.09	162
'Rome Beauty'	18.04	4.10	169
'Braeburn'	15.04	4.10	172
'Granny Smith'	17.04	6.10	172
'Fuji'	16.04	17.10	184

Data supplied by S. Sansavini.

some northern areas with especially favourable microclimates where some of these can be produced.

Short-season cultivars often have a wide climatic tolerance. 'Cox' is grown as an early-season cultivar in New Zealand and 'Gala' is an early-season cultivar in parts of France. In addition, very short-season apples are grown as 'early' apples even in northern climates, e.g. 'Discovery' which is harvested in mid-August in England. These 'early' apples usually do not store well and most are of poor quality compared with main-season apples.

Apple cultivars for mild winter areas

There are many areas with high incident radiation, and with the dry summers which reduce the severity of fungal disease attack, which would be very suitable for apple production but have too little chilling in winter to give good budbreak of most cultivars. This problem can be dealt with to some extent by chemical treatments to induce budbreak and these are described later. However, the use of cultivars with little requirement for winter chilling is beginning to have a major impact. For example, large-scale production of apples in Egypt is based on the growing of 'Anna' and 'Dorsett Golden', which can produce heavy crops in climates with few hours below 7.2 °C.

Apple cultivars for cold-winter areas

Apple production in a number of areas with continental climates is dependent on the use of cultivars which can withstand very low winter temperatures. 'McIntosh' and 'Antonovka' are outstanding for winter hardiness.

Major commercial cultivars of apple

'Delicious' ('Red Delicious') is still the most popular apple in the world although its share of total production has been declining. It was discovered as a seedling tree growing in an orchard in Iowa, USA in 1872 and introduced commercially by Stark Brothers in 1895. The archetypal 'Delicious' is produced in Washington State and its fruits are of the highest quality when grown in this area, or places with similar climate, with high light intensity and warm daytime temperatures coupled with cool nights. These result in good red colour development and an elongated shape (Westwood, 1978). Other major producers of 'Delicious' are China, Italy, Chile, France, Argentina, Spain and Canada. The climate of northern Europe is not ideal because it tends to produce small fruits in cool summers (Bultitude, 1983). There are many distinct clones of 'Delicious', selected from 'bud-sports' which possess advantages in terms of compactness of habit and fruitfulness (spur-types), the extent, deepness and attractiveness of red coloration of their skin and season of harvest. Many of these clones are registered as separate cultivars under plant breeder's rights or plant patent schemes. 'Delicious' has a high chilling requirement but some lower-chill mutants have been found in Mexico (Hauagge and Cummins, 2000).

The original 'Delicious' was a moderately vigorous, upright-spreading and round-headed tree which produced spurs fairly freely. The fruits had a greenish-yellow, becoming golden-yellow, ground cover with a dark crimson flush and short, broken, dark red stripes. They had a very characteristic oblong-waisted shape, flattened at the base, distinctly ribbed and five-crowned at the apex. They were of medium size (69 mm diameter) with very firm, fine-textured juicy and sweet flesh and a tough skin. 'Starking Delicious', discovered as a bud-mutation of 'Delicious' in 1921, is similar to 'Delicious' but with more red skin colour. 'Starkspur Supreme', discovered as a bud-mutation of 'Starking Delicious' in 1967, has bright, cherry-red fruit skin, is harvested about a week after 'Delicious' and its trees are medium-small, upright and winter-hardy spur-types. 'Red Chief', a bud-mutation of 'Starkrimson Delicious' (itself a spur-type, solid-blushed variant of 'Delicious'), was discovered in 1967 and is notable for its very large as well as brightly coloured fruits. These few examples of the many strains of 'Delicious' illustrate the way in which the 'Delicious' apple group encompasses a range from those strains which give medium-sized fruits to those which give large fruits, from those with partially red-coloured skin to those that are red all over and from those with a spreading branch habit to those that are compact and heavily spurred. With modern storage technology the fruits can be kept for 10 months or more.

The 'Delicious' group is susceptible to apple scab, which is a problem in humid regions but less so in dry and hot production areas such as Washington

State, resistant to apple mildew and fire blight so that control is only needed under high disease pressure, and very resistant to cedar apple rust. It is fairly susceptible to winter injury.

'Golden Delicious' originated from a chance seedling found in West Virginia, USA in 1890. It is a dominant cultivar in France, Italy, Spain and the Netherlands and is important in the USA, South Africa and many other countries. The fruits are of medium size (63 mm diameter), have a greenish-yellow ground cover, becoming golden-yellow at maturity although they are sold as pale green fruits on some markets. The flesh is crisp, fine-textured, juicy and sweet with a little acid and a good flavour. A major problem (for some markets only) is a tendency to russeting of the skin. The bulk of the European production is on vigour-controlling rootstocks, especially 'M.9' (see pp. 68–9) so there is relatively less production on spur-type compact clones than is the case with 'Delicious'. Clonal selection has emphasized a reduced tendency to russet, e.g. 'Smoothee', a whole-tree mutant discovered in 1958 and 'Goldenir' ('Lysgolden') which is a radiation mutant of 'Golden Delicious' introduced in France in 1977. 'Goldenir' shows little or no russet, has a tougher, smoother and waxier skin than 'Golden Delicious', very similar flesh and flavour characteristics but better resistance to bruising. It is harvested 10–15 days after 'Golden Delicious' and has a longer storage life (ASHS, 1997). Some clones of 'Golden Delicious' with a lower chilling requirement than the standard type have been identified. 'Golden Delicious' can be stored for about 10 months but has poorer storage characteristics than 'Granny Smith'. It is susceptible to apple scab, fire blight and powdery mildew and very susceptible to cedar apple rust.

'Fuji' was bred in Japan, being a selection made in 1958 from fruit-bearing hybrids of 'Ralls Janet' × 'Delicious' and registered in 1962 (Yoshida *et al.*, 1995). 'Fuji' fruits are medium to large in size, firm, crisp and very sweet with a sugar content of about 17% (Brix 17) when grown in China (Shou-Chun, 1998) and very juicy. The original strain was not highly red-coloured but many red colour sports have been identified and introduced. The trees are large, spreading, vigorous and productive. 'Fuji' is the leading cultivar in Japan and in China. It is increasingly planted in Washington State and California in the USA, in Brazil and in South Africa. In Brazil (Bernardi, 1988) it flowers about a week earlier than 'Starkrimson Delicious', 'Golden Delicious' and 'Granny Smith', which probably indicates a rather lower chilling requirement (Jackson and Bepete, 1995), but ripens three weeks later than 'Golden Delicious' and five days later than 'Starkrimson', but almost a month earlier than 'Granny Smith'. A bud-mutation of 'Fuji', 'Fuji Frey', needs only about 450 chilling units (Hauagge and Cummins, 2000). In Japan, in the north Main Island where 'Delicious' is harvested between 10 and 20 October, 'Fuji' is harvested between 1 and 20 November (Fukuda, 1994). It therefore requires a longer growing season than 'Delicious' and is generally suited to rather warmer climates than 'Delicious' or, especially, 'Golden Delicious'. The fruits of 'Fuji' can be stored successfully for 6 months (187 days) in controlled atmosphere storage or for 8 months at 0 °C in low pressure storage (Yoshida *et al.*, 1995). In Japan 'Fuji' is picked after watercore develops in the flesh since Japanese consumers enjoy the flavour of watercored apples. 'Fuji' foliage is susceptible to Alternaria spot (*Alternaria mali*), apple scab and fire blight.

'Granny Smith' was raised from chance seeds thought to be from an open-pollinated French Crab tree in Australia and fruited in 1868. The fruits are of medium size, round-conical in shape and with a bright green ground colour becoming greenish-yellow at maturity. The fruit flesh is firm, rather coarse-textured, subacid and refreshing. It requires a long growing season. It is the leading apple in South Africa and is important in other southern hemisphere exporting countries, and is also planted in southern Europe and in Washington State and California in the USA. It cannot be grown successfully in the United Kingdom or other northern European countries. Although it requires a warm and long growing season it has a high winter-chilling requirement so is unsuited to regions with very mild winters. 'Granny Smith' apples have outstanding storage potential. Under conditions where (Red) 'Delicious' and 'Golden Delicious' will keep for 9 months 'Granny Smith' will keep for 11 months (van der Merwe, 1996b). 'Granny Smith' is moderately resistant to cedar apple rust, susceptible to apple scab and very susceptible to fire blight and powdery mildew.

'Jonagold' is the product of a cross between 'Golden Delicious' and 'Jonathan' in 1943 at Geneva, NY, USA. It was introduced in 1968. The fruits are large (77 mm diameter), with red striping over 50–80% of the surface and a yellow ground colour. The fruit flesh is firm, subacid and juicy with excellent quality both for eating and processing. The trees are similar to those of 'Golden Delicious' but rather more spreading. In England it flowers two days before 'Golden Delicious' and is harvested about a week earlier. It is a triploid and so cannot be used to pollinate other cultivars. Its intrinsically large fruits and early ripening (compared with 'Delicious', 'Fuji' and 'Granny Smith') enable it to be grown successfully in relatively cool apple producing areas like those of northern Europe. Several sports of 'Jonagold' with a greater extent and depth of red skin colour (e.g. 'Jonagored', 'Jonica' and 'Jored') have been selected. The fruits keep well in storage for about 6 months (ASHS, 1997) or 40 weeks in 1.25% oxygen (Stow, 1987). It is susceptible to apple scab, powdery mildew and fire blight.

'Gala' originated in New Zealand in about 1934 from a cross between 'Kidd's Orange Red' and 'Golden Delicious'. It was introduced in 1960. The fruits are of medium size (67 mm diameter) with pale yellow to golden-yellow glossy skin heavily striped with red. It does not bruise easily and is generally free from skin blemishes and russet. The flesh is creamy yellow, firm, fine-textured, crisp, juicy, fairly sweet and with low acidity and with a good aromatic flavour. The trees are moderately vigorous, upright, fairly wide-spreading and with a spur-bearing habit similar to 'Delicious'. In England it flowers two days after 'Cox's Orange Pippin' and is harvested about a week later, in early October. In Brazil it flowers at about the same time as 'Golden Delicious' but is harvested about three weeks earlier (Bernardi, 1988). It can be grown successfully in most of the world's major apple growing areas, including western Europe, Washington State and the Okanagan Valley of British Columbia as well as New Zealand, although its tendency to produce small fruits is a problem in some climates. A number of 'sports' have been selected, especially for an increase in the proportion of red-coloured skin. 'Royal Gala', introduced by Dr D.W. McKenzie in New Zealand in 1974, was discovered as a bud-mutation in New Zealand in 1971 and is much redder than the parent type, with bright scarlet stripes over a golden-yellow ground colour. 'Gala' cannot be stored for as long as 'Delicious' or 'Golden Delicious' (Combrink, 1996) but has better storage life and a longer shelf life after removal from storage than 'Cox's Orange Pippin' (ASHS, 1997). The mutant 'Imperial Gala' shows better budbreak than 'Royal Gala' under marginal chilling conditions.

'Idared' was raised in Moscow, Idaho by Leif Verner, being selected in 1935 as a cross between 'Jonathan' and 'Wagener' and released in 1942. The fruits are large (82 mm diameter). The skin is almost solid red over a pale greenish-yellow, becoming yellow, ground colour. The flesh is white, tinged with green, firm, crisp, fine-textured, juicy and sweet. The trees are strong, vigorous and upright. In England they flower four days before 'Cox's Orange Pippin' (and five days before 'Golden Delicious') and are harvested in early October, i.e. after 'Cox's Orange Pippin' but before 'Golden Delicious' (Bultitude, 1983). 'Idared' is the predominant apple cultivar in Poland and accounted for 25% of the total Polish apple yield in 1992 (Florkowski *et al.*, 1996). It has a very long storage life. 'Idared' is susceptible to apple scab, cedar apple rust and powdery mildew and very susceptible to fire blight.

'Jonathan' originated in New York State, USA and was first described in 1826. The fruits are of medium size (69 mm diameter) with a pale greenish-white ground colour which becomes pale yellow and is flushed bright crimson over half to three quarters of its surface in scattered broken red stripes. The flesh is white with a greenish tinge, soft, fine-textured and sweet. The trees are weak and semi-weeping and produce spurs freely. Flowering is one day before 'Cox's Orange Pippin' and harvesting about a week later. There are a number of sports with a higher proportion of red-coloured skin, notably 'Jonared', introduced in 1934 following its discovery as a bud-mutation in 1930; 'Blackjon', a bud-mutation of 'Jonathan' selected in 1929 and introduced in 1931 which has earlier colouring and brighter red skin than 'Jonathan' but is otherwise identical; and 'Jonnee', a bud mutation of 'Blackjon' discovered in 1964 which has even earlier developing and more intensive fruit colour than 'Blackjon' and larger fruits. 'Jonathan' and its sports were widely grown, e.g. in the eastern USA and Hungary, but its share of world production has declined in recent years. 'Jonathan' is susceptible to cedar apple rust, fire blight and powdery mildew.

'Rome Beauty' originated in Ohio, USA at some date before 1848 when it was brought to notice. It has medium-large fruits (72 mm diameter) which have a greenish-yellow becoming pale yellow ground colour flushed with red. The flesh is creamy white, coarse-textured and often lacking in flavour although juicy. The trees are vigorous, upright-spreading and become round-headed. They do not produce spurs very readily. Flowering in England is seven days after 'Cox's Orange Pippin' and picking is in the third week of October. There are a number of red sports. The outstanding characteristic of 'Rome Beauty' is its heavy cropping and regular annual bearing, Childers (1983) attributing its economic success to this factor despite inferior quality. Its relative decline since that time probably reflects improved technology of production of high quality cultivars which need more 'management' to crop consistently, and also improvements in the storage and transport of such cultivars. 'Rome Beauty' is susceptible to powdery mildew and very susceptible to apple scab, cedar apple rust and fire blight.

'McIntosh' was discovered in Ontario, Canada in 1796 by John McIntosh and named in about 1870. The fruits are medium-large (74 mm diameter) with a greenish-yellow becoming pale yellow ground colour with half or more of the surface a deep purplish red. The flesh is white with a faint tinge of pink, fine-textured, sweet and juicy but rather soft. The trees are moderately vigorous and spreading and produce spurs freely. In England it flowers four days before 'Cox' and is picked in mid-September, i.e. a week or so before 'Cox' and more than a month before 'Golden Delicious'. This requirement for only a short growing season and its winter hardiness facilitated its production in northerly areas such as British Columbia (Swales, 1971) and Poland (Rejman, 1974) as well as north-eastern USA (Childers, 1983). There are a number of sports of 'McIntosh', e.g. 'Summerland McIntosh' which was introduced in 1929 as a solid red bud mutation, and 'MacSpur' which was discovered as a whole-tree sport of 'Summerland McIntosh' in 1964 and has very compact growth, high yield, high resistance to powdery mildew and is very hardy. This is now the official 'McIntosh' strain for British Columbia and 'McIntosh' is still the second most important apple cultivar in that province (Watson, 1997). The use of 'McIntosh' in England has been restricted by its very great sensitivity to apple canker under UK conditions. It is very susceptible to apple scab, moderately resistant to fire blight and very resistant to cedar apple rust.

'Elstar' was bred at Wageningen in the Netherlands by T. Visser. It originated as a cross between 'Golden Delicious' and 'Ingrid Marie' made in 1955 and it was introduced in 1972. The fruits are medium to large (70–80 mm diameter) and yellow with orange-red stripes. The flesh is white, firm and rather coarse, and acid at harvest but develops a very good flavour after two weeks in store. The trees are very vigorous and precocious. They are harvested three and a half weeks before 'Golden Delicious', i.e. require only a relatively short growing season, and perform best in cool areas. 'Elista' is a natural mutation with much more and brighter red striping of the skin surface. 'Elstar' has rather similar storage life to 'Cox's Orange Pippin'. It is susceptible to apple scab and very susceptible to powdery mildew and to *Phytophthora* fruit rot.

'Braeburn' was discovered in Waiwhero, New Zealand by O. Moran in 1952 as a seedling of unknown parentage. The fruits are medium to large with a glossy skin covered with short stripes of dark crimson three-quarters overlaid with a dark scarlet blush. The flesh is pale cream, very firm, crisp and juicy with a subacid flavour and overall excellent quality. The trees are of moderate vigour, spreading, productive and precocious with a tendency to biennial bearing easily controlled by pruning. The fruit ripens late, just before that of 'Fuji' when grown under the same conditions. It is an important cultivar in New Zealand and is being planted in other countries or regions with long growing seasons. It has fruited heavily in Zimbabwe without application of a dormancy-breaking spray at a site with a long-term average of only 417 chilling hours (below 7.2 °C) under conditions where 'Gala' and 'Jonagold' required a dormancy-breaking chemical spray. 'Hillwell' ('Red Braeburn') is a more highly coloured sport. 'Braeburn' has a longer storage life than 'Cox's Orange Pippin' but shorter than 'Red Delicious' or 'Golden Delicious'. It is subject to Braeburn Browning Disorder in store.

'Cox's Orange Pippin' was raised by Richard Cox in Slough, England in about 1825 and is said to be a seedling of 'Ribston Pippin'. The fruits are of medium size (65 mm diameter) with a light golden-yellow ground colour, one-quarter to three-quarters flushed with brownish orange-red stripes. The flesh is cream in colour and is firm, fine-textured, juicy, slightly acid although with sweetness and with a rich flavour. The trees are moderately vigorous, upright-spreading and produce spurs freely. In England it usually flowers in mid- to late May and is picked in the last week of September. This relatively short season makes 'Cox's Orange Pippin' suitable for production in southern England and the Netherlands but a tendency to produce small fruits in cooler areas limits its potential further north. It is also grown in New Zealand, largely for export to England, but the fruits are of very poor quality if grown in southern Europe. In general yields are lower and more variable than those of, for example, 'Golden Delicious' and storage life is shorter although it can be kept for at least 6 months in controlled atmosphere stores. 'Queen Cox' is a more highly coloured sport. 'Cox's Orange Pippin' is susceptible to bitter pit in storage and to apple scab, fire blight and mildew in the field.

Currently available summaries do not adequately reflect the importance of apples with little requirement for winter chilling, the 'low-chilling-requirement' cultivars, because these are grown primarily for domestic and local markets in areas without enough winter chilling for adequate budbreak of the conventional cultivars. The apple industry of Egypt is based on these as are others in subtropical and tropical climates.

'Anna' is the most widely grown of these 'low-chilling-requirement' apple cultivars. It was bred in Doar Na Shomron, Israel as a cross between the local cv. 'Red Hadassiya' and 'Golden Delicious' made in 1959 by Abba Stein, and introduced in 1963. The fruit is large with red cheeks where exposed to the sun. The flesh is subacid to sweet, juicy and mild flavoured with a smooth texture. It is very precocious and heavy cropping with a compact habit when grown on a 'MM.106' or 'MM.111' rootstock. It requires very little winter chilling in order for budbreak to occur, cropping in Florida in areas with less than 50 hours below 7.2 °C (Childers, 1983) and giving a second, winter, crop in Zimbabwe after a summer without chilling (Jackson, 1990). The period from flowering to harvest is approximately 163–181 days in Brazil (Bernardi, 1988) and is similar in Zimbabwe. 'Anna' is unsuitable for areas with prolonged winters with frost risk, because budbreak may under these circumstances occur while there is still risk of lethal frost. Other cultivars with little requirement for winter chilling include 'Dorsett Golden', a chance seedling of 'Golden Delicious' found in Nassau, Bahamas by Mrs I. Dorsett and introduced commercially in 1964. It resembles 'Golden Delicious' but has poorer colour. 'Maayan' originated in Rehovot, Israel as a selection by Chanan Oppenheimer of a cross between 'Calville St Saveur × Damascus' and 'Delicious' and was introduced in 1967. The fruits are of medium size, the skin yellow with more than 50% of the surface red-coloured. The flesh is juicy and sweet with a taste rather similar to 'Delicious'. The trees are vigorous, cropping on spurs. It has a slightly greater chilling requirement than 'Anna'. 'Princesa' was bred at Cacador, Brazil, by F. Denardi, L.F. Hough and A.P. Camilo from a crossing programme which included 'Anna' in the ancestry. It was selected in 1984. In Brazil the fruits average 160 g in weight and are nearly 100% red over a yellow ground colour. The flesh is whitish-cream, firm, juicy, sweet and subacid. The trees are moderately vigorous. Flowering is 3–4 weeks after 'Anna', probably indicating a slightly greater chilling requirement, and harvesting is also 3–4 weeks after 'Anna'. 'Adina' originated in Stanthorpe, Queensland, Australia, by H. Franklin, and was given a plant patent in 1988. The fruits are large, red to purplish-red, with creamy white, firm juicy and sweet flesh of high eating quality. The trees have a low chilling requirement of 350 hours below 7.2 °C.

Disease-resistant cultivars are a major target of current apple breeding programmes. With the exception of 'Judeline', grown for juice production, they face a problem in competing in dessert quality and handling characteristics with established cultivars, and new cultivars selected primarily for eating, handling and storage quality.

'Florina', selected in Beaucouze, France, by Y. Lespinasse, J.M. Oliver, J. Lespinasse and M. LeLezec, was made available to French growers in 1977. It has 'Golden Delicious', 'Giant Starking', 'Jonathan' and a donor of the *Vf* gene in its ancestry, is resistant to apple scab, tolerant to rosy apple aphid and fire blight but moderately susceptible to powdery mildew. It is early-bearing and productive with medium to large sweet fruits with cream-coloured, medium-firm flesh. The skin is three-quarters purplish-red on a yellow background. The fruit stores well but softens quickly out of storage.

'Freedom', bred in Geneva, New York by Robert C. Lamb was released in 1983. Its ancestry includes 'Machan' (a 'McIntosh' cross), 'Golden Delicious', 'Rome Beauty', 'Antonovka' and *Malus floribunda*. The trees are precocious and productive and the fruits are large with mainly red-coloured skin, creamy coloured firm, juicy and subacid flesh. It is resistant to apple scab, cedar apple rust, fire blight and powdery mildew.

'Liberty' was bred at Geneva, New York, by Robert Lamb with 'Macoun' × PRI 54–12' parentage and released in 1978. The trees are resistant to scab, cedar apple rust, fire blight and powdery mildew. They are very productive, giving a 'McIntosh' type of fruit with a crisp texture.

'Goldrush' ('Co-op 38') was bred at West La Fayette, Indiana by J.A. Crosby, J. Janick, P.C. Pecknold, J. Goffreda and S.S. Korban and introduced in 1993. It had 'Golden Delicious' and 'Co-op 17' as parents. The harvest season is 25 days after 'Delicious', the fruits have good flavour and outstanding storage life and the trees are immune to scab (*Vf* gene), moderately resistant to fire blight and moderately tolerant to mildew but susceptible to cedar apple rust.

'Primicia', released in 1988, was bred in Cacador, Santa Catarina, Brazil by F. Denardi, L.F. Hough and A.P. Camilo. It is resistant to apple scab and moderately resistant to powdery mildew. The trees are precocious and heavy cropping. They flower a few days before 'Gala' and have only a moderate winter-chilling requirement, cropping without the need for a dormancy-breaking spray in Santa Catarina where the mainstream cultivars require such a spray. The fruits are of medium size, firm and ripen three weeks before those of 'Gala'.

'Saturn' was raised at East Malling by Frank Alston and Ray Watkins from a cross between 'PRI 1235' and 'Starkspur Golden Delicious'. It was registered in 1995, has a very high level of scab resistance, moderate mildew resistance and is precocious and heavy cropping. The fruits are large, juicy, crisp and sweet, predominantly bright red in colour with excellent skin finish.

'Pink Lady' is outstanding among the ultra-late cultivars bred by John Cripps in Western Australia. The general expectation is that these cultivars, maturing over a very long season, should be outstanding in firmness and post-harvest storage life. 'Pink Lady', selected in 1979, is a cross between 'Golden Delicious' and 'Lady Williams', a very late chance seedling, and is harvested 10 days after 'Granny Smith'. The fruits are medium to large, pink-blushed with excellent crisp texture and juiciness and a flavour which combines sweetness with acidity. It is now being produced in other areas with long growing seasons such as South Africa.

Despite the apparent continued dominance of some very long established cultivars there has been a steady process of uptake of new strains of these and of new cultivars. This process is greatly facilitated by the use of high-density plantings on dwarfing rootstocks. These high-density orchards can reach their full yield levels within 4 or 5 years from planting or even less, greatly shortening the time from the first launch of a new cultivar to the large-scale production of its fruits.

Apple rootstocks

Selection of rootstock cultivars has played a dominant role in the development of apple orchard systems.

The original reason for using rootstocks was that the scion cultivars which produced desirable fruits could not be produced true-to-type from seed, and in any case seedling apple trees have juvenile characteristics and are generally slow to begin to bear fruits. Moreover most apple scion cultivars are very difficult to root by vegetative means so were not readily produced as clonal plants on their own roots.

Scion cultivars were therefore propagated from very early times by grafting or budding on to rootstocks. The Romans generally used rootstocks raised from cuttings or suckers taken from existing trees (Roach, 1985) but the earliest books on fruit growing in England advised rootstocks raised from seed. Mascall (1575) described how to raise rootstocks from cider apple pomace, i.e. the remains of the apples, containing the seeds, after the juice had been extracted. Seedlings from crab apples (*Malus sylvestris*) were often preferred because of their greater uniformity.

The first mention of dwarfing apple rootstocks in the literature was in the sixteenth and seventeenth centuries when the name 'Paradise' was recorded by Ruellius (1536), and Parkinson (1629) described the way in which the roots of a dwarf 'Paradise' apple sent up many shoots and suckers as a method of increase and noted that whatever fruit (cultivar) was grafted on it would also be dwarfed.

A distinction was later made between the very dwarfing 'Paradise' or 'French Paradise' and the less dwarfing 'Doucin' or 'English Paradise' rootstocks (Tukey, 1964). Rivers (1865) mentioned 14 kinds of 'Paradise' stock but as new forms were introduced into trade there was much confusion over their identity and trueness to type.

Modern selection and breeding programmes

Rootstock selection at East Malling in England was initiated by R. Wellington in 1912 and then carried out by R. Hatton. Hatton gathered 71 collections from 35 sources and found many of them to contain more than one phenotype or to be improperly named. He reclassified them, in his first publication describing nine types numbered I to IX in Roman numerals (Hatton, 1917). Subsequently the number was increased to sixteen (XVI), then to twenty-four (XXIV). These were subsequently given the designations 'EM.', and later 'M.', 1 to 24. To these were added two selections from crosses, 'EM.XXV' (later designated 'M.25') from 'Northern Spy' × 'EM.II' and 'EM.26' (later designated 'M.26') from a cross between 'EM.XVI' and 'EM.IX'. These rootstocks were evaluated first at East Malling (e.g. Preston 1958a,b, 1959) and subsequently world-wide.

Also in England, the John Innes Institute at Merton produced four apple rootstocks from crosses between 'EM.II' and 'Northern Spy', the latter being resistant to the woolly aphid (*Eriosoma lanigerum*), which is a major pest in Australia and New Zealand, but being difficult to propagate. The four rootstocks were designated 'Merton 778', '779', '789' and '793' and were introduced into New Zealand in 1941 for preliminary trial.

A further series of rootstocks, the 'Malling–Merton' or 'MM.' series, were produced by East Malling and John Innes staff, especially H.M. Tydeman and M.B. Crane, collaborating to produce woolly aphid resistant rootstocks. 'Northern Spy' was used as one of the parents and various 'Malling' and other rootstocks as the other parent. Fifteen rootstocks, designated 'MM.101' to 'MM.115', survived initial screening and were taken on for trials (Tydeman, 1953; Tukey, 1964). 'EM.XXV' ('M.25') also arose from this crossing programme but was found to lack the aphid resistance characteristic of the 'MM.' series, so was given an 'EM.' number.

Most recently a new 'AR.' series of rootstocks has been bred at East Malling using a range of parents including some of the 'M.' and 'MM.' series, 'Merton 793', 'Robusta 5J' and 'Ottawa 3'. Some of these have markedly outperformed 'M.9' in initial trials (Webster *et al.*, 1997). One, 'AR.86.1.25', which has a similar effect on scion vigour to 'MM.106' has been released as 'M.116' although it is not part of the original 'M.' series.

In the former Soviet Union, at Michurinsk, V.I. Budagovsky introduced 'Red-Leaved Paradise' or 'Budagovsky 9' in 1946 as a product of 'M.8' × 'Red

Standard' and crossed this with 'Bud. 13–14' to produce the very winter-hardy 'Budagovsky 57–490'. He also produced the winter-hardy 'Budagovsky 57–491'. The Siberian apple, *Malus baccata* Borkh was the original source of winter hardiness in the rootstock breeding programmes at the Michurin Institute (Stepanov, 1974). The best known products of this programme are the 'Budagovsky' ('Bud.' or 'B.') series.

Rootstock breeding in Poland was initiated at Skierniewice in 1954 with the objective of producing winter-hardy rootstocks, because the 'M.' and 'MM.' series were not hardy enough for Polish conditions. The 'P.' series of rootstocks were derived from 'M.9' crossed with the cold-tolerant 'Antonovka'. A second series of crosses was started in the early 1970s using 'P.' series clones and 'M.9', 'B.9', 'M.26', 'MM.106', 'Robusta 5' and 'Ottawa 3' as parents (Zagaja, 1981).

In the United States the 'Michigan Apple Clone' ('MAC.') series originated from open-pollinated seed collected in 1959 from a planting of the 'Malling' rootstocks, 'Alnarp 2' (a winter-hardy Swedish rootstock) and 'Robusta 5' (selected in Canada from seedlings from seed collected in Siberia).

The 'Cornell–Geneva' ('CG.') rootstock series was developed at Geneva, New York with selections starting in 1953 from seedlings from open-pollinated seed of the very dwarfing 'M.8' with 'M.1' to 'M.16' of the 'Malling' series and 'McIntosh' and 'Northern Spy' as likely parents (Cummins and Aldwinkle, 1983). Most of these proved unacceptably susceptible to fire blight and a breeding programme was developed to produce fire blight and *Phytophthora* root rot resistant rootstocks by crossing 'Malling' and other rootstocks with sources of resistance, e.g. 'M.26' with 'Robusta 5' and 'M.27' with 'Beauty Crab'. These crosses resulted in further 'CG.' and 'G.' rootstocks (Robinson *et al.*, 1997).

In Canada a breeding programme at Ottawa started in 1961 and resulted in the introduction of six rootstocks in 1971. This 'OH.' series of rootstocks included ones with cold-hardiness derived from 'Antonovka' and also involved 'Robusta 5' and 'Osman' as parents. A series of winter-hardy clonal rootstocks, the 'O' series 'Ottawa 1' to 'Ottawa 14', was produced from parents including 'M.9' and 'M.7' crossed with 'Robin' crab and *M. baccata* to give cold hardiness. Also in Canada the 'Kentville Stock Clone' ('KSC.') series of rootstocks were selected from seedlings of 'Beautiful Arcade' which had survived wide mid-winter changes in temperature (Ferree and Carlson, 1987). The 'Vineland' ('V.') clonal rootstock series were produced at Vineland, Ontario by crossing 'M.9' with the winter-hardy crab apples 'Dolgo' and 'Kerr' (Hutchinson, 1967). Some of the progeny have acceptable levels of resistance to woolly apple aphid and to fire blight (Cummins and Aldwinkle, 1983).

In Germany the Pillnitz rootstock breeding programme started by crossing the 'EM.' rootstocks and by clonal selection within the progeny. It subsequently included crosses with *M. baccata* (L.) Borkh, *M.* × *floribunda*, 'Antonovka', *M.* × *micromalus*, *M.* × *robusta* and others to introduce resistance to low temperatures, apple scab, apple mildew and woolly aphid. Rootstocks from this programme are available as the Pillnitzer 'Supporter' series (Fischer, 1997). Jork research station, also in Germany, produced 'Jork 9' ('J.9') from seed from open-pollinated 'M.9'.

Selection of sub-clones of rootstock cultivars

In common with the scion cultivars the rootstock cultivars of *Malus*, although clonal, show an appreciable degree of variability and mutability (Baumann, 1981; Engel, 1977, 1986; van Oosten, 1977). This is demonstrated in two main ways. First, either nurserymen have selected within their stocks, especially of 'M.9', or their standard clones have been found to differ from those in other nurseries. Secondly, the process of setting up virus-free or virus-tested rootstock sources inevitably involves the selection of a limited number of shoot tips for heat treatment or meristem culture, and the plants derived from these selections have been found to differ even if all are of similar virus-free status. Some of the apparent differences may be ontogenetic and reflect differing degrees of juvenility, plants from juvenile sources being easier to root but spinier in the nursery and not floriferous. Some such differences may relate to the source material; for example, a rootstock bed in the nursery may be maintained in a juvenile phase by regular cutting back whereas a virus-free rootstock mother-tree may be in the adult phase (van Oosten, 1986b). Such juvenility-related differences may be fairly transient although van Oosten suggested that the observed effect of different sources of 'M.9' on the vigour of scion trees worked on to them could be because juvenile plants root more easily, give larger rootstocks and, in turn, more vigorous maiden trees. There are, however, some differences between sub-clones that are large and long-term in nature. Parry (1980) demonstrated that the 'M.9a' clone of 'Malling 9' was much more dwarfing than the parent type or the later 'M.9 EMLA', and it is also less productive in the nursery. Wertheim (1997) found trees of 'Red Boskoop', 'Cox's Orange Pippin', 'Jonagored' and 'Red Elstar' on 'M.9 Fl 56' to be less vigorous than on other clones of 'M.9'.

Rootstock effects on scion performance

Control of scion vigour

The apple tree growing on its own roots as a seedling or rooted cutting, or an apple scion growing on a genetically similar seedling rootstock, is, in general, too

Table 2.7 *Production inputs into large-tree and dwarfed-tree orchards*

	Rootstock	
	Seedling	'Malling 9'
Trees ha^{-1}	250–400	1200–2800
Pruning hours ha^{-1}	120–180	60–80
Picking hours $tonne^{-1}$	10.0–12.5	5.0–5.5
Spray volume hl ha^{-1} yr^{-1}	600–800	300–350

Data from Italian orchards, taken from Werth (1981). Reproduced with permission.

large for the economic production of high quality fruits. Much of its canopy volume is unproductive and adds to management costs but not to financial returns.

The light intensity required for fruit bud production, fruitlet retention, growth of large fruits and, especially, optimal red skin colour development is much greater than that required for net photosynthesis and vegetative growth (Jackson, 1980). As a consequence only a relatively shallow outer canopy zone, giving a Leaf Area Index (LAI) of less than 2 m^2 leaf per m^2 of orchard floor can produce good quality fruits even under high ambient light conditions. The typical tree on a seedling rootstock can have a vertically-summed LAI of 4 or more (Heinicke, 1964), so much of this canopy volume is too heavily shaded. Moreover, apple trees do not bear fruits on the current season's shoots and their fruit yield is related to light interception by the spur leaves on older wood (Jackson, 1980; Wunsche *et al.*, 1996; Lakso and Robinson, 1997). Excess annual shoot growth and leaf production external to the fruit-bearing zone is therefore counter-productive and has to be prevented or controlled by pruning and branch training. Some effects of tree size on light levels in the tree, fruit size and fruit colour are shown in Figure 2.1.

Large and vigorous trees are not only inefficient in their horticultural productivity but are also expensive to manage. Much of their picking and pruning involves the use of ladders, which slows down operations, and the size and the density of their canopies necessitates the use of high volumes of pesticide and powerful machinery to spray it. The effects of tree size on costs of production were reported by Werth (1981), who compared those for large, 'round-headed' apple trees on seedling rootstocks, filling the available orchard space when planted at 250–400 trees per hectare, with those on dwarfing rootstocks planted at 1200–2800 trees ha^{-1}. The 'round-crown' trees attained a height of 5 m or more, the dwarfed trees only about half of this. The costs of production per kilogram of fruit were more than twice as high in the large-tree orchards as in those with dwarfing rootstocks (Table 2.7).

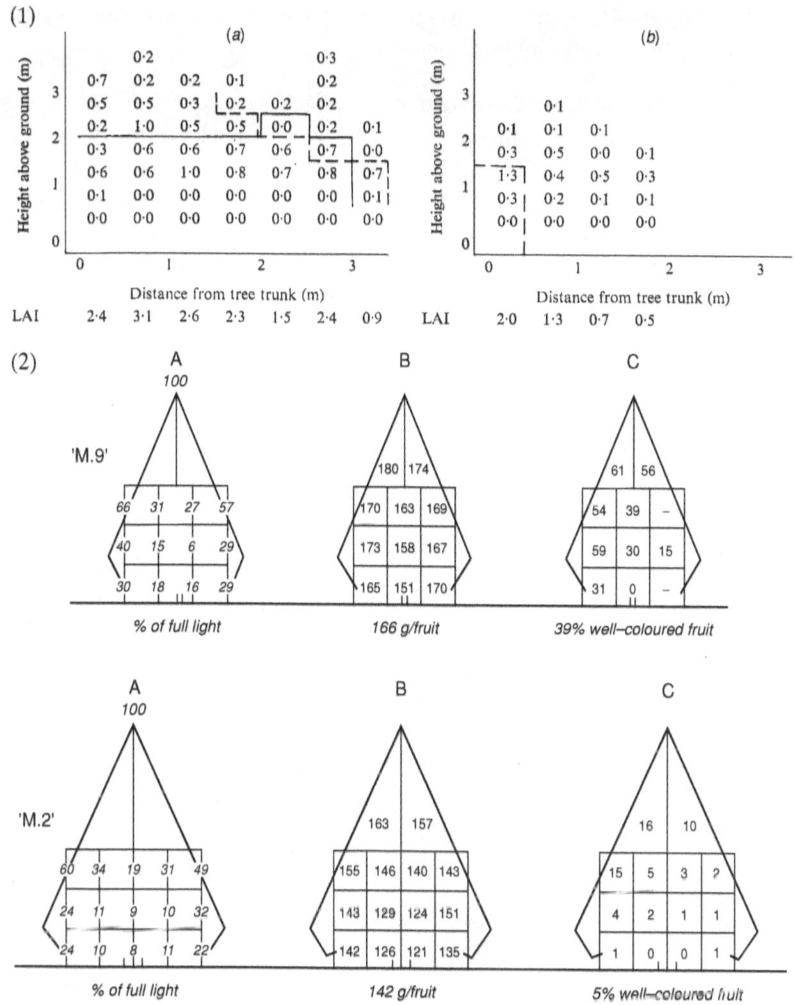

Figure 2.1 Distribution of shade, light, fruit size and fruit colour in large and small apple trees. (1) Bush trees of 'Cox's Orange Pippin' on a semi-dwarfing (a) and a dwarfing (b) rootstock showing leaf areas at different distances from the trunk and ground (measured in half metre grid cubes) and leaf area indices (LAIs) summed vertically. The solid line gives the boundary within which apple fruit weight did not exceed 80 g, the broken line the inner boundary of the zone in which, on average, more than 25% of the surface was red. Taken from Jackson (1970). Reproduced with permission from Blackwell Science. (2) Hedgerows of 'Jonathan' on 'M.9' and 'M.2' rootstocks in section showing (A) light intensity (B) fruit size in each position and on average and (C) percentage of well-coloured fruits in each position and on average. Each grid side is 0.5 m. Reproduced from Verheij and Verwer (1973) with permission from Elsevier Science.

The degree of dwarfing which is needed to achieve the most desirable tree vigour depends on a number of factors:

1 The vigour of the cultivar: a tree of a vigorous cultivar on a dwarfing rootstock may be of similar size to a tree of a compact cultivar on a semi-dwarfing rootstock (Table 2.8). Trees of 'Fuji' on the dwarfing 'M.9' rootstock can be from 45% to 106% larger than trees of 'Redchief Delicious' on the more invigorating 'M.26' rootstock (Costa *et al.*, 1997).
2 The vigour of the site: there are large differences in this, not just because of variations in light and temperature at different latitudes (Wagenmakers, 1995) but also because of much more localized differences in soil fertility. The standard clone of the most widely used dwarfing rootstock, 'M.9', is not sufficiently dwarfing in some areas of the Netherlands where natural soil fertility is high (Wertheim, 1997).
3 The tree training and pruning system and its effects on vigour: in general the training of the main axes into a horizontal or low-angled position, as in horizontal or Y- or V-trellis systems of tree management, will induce fruitfulness and check vegetative growth. Robinson (1997) found that Y-trellis trees were smaller in terms of trunk cross-sectional area than vertical-axis trees.
4 The light intensity: areas with higher light intensities have a greater depth of canopy which is adequately illuminated for the production of fruit buds and good quality fruits. In such areas there may also be a need for more leafy canopies to reduce the risk of fruit sunburn.
5 The light requirements of the cultivar: large-fruited and green-skinned cultivars, for example, may be able to produce good quality fruits over a greater depth of canopy than small-fruited and partially-coloured cultivars.

Now there are rootstocks available which will give any desired degree of apple tree size control. The scion cultivar that at maturity would give a tree more than 6 m high and with corresponding canopy spread if on a seedling rootstock may be restricted to about 1.5 m in height and spread if grown on the 'M.27' rootstock. Other rootstock cultivars give all levels of vigour control in between (Table 2.8, Figure 2.2).

Effects on precocity of cropping

Apple trees are perennial and although the costs of establishing an orchard are incurred in the year of planting or even, with respect to land preparation, in earlier years, returns are not received until the orchard bears fruits. This difference in time has important economic consequences because of the effects of real or imputed interest charges. The returns from the crop of each year need to be adjusted for the timing of these returns in years from planting

Table 2.8 *Trunk cross-sectional area (TCA, cm²) at year 9 for three scion cultivars in trials for supported spindle trees with 23 rootstocks (dwarf trial) and for freestanding central leader trees with 18 rootstocks (vigorous trial). Within groups, rootstocks are listed in estimated order of increasing tree size*

	Cultivar		
Rootstock	'Golden Delicious'	'Granny Smith'	'Redchief Delicious'
Dwarf trial–supported spindle trees			
'M.27 EMLA'	24	16	8
'MAC.9'	23	38	14
'B.146'			16
'P.16'	24	27	10
'P.22'	35	29	
'V.3'	35	20	18
'M.9'	42	47	16
'Mark'	22	36	13
'CG.10'	43	47	18
'V.1'	50		28
'P.2'	41	41	
'M.9 EMLA'	56	60	26
'MAC.39'	57	73	42
'B.9'	63		33
'O.3'	70		35
'C6'	83	120	36
'M.26 EMLA'	84	73	50
'V.2'	85	128	44
'V.7'	88	114	
'M.7A'	105		49
'OAR.1'	112	138	58
'P.1'	119	174	70
'V.4'	145	158	65
$LSD_{0.05}$	14	19	10
Vigorous trial–freestanding central leader trees			
'M.7A'	98	140	40
'A.306'	115	164	70
'MM.106 EMLA'	127	198	57
'M.7 EMLA'	120	163	48
'P.13'	130	181	56
'MM.111 EMLA'	116	165	62
'MM.106'	133	147	70
'B. 490'	137	180	76
'M.2 EMLA'	128	154	72
'M.4'	151	227	80
'MAC.16'	157	208	82
'MAC.1'	185	165	93
'MM.104 EMLA'	158	222	
'B.118'	167	200	93
'P.18'	172	203	103

(cont.)

Table 2.8 (*cont.*)

Rootstock	Cultivar 'Golden Delicious'	'Granny Smith'	'Redchief Delicious'
'MAC.4'	176	189	109
'A.313'	173		104
'MAC.24'	220	256	151
$LSD_{0.5}$	24	38	19

Rootstock codes: 'M.', 'Malling'; 'MAC.', 'Michigan Apple Clone', 'B.', 'Budagovsky'; 'P.', 'P. Series' (Polish); 'V.', 'Vineland'; 'CG.', 'Cornell–Geneva'; 'O.', 'Ottawa'; 'C.', 'Clark'; 'OAR.', 'Oregon Apple Rootstock'; 'A.', 'Alnarp'; 'MM.', 'Malling–Merton'.
From Barritt, *et al.* (1997a). Reproduced with permission.

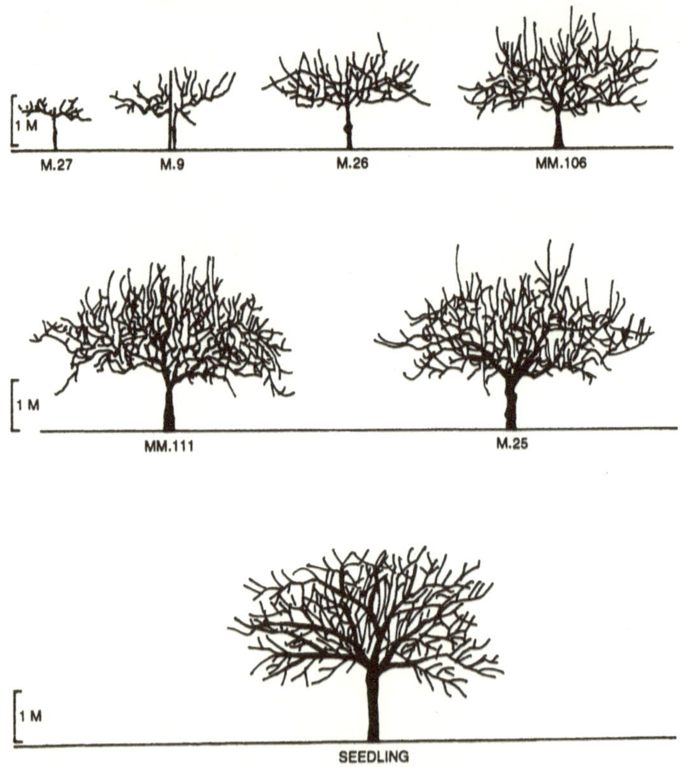

Figure 2.2 Relative size of apple trees when grafted on a seedling rootstock or on different clonal rootstocks. Reproduced by permission of Horticultural Research International, East Malling.

Table 2.9 *The effects of rootstock × planting density combinations on yields following planting in 1959*

Rootstock	Average yields per year (tonnes ha^{-1})					
	Seedling		'M4'		'M9'	
Trees per ha	114		417		1667	
Tree form	Standard		Spindlebush		Hedgerow	
Cultivar	'GD'	'S'	'GD'	'S'	'GD'	'S'
1961–65	5.7	2.5	14.5	6.8	18.7	16.6
1966–70	15.7	9.9	65.9	32.8	57.0	49.3
1971–75	26.7	17.2	49.1	34.7	70.1	50.6

'GD', 'Golden Delicious'; 'S', 'Starking' ('Red Delicious').
Data from Hungary, abstracted from Gyuro (1978).

by discounted cash flow analysis (Jackson, 1985) to evaluate overall orchard system profitability. Moreover the prices which will be obtained in the distant future are always uncertain. Precocious cropping, i.e. cropping early in the life of the tree, therefore has a value much greater than might be inferred from its effect on accumulated yield over the life of the orchard. There are direct effects of rootstock on the precocity of flowering and fruiting at the individual tree level. Tubbs (1974) concluded that precocity may occur independently of a dwarfing influence and there are precocity-inducing rootstocks within each vigour class. 'M.9' has long been regarded as outstanding in this respect but 'MM.106' and 'M.25' also induce early crops. However, the most important effects of rootstock on precocity are at the orchard level. As a result of the much greater number of trees per hectare which can be planted with trees on dwarfing rootstocks, the early yield per hectare is much higher even if yields per tree are similar or lower (Table 2.9). This effect is accentuated when the trees on the more vigorous rootstocks are pruned relatively severely in their early years and defruited to facilitate development of a strong branch framework.

Effects on yield to tree size ratio

The ratio of fruit yield to tree size reflects the efficiency with which the products of photosynthesis are partitioned between crop and vegetative growth, i.e. it is similar to the harvest index of annual crop plants. This ratio, for apple trees, is partly dependent on tree size so the effects of rootstocks on it are not independent of their effects on size itself. This is because once a canopy is of more than a certain depth any further increase in canopy depth does not increase the volume which can produce good quality fruits: it simply increases the amount of excessively shaded and unproductive volume. It is therefore to be expected that trees on dwarfing rootstocks or with dwarfing interstocks

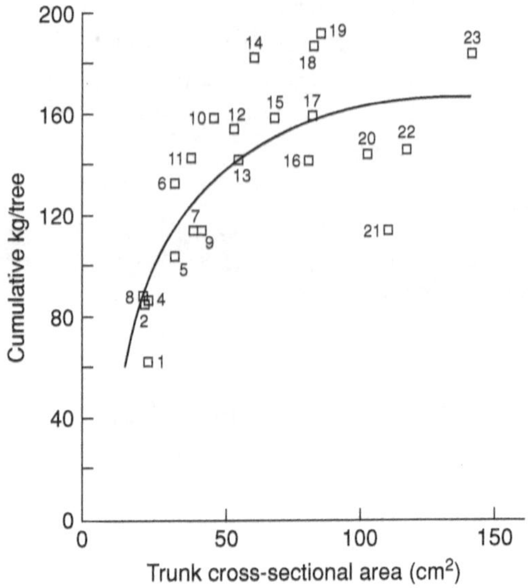

Figure 2.3 The relationship between tree size (measured as trunk cross-sectional area in year 9) and accumulated yield up to year 9. 'Golden Delicious' trees on 22 rootstocks. Rootstock codes: 1 = 'M.27 EMLA', 2 = 'MAC.9', 4 = 'P.16', 5 = 'P.22', 6 = 'V.3', 7 = 'M.9', 8 = 'Mark', 9 = 'CG.10', 10 = 'V.1', 11 = 'P.2', 12 = 'M.9 EMLA', 13 = 'MAC.39', 14 = 'B.9', 15 = 'O.3', 16 = 'C.6', 17 = 'M.26 EMLA', 18 = 'V.2', 19 = 'V.7', 20 = 'M.7A', 21 = 'OAR.1', 22 = 'P.1', 23 = 'V.4'. The equation of the fitted curve is $y = 167(1-e^{(-0.034x)})$ with $n = 22$ and $r^2 = 0.66$. From Barritt *et al.* (1997a). Reproduced with permission.

will have a higher ratio of yield to tree size than more invigorating ones. This is generally the case (Figures 2.3 and 2.4), although there is some evidence of effects of rootstocks on efficiency independent of tree size. For example, Robinson (1997) found that the yield per unit trunk cross-sectional area (TCA) of 'Empire' and 'Jonagold' on 'O.3' rootstock was significantly lower than that on 'M.9 EMLA' (Table 2.10) although the trees were of very similar size (slightly but not significantly smaller TCA). TCA is widely used as a measure of tree size, being highly correlated with the above-ground weight of the tree (Westwood and Roberts, 1970). Although there are some inconsistencies (Parry and Rogers, 1972) a large amount of published data shows linear relationships between TCA and scion weight to apply irrespective of rootstock. Rootstock effects on TCA therefore, in general, give an unbiased estimate of effects on scion weight. Robinson (1997) also found that the trees on both 'O.3' and 'Mark' rootstocks had significantly lower light conversion efficiencies (yield per unit light interception) than trees on 'M.9 EMLA' even though those on 'Mark' were significantly smaller (intercepted less light). Robinson's results also

Table 2.10 *The effects of rootstocks on cropping efficiency*

The effects of six rootstocks averaged over 'Empire' and 'Jonagold' and vertical-axis and Y-trellis training systems planted in 1990 ('M.9 EMLA' = 100).

Rootstock	Trunk CSA yr 6	Cumulative yield to yr 6	Cumulative yield/CSA	Light interception yr 6	Yield per unit light interception
'P.1'	169 a	86 b	50 d	110 ab	79 c
'M.9' / 'MM. 111'[a]	131 b	88 b	67 c	106 ab	88 abc
'M.26 EMLA'	129 b	88 b	68 c	111 a	92 ab
'M.9 EMLA'	100 c	100 a	100 a	100 b	100 a
'O.3'	94 c	87 b	92 b	102 b	79 c
'Mark'	78 d	77 c	97 ab	83 c	85 bc

Means within columns followed by the same letter are not significantly different (n = 16, P = 0.05).
CSA, cross-sectional area (cm^2).
[a] Trees on 'MM.111' rootstock with an 'M.9' interstock.
Data from Robinson (1997). Reproduced with permission.

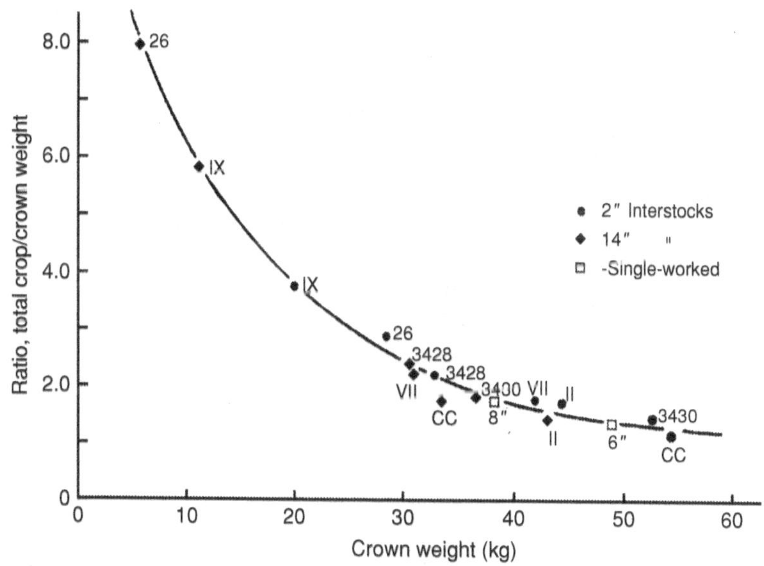

Figure 2.4 The relationship of cropping efficiency to tree size, 'Cox's Orange Pippin' on 'Malling 2' rootstock with size-controlling interstocks of 'M.9' ('IX'), 'M.26', '3428', 'MVII' ('M.7'), 'M.II' ('M.2'), 'Crab C' ('CC'), and '3430'. Reproduced from Parry and Rogers (1972) with permission.

showed unequivocally the higher cropping efficiency of trees on 'M.9 EMLA' in that they gave significantly higher yields than trees on 'P.1', 'M.9'/'MM.111', and 'M.26 EMLA' which were significantly larger.

By the standards of other crops, apple trees on 'M.9' have a very high partitioning efficiency (harvest index). Hansen (1980) showed that fruits represented more than 70% of the total dry matter increment of 4- and 5-year-old trees of 'Golden Delicious' on 'M.9' rootstocks. Barlow and Smith (1971), reporting on a 13-year study of total dry matter production of the cv. 'Laxton's Superb' showed that more than 70% of the dry weight had been in the fruits of trees on 'M.9' as contrasted with 40–50% of that of trees on 'M.16'. The scope for further advance in this respect is clearly limited. However, Callesen (1997) showed that trees of 'Elstar' on 'J. 9' rootstock had a 32% higher ratio of crop to trunk cross-sectional area than similar-sized trees on 'M.9 EMLA', and Hrotko, *et al.* (1997) found that trees of 'Idared' on 'J. 9' had a 20% higher crop to TCA ratio than that of rather smaller trees on 'M.9 T337'.

Data on cropping efficiency must be interpreted with care. Very small trees, with their canopy well illuminated throughout, will generally show a high ratio of crop to tree size and growth. If, however, they are so dwarfed that even at very close spacings they cannot intercept most of the available light, then the orchard yield will be low. Moreover the ratio of crop to tree size may depend on how the latter is measured. Comparative data on widely spaced trees on different rootstocks will give unbiased estimates of their relative cropping efficiencies. However, once the trees have filled the space allocated to them, e.g. in hedgerow planting, they will continue to increase in girth and weight even though their canopy volume and light interception does not increase further and their apparent cropping efficiency, if expressed, e.g. as yield per unit TCA, will decline. Where, as in many experiments, trees on rootstocks of differing vigour are planted at a common within-row spacing, those on the more vigorous rootstocks will fill their available space more quickly, and from then on appear less efficient than they actually are, compared with those on more dwarfing rootstocks. Yield per unit of canopy volume (Figure 2.5) or per unit light interception may then provide a more useful measure especially if used in conjunction with the data on total canopy volume or light interception.

Effects on fruit quality factors

Size

Rootstocks can influence fruit size in three different ways. First, it is well established that shaded areas of apple trees produce smaller fruits, whether the shade is artificially imposed (Jackson and Palmer, 1977) or attributable to position in the canopy (Jackson *et al.*, 1971; Tustin *et al.*, 1989). It is therefore to be expected that trees on dwarfing rootstocks in which most of the canopy is well

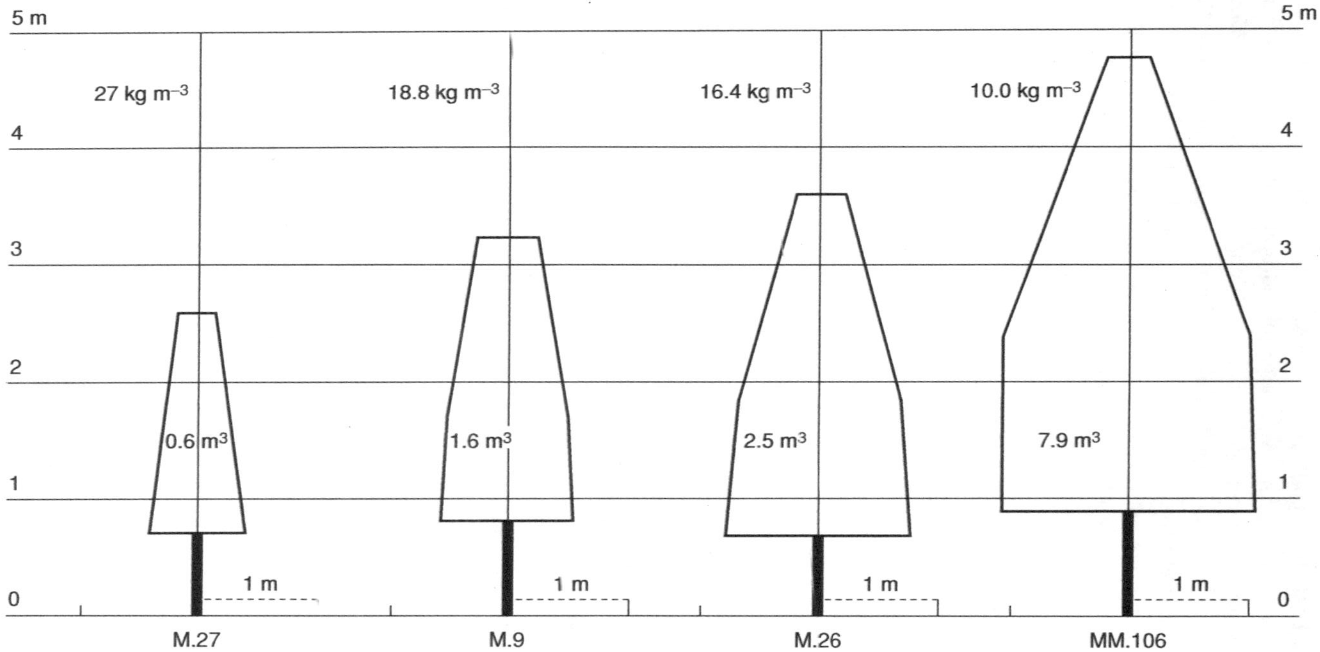

Figure 2.5 The effect of rootstocks 'M.27', 'M.9', 'M.26' and 'MM.106' on the canopy volume (m^3) and yield (kg m^{-3}) of 'Golden Delicious' apple trees, 7 years after planting. Reproduced from Lespinasse and Delort (1986) with permission.

exposed to sunlight will produce, on average, larger fruits than those on more vigorous rootstocks which have a larger proportion of shaded canopy. This is true in some cases; for example, Silbereisen (1981) found that fruits of 'Golden Delicious' and 'Cox's Orange Pippin' on 'M.9' were larger than those on 'M.2' when the trees were grown in planting systems appropriate to their vigour even though the average annual orchard yields were higher on 'M.9'. Similarly, Gyuro *et al.* (1986) found the percentage of fruits (averaged over 'Jonathan', 'Golden Delicious' and 'Starking') of more than 60 mm diameter to be 55, 42 and 29 on dwarfing ('M.9'), semi-vigorous ('M.4') and vigorous (seedling) rootstocks, respectively. However, within each category of rootstock vigour, i.e. very dwarfing, dwarfing, semi-dwarfing, semi-vigorous and vigorous, there are clear differences between individual rootstocks in their effects on fruit size and many cases in which the more dwarfing rootstocks do not give the larger fruits. For example, trees on 'M.26' (semi-dwarfing) and 'M.9' (dwarfing) generally give larger fruits than those on 'M.27' which is very dwarfing (van Oosten, 1986a; Palmer *et al.*, 1989).

Secondly, rootstocks can influence fruit size through their effects on the number of fruits in relation to the size and photosynthetic potential of the tree. In general, as this ratio is increased fruit size is reduced, hence the widely used practice of fruit thinning to increase the size of the remaining fruits. Trees on some very dwarfing rootstocks which have a higher yield per unit volume than those on 'M.9', e.g. on '3426' and 'M.27', can give much smaller fruits (van Oosten, 1986a).

There are also effects of rootstock on fruit size which cannot be explained by differences in either proportion of shaded canopy or cropping level. Callesen (1997) reported larger fruits when on the rootstocks 'M.9 EMLA' and 'Jork 9' than when on 'Bemali' rootstock even though the trees on the latter were of similar size and were lower yielding. 'Gala' trees on 'Mark' rootstock can give smaller fruits than when on 'M.9 EMLA' and 'Bud.9' even after taking cropping levels in relation to tree size into account (Barritt *et al.*, 1997b; Perry, 1997). Hampson *et al.* (1997) also found that trees of 'Summerland McIntosh' on 'Mark' rootstock had significantly smaller fruits than the same cultivar on 'M.9 EMLA' and 'Jork 9' even when all the trees were of very similar size, and those on 'M.9 EMLA' and 'Jork 9' had higher yields, though not significantly so.

Colour

Rootstock influences on fruit colour appear to be mainly a secondary effect of their effects on the vigour of tree growth and within-tree shade. Gyuro *et al.* (1986) recorded fruit colour of apples of 'Jonathan' and 'Starking Delicious' in the outer, middle and inner zones of trees on 'M.9', 'M.4' and seedling rootstocks and on their own roots (Table 2.11). In each case the fruits from the outer zone, defined as 1 m from the external surface, had the greatest

Table 2.11 *Colouring of 'Jonathan' and 'Delicious' fruits in the top zones on different rootstocks, 1979–81*

Average colour grade on a scale of 1 to 5.

Variety	Top zone	Rootstock 'M.9'	'M.4'	Own-rooted	Seedling	Average
'Jonathan'	outer	4.47	4.15	3.54	3.45	3.90
	middle	3.78	3.63	3.20	3.02	3.41
	inner	3.27	3.45	3.22	2.81	3.19
	average	3.84	3.74	3.32	3.09	3.50
	$LSD_{0.05}$	0.41	0.31	0.34	0.37	
'Starking Delicious'	outer	4.62	4.68	4.19	4.02	4.38
	middle	4.48	4.54	3.94	3.70	4.17
	inner	4.22	4.46	3.99	4.10	4.19
	average	4.44	4.56	4.04	3.94	4.25
	$LSD_{0.05}$	0.34	0.17	0.30	0.29	

% of surface red-coloured: Grade 1, 0–20; Grade 2, 20–40; Grade 3, 40–60; Grade 4, 60–80; Grade 5, 80–100.
From Gyuro *et al.* (1986). Reproduced with permission.

proportion of red-coloured surface. Since 97% of the canopy volume of the 'Jonathan' trees on 'M.9' was in this outer zone, compared with only 75%, 69% and 67% of the canopy volume of the trees on 'M.4', 'own roots' and seedling respectively, the better average colour of fruits from the trees on 'M.9' was inevitable. Similar results, showing a decrease in average fruit colour in the sequence 'M.9' > 'M.4' > own root > seedling was shown for 'Starking'. For both cultivars the trees on 'M.9' gave better-coloured fruits within each tree zone. This probably also reflected differences in shade: whereas the outer zone of dwarfed trees is largely 'top', that of larger trees includes lower peripheral areas shaded from direct sunlight for part of the day.

Skin-finish

In general shaded fruits are less prone to russet and cracking than exposed fruits, and it might be expected that fruits from trees on the most dwarfing rootstocks would suffer more from these problems, but results are inconsistent (van Oosten, 1986a).

Sugar and acid content

Sansavini *et al.* (1986) found the percentage of soluble solids in expressed juice of scion cvs. 'Golden Delicious' and 'Clear Red' to decline with increasing rootstock vigour over the range 'M.27', 'M.9', 'M.26' and 'MM.106'. This is in keeping with effects of increasing canopy shade. Ogata *et al.* (1986) found

Table 2.12 *Effects of rootstock on incidence of bitter pit and senescent breakdown, for 'Cox's Orange Pippin', 1974*

	Treatments in 1973								Rootstock means % of total	
	Hand pollinated		Control		Light thinning		Heavy thinning			
Rootstock	b.p.	s.b.	b.p.	s.b.	b.p.	s.b.	b.p.	s.b.	b.p.	s.b.
'M.9'	0.0	0.0	3.2	10.9	0.0	5.8	0.0	17.7	0.8	8.6
'M.26'	15.2	20.2	22.5	22.7	22.9	45.2	34.8	30.3	23.9	30.2
'M.7'	–	–	2.5	12.2	12.5	5.3	10.0	16.0	8.3	11.2
'MM.106'	17.2	12.3	7.5	12.4	2.8	12.8	0.0	14.8	6.9	13.1

b.p., % number of fruits showing bitter pit; s.b., % number of fruits showing senescent breakdown.
From Blasco (1976). Reproduced with permission.

a higher soluble solids content in fruits of 'Red King Delicious' and 'Fuji' on 'M.27' than on 'M.9a' or 'M.26'. Autio *et al.* (1996), in trials of 'Starkspur Supreme Delicious' on 15 rootstocks at four sites, found significant negative correlations between rootstock vigour, expressed as scion trunk cross-sectional area, and soluble solids in the fruit at three of the sites. The effects were large: at the Californian site the soluble solids content ranged from 13.7% on the most dwarfing rootstock ('P.2') to 10.7% on the most invigorating ones ('M.4' and 'P.18'). Sansavini *et al.* (1986) did not find any effect of rootstock on acidity.

Effects on fruit maturity and storage

It has been known for a long time that fruits of trees on 'M.9' ripen on the tree earlier and require earlier picking (MAFF, 1972). Autio *et al.* (1996) found that the time of ripening of 'Starkspur Supreme Delicious' on 15 rootstocks, as measured by the date of the C_2H_4 climacteric and the fruit starch index (see Chapter 10), was a negative function of rootstock vigour. This may indicate that the rootstock effect on fruit maturity is a consequence of its effect on tree size and shade or on vigour-related hormonal factors. Early findings on the effects of rootstock on fruit storage potential were probably confounded with effects on fruit maturity. Harvested at the correct (earlier) time, fruits from trees on dwarfing rootstocks store well.

Rootstocks can also influence storage through their effects on fruit calcium content. Blasco (1976) showed that fruits of 'Cox's Orange Pippin' on 'M.26' rootstock had less calcium per unit fruit weight than when on 'M.9', 'M.7' or 'MM.106', both on average and when only fruits of the same size were considered. This was accompanied by a significant and large increase in storage losses from bitter pit and senescent breakdown in fruits grown on 'M.26' compared with the other rootstocks (Table 2.12).

Other characteristics of apple rootstocks

Resistance to temperature stresses

Soil temperatures above 28 °C may have adverse effects on the growth of trees on most rootstocks. Nelson and Tukey (1956) and Gur *et al.* (1976) found that 'M.7', 'M.16', 'M.25', 'MM.109' and seedling rootstocks were relatively resistant to high soil temperatures but 'M.1', 'M.2', 'M.9' and 'MM.104' performed poorly at temperatures of 25 °C and above. This is very important at the macro-scale level of apple growing (Ferree and Carlson, 1987). Soil temperature is closely related to air temperature, often being higher in the top few centimetres. Diaz and Romo (1988) showed that under Mexican conditions, when air temperatures were 34 °C soil temperatures were about 48 °C at the surface, about 32 °C at 5 cm and then about 30 °C from 20 to 120 cm depth. When air temperatures were 45 °C soil temperatures were about 56 °C at the surface declining to around 34 °C at 40 cm and 30 °C at 100 cm. Since most tree roots are located within the top 120 cm, or even the top 40 cm (cf. Chapter 3), high soil temperatures provide a major constraint on the use of 'M.9' rootstock in warm-climate fruit-growing areas. It is likely that this constraint will apply to many of the newer rootstocks with 'M.9' parentage.

Freezing soil temperatures, and in particular freezing air temperatures just above the ground, can result in winter-freeze death of rootstocks and therefore of trees. In regions prone to severe winter freezes, cold-hardy seedling rootstocks, e.g. 'Antonovka Seedling' in Poland and 'Beautiful Arcade' in Canada, were in general use. Trees on these were too vigorous to fit in with the evolving systems of high-density planting, so the 'Malling' series of clonal rootstocks was evaluated. When it became clear that although these showed a wide range of cold tolerance 'M.9' was sensitive, there was considerable emphasis on the breeding and selection of cold-tolerant dwarfing rootstocks. Some of these proved very difficult to propagate so were used as dwarfing interstocks between a cold-resistant but vigorous rootstock and the scion cultivar which it was desired to dwarf. Observations on cold-hardiness have given some variable results, possibly because there are a number of different aspects to hardiness (Ferree and Carlson, 1987). However, there is ample data that 'M.9' has poor tolerance of winter-freezing conditions whereas 'M.26' and 'MM.106' are as tolerant as, or more tolerant than, 'Beautiful Arcade' seedling rootstock (Figure 2.6, from Privé and Embree, 1997). Quamme and Brownlee (1997) also found 'M.7' and 'M.9' to be sensitive to freezing injury in terms of both field performance and exposure to different freezing temperatures, while 'J.9' was hardier than either and the equally dwarfing 'B.9' and 'P.2' were as hardy as the vigorous Swedish 'Alnarp 2' (Table 2.13). In Poland, in a year without snow cover with a soil temperature of −11 °C at a depth of 5 cm, only 15–20%

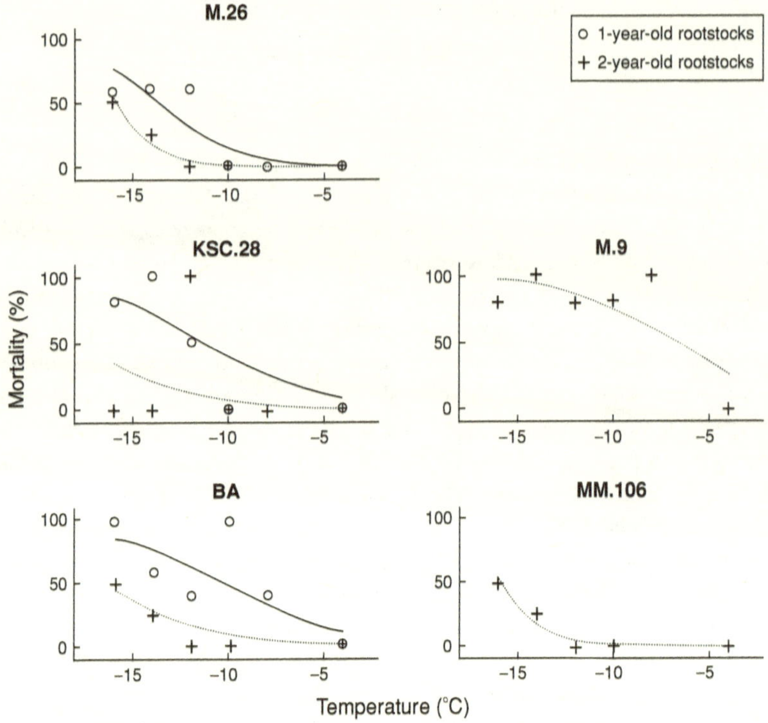

Figure 2.6 Percentage tree mortality of 1- and 2-year-old apple rootstocks 'M.9', 'M.26', 'MM.106', 'Beautiful Arcade' ('BA'), and 'Kentville Selected Clone 28' ('KSC.28') of 'Beautiful Arcade' following low temperature treatments. Reproduced from Privé and Embree (1997) with permission.

of trees on 'M.26' and 'B.9' were killed although nearly 75% of trees on 'M.9' died (Czynczyk and Holubowicz, 1984).

Although 'M.9' is regarded as moderately sensitive to winter freezing, it has been grown in British Columbia, Canada (mainly in the Okanagan Valley) for 35 years with few losses from winter injury (Quamme *et al.*, 1996).

Resistance to soil moisture stress

The most important stress associated with soil moisture is that due to waterlogging and anaerobiosis. This is particularly dangerous if it occurs in the non-dormant period after the trees leaf out. In general the 'Merton Immune' and 'MM.' rootstock series, descended from 'Northern Spy' and in many cases 'M.2', which are sensitive, range from sensitive to very sensitive to waterlogging (Cummins and Aldwinkle, 1983), and 'M.1', 'M.13' and 'M.16' show field

Table 2.13 *Comparative low temperature hardiness of rootstocks*

Hardiness as measured by minimum survival temperature with no effect on growth (MST) and field hardiness.

Rootstock	Mean MST (°C)	Hardiness class
'M.7'	−7.5 a	Very sensitive
'M.9'	−9.6 b	Moderately sensitive
'A.2'	−11.2 c	Hardy
'J.9'	−11.8 cd	Moderately hardy
'B.9'	−12.3 cd	Hardy
'O.3'	−13.2 cde	Hardy
'P.2'	−13.6 e	Hardy

Means followed by the same letter are not significantly different by the DMR test at the 0.05 level.
Hardiness classes on the scale developed from Quamme (1990). From Quamme and Brownlee (1997). Reproduced with permission.

tolerance. Rom and Brown (1979) found 'M.7' and 'M.26' to be relatively tolerant and 'MM.111' least tolerant.

Although there are differences in drought sensitivity between rootstocks, with 'MM.111' being relatively tolerant and 'M.9', 'M.26' and 'MM.106' being intolerant, this is not usually an important factor in rootstock choice because of the widespread use of irrigation. However, it does influence the choice of planting density because trees on 'MM.106' are much smaller, relative to those on other rootstocks, when grown on sandy soils under rainfed conditions.

Resistance to pests and diseases

Different plant diseases are of variable importance in different apple growing regions depending on local environmental conditions. Fire blight, caused by the bacterium *Erwinia amylovora* (see Chapter 13, p. 451), is of major importance in warm, humid apple growing areas such as the eastern United States. Crown rot and collar rot, caused by species of *Phytophthora* fungi especially *cactorum* and *syringae*, are most severe where there are recurrently wet and waterlogged soils. Each of these diseases can be of such severity that trees on intolerant (sensitive) rootstocks die. Differences between rootstocks in their sensitivity to latent viruses are particularly important in cases where the scion graftwood may be contaminated and graft unions fail, and where the virus can be spread by nematodes as is the case with apple union necrosis and decline due to tomato ringspot virus (Ferree and Carlson, 1987).

Table 2.14 *Relative susceptibility of various apple rootstocks to selected diseases and insects*[a]

Rootstock	Crown rot	Fire blight	Apple scab	Powdery mildew	Latent viruses	Woolly aphid
'A.2'	MS	VS	MR	MS	T	S
'B.9'	VR	S	M	MS	T	S
'B.490'	MR	M	M	S	T	MS
'B.491'	MS	S	M	MS	NT	S
'M.2'	MR	MR	M	MR	T	S
'M.4'	R	MR	M	M	T	S
'M.7'	MR	R	M	MR	T	S
'M.9'	R	S	M	MR	T	S
'M.13'	R	M	M	MR	T	S
'M.26'	MS	S	M	MR	MS	S
'M.27'	R	MS	M	MR	MS	S
'MM.104'	S					R
'MM.106'	MS	M	M	M	T	R
'MM.111'	M	M	M	MS	T	R
'Novole'	R	R	R	MR	S	MR
'O.3'	R	MS	M	MR	S	VS
'P.1'	MR	MS	NT	MR	NT	MS
'P.2'	R	MS	NT	MR	NT	MS
'P.18'	R	MR	NT	MR	NT	S
'P.22'	R	MS	NT	MR	NT	MS
'Robusta 5'	MR	R	R	MR	M	VR

Based largely on Cummins and Aldwinkle (1982). Rootstock abbreviations as in Table 2.8.
[a] Rating system: VS, very susceptible; S, susceptible; MS, moderately susceptible; M, intermediate; MR, moderately resistant; R, resistant; VR, very resistant; NT, not tested; T, tolerant.
From Ferree and Carlson (1987). Reprinted by permission of John Wiley & Sons Inc.

The most important insect pest of rootstocks is the woolly apple aphid (*Eriosoma lanigerum*; see Chapter 13, p. 461), which can stop tree growth in warm-climate production areas. The 'M.I' and 'MM.' series of rootstocks were bred specifically for resistance to this pest.

The relative resistances of the different rootstocks to these pests and diseases are given in Tables 2.14 and 2.15. These ratings must be treated with caution. Quamme *et al.* (1996) concluded that 'M.9' is among the most resistant of rootstocks to *Phytophthora cactorum*, in agreement with the results of Cummins and Aldwinkle (1982), Sewell and Wilson (1959) and general field experience. However, whereas Sewell and Wilson (1959) found 'M.7' and 'MM.106' to be resistant and 'M.26' to be susceptible, Ferree and Carlson (1987) classed 'M.7' as moderately resistant and 'M.26' and 'MM.106' as moderately susceptible; and Quamme *et al.* (1996) classed 'M.7' as moderately susceptible, 'M.26' as

Table 2.15 *Resistance to* Phytophthora cactorum *as measured by the lesion length on cut shoots placed in inoculated agar medium*

Rootstock	Resistance classification	Mean lesion length (mm)[a]
'J.9'	–	15.8 a
'M.9'	R	18.5 ab
'P.2'	–	19.8 ab
'O.3'	–	25.0 abc
'A.2'	–	25.8 abc
'P.22'	–	26.0 abc
'P.1'	–	26.8 abc
'M.4'	R	28.3 bc
'B.9'	–	33.0 cd
'M.26'	MR	35.5 cd
'M.7'	MS	39.5 d
'MM.111'	MS	43.3 d
'MM.106'	S	45.0 c

[a] Mean separation by Duncan's Multiple Range test ($P < 0.05$).
From Utkhede and Quamme (1988). Reproduced with permission.

moderately resistant and 'MM.106' as susceptible. Similarly, 'B.9' was less resistant to *P. cactorum* in tests carried out by Utkhede and Quamme (1988) than was 'M.9', although classed as very resistant by Cummins and Aldwinkle (1982). A possible reason for these discrepancies lies in the different pathogenicities among isolates of *P. cactorum* and different responses of the different rootstocks to these (Sewell and Wilson, 1959). Although the resistance to woolly apple aphid transferred from 'Northern Spy' to the 'Merton' and 'MM.' rootstocks has been immensely valuable, some biotypes of woolly apple aphid capable of colonizing 'Northern Spy' and the 'M.I' and 'MM.' stocks have been identified (Cummins and Aldwinkle, 1983). A number of 'CG.' and 'G.' series rootstocks are resistant to fire blight and to *Phytophthora* root rots (Robinson *et al.*, 1997). 'M.116' ('AR.86.1.25') is resistant to *Phytophthora* crown and collar rots, mildew, woolly apple aphid and the specific apple replant disease common in the UK.

Apple rootstocks also influence the incidence of diseases of the scion through mechanisms distinct from those of rootstock resistance. Sewell and Wilson (1973) found that the rootstock effect on scion resistance to *P. cactorum* was inversely related to the effect on tree vigour and not related to inherent rootstock resistance. The degree of scion susceptibility to fire blight is greater the more dwarfing the rootstock and is influenced by rootstock effects on precocity of flowering (Cummins and Aldwinkle, 1983).

Table 2.16 *Stooled layer production of rootstocks at Summerland, British Columbia*

Rootstock	Number of rooted shoots m^{-1}
'M.26'	37
'B.118'	35
'M.4'	27
'M.27'	22
'M.9'	21
'M.7'	18
'J.9'	16
'P.16'	12
'P.22'	9
'B.9'	8
'P.2'	8
'P.1'	7
'O.3'	6

From Quamme and Brownlee (1990). Reproduced with permission.

Nursery productivity

Although the orchard performance of trees grafted on to the different rootstocks determines the demand for them, differences in the ease with which they can be propagated determines the supply. Some very desirable rootstocks are very difficult to multiply by conventional nursery techniques (Table 2.16). Those which produce the fewest rooted shoots suitable for being grafted or budded with scions may instead be used as interstocks. Micropropagation methods may help overcome the difficulties in producing rooted propagules.

Major apple rootstocks

'Malling 9' ('M.9') is the most widely used dwarfing rootstock. It is the basis of the western European apple industry and is rapidly gaining in use in north America. Trees on 'M.9' are relatively easy to maintain at a height of between 2 and 3 metres, depending on cultivar, soil and climate, although they can be taller if required. The spacings between trees on 'M.9' range from about 4 m × 2 m or closer in single rows to 1.75 m × 1.25 m in 'full-field' systems (Wertheim *et al.*, 1986). Trees on 'M.9' give larger fruits than those on most other rootstocks. Yield per unit of tree size is outstandingly high and the trees start to crop within a year or so of planting; indeed, in many areas the early fruiting has to be reduced to prevent an excessive check to growth. 'M.9' is

propagated by stooling: it is less productive in the nursery than many other 'Malling' rootstocks. Trees on 'M.9' are poorly anchored and, because of this and the heavy early crops, need support by individual stakes or wire trellises. 'M.9' is resistant to collar rot but susceptible to fire blight, woolly apple aphid, crown gall and mildew. It does not tolerate high soil temperature. The main clones (sub-clones) are all virus-free selections, as follows:

'M.9 EMLA' was produced by East Malling and Long Ashton Research Stations. Trees grafted or budded on it were much more vigorous than those on the virus-tested (latent virus-infected) 'M.9a' clone which preceded it and the 'old' 'M.9' infected with latent viruses (Campbell, 1981). They gave 60% more crop up to 12 years of age than those on the virus-infected stocks as a consequence of their larger size.

'M.9 T337' ('NAKB 337') is a Dutch selection which gives trees of similar or rather smaller size than 'M.9 EMLA' (Webster, 1997) and of similar cropping efficiency.

'M.9 Fleuren 56' is a Dutch selection giving trees about 10–20% smaller than those on 'M.9 T337' (Wertheim, 1997). It induces a higher yield per unit of tree volume and, possibly as a result, reduces fruit size slightly.

'M.9 Pajam 1' ('Lancep') is a French clone which has been about 10% more dwarfing than 'M.9 EMLA' in some trials but less dwarfing in others (Callesen, 1997).

'M.9 Pajam 2' ('Cepiland') is similar to, or slightly more invigorating than, 'M.9 EMLA' or 'M.9 T337'.

'M.9 RN29' is a Belgian clone, similar in its effects on vigour and cropping to 'M.9 T337'.

'Malling 27' ('M.27') originated at East Malling from a cross made in 1929 between 'M.13' and 'M.9'. It was selected in 1934 (as 'Malling 3431') and patented as 'M.27' in 1975. It is much more dwarfing than either of its parents, scion cultivars grafted on it giving trees only about half of the size of those on 'M.9'. It is suitable only for orchards of very closely spaced trees or with very vigorous scion cultivars. It induces precocious cropping but trees on it tend to have small fruits.

'Malling 26' ('M.26') originated at East Malling from a cross between 'M.16' and 'M.9'. It is better anchored than 'M.9' but is less dwarfing and is unsuited for intensive planting systems, excepting possibly with spur-type scions. It induces precocious cropping and large fruit size but has a less positive effect on cropping efficiency than 'M.9'. It is the most tolerant of low winter temperatures of all the 'Malling' rootstocks in commercial use, but is susceptible to fire blight and to woolly apple aphid and moderately susceptible to crown rot, although resistant to collar rot.

'Malling 7' (*'M.7'*) was selected at East Malling in 1912. It is semi-dwarfing, producing trees 55–65% of the size of those on apple seedling rootstocks. It is widely used in North America but by the mid-1990s more trees were planted on the more dwarfing rootstocks. 'M.7' induces precocious cropping on a per tree basis but is unsuited to high-density plantings. It tolerates a wide range of soil temperatures and is resistant to fire blight. It tends to form root suckers.

'Malling 25' (*'M.25'*) was selected at East Malling in 1930, its parents being 'Northern Spy' and 'M.2', and was introduced into commerce in 1952. It gives trees almost as vigorous as those on seedling rootstocks but they are much more precocious and productive. Although it is not resistant to woolly apple aphid it is used in South Africa on poor soils or where there is a 'replant' problem which checks growth.

'Malling–Merton 106' (*'MM.106'*) was selected from crossing 'Northern Spy' and 'M.1' between 1930 and 1932 and introduced in 1952. It is semi-dwarfing, giving trees rather larger than those on 'M.7', and is outstanding for its effects on precocity and fruitfulness. The trees do not need staking. It is widely used in western Europe, especially in semi-intensive 'hedgerow' orchards, and in New Zealand and South Africa where its resistance to woolly apple aphid is important. Its susceptibility to *Phytophthora cactorum* limits its use in North America and parts of Europe. It is very easy to propagate.

'Malling–Merton 109' (*'MM.109'*) originated at Merton, England from 'M.2' × 'Northern Spy' parentage. It was selected in 1930 and 1931 and introduced into commerce in 1952. Trees on it are as vigorous as those on seedling rootstocks but it is resistant to woolly apple aphid and gives heavier cropping. The trees are poorly anchored but it is still used in South Africa on soils of low growth potential. It is easily propagated by stooling.

'Malling–Merton 111' (*'MM.111'*) was selected at Merton, England in 1932, its parents being 'Northern Spy' and 'Merton 793'. It gives trees 70–80% of the size of those on seedling rootstocks, but they are much more precocious and productive than those on seedlings or on the older 'M.2' rootstock of similar vigour. It is particularly tolerant of drought conditions and sandy soils (MAFF, 1972). It is resistant to woolly apple aphid and is still used in South Africa. It is easily propagated.

'M.116' (*'AR.86.1.25'*) was selected at East Malling from crosses made in 1964 between 'MM.106' and 'M.27' and introduced in 2001. It gives trees similar in size to, or appreciably smaller than, those on 'MM.106', has similar effects on cropping efficiency and fruit size and is resistant to collar rot, mildew, woolly apple aphid and the specific apple replant disease common in UK orchards. It roots well from hardwood cuttings.

'Merton 793' (*'M.793'*) was selected from seedlings of the cross 'Northern Spy' × 'M.2'. It gives trees of similar size to those on 'MM.111' and is resistant

to woolly apple aphid and collar rot. It is still used in South Africa but has been largely replaced by other rootstocks.

'Jork 9' ('J.9') was selected from a population of open-pollinated seedlings of 'M.9' at Jork in Germany and introduced as a clonal rootstock in 1981. Trees on it are of similar vigour or slightly smaller than those on 'M.9 EMLA'. It is rather more resistant to low winter temperatures than 'M.9', has a similar or even more positive effect on productivity and its use results in larger fruits. It is very susceptible to fire blight and woolly apple aphid. It is easy to propagate.

'Mark' ('MAC. 9') originated in East Lansing, Michigan, being selected by R.F. Carlson from seedlings from open-pollinated seeds of 'M.9' and introduced in 1979. It gives trees similar in size to those on 'M.9 EMLA' and with similar productivity. It is tolerant of *Phytophthora* and common latent viruses and more winter-hardy than 'M.9' but is susceptible to fire blight and woolly apple aphid. Fruits from trees on 'Mark' tend to be smaller than from those on 'M.9' and tissue proliferates at the soil line in a way which might be deleterious.

'Maruba-kaido N-1' is a clone of a weeping form of 'Maruba-kaido' (*Malus prunifolia ringo*) selected by Koike and Tsukahara. It is semi-vigorous, winter-hardy, resistant to woolly apple aphid and collar rot but susceptible to chlorotic leaf spot virus and stem grooving virus. It is usually used with a dwarfing interstem, originally 'M.26', but now that clones of 'M.9' free from latent chlorotic leaf spot virus (which caused incompatibility) are available these and some 'CG.' series interstems are often used. This is the most used rootstock in Japan (Ogata *et al.*, 1989) although 'M.9' is increasingly used to obtain fruit size and colour.

Malus prunifolia and *M. sieversii* seedlings are also widely used in China for their drought and cold tolerance with 'M.26' as a dwarfing interstem.

'Antonovka' seedlings are still used in eastern Europe and the countries of the former USSR for their cold tolerance, either giving large trees or being used with a dwarfing interstem of the 'B.' or 'P.' series.

M. × *domestica* seedlings, from fruits of the main scion cultivar, are still used in South Africa on soils of very low fertility; but in general such seedlings are passing out of commercial use.

Cold-tolerant dwarfing and semi-dwarfing rootstocks are still being actively sought in a number of breeding programmes.

'Budagovsky 9' ('B.9') has 'M.8' and 'Red Standard' as parents and was introduced in 1946 by V.I. Budagovsky of the Michurinsk College of Agriculture. It is slightly more dwarfing than 'M.9', induces precocious and heavy cropping, is very resistant to *Phytophthora* and is tolerant of the common latent viruses but is susceptible to fire blight and woolly apple aphid. It has been successfully used as a dwarfing interstem especially in Poland.

'Budagovsky 57–490' (*'Bud. 490'*) originated at Michurinsk as a cross between 'B.9' and 'Bud. 13–14'. It is a semi-vigorous clonal rootstock, with vigour similar to 'MM.111' but inducing greater precocity. It is very winter-hardy, equivalent to 'Common Antonovka', and resistant to crown rot and is very easy to propagate, even by hardwood cuttings.

'Budagovsky 57–491' (*'Bud. 491'*) also originated at Michurinsk and gives tree vigour control equivalent to 'M.27'. It induces early and heavy cropping, is easy to propagate and is even more winter-hardy than 'Common Antonovka'. It is very susceptible to fire blight and woolly apple aphid and is not resistant to *Phytophthora.*

'P.1' originated in Skierniewice, Poland, being bred by S. Zagaja and A. Czynczyk from 'M.9' × 'Common Antonovka'. It has proved dwarfing in some trials (Hrotko *et al.*, 1997) but more commonly has had similar effects on growth to 'M.26' or 'MM.106' (Barritt *et al.*, 1997a,b; Callesen, 1997). 'P.1' is slightly less cold-resistant than 'Common Antonovka' but still very winter-hardy. It is also crown-rot resistant and easily propagated but is moderately susceptible to fire blight, *Phytophthora* and woolly apple aphid and has many burr-knots just below the graft union.

'P.2' originated in Skierniewice, Poland, also bred by S. Zagaja and A. Czynczyk from 'M.9' and 'Common Antonovka'. It is about as dwarfing as 'M.9 EMLA' but trees on it have a lower cropping efficiency. It is winter-hardy and crown-rot resistant but is moderately susceptible to fire blight and susceptible to woolly apple aphid.

'P.16', also bred by S. Zagaja and A. Czynczyk from 'M.9' × 'Common Antonovka' is more dwarfing than 'M.9' and in some trials (Callesen, 1997) has induced very high productivity combined with large fruit size. It is easier to propagate than the other 'P.' series rootstocks, is susceptible to fire blight and woolly apple aphid and is said to have winter-hardiness only as good as 'M.9'.

'P.22', also bred by S. Zagaja and A. Czynczyk from 'M.9' and 'Common Antonovka', is about as dwarfing as 'M.27' in some trials but equivalent to 'M.9 EMLA' in others. It is winter-hardy and resistant to *Phytophthora* but moderately susceptible to fire blight and very susceptible to woolly apple aphid. It is difficult to propagate.

'Ottawa 3' was bred in Ottawa, Ontario, Canada by L. Spangelo, S.O. Fejer, S.J. Leuty and R.L. Granger, with 'M.9' and 'Robin Crab' as parents. It was introduced in 1974. It is rather less dwarfing than 'M.9' but more dwarfing than 'M.26' and induces precocious and heavy cropping. It is cold-hardy and resistant to *Phytophthora* but is susceptible to fire blight, woolly apple aphid and stem grooving virus, and has poor productivity in the nursery although it is easy to micropropagate.

Apple rootstock × site interactions

In relative terms rootstock effects on scion growth and cropping are fairly consistent over a wide range of environments. Rootstock × site interactions, although often statistically significant, are usually small compared with rootstock effects especially when a wide range of rootstocks is being tested. They may, however, be important where they reflect site variations in major disease or climatic stresses and in rootstock tolerance of these. Parry (1977) found that even within the limited confines of English apple growing areas trees on 'MM.106' varied more in relation to soil type than those on 'M.7a', and on sites with replant disease the vigour of trees on 'M.9a' was reduced more than on other rootstocks. Other site × rootstock interactions could not be attributed to any specific cause.

Apple rootstock × scion interactions

In broad terms the direction of rootstock effects is consistent over the range of scion cultivars with respect to vigour of growth, cropping, fruit size, etc. There are, however, numerous reports of significant rootstock × scion interactions; for example, Parry (1977) found large differences between effects of rootstocks on the cropping efficiency of 'Cox' and on that of 'Worcester Pearmain', and Barritt *et al.* (1997a) found that as rootstock vigour increased, the incidence of biennial bearing of 'Golden Delicious' and of 'Granny Smith' increased but that of 'Redchief Delicious' decreased.

Pear scion cultivars

European-type pear scion cultivars

The dominant cultivars of European pears (*Pyrus communis*) were selected in the eighteenth and nineteenth centuries. A relatively small number of 'sports' have also been selected from these.

'Williams' Bon Chrétien' ('Bartlett') was bred by Stair, in Berkshire, England, distributed by Williams of Middlesex and taken to America in 1797. It was later distributed under the name of 'Bartlett' in the USA and became the leading cultivar there both for fresh consumption and, especially, for canning. It is primarily grown in Washington State and California and is also widely grown in the most important Italian pear growing region, the lowlands of Emilia-Romagna and Veneto. The fruits are medium-large, yellow blushed with brownish-red and of excellent quality. 'Bartlett' is a prolific and heavy cropper, requires a shorter growing season than most major pear cultivars (105 days from full bloom to harvest in South Africa) and can succeed on a range of soil types. However, like other important pear cultivars it flowers before most apples so is subject to more risk of frost damage. It has a fairly high winter-chilling requirement (*c*. 800 h), is sensitive to fire blight and produces fruits of the best quality only under conditions of high temperature for the 2 months preceding harvest. 'Bartlett' is self-sterile but can be self-fruitful through the production of parthenocarpic fruits. A number of different clones, especially with red skin colour, have been selected. Some have slightly lower chilling requirements than the parent type.

'D'Anjou' pear was introduced into England in the nineteenth century by Thomas Rivers and is the most important winter pear in the Pacific Northwest of the USA, being grown mainly in Washington and Oregon. The fruits are large and green. The trees are vigorous and consistent in cropping and require cross-pollination, usually by 'Bartlett' or 'Bosc'. It is more cold-tolerant than 'Bartlett' but has a similar chilling requirement. It ripens 3–5 weeks after 'Bartlett' in the USA and, relative to 'Bartlett', is even later in Italy when it is harvested between 10 October and 10 November. It is moderately resistant to fire blight and has a very long storage life. One sport, 'Columbia Red Anjou', has bright red fruits with lower rates of respiration and ethylene production than the standard 'Anjou', giving even longer storage life.

'Beurré Hardy' was bred by M. Bonnet at Boulogne in about 1820. It ripens soon after 'Bartlett' in the Pacific Northwest and about 19 days after 'Bartlett' in South Africa. It has fairly large fruits with a golden-brown skin speckled with pink. It tends to form fruiting spurs instead of lateral long shoots along the fruiting branches. Its wood unites well with that of quince and it is frequently budded on quince rootstocks and then budded or grafted with other cultivars less compatible with quince.

'Beurré Bosc', also grown in Oregon, is slightly later ripening than 'Beurré Hardy'. It has large fruits and a yellow skin almost covered with a bronze russet. It is more frost-hardy than 'Bartlett' but has a slightly greater chilling requirement, requires a pollenizer and is very susceptible to fire blight, and to stony pit virus.

'Doyenné du Comice' first fruited at Angers, France, in 1849. It has large, light greenish-yellow fruits with probably the best flavour of all pears. It ripens three-and-a-half to four-and-a-half weeks after 'Bartlett' and has good storage quality but comes into bearing slowly and is an erratic cropper. It is grown in Oregon and coastal California as well as being an important cultivar in western Europe from southern England to Italy. It is reported to have a chilling requirement of only about 600 hours.

'Conference' was introduced by Rivers in England in 1894. The fruits are medium to large, long-pyriform in shape, with a smooth, unevenly russeted, green skin. It is a late-maturing pear with a very long storage life. It can produce parthenocarpic fruits and is a heavy and regular cropper but is susceptible to fire blight. It is the predominant cultivar in the European Union.

'Packham's Triumph' was selected in New South Wales, Australia from a cross made in 1896 or 1897 between 'Uvedale's St Germain' and 'Williams' Bon Chrétien' ('Bartlett'). The fruits are large and obtusely pyriform with a pale green skin becoming lime-yellow when ripe. It matures about 30 days later than 'Williams'' in South Africa. The fruits are of excellent eating and storage quality. The trees are good, consistent croppers. They have a rather lower chilling requirement than other major European cultivars so 'Packham's' is particularly important in areas with marginally inadequate winter chilling for temperate tree fruit production such as South Africa. Some 'low-chilling-requirement' strains, e.g. 'Africana' have been selected (Hauagge and Cummins, 2000). 'Packham's Triumph' is very susceptible to fire blight.

'Coscia' originated in Italy prior to 1800. It is grown in North Africa where it is considered to have a low chilling requirement.

Asian pear cultivars

The impact of modern plant breeding has been particularly great with respect to Japanese pears (Kajiura, 1994).

'Kosui' is the leading pear in Japan. It was released by the National Fruit Tree Research Station in 1959. The trees are vigorous, moderately productive and resistant to black spot (*Alternaria alternata*). The fruits are medium-sized, yellow to golden-brown and russeted with very sweet flesh.

'Hosui', the second in importance in Japan, was introduced in 1972. It has larger fruits (300–350 g) and ripens about two weeks later than 'Kosui'. The fruits are russeted and golden to golden-brown with sweet, crisp and juicy flesh.

'Nijisseiki', the leading cultivar in the past, was introduced in 1898. The fruits are green, medium-sized, crisp, juicy and sweet with a longer storage life than 'Kosui' or 'Hosui'. It is, however, difficult to grow and susceptible to black spot. An irradiation-induced mutant, 'Gold Nijisseiki', is resistant to black spot.

'Niitaka', an early breeding programme product, is a late-season pear with very large orange-brown russeted fruits that can be stored for 5–6 months.

'Shinko', another late-season pear, is resistant to black spot with russet brown, medium to large fruits, and good eating and storage quality.

'Tse-Li' and 'Ya-Li' are traditional Chinese pears with large fruits that can be stored for 24 weeks. They are cold-hardy and have lower chilling requirements than the main Japanese cultivars.

Pear rootstocks

Rootstocks for European pears

Clonal quince rootstocks, which can induce different degrees of dwarfing relevant to high-density planting, are used in areas with mild winters, especially in western Europe, but are generally sensitive to fire blight and to lime-induced chlorosis.

'Quince C' ('EMC'), is dwarfing as a result of the heavy and precocious cropping that it induces. It was selected and produced in virus-free form at East Malling.

'Quince A' ('EMA'), also from East Malling, is semi-dwarfing. It also induces high productivity in relation to tree size but does not induce such precocious cropping as 'Quince C'.

'EMH' ('QR 193–16') is a much more recent quince rootstock introduced by East Malling in 2001 following field trials starting in 1980/81. It is more dwarfing than 'Quince A' but usually rather more invigorating than 'Quince C', while consistently inducing larger fruits of pear scion cultivars than either of these.

'Adams 332' from Belgium is intermediate between 'Quince C' and 'Quince A' in its effect on scion vigour. It induces precocious cropping.

'Sydo', from France, is of similar vigour to 'Quince A' and is particularly recommended for 'Comice'.

'BA29', also from France, is rather more invigorating than 'Quince A' and is widely used on poorer soils. It has also been recommended for inherently weak-growing and precocious cultivars, e.g. 'Passe Crassane' (Carrera and Ortiz, 1984).

Some clonal *Pyrus* rootstocks are also used:

'Old Home × Farmingdale' ('OHF') clones 51 and 333 are more invigorating than 'BA29', but are resistant both to low temperature and to fire blight.

'BP1', from South Africa, gives some dwarfing on poor soils but is too invigorating for high-density planting systems on good soils.

Pyrus communis seedlings, especially of the main commercial cultivars such as 'D'Anjou' and 'Bartlett', are still widely used although trees on them can be excessively vigorous and they are susceptible to fire blight.

Rootstocks for Asian pears

Seedling rootstocks of *Pyrus pyrifolia* are generally used. Japanese pear trees generally have their vigour controlled by being trained on a horizontal trellis and therefore developing a dwarf rootstock has not been a priority. Strains of *P. calleryana* tolerant of flooding have been selected in Japan.

Pear rootstock × scion interactions

Although the different quince rootstocks have consistent effects on the size of pear cultivars grafted on them, their effect on cropping per unit of tree size varies with the scion cultivar (Carrera and Ortiz, 1984).

References

ASHS (1997). *The Brooks and Olmo Register of Fruit & Nut Varieties*, 3rd edn. Alexandria: American Society for Horticultural Science.

Autio, W.R., Hayden, R.A., Micke, W.C. and Brown, G.R. (1996). Rootstock affects ripening, colour and shape of 'Starkspur Supreme Delicious' apples in the 1984 NC-140 co-operative planting. *Fruit Varieties Journal* **50**, 45–53.

Barlow, H.W.B. and Smith, J.G. (1971). Effect of cropping on growth of the apple tree. *Report of the East Malling Research Station for 1970*, 52.

Barritt, B.H., Barden, J.A., Cline, J., Granger, R.L., Kushad, M.M., Marini, R.P., Parker, M., Perry, R.L., Robinson, T., Unrath, C.R. and Dilley, M.A. (1997b). Performance of 'Gala' at year 5 with eight apple rootstocks in an 8-location North American NC-140 trial. *Acta Horticulturae* **451**, 129–35.

Barritt, B.H., Konishi, B.S. and Dilley, M.A. (1997a). Tree size, yield and biennial bearing relationships with 40 apple rootstocks and three scion cultivars. *Acta Horticulturae* **451**, 105–12.

Baumann, G. (1981). Investigations on virus-indexed and virus-free plants of different clones of M.9 apple rootstock. *Acta Horticulturae* **114**, 171–84.

Bell, R.L., Quamme, H.A., Layne, R.E.C. and Skirvin, R.M. (1996). Pears. In *Fruit Breeding*, Vol. I. *Tree and Tropical Fruits*, ed. J. Janick and J.N. Moore, pp. 441–514. New York: John Wiley and Sons.

Belrose (1996). *World Apple Review 1996*. Pullman, WA: Belrose Inc.

Belrose (1997). *World Apple Review 1997*. Pullman, WA: Belrose Inc.

Bernardi, J. (1988). Behaviour of some apple cultivars in the subtropical region of Santa Catarina, Brazil. *Acta Horticulturae* **232**, 46–50.

Blasco, A.B. (1976). Rootstock effects on growth and cropping of apples with special reference to fruit quality. PhD thesis, University of London.

Bultitude, J. (1983). *Apples. A Guide to the Identification of International Varieties*. London: Macmillan.

Callesen, O. (1997). Testing 20 apple rootstocks. *Acta Horticulturae* **451**, 137–45.

Campbell, A.I. (1981). The effects of viruses on the growth, yield and quality of three apple cultivars on healthy and infected clones of four rootstocks. *Acta Horticulturae*, **114**, 185–91.

Carrera, M. and Ortiz, L. (1984). Performance of three quince rootstocks for pear. *Acta Horticulturae* **161**, 231–4.

Childers, N.F. (1983). *Modern Fruit Science*. Gainesville, FL: Horticultural Publications.

Clapham, A.R., Tutin, T.G. and Warburg, E.F. (1952). *Flora of the British Isles*. Cambridge: Cambridge University Press.

Combrink, J.C. (1996). Cold-storage requirements of specific commodities. In *Integrated Management of Post-harvest Quality*, ed. J.C. Combrink, pp. 120–9. Stellenbosch: Infruitec.

Combrink, J.C. and Von Mollendorff, L.J. (1996). Fruit quality. In *Integrated Management of Post-Harvest Quality*, ed. J.C. Combrink, pp. 2–7. Stellenbosch: Infruitec.

Costa, G., Beltrame, E., Zerbeni, P.E. and Pianezzola, A. (1997). High density planted apple orchards: effects on yield, performance and fruit quality. *Acta Horticulturae* **451**, 505–12.

Cummins, J.N. and Aldwinkle, H.S. (1982). New and forthcoming apple rootstocks. *Fruit Varieties Journal* **36**, 66–73.

Cummins, J.N. and Aldwinkle, H.S. (1983). Breeding apple rootstocks. *Plant Breeding Reviews* **1**, 294–394.

Czynczyk, A. and Holubowicz, T. (1984). Hardy, productive apple rootstocks used in Poland. *Compact Fruit Tree* **17**, 19–31.

Dayton, D.F. (1959). Red colour distribution in apple skin. *Proceedings of the American Society for Horticultural Science* **74**, 72–81.

Diaz, D.H. and Romo, R. (1988). Nutritional status and root system of Anna apple under alkaline soil in a warm dry climate of Mexico. *Acta Horticulturae* **232**, 177–86.

Engel, G. (1977). Comparisons of M.9 clonal selections with respect to growth and cropping of three apple varieties. *Acta Horticulturae* **75**, 163–7.

Engel, G. (1986). Virus free and virus tested M.9 selections. *Acta Horticulturae* **160**, 79–82.

Ferree, D.C. and Carlson, R.F. (1987). Apple rootstocks. In *Rootstocks for Fruit Crops*, ed. R.C. Rom and R.F. Carlson, pp. 107–43. New York: John Wiley and Sons.

Findlay, J.S. and Combrink, J.C. (1996). Cold chain management. In *Integrated Management of Post-Harvest Quality*, ed. J.C. Combrink, pp. 79–90. Stellenbosch: Infruitec.

Fischer, M. (1997). The Pillnitz apple rootstock breeding methods and selection results. *Acta Horticulturae* **451**, 89–97.

Florkowski, W.J., Majewski, E. and Bagel, M. (1996). Opportunities in production and marketing of integrated fruit production (IFP) certified apples in Poland. *Journal of Tree Fruit Production* **I**, 15–31.

Fukuda, H. (1994). Apple. In *Horticulture in Japan*, ed. K. Konishi, S. Iwahori, H. Kitagawa and T. Yakuwa, pp. 23–7. Tokyo: Asakura Publishing Co. Ltd.

Gonzalez-Cepeda, I.A. (1992). The Mexican apple industry as it exists today. *Compact Fruit Tree* **25**, 95–8.

Gotz, G. and Silbereisen, R. (1989). *Obstsorten Atlas*. Stuttgart: Ulmer.

Gur, A., Mizrahi, Y. and Samish, R. (1976). The influence of root temperature on apple trees. II. Clonal differences in susceptibility to damage caused by supraoptimal root temperatures. *Journal of Horticultural Science* **51**, 195–202.

Gyuro, F. (1978). High density apple planting in Hungary. *Acta Horticulturae* **65**, 53–60.

Gyuro, F., Gondor-Pinter, M. and Dib, K.A. (1986). The effect of rootstock and variety on the productive surface of apple trees. *Acta Horticulturae* **160**, 83–96.

Hampson, C.R., Quamme, H.R. and Brownlee, R.T. (1997). Performance of dwarfing apple rootstocks in five trials in British Columbia, Canada. *Fruit Varieties Journal* **51**, 183–91.

Hansen, P. (1980). Crop load and nutrient translocation. In *Mineral Nutrition of Fruit Trees*, ed. D. Atkinson, J.E. Jackson, R.O. Sharples and W.M. Waller, pp. 200–12. London: Butterworths.

Hatton, R.G. (1917). Paradise apple stocks. *Journal of the Royal Horticultural Society* **42**, 361–99.

Hauagge, R. and Cummins, J.N. (2000). Pome fruit genetic pool for production in warm climates. In *Temperate Fruit Crops in Warm Climates*, ed. A. Erez, pp. 267–303. Dordrecht: Kluwer.

Heinicke, D.R. (1964). The microclimate of fruit trees. III. The effect of tree size on light penetration and leaf area in Red Delicious apple trees. *Proceedings of the American Society for Horticultural Science* **85**, 33–41.

Hong, Y.P., Choi, J.H. and Lee, S.K. (1997). Optimum CA conditions for four apple cultivars grown in Korea. In *CA 1997 Proceedings*, Vol. 2. Apples and Pears, ed. E.J. Mitcham, pp. 241–5.

Hrotko, K., Magyar, L. and Hanusz, B. (1997). Apple rootstock trials at the Faculty of Horticulture UHF Budapest. *Acta Horticulturae* **451**, 153–9.

Hutchinson, A. (1967). Inheritance of stature, stooling, rooting ability and other characteristics in progenies of Malling IX. *Annual Report of the Horticultural Research Institute of Ontario 1966*, 76–82.

Jackson, J.E. (1970). Aspects of light climate within apple orchards. *Journal of Applied Ecology* **7**, 207–16.

Jackson, J.E. (1980). Light interception and utilisation by orchard systems. *Horticultural Reviews* **2**, 208–67.

Jackson, J.E. (1985). Future fruit orchard design: economics and biology. In *Attributes of Trees as Crop Plants*, ed. M.G.R. Cannell and J.E. Jackson, pp. 441–59. Monks Wood, UK: Institute of Terrestrial Ecology, Natural Environment Research Council.

Jackson J.E. (1990). Cropping and multiple cropping of apples in medium altitude tropical Zimbabwe. In *Abstracts of the 23rd International Horticultural Congress.* Florence: International Society for Horticultural Science, Abstract 4325.

Jackson, J.E. and Bepete, M. (1995). The effect of hydrogen cyanamide (Dormex) on flowering and cropping of different apple cultivars under tropical conditions of sub-optimal winter chilling. *Scientia Horticulturae* **60**, 293–304.

Jackson, J.E. and Palmer, J.W. (1977). Effects of shade on the growth and cropping of apple trees. II. Effects on components of yield. *Journal of Horticultural Science* **52**, 253–66.

Jackson, J.E., Sharples, R.O. and Palmer, J.W. (1971). The influence of shade and within-tree position on apple fruit size, colour and storage quality. *Journal of Horticultural Science* **46**, 277–87.

Janick, J., Cummins, J.N., Brown, S.K. and Hemmat, M. (1996). Apples. In *Fruit Breeding*, Vol. I. *Tree and Tropical Fruits*, ed. J. Janick and J.N. Moore, pp. 1–77. New York: John Wiley and Sons.

Juniper, B.E., Robinson, J., Harris, S.A. and Watkins, R. (2001). Origin of the apple (*Malus domestica*, Borkh.). In *Encyclopedia of Genetics*, ed. E.C.R. Reeve, pp. 674–7. London: Fitzroy Dearborn.

Kajiura, I. (1994). Nashi (Japanese pear). In *Horticulture in Japan*, ed. K. Konishi, S. Iwahori, H. Kitagowa and T. Yakuwa, pp. 40–7. Tokyo: Asakura Publishing Co. Ltd.

Korban, S.S. and Skirvin, R.M. (1984). Nomenclature of the cultivated apple. *HortScience* **19**, 177–80.

Lacey, C.N.D. and Campbell, A.I. (1987). Selection, stability and propagation of mutant apples. In *Improving Vegetatively Propagated Crops*, ed. A.J. Abbott and R.K. Atkin, pp. 351–61. New York: Academic Press.
Lakso, A.N. and Robinson, T.L. (1997). Principles of orchard systems management optimising supply, demand and partitioning in apple trees. *Acta Horticulturae* **451**, 405–15.
Lespinasse, J.M. and Delort, J.F. (1986). Apple tree management in vertical-axis: appraisal after ten years of experiments. *Acta Horticulturae* **160**, 139–55.
Lombard, P.B. and Westwood, M.N. (1987). Pear rootstocks. In *Rootstocks for Fruit Crops*, ed. R.C. Rom and R.F. Carlson, pp. 145–83. New York: John Wiley and Sons.
Luton, M.T. and Hamer, P.J.C. (1983). Predicting the optimum harvest dates for apples using temperature and full-bloom records. *Journal of Horticultural Science* **58**, 37–44.
MAFF (1972). *Apples. Bulletin 207. Ministry of Agriculture, Fisheries and Food.* London: HMSO. p. 20.
Mascall, L. (1575). *The booke of Arte and Maner, howe to plant and graffe all sortes of Trees.* London.
Mattheis, J. and Fellman, J. (1997). Factors influencing 'Gala' apple volatile emission following dynamic atmosphere storage. In *CA '97 Proceedings*, Vol. 2. *Apples and Pears*, ed. E.J. Mitcham, p. 112.
Mukherjee, S.K., Randhawa, S.S., Gosami, A.M. and Gupta, P.H. (1969). Description of wild species of pome and stone fruits. 1. *Pyrus. Indian Journal of Horticulture* **26**, 130–3.
Nelson, S.H. and Tukey, H.B. (1956). Effects of controlled root temperatures on the growth of East Malling rootstocks in water culture. *Journal of Horticultural Science* **31**, 55–63.
Ogata, R., Goto, K., Kumisawa, T. and Harada, R. (1986). Productivity and fruit quality of four apple cultivars on three different rootstocks and at different planting densities. *Acta Horticulturae* **160**, 97–104.
Ogata, R., Koike, H. and Tsukahara, K. (1989). Apple tree management on dwarf rootstocks in Japan. *Acta Horticulturae* **243**, 269–78.
O'Rourke, A.D. (1994). *The World Apple Market.* New York: Food Products Press.
Palmer, J.W., Bunemann, G., Sansavini, S., Wagenmakers, P.S. and Winter, F. (1989). The International Planting Systems Trial. *Acta Horticulturae* **243**, 231–41.
Parkinson, J. (1629). *Paradisi in Sole.* London.
Parry, M.S. (1977). Field comparisons of 'M.26' and other dwarfing apple rootstocks on a diversity of sites. *Journal of Horticultural Science* **52**, 59–73.
Parry, M.S. (1980). Evidence of clonal variation and latent virus effects on the vigour of Cox's Orange Pippin apple trees on M.9 rootstocks. *Journal of Horticultural Science* **55**, 439–40.
Parry, M.S. and Rogers, W.S. (1972). Effects of interstock length and vigour on the field performance of Cox's Orange Pippin apples. *Journal of Horticultural Science* **47**, 97–105.
Perry, R.L. (1997). The performance of 'Jonagold' and 'Empire' apple in an NC-140 rootstock and orchard systems trial in Michigan. *Acta Horticulturae* **451**, 453–8.
Pieniazek, S.A. (1966). Fruit production in China. *Proceedings of the 17th International Horticultural Congress* **4**, 427–56.
Pratt, C. (1988). Apple flower and fruit: morphology and anatomy. *Horticultural Reviews* **10**, 273–308.

Preston, A.P. (1958a). Apple rootstock studies: thirty-five years' results with Lane's Prince Albert on clonal rootstocks. *Journal of Horticultural Science* **33**, 29–38.
Preston, A.P. (1958b). Apple rootstock studies: thirty-five years' results with Cox's Orange Pippin on clonal rootstocks. *Journal of Horticultural Science* **33**, 194–201.
Preston, A.P. (1959). Thirty-five years' results with Worcester Pearmain on clonal rootstocks. *Journal of Horticultural Science* **34**, 32–8.
Privé, J-P. and Embree, C. (1997). Low temperature injury and recovery of apple rootstocks. *Acta Horticulturae* **451**, 179–86.
Quamme, H.A. (1990). Cold hardiness of rootstocks. Compact Fruit Tree **23**, 11–16.
Quamme, H.A. and Brownlee, R.T. (1990). Stool layering ability of thirty one apple rootstock cultivars. *Fruit Varieties Journal* **44**, 165–9.
Quamme, H.A. and Brownlee, R.T. (1997). Cold hardiness evaluation of apple rootstocks. *Acta Horticulturae* **451**, 187–93.
Quamme, H.A., Brownlee, R.T. and Hampson, C.R. (1996). Apple rootstock performance in British Columbia. *Compact Fruit Tree* **29**, 12–18.
Rehder, A. (1967). *Manual of Cultivated Trees and Shrubs*, 2nd edn. New York: Macmillan.
Rejman, A. (1974). Apple varieties for the Polish climate. In Sad nowoczesny. The modern orchard. *Special issue for the XIX International Horticultural Congress*, ed. in chief K. Palinski, pp. 11–14. Warsaw: State Agriculture and Forestry Publishing House.
Rivers, T. (1865). *The Miniature Fruit Garden*. London.
Roach, F.A. (1985). *Cultivated Fruits of Britain: their Origin and History*. Oxford: Blackwell.
Robinson, T.L. (1997). The interaction of tree form and rootstock on light interception, yield and efficiency of 'Empire', 'Delicious', and 'Jonagold' apple trees trained to different systems. *Acta Horticulturae* **451**, 427–36.
Robinson, T.L., Cummins, J.N., Hoying, S.A. and Smith, W.H. (1997). Commercial orchard evaluation of the new Cornell-Geneva apple rootstocks. *Acta Horticulturae* **451**, 113–9.
Rom, R.C. and Brown, S.A. (1979). Water tolerance of apples on clonal rootstocks and peaches on seedling rootstocks. *Compact Fruit Tree* **12**, 30–3.
Ruellius, J. (1536). *De Natura Stirpium libri tress*. Paris.
Sansavini, S., Marangoni, B., Buscaroli, C., Corelli, L. and Tazzari, G. (1986). The relationship between spacing and rootstock effects in an intensive planting trial of two apple cultivars. *Acta Horticulturae* **160**, 23–37.
Sewell, G.W.F. and Wilson, J.F. (1959). Resistance trials of some apple rootstock varieties to *Phytophthora cactorum* (L. & C.) Schroet. *Journal of Horticultural Science* **34**, 51–8.
Sewell, G.W.F. and Wilson, J.F. (1973). *Phytophthora* collar rot of apple: influence of the rootstock on scion variety resistance. *Annals of Applied Biology* **74**, 159–69.
Shewfelt, R.L. (1993). Measuring quality and maturity. In *Post-harvest Handling. A Systems Approach*, ed. R.L. Shewfelt and S.E. Prussia, pp. 99–124. New York: Academic Press.
Shou-Chun, Q. (1998). Cited in: China pushes Fuji into overseas markets. *Asiafruit* January/February 1998, p. 22.
Silbereisen, R. (1981). Productivity, fruit quality and physiology aspects of fruiting at different planting densities and with different rootstocks with apples. *Acta Horticulturae* **114**, 137–8.

Smith, S.M. (1984). Improvement of aroma of Cox's Orange Pippin apples stored in low oxygen atmospheres. *Journal of Horticultural Science* **59**, 515–22.

Stepanov, S. N. (1974). Species and forms of fruit trees and bushes of the USSR and their utilization in plant breeding. *Proceedings of the XIX International Horticultural Congress*, Volume II. Warsaw, 1–12.

Stow, J. (1987). Storage of 'Jonagold' apples. *Scientia Horticulturae* **31**, 245–51.

Swales, J.E. (1971). *Commercial Apple-growing in British Columbia*. Department of Agriculture, Victoria, BC.

Tubbs, F.R. (1974). Rootstock/scion relations in horticultural crop physiology. *Scientia Horticulturae* **2**, 221–30.

Tukey, H.B. (1964). *Dwarfed Fruit Trees*. New York: Macmillan.

Tustin, S., Hurst, P., Warrington, I. and Stanley, J. (1989). Light distribution and fruit quality through multi-layered trellis apple canopies. *Acta Horticulturae* **243**, 209–12.

Tydeman, H.M. (1953). A description and classification of the Malling–Merton and Malling XXV apple rootstocks. *Report of the East Malling Research Station for 1952*, 55–63.

Utkhede, R.S. and Quamme, H.A. (1988). Use of the excised shoot assay to evaluate resistance to *Phytophthora cactorum* of apple rootstock cultivars. *Canadian Journal of Plant Science* **68**, 851–7.

van der Merwe, J.A. (1996a). Maturity standards for deciduous fruit. In *Integrated Management of Post-harvest Quality*, ed. J.C. Combrink, pp. 25–32. Stellenbosch: Infruitec.

van der Merwe, J.A. (1996b). Controlled and modified atmosphere storage. In *Integrated Management of Post-harvest Quality*, ed. J.C. Combrink, pp. 104–12. Stellenbosch: Infruitec.

van Oosten, H.J. (1977). Growth and yield of Golden Delicious on M.9a and on three other M.9 rootstock sources. *Acta Horticulturae* **75**, 157–61.

van Oosten, H.J, (1986a). Effects of some new rootstocks on orchard behaviour of apple trees. *Acta Horticulturae* **160**, 39–46.

van Oosten, H.J, (1986b). Intraclonal variation in M.9 rootstock and its effects on nursery and orchard performance. *Acta Horticulturae* **160**, 67–78.

Verheij, E.W.M. and Verwer, F.L.J.A.W. (1973). Light studies in a spacing trial with apple on a dwarfing and a semi-dwarfing rootstock. *Scientia Horticulturae* **1**, 25–42.

Wagenmakers, P.S. (1995). Light relations in orchard systems. Thesis, Wageningen.

Watson, T. (1997). Fruit growing in the Okanagan Valley. *Acta Horticulturae* **451**, 45–50.

Way, R.D., Aldwinkle, H.S., Lamb, R.C., Rejman, A., Sansavini, S., Shen, T., Watkins, R., Westwood, M.M. and Yoshida, Y. (1990). Apples (*Malus*). In *Genetic Resources of Temperate Fruit and Nut Crops*, ed. J.N. Moore and J.R. Ballington. International Society for Horticultural Science, *Acta Horticulturae* **290**, 1–67.

Webster, A.D. (1997). A review of fruit tree rootstock research and development. *Acta Horticulturae* **451**, 53–73.

Webster, A.D., Tobutt, K.R., James, D.J., Evans, K.M. and Alston, F.A. (1997). Rootstock breeding and orchard testing at Horticultural Research International East Malling. *Acta Horticulturae* **451**, 83–8.

Werth, K. (1981). Development and current achievement of high density plantings in Italy, Switzerland, Austria and Yugoslavia. *Acta Horticulturae* **114**, 295–9.

Wertheim, S.J. (1997). Useful differences in growth vigour between subclones of the apple rootstock M.9. *Acta Horticulturae* **451**, 121–8.

Wertheim, S.J., de Jager, A. and Duyzens, M.J.J.P. (1986). Comparison of single-row and multi-row planting systems with apple with regard to productivity, fruit size and colour and light conditions. *Acta Horticulturae*, **160**, 243–58.

Westwood, M.N. (1978). *Temperate-zone Pomology*. San Francisco: Freeman.

Westwood, M.N. (1982). Pear germplasm of the new national clonal repository: its evaluation and uses. *Acta Horticulturae* **124**, 57–65.

Westwood, M.N. (1993). *Temperate Zone Pomology, Physiology and Culture*. Portland, OR: Timber Press.

Westwood, M.N. and Roberts, A.N. (1970). The relationship between trunk cross-sectional area and weight of apple trees. *Journal of the American Society for Horticultural Science* **95**, 28–30.

Wunsche, J.N., Lakso, A.N., Robinson, T.L., Lenz, F. and Denning, S.S. (1996). The bases of productivity in apple production systems: the role of light interception by different shoot types. *Journal of the American Society for Horticultural Science* **121**, 886–93.

Yoshida, Y., Fan, X. and Patterson, M. (1995). 'Fuji' Apple. *Fruit Varieties Journal* **49**, 194–7.

Zagaja, S.W. (1981). Performance of two apple cultivars on P. series dwarf rootstocks. *Acta Horticulturae* **114**, 162–9.

Zeven, A.C. and Zhukovsky, M. (1975). *Dictionary of Cultivated Plants and their Centres of Diversity*. Wageningen: Centre for Agricultural Publishing and Documentation.

3

Apple and pear root systems: induction, development, structure and function

Introduction

In field crops grown from seed the root system develops, in general, in an uncomplicated way so as fully to exploit the soil to a depth characteristic of the crop. In crop physiological analysis, and for irrigation scheduling, the root system is considered to be defined fully in terms of rooting depth and density, the roots usually being considered to be evenly distributed.

Root development and distribution is much more complex in apple and pear trees as grown commercially. It involves, and relates to, very distinctive technologies of propagation, tree establishment, and nursery and orchard soil management and irrigation. Prior to consideration of the biological basis of these technologies it is best to consider the general anatomy and structure of the roots of the orchard tree.

The essential root system of the mature tree consists, as a rule, of an underground rootstock stem from which arises a system of permanent, thickened, scaffold roots spreading almost horizontally, usually less than 50 cm from the surface, and numerous more or less vertical 'sinkers' descending as a rule to either an impermeable layer or a water table (Rogers and Head, 1966; Atkinson, 1980). These woody 'skeletal' roots are long-lived, provide anchorage and form the framework which bears the fine (fibrous) roots. New roots can emerge from the underground rootstock stem, from coarse roots (>2 mm diameter) or from fine roots. They are initially white and most are ephemeral with only a proportion thickening and becoming perennial roots. Figure 3.1 illustrates the root system of an apple tree. Transverse sections of pear roots are shown in Figure 3.2.

There are three main stages in the development of such a root system: the initiation of roots on rootstock shoots, the development of roots on lined-out rootstocks and grafted or budded trees in the nursery, and the development of a root system after transplanting into the orchard.

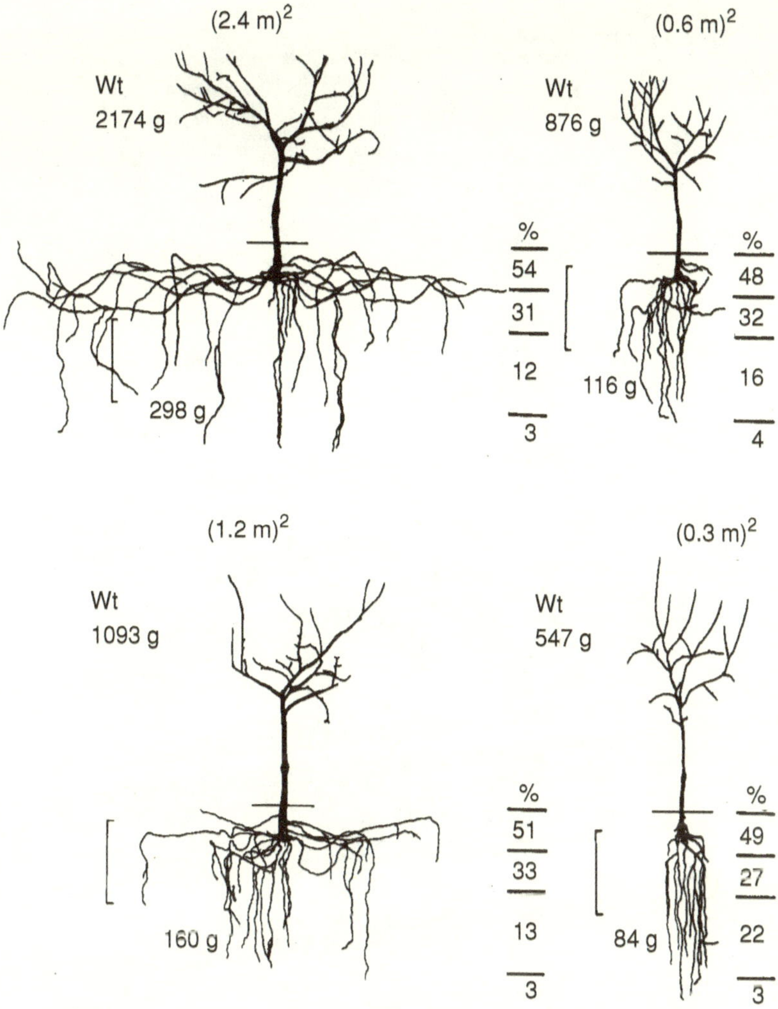

Figure 3.1 The form of an apple tree root system ('Golden Delicious' on 'M.9' rootstock) as affected by tree spacing showing the weight of roots and shoots and the distribution of roots with depth. The vertical brackets represent 0.5 m. From Atkinson (1978). Reproduced with permission.

Root initiation

With the exception of the roots of seedling rootstocks, all the roots of apple and pear trees are adventitious in origin, i.e. arise from plant parts other than by the normal development of the seedling root and its branches. The

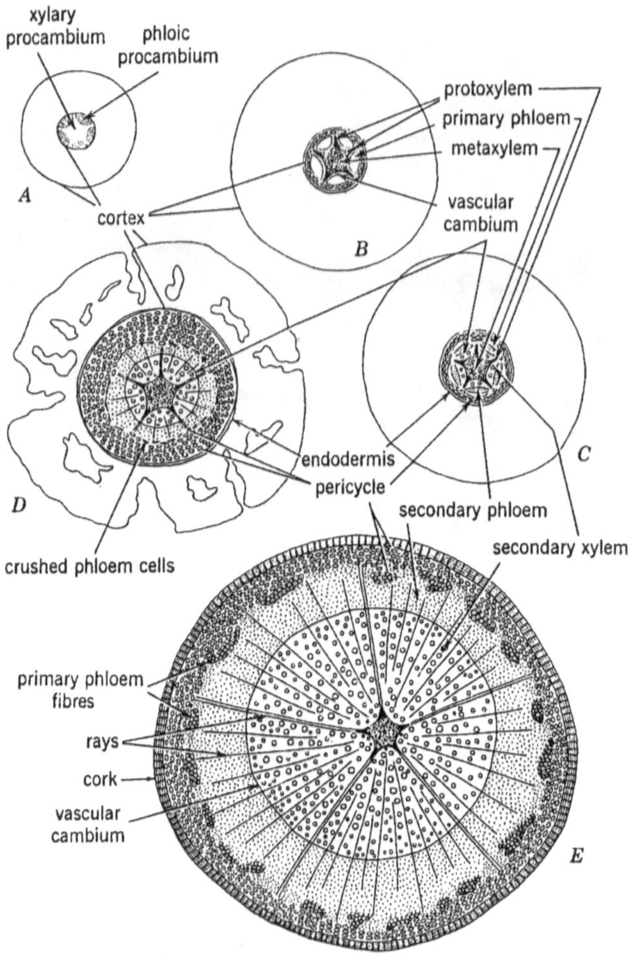

Figure 3.2 Development of *Pyrus* (pear) root. Transections. A, vascular cylinder in procambial state. B, primary growth completed. C, strips of vascular cambium between phloem and xylem have produced some secondary vascular tissues. D, vascular cambium, now a cylinder, has produced additional secondary tissues; pericycle has undergone periclinal divisions; endodermis partly crushed; cortex breaking down. E, secondary growth has progressed further; periderm has appeared, cortex has been shed. Cambium opposite protoxylem poles has formed wide rays (D.E.). All × 26. From Esau (1965). Reprinted by permission of John Wiley & Sons Inc.

name 'adventitious', meaning 'coming from without, accidental, casual, in an unusual position' (*Oxford English Dictionary*), is misleading. Such roots are in fact internal in origin and, far from being exceptional, are the norm for most of the plant kingdom (Groff and Kaplan, 1988; Harper *et al.*, 1991). Most of the so-called adventitious roots of apple and pear are more properly described as shoot-borne roots, although 'adventitious' roots can also be regenerated from root-pruned seedlings. The capacity to produce such roots is dependent on two basic characteristics of plant cells: totipotency, i.e. the fact that each living cell contains the genetic information needed to reconstitute all plant parts and their functions; and the ability of previously developed, differentiated, cells to re-differentiate and develop a new growing point.

Shoot-borne roots of apple and pear arise in two ways:

1 As pre-formed roots which develop naturally on stems. These generally lie dormant until they are placed in environmental conditions suitable for development from primordia into actual roots. They may, however, develop into swellings called burr-knots and show evidence of root growth even when still above ground.
2 As wound roots which develop *de novo* when a shoot is severed from the parent plant and used as a cutting. In this case there is a clear sequence in which previously differentiated cells de-differentiate, root initials form from de-differentiated, newly meristematic cells near vascular bundles, these develop into root primordia and these primordia grow outwards through the stem tissue to emerge while also forming vascular connections with the vascular tissues of the stem (Hartmann *et al.*, 1990). The term de-differentiation is used here to describe a specific stage of differentiation into a different anatomical structure.

There are many features in common between the development of pre-formed roots and wound roots, but it is simplest to consider them separately.

Pre-formed shoot-borne roots are developed by the nursery practices of stooling and layering. In these the stems of rootstock cultivars are covered in soil while still attached to the parent plant and harvested as rooted shoots suitable for budding or grafting with a scion cultivar. Most of the commercial apple rootstocks are rooted in this way. They also develop from that part of the rootstock stem which is buried in the soil when the composite tree is planted in the orchard and also from any part of the scion trunk which becomes buried. The new roots which develop from the below-ground rootstock stem in the orchard are of particular benefit with respect to tree anchorage if the trees are deep-planted (Rogers and Parry, 1968; Parry, 1974): the new roots which may emerge from scions if these are partially buried are usually undesirable because such an 'own-rooted' tree loses the specific benefits conveyed by the rootstock.

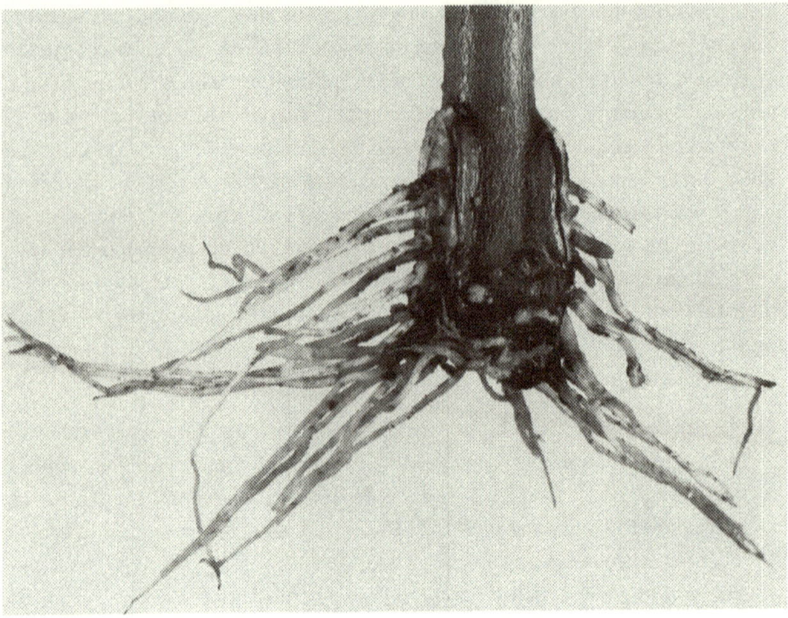

Figure 3.3 The effect of splitting the base of a cutting on root emergence: roots emerging in ranks from the wound. From Howard *et al.* (1984). Reproduced with permission.

Root emergence from layered shoots of *Malus* rootstocks is largely confined to nodal positions near lateral buds (Doud and Carlson, 1977). It is closely associated with parenchymatous, starch-rich tissues in bud and leaf gaps in the stem.

Wound-induced roots are produced by cuttings. In some cases cuttings already have pre-formed roots. This is so with quince (*Cydonia oblonga*), where a small slice of older wood (a 'heel') with pre-formed root initials is retained at the base of the cutting to obtain maximum rooting (Hartmann *et al.*, 1990). In other cases the presence of nodal rooting sites on the cutting is important. Rooting potential of apple cuttings is highest if they are from the basal part of the annual shoot as the result of a rosette of buds with their associated nodal sites (Howard, 1971, 1987). In general, however, rooting of cuttings is stimulated by wounding, which is especially important when the cutting base is of internodal tissue. Outward-pointing cambial salients develop in the callus which arises following stem wounding. The wounding which stimulates this is the cut across the base of the hardwood (winter) cutting, which has separated it from the parent plant, or a vertical cut which splits the base of the cutting (Figure 3.3).

Lateral roots arise in the pericycle of the parent root some distance from the apical meristem of this and grow out through the cortex. In apple and pear root systems the first laterals produce secondary laterals and this process continues to give up to eight 'orders' of laterals within a single season (Rom, 1987).

Factors influencing root initiation and emergence

JUVENILITY FACTORS

These are of particular importance because the rootstock cultivars which have the most desirable effects on the scions grafted on them are quite often difficult to root.

In all woody plants there is a juvenile phase during which flowering does not occur. When flowering occurs the tree is said to have attained maturity. The phase change from juvenility to maturity is accompanied by changes in a number of characteristics, one of which is a reduction in the ability to produce adventitious roots from the shoots (Gardner, 1929; Hackett, 1989).

In apple the juvenile phase, defined as time from seed to flowering, is short compared with that of many trees. It ends when the tree reaches a characteristic height or number of nodes and so is shortened by good management, improvement in growing conditions and length of growing season. Visser *et al.* (1976) reported that through the course of 20 years the juvenile period of comparable groups of seedlings was reduced from 7.4 to 4.3 years in apple and 9.2 to 6 years in pear.

The period of high rooting ability is even shorter. Gardner (1929) reported that softwood cuttings from the tops of apple seedlings during the first year of growth could be rooted easily whereas cuttings from 2- to 4-year-old seedlings rooted poorly. The technology of propagating apples through the use of rooted cuttings from vegetatively propagated plants, some centuries away from seedling progenitors, therefore requires a detailed understanding of methods of maintaining and restoring the ready rooting normally associated with juvenility in the face of apparently unpromising circumstances. Although it is conventional to define juvenility in relation to flowering there is evidence that juvenile behaviour as regards rooting may be separable from juvenile behaviour as regards flowering ability (Hackett, 1989). In the following discussion an effect on rooting will be considered as a juvenility effect if it is associated with effects on other morphological, anatomical and physiological traits typical of a juvenile plant or part of a plant. Such traits include the presence of many spines (short axillary laterals) on the stems, thin leaves, prolonged leaf retention in autumn, reduced leaf and stem pubescence, and abundant anthocyanin production in leaves and stems (Stoutemyer, 1937; Howard, 1987). Beakbane (1961) also noted that in seedling apple trees the basal juvenile shoots

Table 3.1 *The effect of different degrees of severity of pruning of apple rootstock hedges on the percentage rooting of cuttings taken from them*

Pruning	Rootstock		
	'M.26'	'MM.106'	'MM.111'
Ultrasevere (23 cm framework)	88	48	63
Severe (45 cm framework)	75	48	60
Normal (100 cm framework)	42	25	58
Light (100 cm framework and one-third of annual shoot length retained)	23	25	33

From Howard, (1987). Reprinted by permission of John Wiley & Sons Inc.

are relatively free from sclerotic cells in the primary phloem compared with lateral shoots from fruiting branches close to the apex.

Seedling apple and pear trees exhibit juvenile characteristics throughout when very young. As the tree matures the plant tissues in the lower part of the tree remain juvenile, Beakbane (1961, 1969) finding typical juvenile shoots at the base of 10-year-old seedling trees which were fruiting at the top. Juvenile characteristics are also found in shoots developing from adventitious buds in roots (root suckers), watersprouts from latent buds in the basal parts of trees (epicormic shoots) or sphaeroblasts, which are masses of shoots from adventitious buds in the trunks of trees and can be induced to develop by heavily cutting back stock plants (Stoutemyer, 1937; Wellensiek, 1952; Baldini and Mosse, 1956; Beakbane, 1961; Rom and Brown, 1979).

Cuttings from severely pruned apple rootstock hedges, stools and layer beds, which all show juvenile characteristics, have enhanced rooting ability (Howard, 1987; Table 3.1), as do cuttings of shoots developing from adventitious buds on the roots of mature apple trees (Robinson and Schwabe, 1977) and shoots developed from sphaeroblasts on apple (Stoutemyer, 1937).

In vitro culture can be used as a technique for rejuvenation. Although propagation of apple trees *in vitro* is achieved most readily with explants from young nursery trees (Jones, 1983), it can be fully successful with shoot tips in either the juvenile or the adult phase (Welander and Huntrieser, 1981). Rejuvenation takes place during *in vitro* subculture and the capacity of the *in vitro* shoots to initiate roots increases with the number of subcultures (Figure 3.4). Shoots of 'M.7' apple rootstock required 2 months of subculture before more than 70% of them could be rooted, while the corresponding figure for the apple scion cv. 'Greensleeves' was 5 months (Jones *et al.*, 1982), and the cvs. 'Jonathan' and 'Red Delicious' required 6 and 36 months of subculture, respectively, before rooting was achieved (Sriskandarajah *et al.*, 1982). Apple trees produced by grafting scions and rootstocks produced by micropropagation show

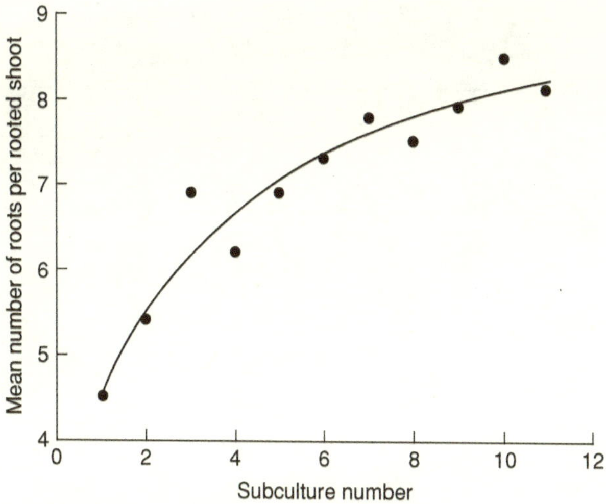

Figure 3.4 Rejuvenation with increasing subculture of initially poor-rooting somaclones of 'Greensleeves' apple. From Donovan *et al.* (1994). Reproduced with permission.

juvenile-like characteristics when grown in the field (Jones and Hadlow, 1989), and conventional cuttings taken from micropropagated plants (plants produced *in vitro*) showed improved rooting for both apple and pear (Table 3.2; James *et al.*, 1987; Jones and Webster, 1989; Webster and Jones 1992). *In vitro* culture may lead to excessive development of burr-knots and suckers (both characteristic of juvenile, ready-rooting material) when the rootstocks are planted directly from micropropagation into the field. This problem is reduced if the rejuvenated plants are used as source material for conventional propagation (Jones and Webster, 1993).

POLARITY

Stem cuttings form shoots at the distal end, i.e. nearest the shoot tip, and roots at the proximal end, i.e. nearest the original root system. This polarity is correlated with auxin movement (Robinson and Schwabe, 1977). Rooting potential is higher in the proximal parts of annual shoots of apple rootstocks, especially in the basal region (Howard, 1987; Howard *et al.*, 1983).

SCLERIFICATION

The production and emergence of shoot-borne roots is negatively related to the continuity of the sclerenchymatous sheath arising from the primary phloem. This relationship holds with respect to the rooting of dormant winter cuttings (hardwood cuttings) of a range of apple rootstocks and apple and pear

Table 3.2 *Rooting of softwood shoot cuttings from micropropagated (M) and conventionally (C) propagated rootstocks*

	Shoot cuttings		
Rootstock	Source	Number	% rooted after 4 weeks
'M.9'	M (a)	10	70
	M (b)	7	71
	C	10	30
'B655'	M	10	80
	C	10	0
'B657'	M	10	60
	C	10	0
'B660'	M	6	17
	C	10	0
'B661'	M (c)	10	40
	M (d)	10	80
	C	10	40
'B654'	M	4	0
	C	10	20

a, micropropagated plants from a 3-year-old shoot culture line; b, micropropagated plants from an 8-year-old shoot culture line; c, micropropagated plants from a 2-year-old shoot culture line; d, micropropagated plants from a 3-year-old shoot culture line. All other micropropagated plants from 2-year-old shoot culture lines.
From James *et al.* (1987). Reproduced with permission.

scion cultivars (Table 3.3; Beakbane, 1961, 1969), of juvenile as compared with mature shoots, of etiolated as compared with un-etiolated layer-bed shoots of apple rootstocks (Doud and Carlson, 1977), and of wounded or split-based apple rootstock cuttings as compared with controls (MacKenzie *et al.*, 1986; Pontikis *et al.*, 1979). The effect appears to be on root initiation and differentiation, with a continuous sheath of phloem fibres preventing the development of outward-pointing meristematic masses such as occur at the lateral margins of leaf traces or after wounding (MacKenzie *et al.*, 1986). Wounding also facilitates the penetration of IBA (indolebutyric acid) when this is applied to aid the rooting of cuttings.

WOUNDING

Wounding is intrinsic to the process of taking cuttings and the production of roots from a cutting is, at least initially, mainly from tissues close to the wound. The wound surface can be increased by scoring or splitting the base

Table 3.3 *Relation between propagation capacity and primary phloem structure in stems*

Cultivar	Propagation capacity	Continuity of fibre ring[a] %	Continuity of fibre ring[a] Mean
Apple rootstocks			
'M.13'		28	
'M.5'	Excellent	43	41
'M.11'		52	
'M.4'		59	
'M.16'	Good	64	64
'M.2'		69	
'M.12'		69	
'M.25'	Fair	71	75
'M.8'		85	
'Crab C'		82	
'Northern Spy'	Poor	84	84
'Ivory's Double Vigour'		86	
Scion cvs.			
Apple			
'Bramley's Seedling'		86	
'Cox's Orange Pippin'	Poor	83	80
'Lane's Prince Albert'		70	
Pear			
'Conference'		99	
'Williams''	Poor	96	98

[a] Proportion of radii through the primary phloem blocked by sclerenchyma.

From Beakbane (1961). Reprinted with permission from *Nature*, Macmillan Magazines Limited.

of a cutting. Characteristically there is callus formation and production of root primordia along the margins of the wound (MacKenzie *et al.*, 1986). The wounded surface permits penetration of synthetic auxin rooting compounds and also there is probably a natural accumulation of auxins and carbohydrates in the wounded area. Wounding by disrupting the sclerenchymatous sheath in winter (hardwood) cuttings permits the formation of cambial salients, as noted earlier.

EFFECTS OF BUDS AND LEAVES

Hartmann *et al.* (1990) present photographic evidence (in their publication, courtesy W. Chantarotwong) that the presence of leaves and treatment of the cutting base with auxin are both needed to induce vigorous rooting of leafy (summer) cuttings of 'Old Home' pear. Removal of the leaves prevented rooting even with auxin treatment, and reduction of the leaf area to one quarter had a major negative effect, but bud removal had no apparent effect.

Howard (1987) concluded that the negative effect of disbudding on the rooting of leafless winter cuttings of apple was primarily due to desiccation through disbudding wounds.

The need for leaves on summer but not winter cuttings may reflect differences in carbohydrate reserves.

CARBOHYDRATE RESOURCES

All growth requires energy. A carbohydrate source, usually sucrose, is an essential constituent of tissue-culture media and the regeneration potential and survival rate of root cuttings is highest if they are taken at the time of maximum accumulated reserves of storage polysaccharides (Robinson and Schwabe, 1977). Summer cuttings (see above) require the presence of leaves to root even if auxin is supplied. Leafless winter cuttings are much larger and have the benefit of carbohydrate exported from the leaves during and at the end of the growing season. Their ability to root is not necessarily related to initial differences in carbohydrate content. Rooting and subsequent establishment of these cuttings may, however, be reduced by factors such as high temperatures and bud growth which lead to increased respirational losses of carbohydrate or to its redistribution away from the rooting zone (Cheffins and Howard, 1982a,b).

PLANT GROWTH SUBSTANCES

Auxins, co-factors and inhibitors

Following work on other plants which showed that auxin was involved in adventitious root formation, Pearse and Garner (1937) found that immersion of the bases of softwood (leafy) cuttings of *Pyrus communis* rootstock in 40 ppm NAA (naphthaleneacetic acid) for 12 hours was very effective in inducing rooting. The control cuttings did not root and many subsequent studies have shown synthetic auxins to be effective in inducing rooting of cuttings which would not otherwise do so. IBA (indolebutyric acid) and NAA are more effective than the naturally-occurring indoleacetic acid (IAA). IBA is now generally used to root apple and pear cuttings, being preferred to NAA because of its ability to induce a fibrous root system and not to translocate in the stem and inhibit bud growth to the same extent as NAA (Howard, 1987). IBA is applied to the bases of cuttings as the acid dissolved in ethanol or acetone, as the potassium salt dissolved in water, or as a powder formulation. In the United Kingdom a 'quick dip' of the base of the cutting for 5 seconds to a depth of 1 cm in 2500 ppm IBA in a 50% aqueous solution of acetone is routinely used for the apple rootstocks 'M.26', 'MM.111' and 'M.27' (although only a small proportion of these rootstocks are propagated by winter cuttings). A concentration of 1000 ppm is used for quince rootstocks if they do not bear pre-formed roots (Howard, 1987).

Treatment with IBA of the roots of dormant pear trees in the nursery prior to planting out can lead to increased root production after this (Looney and McIntosh, 1968).

Divisions of the first root initial cells of a range of plants have been shown to be dependent on either applied or endogenous auxin (Hartmann *et al.*, 1990). The process of root elongation which follows initiation is either insensitive to, or inhibited by, auxin. James (1983) showed that the auxin-sensitive phase for apple cuttings *in vitro* lasts for only 6 days, during which time root initiation is very sensitive to incident light.

The effect of auxin on rooting is generally enhanced by the action of rooting co-factors (Hess, 1961). These have been studied in both *in vitro* (micropropagation) and large cutting (macropropagation) systems.

Jones and Hatfield (1976) showed that phloroglucinol, a phenolic compound, and auxin together induce rooting of apple rootstocks *in vitro*, and James and Thurbon (1981) found that phloroglucinol acted as an auxin synergist in root initiation in 'M.9' apple rootstock shoot cultures over a range of concentrations from 16.2 to 1620 mg l^{-1}.

Challenger *et al.* (1965) used the mung bean bioassay to find rooting co-factors in 'Crab C' and 'M.26' apple rootstocks, and Ashiru and Carlson (1968) found strong root promoting factors in the easy-to-root 'MM.106', and lower amounts, and also rooting inhibitors, in the difficult-to-root 'M.2'. Fadl and Hartmann (1967a, b) isolated an endogenous root promoting factor in basal sections of the easy-to-root 'Old Home' pear cultivar cuttings after IBA treatment but not in similar cuttings of the difficult-to-root 'Bartlett' cultivar. This rooting factor was possibly a condensation product of the applied auxin and a phenolic substance produced by the buds. Bassuk *et al.* (1981) showed a positive correlation between endogenous root-inducing co-factor activity in sap and seasonal changes in rooting of 'M.26' winter cuttings and Bassuk *et al.* (1981) found an apparent involvement of polyphenol oxidase and phloridzin in the production of apple rooting co-factors.

Cytokinins

Although in general cytokinins are inhibitory to adventitious root formation (McCown, 1988), they may have a positive indirect effect through rejuvenation by repeated subculturing *in vitro* (Sriskandarajah *et al.*, 1982).

Gibberellins

Takeno *et al.* (1983) found a negative relationship between endogenous gibberellin level and the *in vitro* rooting ability of *Malus*.

SOURCE OF PROPAGATING MATERIAL

The conventional production of clonal apple rootstocks is by the stoolbed technique in which one-year-old rooted shoots are planted in rows and established for one season before being cut close to ground level (stooled). Shoots are produced from the stumps and encouraged to root during summer by mounding soil around them. The rooted shoots are then removed by cutting below the new root system during the following winter. This stooling process is repeated annually with each new flush of shoots. Howard (1987) found that the use of good quality shoots (>60 cm in length) to establish the stoolbeds led to larger permanent root systems and many more rooted shoots (liners) being produced each year with the difficult-to-root 'M.9a' rootstock, although effects on the easy-rooting 'MM.111' were slight.

Where propagation is by winter cuttings (hardwood cuttings), as with the easy-to-root 'M.26' apple rootstock, the cuttings taken at the base of the annual shoot root most readily. After the removal of the winter cuttings it is important to prune the framework shoots of the hedges to three or four buds. This stimulates the production of shoots with a high rooting potential compared with shoots from un-pruned or lightly pruned bushes (Howard, 1987; Table 3.1).

For quince, better rooting of winter cuttings is achieved if they have a 'heel' of older wood, with pre-formed root initials, retained at the base of the cutting.

Leafy summer cuttings (softwood cuttings) are not generally used to produce apple and pear rootstocks. Their small size at rooting, and the need for mist to prevent them from desiccating, results in them being much more expensive than winter hardwood cuttings to grow to a size suitable for budding or grafting. They can, however, be used when very rapid multiplication is required. Cuttings are generally taken from the tips of shoots, which probably have more cells capable of becoming meristematic and less sclerified tissue than older shoots. Lateral shoots, about 3 inches (7.5 cm) in length and not in active growth, are preferred. They are best obtained from bushy stock plants established in pots and brought into a warm house in late winter. These provide excellent cuttings in May and early June which can quickly be induced to root and are fit to be transferred to open ground by early July (Garner 1944, 1988).

SEASON OF COLLECTION OF WINTER CUTTINGS

The rooting potential of apple cuttings is at a peak in late winter and early spring (Table 3.4). These changes in rooting potential are associated with changes in rooting co-factor activity (Bassuk and Howard, 1981). Rootstocks of quince, which leaf out early in spring, are best propagated in autumn when their buds are fully dormant (Table 3.5).

Table 3.4 *The effect of time of taking cuttings of 'M.26' apple rootstock and of the propagation environment on the percentage rooting*

	Percentage rooting	
Dates (1983/84)	RH 80–90%	RH 50–60%
9 November	71	32
14 December	30	69
18 January	15	69
22 February	61	78
28 March	90	51

From Howard (1987). Reprinted by permission of John Wiley & Sons Inc.

Table 3.5 *Establishment of Quince cuttings related to season of direct field planting or prior stimulation with basal heat*

	Collection date		
	Mid-Nov	Mid-Jan	Mid-Feb
Prior treatment for 2 wks at 21 °C	76	85	12
Directly planted into field	61	8	2

From Howard (1987). Reprinted by permission of John Wiley & Sons Inc.

ETIOLATION AND BLANCHING

Strictly speaking, etiolation is the development of plant parts in the absence of light, but horticulturalists use the term to include development under heavy shade. Banding is localized light exclusion. Bands of black adhesive tape or Velcro are applied both to etiolated shoots and to light-grown shoots in the softwood stage to exclude light from the portion of a stem that will be used as the cutting base. When applied to non-etiolated shoots it is said to blanch the relevant part of these. Shading refers to reducing the light level under which stock plants are grown.

These treatments 'precondition' apple and pear shoots to root when applied at a number of different stages of development. Covering 'M.9' apple rootstock hedges with black polythene structures for a month from budburst raised the proportion of subsequent summer cuttings rooting from 14% to 84% (Harrison-Murray and Howard, 1981). Although total light exclusion during the period of shading stock plants of 'M.9' in the field maximized the improvement in rooting of their cuttings, compared with those from unshaded plants, even reduction of light to 70% of ambient had a substantial effect on both the percentage of cuttings rooted and the number of roots per cutting (Howard, 1983). Blanching 'MM.106' rootstock shoots by application of a 2.5 cm wide black adhesive tape to a de-leafed node just behind the shoot tip greatly increased rooting of cuttings subsequently taken in winter with blanched bases, the rooting percentage of these blanched distal winter cuttings being 81% compared with 24% for unblanched distal and 77% for unblanched basal cuttings (Heighton and Howard, 1984). In these studies IBA was applied to the cut surfaces of the cuttings as is conventional for summer and winter cuttings. The 'pre-etiolation' effect can, however, be demonstrated in the absence of IBA treatment. Harrison-Murray *et al.* (1985) found that etiolation of sucker shoots of 'M.9' resulted in 92% rooting of summer cuttings even without IBA application. Banding also induces root primordia emergence both on young apple stems still on the stock plant (Gardner, 1937), and on much more mature rootstock stems below the graft union during their maiden year in the nursery (Harrison-Murray *et al.*, 1984; Jackson *et al.*, 1984). In the latter case the portion of rootstock stem which had been banded while above ground in the nursery gave rise to a new root system when deep-planted in the orchard.

The rooting response to etiolation appears to be complex: the associated effects range from changes in stem anatomy which are correlated with ease of rooting to changes in the light control of auxin metabolism, factors affecting the activity of IAA-O and in the activity of rooting co-factors (Maynard and Bassuk, 1988).

TEMPERATURE

In cool-winter climates such as that of the United Kingdom winter (hardwood) cuttings of apple do not root adequately if planted directly in the field, so are rooted in cutting bins with the benefit of base heat and transplanted after initiation of rooting. A temperature of 20 °C at the cutting base has been found to be suitable for rootstocks (Howard, 1987) and is generally used although higher temperatures, up to 30 °C, are more effective especially with difficult-to-root material such as 'Cox' apple scionwood cuttings (Fenlon, 1984). The period of exposure to bottom heat is usually limited to about 4 weeks: longer periods result in poor establishment, probably as a result of depletion of carbohydrates (Howard, 1978). Appreciable rooting is stimulated at 17 °C (Howard, 1968) and in countries with warm winters and relatively high soil temperatures in the dormant season apple rootstock cuttings treated with IBA can be rooted directly in the field.

HUMIDITY AND INTERNAL WATER STATUS

Although winter cuttings of apple and quince are leafless their rooting is influenced by the relative humidity of their environment and the water content of the rooting medium. The effect of relative humidity varies with time of season; excessive water content in the rooting medium leads to rotting instead of rooting.

Root growth of the rooted cutting and during tree establishment

The growth of roots of rooted cuttings and of orchard trees can be best understood in terms of the environmental factors which affect root growth *per se* and of those involving the functional equilibrium between roots and shoots.

This concept of functional equilibrium (Brouwer, 1963, 1977) is that shoots and roots are interdependent with each supplying essential materials for the growth of the other. Thus a reduction in the function of the one generally results in a reduction in the growth rate of the other (Brouwer and De Wit, 1969), if all other factors remain constant. The general relationship has been reviewed by Klepper (1991). Root/shoot ratios tend to be higher when soils are deficient in water and nutrients than when these are readily available but under any given set of conditions plants of all species tend to restore the ratio between root and shoot when part of either is removed. For example, relative shoot growth increases and the proportion of dry matter increment going to root decreases after grazing or defoliation of grasses.

Root growth in the nursery

Following rooting or root initiation in stoolbeds or cutting bins the rootstocks are transplanted to the site at which they will be budded or grafted with the scion cultivar.

In the stooling system the rooted shoots are harvested by hand or mechanically from the stoolbed (Howard, 1987). They usually have numerous roots but the proportion of root to shoot dry weight is very low. Vyvyan (1934) found the root/shoot ratio of typical rooted shoots taken from a stoolbed of 'M.13' to be 0.031 at the time of harvesting and planting out in April. By December the stem weight had approximately doubled but root weight had increased more than 10-fold to give a root/shoot ratio of 0.232, which is within the range normal for young apple trees (Avery, 1970).

In the winter-cutting system, cuttings of easy-to-root subjects may be planted in the nursery field after only two or three weeks' basal heat treatment in the cutting bins, with most of them showing only good callus development and only a few showing roots (Howard, 1987). Slower-to-root subjects are given longer exposure to heat and require the presence of a few roots on each cutting prior to transplanting, but even with these the root to shoot ratio is very low. In most rootstocks a proportion of the roots developed in the cutting bin survive transplanting and further roots emerge from the buried rootstock stem during the growing season. Newly rooted hardwood cuttings establish just as well if the roots are removed before planting as if the roots are retained (Howard, 1971).

Rooting during orchard establishment

A rootstock planted in winter in a nursery is usually budded in the following summer, when in rapid growth, and the bud remains dormant until the following spring. If the rootstock is grafted with dormant scionwood this is done in the second spring after planting. The rootstock shoot above the bud or graft is removed, temporarily increasing the root to shoot ratio, but the balance is restored by rapid growth of the scion in the nursery. It is reduced again by the process of lifting trees from the nursery, transporting and transplanting them. This is done during the winter prior to budbreak.

Hatton *et al.* (1924) reported on a number of studies involving excavation of apple root systems within 2 years of planting. They commented that their results 'confirmed, in a modified form, the oft-repeated assertion that, after removal from the nursery and replanting, the young tree has practically to start afresh in its root-making'.

Tamasi (1986) reported that during the first 2 years after planting apple trees on seedling rootstocks, new roots grew almost exclusively from thick 'skeletal' roots cut back at planting and after 3 years roots began to develop from the rootstock neck.

Abod and Webster (1989) assessed the mortality of the transplanted root systems of one-year-old transplanted trees of 'Spartan' on 'MM.106' and 'M.9' rootstocks by measuring total 'old' root length on trees at the date of planting and at excavation at 15-day intervals after this. This length of 'old' roots declined to 80%, 60%, 40% and 25% of the initial length at 15, 30, 45 and 120 days after planting, respectively. No new roots were regenerated in the first 15 days from planting but by 30 days the trees on 'M.9' had 35 m of new roots and those on 'MM.106' had 12 m. By 120 days they had 133 m and 174 m, respectively. Fifty-eight per cent of the new 'M.9' roots originated directly from the rootstock stem, 27% from old coarse roots (>2 mm diameter) and 15% from old fine roots. For 'MM.106' the corresponding figures were 23%, 58% and 19%.

This rapid root regeneration can lead to excellent new root systems being produced within 2 years even when all existing roots are trimmed off at the time of planting into the orchard (Hatton *et al.*, 1924; Figure 3.5). This new root establishment is, however, accompanied by a severe check to shoot growth (Hatton *et al.*, 1924; Preston, 1972; Young and Werner, 1982). The practice of root pruning before planting to enable cheap and rapid planting into holes drilled into the ground (called Stringfellowing after its originator) is therefore not widely employed. A secondary problem is that the mechanical drilling of

(*a*) (*b*)

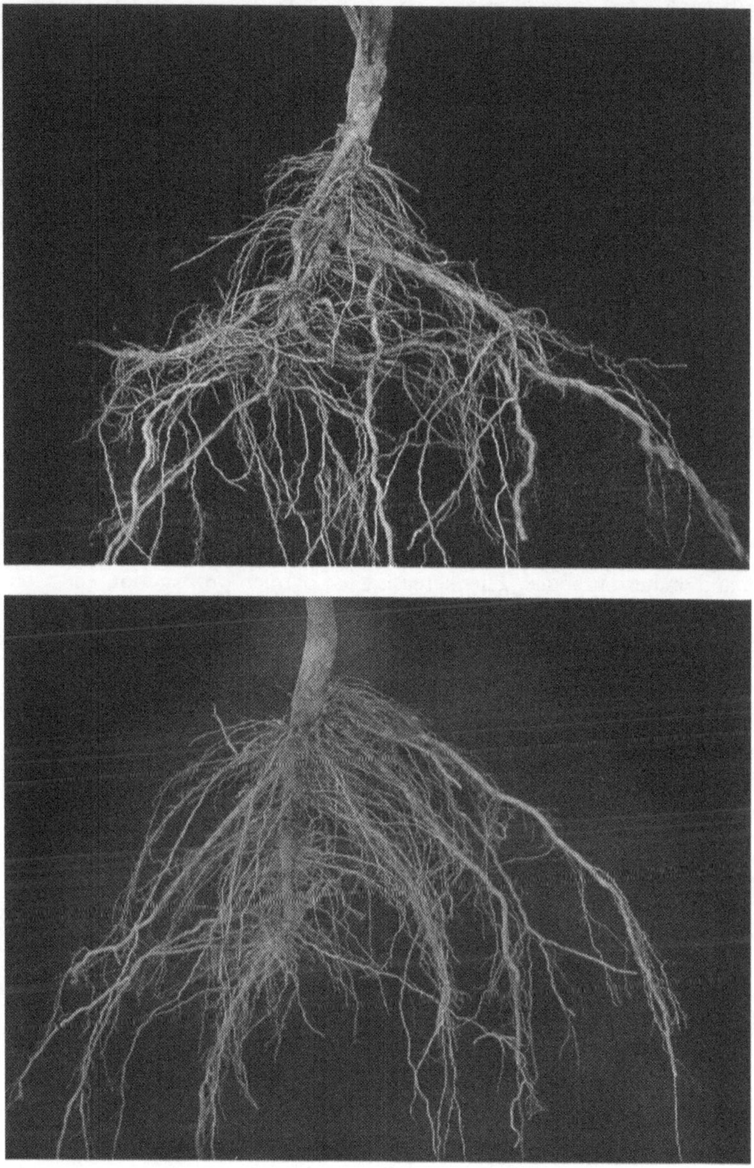

Figure 3.5 Comparison of an apple tree root systems two years after planting, (*a*) with roots intact and (*b*) with all roots cut away. From Hatton *et al.* (1924). Reproduced with permission of Horticultural Research International, East Malling.

the holes may lead to glazing of their inner surfaces with adverse effects on root penetration (Auxt *et al.*, 1980).

Mycorrhizal infection and initial root growth

Apple root systems in the orchard are commonly infected with vesicular-arbuscular mycorrhizal (VAM) fungi (Atkinson, 1983), and pear roots readily become mycorrhizal (Gardiner and Christensen, 1991). The VAM fungi form branched haustorial structures within the root cortex cells and have mycelium or hyphal strands which extend well into the surrounding soil. Water and nutrients absorbed by this mycelium pass back into the root so that the effective absorbing surface of the latter is greatly increased. This fungal infection has beneficial effects on phosphorus nutrition, trace element and water uptake, hormone production, nitrogen fixation and resistance to root disease, although it results in a drain on carbohydrate resources (Gianinazzi-Pearson and Gianinazzi, 1983; Marschner, 1995).

Infection is likely to occur naturally in the course of traditional nursery and orchard practice. The increasing use of micropropagation, the rooting of cuttings in sterilized media and soil fumigation in the nursery and orchard, reduce the likelihood of this. Delays in mycorrhizal infection may have serious adverse effects. Mosse (1957), using apple seedlings, found mycorrhizal plants to have 38% more root dry weight and 25% more shoot dry weight than non-mycorrhizal ones and to have higher concentrations of K, Ca and Fe and lower concentrations of Mn. Sewell *et al.* (1985) also found apple leaf Mn concentrations to be reduced in the presence of a mycorrhizal fungus and found leaf P levels to be increased in the presence of VA mycorrhiza and grass.

Sewell and Roberts (1985) showed that in sterilized soil containing only 17 mg P g^{-1} apple seedlings inoculated with *Glomus caledonicum* grew more than eight times as rapidly as did non-mycorrhizal seedlings. In high-phosphorus soils responses to mycorrhizal infection are less dramatic, but still large (Sewell and Roberts, 1985; Morin *et al.*, 1994).

Rapparini *et al.* (1994) inoculated micropropagated plants of OH × F51 clonal pear rootstock with *Glomus* sp. This resulted in a three-fold increase in shoot growth and a smaller and non-significant increase in root growth by 7 months after inoculation, and also a greater development in the second growing season after overwintering.

Orchard tree root systems

The mature tree root system includes roots which differ in age, thickening and suberization.

The young root is initially white and succulent with short root hairs usually 0.25–0.50 mm in length (Rogers, 1939; Rogers and Head, 1966). It grows at up to 1 cm per day. After 1–4 weeks it begins to turn brown and the root hairs shrivel. The browning takes 2–3 weeks in summer and up to 12 weeks in winter (Head, 1967) and is followed by decay and disintegration of the cortex due largely to the feeding of soil fauna. Some of the roots become secondarily thickened and form part of the perennial root system (Head, 1968b), others remain unthickened or disappear.

Casparian strips and lignified thickenings differentiate in the anticlinal walls of all endodermal and *phi* layer cells respectively, 4–5 mm from the root tips in apple. Suberin lamellae are laid down, on the inner surface of endodermis cell walls only, 16 mm from the tip and an additional cellulosic layer about 35 mm from the tip. Browning, phellogen development, and sloughing off of the cortex commences 100 mm from the tip. Plasmodesmata traverse the suberin and cellulose layers of the endodermis and are especially frequent in the *phi* layer (MacKenzie, 1979).

The growth of the root system in the orchard is controlled by temperature, soil moisture and nutrients, the supply of carbohydrates and plant growth substances from the shoot, and by within-plant and between-plant competition for these resources.

Slow growth of extending roots may continue through the winter but rapid new growth in the spring usually begins when the soil has warmed to about 43 °F (6.1 °C) (Rogers and Head, 1966). The roots can spread to 6 feet (almost 2 m) on either side of the trunk in the first year after planting. Rogers (1939) concluded from earlier studies that the root system is generally more extensive than the shoot system and will grow in any soil area which supplies what it requires, subject to limiting factors. Atkinson (1980) cites studies in which the roots of 'M.9' apple rootstocks grafted with different cultivars extended to 4 m in depth and *Malus sieversii* rootstock roots to 8.6 m, and the radial spread of roots of a range of rootstocks grafted with 'Cox' covered up to 34 m^2 and those of 'Papirovka' covered 104 m^2. This extensive root distribution greatly exceeds the zones in which most of the roots are found in modern orchard systems and should be regarded as showing the potential. In New Zealand the roots of 'average' mature apple trees on 'MM.106' rootstock under good growing conditions can completely explore, although not fully occupy, the soil volume between trees spaced at 5 m × 3.8 m to a depth of at least 1.8 m (Hughes and Gandar, 1993).

The actual root distribution can best be understood by considering root systems as highly plastic entities, the form of which reflects the availability of water and nutrients. Where the soil is highly heterogeneous in these respects roots tend to proliferate where conditions are favourable for growth, i.e. the roots may be few and far between when passing through infertile or intermittently dry soil and the density of rooting increases dramatically where the roots enter a nutrient and moisture-rich layer (Atkinson, 1973a). Where the soil is so managed that certain zones provide better conditions for root growth than others, the bulk of the roots will be found in these regions, so the trees are effectively rooted in a medium much richer than the bulk soil.

Factors limiting orchard tree root growth

SOIL FACTORS

In general the depth of rooting increases from sand to loam to clay (Rogers and Vyvyan, 1934; Coker, 1958). It can be limited by the presence of a seasonal water table or mechanical impedance due to hard rock or an impermeable hard pan and influenced by soil nutrient and water status, temperature and grass and soil management.

Anaerobiosis

The importance of high aeration to rooting in artificial media has been quantified by Harrison-Murray and Fieldsend (1982) and Harrison-Murray (1982). Rooting percentage of 'M.26' apple rootstock increased to about 80% at an air content of 70% (v/v) while below 15% air few cuttings rooted and most rotted. There was no evidence that shortage of water limited rooting at a water content of 7%, the lowest in the trial. In view of the importance of new root formation in the orchard these results are of relevance to orchard trees, although the coarseness of the media used restricts the range of soils to which the results are directly relevant. Wiersum (1980) cites Stolzy and Letey (1964) in arguing that the oxygen diffusion rate (ODR) measured with a platinum electrode should be above $30–40 \times 10^{-8}$ g O_2 cm^{-2} min^{-1} to ensure good root growth. Webster (1978) found that the abundance of small apple roots (<5 mm diameter) was greatly reduced if soil porosity was less than 29–39% and if less than 10% of the soil volume was air-filled at -10 kPa tension. The most common causes of anaerobiosis in soils used for fruit growing are seasonal water tables (Rogers and Vyvyan, 1934), impeded drainage (Weller, 1971) and excessive water supply to a part of the soil profile as a result of inappropriate irrigation technology (Levin *et al.*, 1980; Huguet and Fourcade, 1980).

Where water tables are permanently high, shallow rooting results. Such shallow root zones may be adequate. Temporary saturation as a result of heavy rain or irrigation combined with locally poor soil structure and drainage can, however, lead to local oxygen deficiency and death of roots in the intermittently saturated zones. This is especially likely in summer when respiration rates are high and can lead to nutritional problems accompanying root death (Wiersum, 1980).

Anaerobiosis also has a number of important effects on root function. The accompanying check to respiration and availability of metabolic energy prevents or reduces energy-requiring nutrient uptake. As this is reduced to a greater extent than shoot growth the mineral ion concentration in leaves is reduced. It also reduces hydraulic conductivity of the roots, which can result in leaves wilting as water uptake falls below transpirational loss, and can trigger ABA (abscisic acid) production leading to a check to leaf growth and reduced stomatal conductance. In orchard trees the effect of waterlogging is minimal or not evident in the dormant season. Non-lethal waterlogging in the growing season has greater adverse effects on root than on shoot growth (Olien, 1987).

Oxygen diffusion to deep soil layers may also be impeded in very dense soils and soil crusts and so limit root growth even in the absence of flooding.

Mechanical impedance

The rooting zone is often limited in depth by the shallowness of the soil layer. At East Malling in Kent sandy loam soil overlies ragstone rock at depths from approximately 45 cm to 1.8 m (Rogers and Head, 1966) and although in England it is recommended that the soil should have a depth of at least 45 cm for apple production (MAFF, 1972) the apple soils of the Western Cape in South Africa are usually only 30–50 cm in depth (Huysamer, 1997). Pans of different types may also provide a barrier to root penetration, e.g. the caliche layer of lime or lime-silica cemented hardpan found in some central Washington soils (Dow, 1982) and frangipans in Ohio (Fernandez *et al.*, 1995). Compacted soil may also provide a mechanical barrier. Roots cannot penetrate soil if its strength is greater than 2000 kPa, such compaction arising either from natural processes or use of heavy vehicles (Kotzé, 1996). Root growth is often facilitated by soil cracks, earthworm channels and old root channels. Auxt *et al.* (1980) found that use of a tractor-mounted soil auger to create a planting hole could so compact the sides of the hole as to prevent its penetration by secondary and tertiary roots.

A special case of mechanical impedance is that provided by the use of woven nylon root restriction membranes to contain the root system within a limited volume. The objective is to achieve a consequent reduction in shoot growth. This is achieved partly by concomitant reduction in availability of water. Part of the reduction in growth is, however, found even when irrigation maintains high soil water levels within the restriction membrane and maintains leaf water potential at a similar level to those of control trees with unrestricted roots. It has been suggested that this component of the restriction of root growth is due to effects on root production of plant growth substances. Roots appear to respond to physical impedance, including that caused by root restriction, by producing a transportable chemical signal (i.e. abscisic acid, ABA) which can reduce shoot growth and leaf expansion. This signal, which is also produced in response to soil drying, is discussed in Chapter 12. Roots are also important sources of cytokinins and gibberellins. Root restriction reduces apple fruit growth and size (Atkinson *et al.*, 1997, 2000; Webster *et al.*, 2000) as well as vegetative growth.

Nutrient deficiencies

Rogers and Head (1966) attributed very shallow root systems on light sandy soil at Wisley, where more than three-quarters of the root weight was in the top 30 cm of soil, to lack of nutrients at greater depths. Head (1969b) showed that in the absence of applied fertilizers, 'Worcester'/'MM.104' apple trees did not show any root growth in spring and Weller (1966) found that adding mineral fertilizer close to the tree trunk increased root density there at the expense of root growth elsewhere.

Soil moisture deficits

In general soil moisture appears to be available over the full range from field capacity to permanent wilting point. Root production and extension tend to be greater when water is limiting than under conditions of ample water supply (Jones *et al.*, 1985). Indeed with other plants the increasing root growth rate in water-stressed plants is maintained up to the point where the leaf water potential decreases to a value at which stomata close and net photosynthesis is zero (Cruz Romero and Ramos, 1980). This is clearly an adaptive feature leading to tolerance of dry conditions. However, there is also evidence that apple and pear root growth is reduced under dry conditions. Rogers (1939) found reduced root growth at soil water potentials of −40 to −50 kPa or lower and Afensev (1959) found soil moisture below 16% in summer and autumn very harmful to root growth. Roots proliferate and concentrate where water is available but do not specifically grow towards it. Tamasi (1964, 1986) found that apple trees on 'M.4' rootstocks planted in a sandy soil 100–120 cm above the soil water level produced a more superficial and much more dense root system compared with those on a similar sandy soil with identical cultural practices but planted 200 cm above the water table. The roots of the latter trees did not, in general, reach the soil water table or the layer dampened by it and relied on precipitation and irrigation. When a limited part of a soil with apple or pear roots is irrigated the roots proliferate mainly in the wetted soil. Huguet (1976) found that drench irrigation limited rooting to a superficial zone. Levin *et al.* (1979, 1980) found that the size of apple tree root systems and root distribution pattern responded very quickly to changes in irrigation pattern. Trees growing in a dry Mediterranean climate which had been irrigated by surface irrigation for many years and had widespread root systems adjusted their roots to the small volume wetted by drip irrigation within one season. The soil volume wetted by the tricklers was about 30–50% of that wetted by surface irrigation and the density of roots became related to distance from the tricklers rather than the tree trunk, the highest concentration being within

60 cm of the drip points with very few roots more than 150 cm away. Goode *et al.* (1978) also found, after the introduction of trickle irrigation, that roots proliferated in the wetted zones and there was a reduction in rooting distant from the point of irrigation. The mutability of the root system in response to relative water availability is a very important consideration in the introduction and long-term use of drip irrigation and other high-frequency irrigation methods.

Soil temperature

In apple the onset of root growth seems to occur at 6.2 °C (Rogers, 1939) and the rate of root growth to be temperature dependent, in England reaching its maximum in June and July at soil temperature of about 21–24 °C (Rogers and Booth, 1959–60). The resultant seasonal pattern of root growth may, however, be modified by concomitant changes in soil moisture availability and also by competition from shoots and fruits for resources. Gur *et al.* (1976a) found adverse effects of temperatures above 25 °C although there are differences in rootstock tolerance of high temperature (Nelson and Tukey, 1956; Gur *et al.*, 1976b). Rom (1987), in a review, concluded that new root formation in *Malus* ceases at 35 °C and above 24 °C roots lack succulence and suffer from early death of the cortex. The number of active roots in the top 30 cm of soil may be limited by high temperatures in hot-desert climates. In cold-winter climates surface roots may be killed by winter freezing in lighter textured and drier soils without snow cover: differences in freezing tolerance of different rootstocks have been discussed earlier (pp. 63–5)

Effects of grass competition and soil management

The lateral spread of apple root systems is checked by competition from grass. Where trees are grown in herbicide-treated strips separated by grassed alleyways most of the roots are confined to the herbicide-treated strip (Figure 3.6; Atkinson and White, 1980).

Under the older management system of cultivating the orchard, surface roots growing in the top 12 cm of soil were pruned annually. When the orchards were grassed tree roots grew nearer the surface but were subject to direct competition for water and nutrients (Coker, 1959). Eliminating grass and weed competition either in herbicide-treated squares of 1.44 m^2 around the base of the trunk (White and Holloway, 1967) or in herbicide-treated strips along the tree rows, or when the entire orchard surface was herbicide-treated (Atkinson and White, 1976, 1980), resulted in more apple roots in the surface layers of the soil. Apple trees grown with straw mulch over the soil surface produced more roots than those with grass, cultivation or herbicide soil management. Reckruhm (1974) found more pear roots under mulch than under grass.

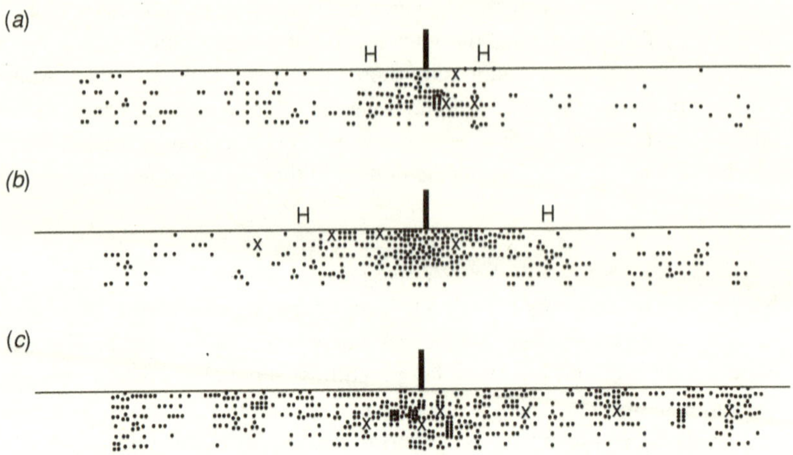

Figure 3.6 The root distribution between the surface and 30 cm depth in a mature apple orchard (•, roots <2 mm diameter; X, roots >2 mm diameter) in relation to the area treated with herbicide (a) Narrow (0.5 m) herbicide strip (b) wide (1.5 m) herbicide strip (c) entire orchard surface herbicided. H shows the boundaries of the area herbicided to each side of the tree row. From Atkinson and White (1980). Reproduced with permission.

Effects of tree to tree competition

Atkinson (1978) showed that the closer the spacing between trees the less the horizontal spread of their roots and the greater the proportion of their roots found below 50 cm in depth (Figure 3.1).

'Replant' effects

Planting apples or pears on land previously used for either of these crops may lead to very adverse effects on root growth and, consequently, on total tree growth and cropping. This will be discussed in detail in Chapter 13.

EFFECTS OF SHOOTS AND FRUITS

The above-ground parts of the tree supply carbohydrates produced in photosynthesis to the root system, re-export mineral nutrients to it and compete with it for these resources. There is experimental evidence of the immediate dependency of root growth on current photosynthesis, although the autumn and spring root growth, which is very important in young trees, obviously depends on reserves. Defoliation even 4–6 weeks before natural leaf fall greatly reduces apple root growth within two weeks of treatment (Head, 1969a), and removal of a ring of bark in September also reduces autumn and winter root growth (Priestley, 1964). Reducing the light received by the leaves of one-year-old pear trees reduces root growth in direct proportion and the effect is reversible within 2 or 3 days (Lucic, 1967).

Roots are, however, a relatively weak sink for assimilates and root growth can be greatly reduced by 'internal' competition from shoots and fruits. Head (1967) and Rogers and Head (1969) confirmed the general pattern of a peak of root growth in May and June, ending at the time of vigorous shoot growth and a second peak in August through October after shoot growth had ended. Quinlan (1965) found that when $^{14}CO_2$ was supplied to individual leaves on a rooted apple shoot, labelled assimilate from the youngest (upper) 8 leaves moved to the stem apex and young leaves. That from the lower leaves moved to the roots, each leaf supplying a specific part of the root system. Pruning apple and pear trees stimulates shoot growth and delays the onset of root growth (Head, 1967, 1968a).

Even light crops of fruit reduce root growth (Head, 1969a). Avery (1970) found that fruiting could lead to a net decrease in root volume of a very dwarfing apple rootstock, presumably because lost roots were not replaced, and Hansen (1980) showed that whereas roots accounted for approximately 20% of the annual growth increment of deblossomed trees of 4- and 5-year-old pot-grown trees of 'Golden Delicious'/'M.4' they accounted for less than 2% of the growth increment of cropping trees. Palmer (1988) found that in heavily cropping field-grown trees of 'Crispin'/'M.27' root weight actually decreased from the second year in the orchard.

Genotype effects

The more vigorous the stock/scion combination, the larger the root system. Data from Hatton (1935), from trials involving 19 rootstocks, shows a positive, linear, relationship between the mass of the root system and that of the tree trunk and branches in a ratio of about 1 to 4, although with wide variations. Any given rootstock had more root mass when combined with the vigorous 'Bramley' scion than with the less vigorous 'Worcester'. Fernandez *et al.* (1995) found that the vigour of trees of 'Starkspur Supreme Delicious' on nine modern apple rootstocks was linearly correlated with the number of rootstock roots. The slope of the relationship varied with soil type and there were some deviations from expectation, e.g. 'M.7 EMLA' had relatively few roots in relation to the vigour it induced in the scion. Even dwarfing rootstocks can, however, exploit the available soil volume at maturity: Rogers and Head (1966) reported 'M.9' roots at 9 feet (2.75 m) in a pocket of deep soil.

In general the root systems of pear seedling stocks are much more vertically oriented than those of quince or of the clonal apple rootstocks. Within apple rootstocks a number of differences in root distribution have been recorded, but effects are inconsistent and may in part reflect differences in tree management (Atkinson, 1980).

Density of rooting

Apple trees, and to a lesser extent pear trees, have very sparse root systems. Reported values of L_A (cm root per cm^2 soil surface) for apple are in the range of 0.8 to 23.8 and for pear from 7 to 69 (Atkinson, 1980). The corresponding range of L_v values (cm root per cm^3 soil volume) is from 0.01 to 0.20 for apple and 0.12 to 0.56 for pear. The very low values may reflect limited exploitation of soil well away from the trunk but even the highest values, obtained from very closely planted trees, are very low compared with those of conifers (L_A for Scots Pine = 126) or *Gramineae* (L_A values of 100–2000).

Density of rooting can be affected by soil type (Fernandez *et al.*, 1995) as well as being greatly increased in the vicinity of trickle irrigation emitters and localized nutrient-rich pockets.

Root death and root system renewal

Root systems are constantly in a process of death and replacement of individual roots. Following the browning of the cortex of young roots this generally decays and disintegrates, largely due to the feeding of soil fauna. Some roots then thicken and become part of the perennial root system but many of the lateral roots, especially, simply disappear (Head, 1968b). Three types of root death are distinguished (Rom, 1987).

1 A systematic dying of short laterals on active roots, following suberization of the roots from which they arose, after 1–2 weeks.
2 Tips of main roots and those of higher branching orders often die back while lateral roots continue to grow.
3 Fibrous roots more than 5 years old may die back completely, to be replaced by new fibrous roots.

Even skeletal or semi-skeletal roots may die, particularly after age five. Any environmental stress increases root shedding and this is also particularly obvious at budbreak and early in the growing season.

This pattern of root death makes an appreciable contribution to soil organic matter status.

All parts of the root system except for the tips appear capable of producing lateral roots from primordia in the pericycle while the tip itself exerts apical dominance on branching. Death of root tips is thus likely to encourage new branching. Tamasi (1986) found that when roots are cut they develop callus and produce many new roots near the cut surface, presumably associated with increased auxin activity such as that demonstrated by Carlson and Larson (1977) after pruning oak roots. Tamasi found that when 58 roots were cut back a total of 160 new roots developed over 4 years from near the cut surfaces and

Faust (1980) found the volume of regenerated roots in solution culture to be twice as high following root pruning as in unpruned controls.

Root/shoot ratios

These are highly variable depending on soil type, tree age, rootstock and the supply of water and nutrients. Rogers and Vyvyan (1934) found the ratio of root to shoot weight of 11-year-old trees to range from 1–1.4 to 1 on sandy soils but from only 0.4–0.5 to 1 on loam soils. Very young trees may have a ratio of only 0.15 to 1 (Avery, 1970). Cripps (1971) found that moisture stress, fluctuating soil moisture availability and waterlogging, which all checked total tree growth, led to increased root/shoot ratios. Rogers and Head (1966) concluded that the ratio is a rough measure of the fertility of a soil. The poorer the soil in terms of water and nutrient supply, the higher the ratio of roots to shoots. Apple trees growing in a high-light environment have much higher ratios of root to shoot than when grown under shade or in a naturally lower-light climate (Cripps, 1972).

Atkinson (1973b) showed that the root/shoot ratio increased progressively with increasing phosphorus and nitrogen deficiencies.

Functions of roots

Anchorage

The firmness of tree anchorage in the soil is dependent on the depth and distribution of the root system, the mechanical strength of the roots and the forces acting on tree stability, primarily wind and the weight of crop.

Apple seedling rootstocks and the clonal rootstocks 'M.25', 'MM.111' and 'MM.106' anchor trees firmly so that they do not require staking. The use of 'MM.109' was largely discontinued due to poor anchorage combined with excessive vigour (Ferree and Carlson, 1987). Apple trees on 'M.7', 'M.4' and 'M.2' often show poor anchorage and tree leaning but not to an extent which justifies all trees being staked. 'Malling 9' and most of its dwarfing derivatives have brittle roots with short fibres and a high proportion of root bark to woody tissues. Under severe windy conditions even their large structural roots will break off cleanly at the base of the trunk. This problem is exacerbated by the very high ratio of crop weight to tree framework, including roots, in trees on dwarfing rootstocks (cf. the general negative relationship between crop yield per unit of tree size and tree size itself, discussed earlier). All apple trees on dwarfing rootstocks, with the possible exception of those on 'MAC.9' ('Mark'), are usually supported either by an individual stake or by a post-and-wire trellis system. This is, however, not solely for reasons of anchorage. It is also because the mechanical supports can be used in tree-training to give the desired branch angles and also to support heavy crops without branch breakage. The alternative to such support would generally involve pruning to stiffen branches, which inevitably delays cropping. An appreciable improvement in anchorage *per se* can be achieved by budding high on the rootstock stem in the nursery and then planting to a greater depth in the orchard. This results in a second root system arising from the newly buried stem (Rogers and Parry, 1968; Parry, 1974), i.e. in root systems distributed over a greater depth.

Almost all of the seedling *Pyrus* rootstocks for pear give good anchorage whereas trees on clonal quince rootstocks require support (Lombard and Westwood, 1987).

Water uptake

Only a very small part of the water uptake by plants is needed to meet their direct metabolic requirement. Most of it is to replenish evaporative water loss through the leaves. This loss is a consequence of the need to have open stomata in the leaves and an internal leaf structure permitting ready gas exchange (including water loss), if carbon dioxide is to be taken up for photosynthesis.

Van den Honert (1948) considered water flux through a plant to be governed by an Ohm's-type law, which can be written

$$F = \frac{\Psi_{\text{soil}} - \Psi_{\text{leaf}}}{r_{\text{p}}} \tag{3.1}$$

where F = transpirational flux, Ψ_{soil} is the water potential at the soil–root interface, Ψ_{leaf} is the leaf water potential and r_{p} is the resistance to liquid flow in the plant.

If the soil is saturated with water, Ψ_{soil} is 0. Ψ_{leaf} is negative when transpiration is taking place but seldom falls below −2.5 MPa (−25 bar) unless there are very severe soil water deficits (Jones *et al.*, 1985).

The water potential of plant cells is in a state of dynamic equilibrium with the transpirational water flux in the xylem. The water potential of an individual plant cell (ignoring matric potential which is usually small) depends on the osmotic potential of its sap (Ψ_{s}) and its (hydrostatic) turgor pressure potential (Ψ_{p}):

$$\Psi_{\text{cell}} = \Psi_{\text{s}} + \Psi_{\text{p}} \tag{3.2}$$

Ψ_{s} is negative because the presence of solutes lowers water potential. If the water potential of the transpiration stream is lowered, i.e. becomes more negative, as when greater tensions develop in the xylem resulting from an increase in transpiration rate, water tends to move from the living cells into the xylem. This is very obvious at mid-day in apple, with actual shrinkage of tree trunks and fruits (Tukey, 1959, 1963), and has a number of negative effects which are discussed later. There are also adverse effects of low tissue water potential (high negative values) on many physiological processes, e.g. cell growth, wall synthesis, protein synthesis and production of new leaf. It follows from equation 3.1 that plant water potential at any given soil water potential and evaporation (transpiration) rate will be more negative the higher the plant resistance to flow. The main source of resistance to water flow in the apple tree is found in the roots; indeed, root resistance is the main resistance to flow in the complete soil–plant pathway with soil resistance only becoming significant at low soil water content (Landsberg and Jones, 1981). This high root resistance can result in severe mid-day water stress even when soil water is not limiting: such stress cannot be alleviated by supplying water to the roots but only by shading the leaves or cooling them by use of over-tree irrigation. In both apple and pear, unlike some other plant species, root resistance appears to be independent of flow rate, i.e. transpiration (Jones *et al.*, 1985), although there is contradictory evidence on this. Dwarfing rootstocks and their graft unions with scions have higher resistance to flow than more invigorating ones (see Chapter 12). Root resistance to hydraulic flow is increased at low temperature and under anaerobic conditions.

The sparsity of apple and pear roots suggests that the soil component of the hydraulic resistance can become significant at higher soil water potentials than is the case for herbaceous species (Landsberg and Jones, 1981). It is probable that both white and woody roots can be effective in water uptake, as they are in cherry (Atkinson and Wilson, 1980).

Although the majority of apple and pear roots are usually within the upper 50 cm of the soil profile, i.e. do not draw water from a greater soil depth than those of many field crops, the frequent presence of at least some deep roots enables apple and pear trees to survive droughts for longer than species which do not have these. Most of the world's high-yielding orchards are, however, irrigated. In this context the ability of apple and pear roots to become concentrated in relatively limited soil volumes maintained at high water potential by drip, trickle or microjet irrigation is particularly important.

Nutrient uptake

Russell (1972, 1977) concluded that uptake of nutrients by intact growing plants is often controlled predominantly by the metabolic demands of the plant as a whole, and Tromp (1980) found a linear relationship between uptake of potassium and total plant growth for apple. Faust (1980) found calcium uptake by apple seedlings to be dependent on the presence of healthy young root tips and either a carbohydrate supply from photosynthesis or a direct supply of sugar. Atkinson (1974) found that the uptake of ^{32}P from soil showed a periodicity which corresponded with that of new root growth. There is, however, evidence that calcium uptake by white and woody roots per unit of root surface area is similar (0.06 and 0.07 nmol Ca^{2+} mm^{-2} h^{-1} respectively), and that at some times of year brown roots make up 100% of the root length of young fruit trees and so must provide the absorbing surface (Atkinson and Wilson, 1980). The zone just behind the root tip, with its root hairs, both provides the greatest surface per unit length of root and also is in undepleted soil. The older secondarily-thickened roots are likely to have better root–soil contact than white roots (Wilson and Atkinson, 1979). These factors are likely to influence the importance of different parts of the root system in nutrient uptake. Uptake by the roots from the soil is frequently ineffective in supplying enough calcium to fruits to ensure optimal storage life, and other nutrients sometimes have to be applied as foliar sprays. This is discussed in Chapters 10 and 11.

Storage of reserves

Growth and fruiting of apple and pear trees in the spring must depend, in their early stages, on mobilization of reserves accumulated in the previous season. Many of these are held in the roots.

CARBOHYDRATES

Hansen (1967), Quinlan (1969) and Hansen and Grausland (1973) found that labelled $^{14}CO_2$ supplied to apple leaves in the autumn was mainly transported down to the roots. In the spring ^{14}C was detected in all new leaf and shoot growth, with the greatest activity in the first-formed leaves. Hansen and Grauslund (1973) found 88% of the ^{14}C in the trees on 20 November to be in the roots and a further 6% in the rootstock. There was then a rapid decline in the amount of ^{14}C in these tissues, especially between 10 January and 1 May when it was only about 40% of the November value. Most of this loss was attributed to respiration, about 25% of it to building materials for new growth.

The quantities of carbohydrate stored in the roots are large. Murneek (1942) found that the roots of 18–year-old apple trees in mid-October contained up to 44% carbohydrates compared with 31% for the above-ground parts. Hemicellulose was the major carbohydrate fraction but there is controversy as to whether this functions as reserve as well as structural material. Starch was the main storage carbohydrate, making up 11% of the dry weight, and sorbitol the main soluble carbohydrate. Sucrose, glucose and fructose are also present (Hansen and Grauslund, 1973).

NITROGEN

Total N concentration in winter is about 1.5% in the roots of well-fertilized young apple trees and about 0.8% is mobilized in spring (Tromp, 1983). The stored nitrogen is mainly in soluble form with the amide asparagine always being an important constituent.

OTHER MINERAL ELEMENTS

Mason and Whitfield (1960) showed that the apple root bark concentration of K decreased from 0.5% in winter to 0.3% in early May. Terblanche *et al.* (1979) found that Ca is redistributed from roots after resumption of growth in spring.

Conversion or synthesis of growth regulators

Root-produced abscisic acid (ABA) is an important signal of soil moisture stress and of impedance to root growth. It helps adapt the plant to these by inducing stomatal closure and checking leaf and shoot growth (see Chapter 12).

Root-produced cytokinins may have important regulatory effects on shoot budbreak, in relation to release from seasonal dormancy and apical dominance, and on other processes involving cell division.

It is possible that root responses to, or production of, plant growth substances are involved in rootstock effects. Movement of labelled IAA from scion leaves into roots of 'MM.106' and 'MM.111' was much greater than that into the dwarfing 'M.9' and 'M.27'. The content of the cytokinin (zeatin + zeatin riboside) in the xylem sap of rootstocks increases with rootstock vigour (Kamboj *et al.*, 1997).

Special features of apple and pear seedling roots

Seedling rootstocks are still widely used for pear and are used to a limited extent for apple.

Production of seedling rootstocks

The seeds used for seedling production are generally those which are most readily available as a result of separating the seeds from the rest of the fruit in the course of fruit processing, e.g. canning or juice production. 'Delicious' is the most commonly used cultivar. Apple and pear seeds are subject to inhibitory dormancy from the fruit and its juices and from seed-coat dormancy. The seeds are thoroughly soaked and washed but with the predominant temperate-zone cultivars there is then still a need for an appreciable period at low temperature under moist conditions (Abbott, 1955). This need for low temperature 'stratification', usually at 4 °C for 90 days (Howard, 1987), is due to physiological deep dormancy (Hartmann *et al.*, 1990) which is overcome by meeting the 'chilling requirement'. Such dormancy parallels the bud dormancy of temperate-zone fruit trees. It is an adaptive characteristic for which there would be natural selection pressure in cold-winter environments in which early seed germination would lead to seedling death by winter freezing. Westwood and Bjornstad (1968) and Westwood (1995) found that pear species from warm-winter environments required less chilling to break seed dormancy than those from cold-winter environments. This suggests the possibility of very different seedling production systems.

Root production in the nursery

Seedlings produce a tap root. If a branched root system is desired this is achieved by under-cutting the seedlings when small.

Seedling root system growth

Tamasi (1986) found that intensive root production by seedling rootstocks after planting in the orchard began only at 4 years of age under conditions where this happened with 'M.4' at 3 years of age. He found that over the years the dominance of the tap root was lost as the lateral roots became better developed.

References

Abbott, D.L. (1955). Temperature and the dormancy of apple seeds. *Report of the 14th International Horticultural Congress Section 2a*, 746–53.

Abod, S.A. and Webster, A.D. (1989). Root and shoot growth of newly-transplanted apple trees as affected by rootstock cultivar, defoliation and time after transplanting. *Journal of Horticultural Science* **64**, 655–66.

Afensev, G.K. (1959). Growth of the active roots of apple trees in the autumn-, winter- and early spring period [in Russian]. *Trudii N.1.1. Sad. Vin i Vin*, **23**: 55–63.

Ashiru, G.A. and Carlson, R.F. (1968). Some endogenous rooting factors associated with rooting of East Malling II and Malling Merton 106 apple clones. *Proceedings of the American Society for Horticultural Science* **92**, 106–12.

Atkinson, C.J., Webster, A.D., Vaughan, S.J. and Lucas, A.S. (1997). Effects of root restriction on the physiology of apple tree growth. *Acta Horticulturae* **451**, 587–95.

Atkinson, C.J., Webster, A.D., Vaughan, S.J., Taylor, L. and Kingswell, G. (2000). Interactions between root restriction, irrigation and rootstock treatments on 'Queen Cox' apple trees: effects of soil and plant water relations. *Journal of Horticultural Science and Biotechnology* **75**, 376–82.

Atkinson, D. (1973a). The root system of Fortune/M.9. *Report of the East Malling Research Station for 1972*, 72–8.

Atkinson, D. (1973b). Some general effects of phosphorus deficiency on growth and development. *New Phytologist* **72**, 101–11.

Atkinson, D. (1974). Some observations on the distribution of root activity in apple trees. *Plant and Soil* **40**, 333–42.

Atkinson, D. (1978). The use of soil resources in high density planting systems. *Acta Horticulturae* **65**, 79–89.

Atkinson, D. (1980). The distribution and effectiveness of the roots of tree crops. *Horticultural Reviews* **2**, 424–90.

Atkinson, D. (1983). The growth, activity and distribution of the fruit tree root system. *Plant and Soil* **71**, 23–35.

Atkinson, D. and White, G.C. (1976). The effect of the herbicide strip system of management on root growth of young apple trees and the soil zones from which they take up mineral nutrients. *Report of the East Malling Research Station for 1975*, 165–7.

Atkinson, D. and White, G.C. (1980). Some effects of orchard soil management on the mineral nutrition of apple trees. In *Mineral Nutrition of Fruit Trees*, ed. D. Atkinson, J.E. Jackson, R.O. Sharples and W.M. Waller, pp. 241–54. London: Butterworths.

Atkinson, D. and Wilson, S.A. (1980). The growth and distribution of fruit tree roots: some consequences for nutrient uptake. In *Mineral Nutrition of Fruit Trees*, ed. D. Atkinson, J.E. Jackson, R.O. Sharples and W.M. Waller, pp. 137–50. London: Butterworths.

Auxt, T., Blizzard, S. and Elliot, K. (1980). Comparison of apple planting methods. *Journal of the American Society for Horticultural Science* **105**, 468–72.

Avery, D.J. (1970). Effects of fruiting on the growth of apple trees on four rootstock varieties. *New Phytologist* **69**, 19–30.

Baldini, E. and Mosse, B. (1956). Observations on the origin and development of sphaeroblasts in apple. *Journal of Horticultural Science* **31**, 156–62.

Bassuk, N.L. and Howard, B.H. (1981). A positive correlation between endogenous root-inducing cofactor activity in vacuum-extracted sap and seasonal changes in rooting in M.26 winter apple cuttings. *Journal of Horticultural Science* **56**, 301–12.

Bassuk, N.L., Hunter, L.D. and Howard, B.H. (1981). The apparent involvement of polyphenol oxidase and phloridzin in the production of apple rooting cofactors. *Journal of Horticultural Science* **56**, 313–22.

Beakbane, A.B. (1961). Structure of the plant stem in relation to adventitious rooting. *Nature* **192**, 954–5.

Beakbane, A.B. (1969). Relationships between structure and adventitious rooting. *Proceedings of the International Plant Propagators Society*, 192–201.

Brouwer, R. (1963). Some aspects of the equilibrium between overground and underground plant parts. *Jaarb, I.B.S., Wageningen*, 1963, 31–9.

Brouwer, R. (1977). Root functioning. In *Environmental Effects on Crop Physiology*, ed. J.J. Landsberg and C.V. Cutting, pp. 229–45. London: Academic Press.

Brouwer, R. and De Wit, C.T. (1969). A simulation model of plant growth with special attention to root growth and its consequences. In *Root Growth*, ed. W.J. Whittington, pp. 224–44. London: Butterworths.

Carlson, W.C. and Larson, M.M. (1977). Changes in auxin and cytokinin activity in roots of red oak (*Quercus rubra*) seedlings during lateral root formation. *Physiologia Plantarum* **41**, 162–6.

Challenger, S., Lacey, H.J. and Howard, B.H. (1965). The demonstration of root promoting substances in apple and plum rootstocks. *Report of the East Malling Research Station for 1964*, 124–8.

Cheffins, N.J. and Howard, B.H. (1982a). Carbohydrate changes in leafless winter apple cuttings. I. The influence of level and duration of bottom heat. *Journal of Horticultural Science* **57**, 1–8.

Cheffins, N.J. and Howard, B.H. (1982b). Carbohydrate changes in leafless winter apple cuttings. II. Effects of ambient air temperature during rooting. *Journal of Horticultural Science* **57**, 9–15.

Coker, E.G. (1958). Root studies XII. Root systems of apple on Malling rootstocks on five soil series. *Journal of Horticultural Science* **33**, 71–9.

Coker, E.G. (1959). Root development of apple trees in grass and clean cultivation. *Journal of Horticultural Science* **34**, 111–21.

Cripps, J.E.L. (1971). The influence of soil moisture on apple root growth and root:shoot ratios. *Journal of Horticultural Science* **46**, 121–30.

Cripps, J.E.L. (1972). The effect of shading and Alar application on apple root:shoot ratios in Western Australia. *Journal of Horticultural Science* **47**, 291–9.

Cruz Romero, G. and Ramos, C. (1980). Growth and hydraulic resistance of roots as affected by water stress and salinity. *Seminaires sur l'irrigation localisée. III. Influence de l'irrigation localisée sur la morphologie et la physiologie des racines.* CNR/University of Naples, 163–70.

Donovan, A.M., Ridout, M.S. and James, D.J. (1994). Assessment of somoclonal variation in apple. II. Rooting ability and shoot proliferation *in vitro*. *Journal of Horticultural Science* **69**, 115–22.

Doud, S.L. and Carlson, R.F. (1977). Effects of etiolation, stem anatomy and starch reserves on root initiation of layered *Malus* clones. *Journal of the American Society for Horticultural Science* **102**, 487–91.

Dow, A.I. (1982). The soil survey as it relates to orchard water management. In *Tree Fruit Water Management and Irrigation. The Proceedings of the 1982 Pacific Northwest Tree Fruit Short Course*, ed. R.B. Tukey, pp. 29–60. Pullman: Washington State University.

Esau, K. (1965). *Plant Anatomy*, 2nd edn. New York: John Wiley and Sons.

Fadl, M.S. and Hartmann, H.T. (1967a). Isolation, purification and characterization of an endogenous root-promoting factor obtained from the basal sections of pear hardwood cuttings. *Plant Physiology* **42**, 541–9.

Fadl, M.S. and Hartmann, H.T. (1967b). Relationship between seasonal changes in endogenous promoters and inhibitors in pear buds and cutting bases and rooting of pear hardwood cuttings. *Proceedings of the American Society for Horticultural Science* **91**, 96–112.

Faust, M. (1980). Interaction between nutrient uptake and photosynthesis. In *Mineral Nutrition of Fruit Trees*, ed. D. Atkinson, J.E. Jackson, R.O. Sharples and W.M. Waller, pp. 193–9. London: Butterworths.

Fenlon, C.A. (1984). Rooting Cox cuttings. *Report of the East Malling Research Station for 1983*, 84.

Fernandez, R.T., Perry, R.L. and Ferree, D.C. (1995). Root distribution patterns of nine apple rootstocks in two contrasting soil types. *Journal of the American Society for Horticultural Science* **120**, 6–13.

Ferree, D.C. and Carlson, R.F. (1987). Apple Rootstocks. In *Rootstocks for Fruit Crops*, ed. R.C. Rom and R.F. Carlson, pp. 107–43. New York: John Wiley and Sons.

Gardiner, D.T. and Christensen, N.W. (1991). Pear seedling response to phosphorus, fumigation and mycorrhizal inoculation. *Journal of Horticultural Science* **66**, 775–80.

Gardner, F.E. (1929). The relationship between tree age and the rooting of cuttings. *Proceedings of the American Society for Horticultural Science* **26**, 101–4.

Gardner, F.E. (1937). Etiolation as a method of rooting apple variety stem cuttings. *Proceedings of the American Society for Horticultural Science* **34**, 323–9.

Garner, R.J. (1944). Propagation by cuttings and layers. Recent work and its application, with special reference to pome and stone fruits. *I.A.B. Technical Communication No. 14.*

Garner, R.J. (1988). *The Grafter's Handbook.* London: Cassell.

Gianinazzi-Pearson, V. and Gianinazzi, S. (1983). The physiology of vesicular-arbuscular mycorrhizal roots. *Plant and Soil* **71**, 197–209.

Goode, J.E., Higgs, K.H. and Hyrycz, K.J. (1978). Trickle irrigation of apple trees and the effect of liquid feeding with NO_3^- and K^+ compared with normal manuring. *Journal of Horticultural Science* **53**, 307–16.

Groff, P.A. and Kaplan, D.R. (1988). The relation of root systems to shoot systems in vascular plants. *Botanical Review* **54**, 387–422.

Gur, A., Bravdo, B. and Mizrahi, Y. (1972). Physiological responses of apple trees to supraoptimal root temperature. *Physiologia Plantarum* **27**, 130–8.

Gur, A., Hepner, J. and Mizrahi, Y. (1976a). The influence of root temperature on apple trees. I. Growth responses related to the application of potassium fertilizer. *Journal of Horticultural Science* **51**, 181–93.

Gur, A., Mizrahi, Y. and Samish, R.M. (1976b). The influence of root temperature on apple trees. II. Clonal differences in susceptibility to damage caused by supraoptimal root temperature. *Journal of Horticultural Science* **51**, 195–202.

Hackett, W.P. (1989). Donor plant maturation and adventitious root formation. In *Adventitious Root Formation in Cuttings*, ed. T.D. Davis, B.E. Haissig and N. Sankhla, pp. 11–28. Portland, OR: Dioscorides Press.

Hansen, P. (1967). ^{14}C studies on apple trees. III. The influence of season on storage and mobilization of labelled compounds. *Physiologia Plantarum* **20**, 1103–11.

Hansen, P. (1980). Crop load and nutrient translocation. In *Mineral Nutrition of Fruit Trees*, ed. D. Atkinson, J.E. Jackson, R.O. Sharples and W.M. Waller, pp. 201–12. London: Butterworths.

Hansen, P. and Grauslund, J. (1973). ^{14}C studies on apple trees. VIII. The seasonal variation and nature of reserves. *Physiologia Plantarum* **28**, 24–32.

Harper, J.L., Jones, M. and Sackville-Hamilton, N.R. (1991). The evolution of roots and the problems of analyzing their behaviour. In *Plant Root Growth. An Ecological Perspective*, ed. D. Atkinson, pp. 3–22. Oxford: Blackwell.

Harrison-Murray, R.S. (1982). Air/water relations of rooting media for leafless winter cuttings. *ISHS XXIst International Horticultural Congress*, Hamburg, Abstract 1787.

Harrison-Murray, R.S. and Fieldsend, M. (1982). Air/water relations of cutting composts. *Report of the East Malling Research Station for 1981*, 64–5.

Harrison-Murray, R.S. and Howard, B.H., (1981). Mechanisms of cutting propagation: etiolation. *Report of the East Malling Research Station for 1980*, 60.

Harrison-Murray, R.S., Avanzato, D., Tal, E. and Aylott, J.W. (1985). Pre-etiolation of M9 cuttings. *Report of the East Malling Research Station for 1984*, 131.

Harrison-Murray, R.S., Jackson, J.E., McNeil, J.C. and Hicks, S.M. (1984). Pre-treatment of apple trees for deep planting. *Report of the East Malling Research Station for 1983*, 79.

Hartmann, H.T., Kester, D.E. and Davies, F.T. (1990). *Plant Propagation, Principles and Practices.* New Jersey: Prentice Hall.

Hatton, R.G., (1935). Apple rootstock studies: effect of layered stocks upon the vigour and cropping of certain scions. *Journal of Pomology and Horticultural Science XIII*, 293–353.

Hatton, R.G., Grubb, N.H. and Amos, J. (1924). Some factors influencing root development. *Report of the East Malling Research Station for 1923*, 110–19.

Head, G.C. (1967). Effects of seasonal changes in shoot growth on the amount of unsuberized root on apple and plum trees. *Journal of Horticultural Science* **42**, 169–80.

Head, G.C. (1968a). Studies of growing roots by time-lapse cinematography. *Proceedings IX International Congress of Soil Science* **1**, 751–8.

Head, G.C. (1968b). Seasonal changes in the diameter of secondarily thickened roots of fruit trees in relation to growth of other parts of the tree. *Journal of Horticultural Science* **43**, 275–82.

Head, G.C. (1969a). The effects of fruiting and defoliation on seasonal trends in new root production in apple trees. *Journal of Horticultural Science* **44**, 175–81.

Head, G.C. (1969b). The effects of mineral fertilizer on seasonal changes in the amount of white root on apple trees in grass. *Journal of Horticultural Science* **44**, 183–7.

Heighton, H.R. and Howard, B.H. (1984). Rooting in relation to cutting position within apple shoots. *Report of the East Malling Research Station for 1983*, 84–5.

Hess, C.E. (1961). The physiology of root initiation in easy and difficult-to-root cuttings. *Hormolog* **3**, 3–6.

Howard, B.H. (1968). The influence of 4(indolyl-3) butyric acid and basal temperature on the rooting of apple rootstock hardwood cuttings. *Journal of Horticultural Science* **43**, 23–31.

Howard, B.H. (1971). Propagation techniques. *Scientific Horticulture* **23**, 116–26.

Howard, B.H. (1978). Field establishment of apple rootstock hardwood cuttings as influenced by conditions during a prior stage in heated bins. *Journal of Horticultural Science* **53**, 31–7.

Howard, B.H. (1983). Plant propagation. *Report of the East Malling Research Station for 1982*, 59–75.

Howard, B.H. (1987). Propagation. In *Rootstocks for Fruit Crops*, ed. R.C. Rom and R.F. Carlson, pp. 29–77. New York: John Wiley and Sons.

Howard, B.H., Harrison-Murray, R.S. and Fenlon, C.A. (1983). Effective auxin treatment of leafless winter cuttings. In *Growth Regulators in Root Development*, ed. M.B. Jackson and A.D. Stead, pp. 73–85. British Plant Growth Regulator Group Monograph No. 10.

Howard, B.H., Harrison-Murray, R.S. and MacKenzie, K.A.D. (1984). Rooting responses to wounding winter cuttings of M.26 apple rootstock. *Journal of Horticulural Science* **59**, 131–9.

Hughes, K.A. and Gandar, P.W., (1993). Length densities, occupancies and weights of apple root systems. *Plant and Soil* **148**, 211–21.

Huguet, J.G. (1976). Influence of a localized irrigation on the rooting of young apple trees [in French]. *Annales Agronomiques*, N.S. **27**, 343–61.

Huguet, J.G. and Fourcade, P. (1980). Distribution des racines de pommiers en relation avec la position des points d'irrigation. *Seminaires sur l'irrigation localisée. III. Influence de l'irrigation localisée sur la morphologie et la physiologie des racines.* CNR/Universite de Naples, Sorrento, 43–50.

Huysamer, M. (1997). Integrating cultivar, rootstock and environment in the export-driven South African deciduous fruit industry. *Acta Horticulturae* **451**, 755–60.

Jackson, J.E. Harrison-Murray, R.S. and Hicks, S.M. (1984). Reducing staking costs and improving tree establishment. *Report of the East Malling Research Station for 1983*, 31–2.

James, D.J. (1983). Adventitious root formation '*in vitro*' in apple rootstocks (*Malus pumila*). I. Factors affecting the length of the auxin-sensitive phase in M.9. *Physiologia Plantarum* **57**, 149–53.

James, D.J. and Thurbon, I.J. (1981). Phenolic compounds and other factors controlling rhizogenesis *in vitro* of the apple rootstocks M.9 and M.26. *Zeitschrift für Pflanzenphysiologie* **105**, 11–20.

James, D.J., Jones, O.P., Hadlow, W.C.C., MacKenzie, K.A.D., Marks, T.R., Murphy, H., Passey, A.J., Webster, A.D. and Webster C.A. (1987). Rejuvenation through sequential subculture. *Report of the East Malling Research Station for 1986*, 118.

Jones, H.G., Lakso, A.N. and Syvertsen, J.P. (1985). Physiological control of water status in temperate and subtropical fruit trees. *Horticultural Reviews* **7**, 301–44.

Jones, O.P. (1983). In vitro propagation of tree crops. In *Plant Biotechnology*, ed. S.H. Mantell and H. Smith, pp. 139–59. S.E.B. Seminar Series **18**. Cambridge University Press.

Jones, O.P. and Hadlow, W.C.C. (1989). Juvenile-like character of apple trees produced by grafting scions and rootstocks produced by micropropagation. *Journal of Horticultural Science* **64**, 395–401.

Jones O.P. and Hatfield, S.G.S. (1976). Root initiation in apple shoots cultured *in vitro* with auxins and phenolic compounds. *Journal of Horticultural Science* **51**, 495–99.

Jones, O.P. and Webster, C.A. (1989). Improved rooting from conventional cuttings taken from micropropagated plants of *Pyrus communis* rootstocks. *Journal of Horticultural Science* **64**, 429–34.

Jones, O.P. and Webster, C.A. (1993). Nursery performance of 'Cox' apple trees with rootstocks of M9 from either micropropagation or improved conventional propagation from micropropagated plants. *Journal of Horticultural Science* **68**, 763–6.

Jones, O.P., Marks, T.R. and Waller, B.J. (1982). Propagation *in vitro*. *Report of the East Malling Research Station for 1981*, 159.

Kamboj, J.S., Blake, P.S., Quinlan, J.D., Webster, A.D. and Baker, D.A. (1997). Recent advances in studies on the dwarfing mechanism of apple rootstocks. *Acta Horticulturae* **451**, 75–82.

Klepper, B. (1991). Root–shoot relations. In *Plant Roots. The Hidden Half*, ed. Y. Waisel, A. Eshel and U. Kafkafi, pp. 265–86. New York: Dekker.

Kotzé, W.A.G. (1996). Soil and nutritional factors affecting apple fruit quality. In *Integrated Management of Post-harvest Quality*, ed. J.C. Combrink, pp. 14–24. Stellenbosch: Infruitec.

Landsberg, J.J. and Jones, H.G. (1981). Apple orchards. In *Water Deficits and Plant Growth*, Vol. VI, ed. T.T. Kozlowski, pp. 419–69. New York: Academic Press.

Levin, I., Assaf, R. and Bravdo, B. (1979). Soil moisture and root distribution in an apple orchard irrigated by tricklers. *Plant and Soil* **52**, 31–41.

Levin, I., Assaf, R. and Bravdo, B. (1980). Irrigation, water status and nutrient uptake in an apple orchard. In *Mineral Nutrition of Fruit Trees*, ed. D. Atkinson, J.E. Jackson, R.O. Sharples and W.M. Waller, pp. 255–64. London: Butterworths.

Lombard, P.B. and Westwood, M.N. (1987). Pear rootstocks. In *Rootstocks for Fruit Crops*, ed. R.C. Rom and R.F. Carlson, pp.145–83. New York: John Wiley and Sons.

Looney, N.E. and McIntosh, D.L.(1968). Stimulation of pear rooting by pre-plant treatment of nursery stock with indole-3-butyric acid. *Proceedings of the American Society for Horticultural Science* **92**, 150–4.

Lucic, P. (1967)). The growth of the root system of the pear. *Pyrus communis* var. *silvestris* under different illumination of the assimilation organs. *Contemp. Agr.-Novi Sad* **15**, 31–7.

MacKenzie, K.A.D. (1979). The development of the endodermis and *phi* layer of apple roots. *Protoplasma* **100**, 21–32.

MacKenzie, K.A.D., Howard, B.H. and Harrison-Murray, R.S. (1986). The anatomical relationship between cambial regeneration and root initiation in wounded winter cuttings of the apple rootstock 'M.26'. *Annals of Botany* **58**, 649–61.

MAFF (1972). *Apples.* Bulletin 207, Ministry of Agriculture, Fisheries and Food. London: Her Majesty's Stationery Office.

Marschner, H. (1995). *Mineral Nutrition of Higher Plants*, 2nd edn. London: Academic Press.

Mason, A.C. and Whitfield, A. B. (1960). Seasonal changes in the uptake and distribution of mineral elements in apple trees. *Journal of Horticultural Science* **35**, 34–55.

Maynard, B.K. and Bassuk, N.L. (1988). Etiolation and banding effects on adventitious root formation. In *Adventitious Root Formation in Cuttings*, ed. T.D. Davis, B.E. Haissig and N. Sankhla, pp. 29–46. Portland, OR: Dioscorides Press.

McCown, B.H. (1988). Adventitious rooting of tissue cultured plants. In *Adventitious Root Formation in Cuttings*, ed. T.D. Davis, B.E. Haissig and N. Sankhla, pp. 289–302 Portland, OR: Dioscorides Press.

Morin, F., Fortin, J.A., Hamel, C., Granger, R.L. and Smith, D.L. (1994). Apple rootstock response to vesicular-arbuscular mycorrhizal fungi in a high-phosphorus soil. *Journal of the American Society for Horticultural Science* **119**, 578–83.

Mosse, B. (1957). Growth and chemical composition of mycorrhizal and non-mycorrhizal apples. *Nature* **179**, 922–4.

Murneek, A.E. (1942). Quantitative distribution of nitrogen and carbohydrates in apple trees. *Research Bulletin of the Missouri Agricultural Experiment Station*, **348**.

Nelson, S.H. and Tukey, H.B. (1956). Effects of controlled root temperatures on the growth of East Malling rootstocks in water culture. *Journal of Horticultural Science* **31**, 55–63.

Olien, W.C. (1987). Effect of seasonal soil waterlogging on vegetative growth and fruiting of apple trees. *Journal of the American Society for Horticultural Science* **112**, 209–14.

Palmer, J.W. (1988). Annual dry matter production and partitioning over the first 5 years of a bed system of Crispin/M.27 apple trees at four spacings. *Journal of Applied Ecology* **25**, 569–78.

Parry, M.S. (1974). Depth of planting and anchorage of apple trees. *Journal of Horticultural Science* **49**, 349–54.

Pearse, H.L. and Garner, R.J. (1937). A note on the use of α-naphthalene acetic acid for rooting soft-wood cuttings of fruit tree stocks. *Journal of Pomology and Horticultural Science* **XV**, 248–51.

Pontikis, C.A., MacKenzie, K.A.D. and Howard, B.H. (1979). Establishment of initially unrooted stool shoots of M27 apple rootstock. *Journal of Horticultural Science* **54**, 79–85.

Preston, A.P. (1972). Apple rootstock studies: growth of trees on three clonal rootstocks planted with and without roots. *Journal of Horticultural Science* **47**, 329–35.

Priestley, C.A. (1964). The importance of autumn foliage to carbohydrate status and root growth of apple trees. *Report of the East Malling Research Station for 1963*, 104–6.

Quinlan, J.D. (1965). The pattern of distribution of ^{14}carbon in a potted apple rootstock following assimilation of ^{14}carbon dioxide by a single leaf. *Report of the East Malling Research Station for 1964*, 117–18.

Quinlan, J.D. (1969). Mobilization of ^{14}C in the spring following autumn assimilation of $^{14}CO_2$ by an apple rootstock. *Journal of Horticultural Science* **44**, 107–10.

Rapparini, F., Baraldi, R., Bertazza, G., Branzanti, B. and Predieri, S. (1994). Vesicular-arbuscular mycorrhizal inoculation of micropropagated fruit trees. *Journal of Horticultural Science* **69**, 1101–9.

Reckruhm I. (1974). Direct root analysis in orchards with different soil management systems [in German]. *Archiv für Gartenbau* **22**, 17–26.

Robinson, J.C. and Schwabe, W.W. (1977). Studies on the regeneration of apple cultivars from root cuttings. II. Carbohydrate and auxin relations. *Journal of Horticultural Science* **52**, 221–33.

Rogers, W.S. (1939). Root studies. VIII. Apple root growth in relation to rootstock, soil, seasonal and climatic factors. *Journal of Pomology and Horticultural Science* **17**, 99–130.

Rogers, W.S. and Booth, G.A. (1959–60). The roots of fruit trees. *Scientific Horticulture* XIV, 27–34.

Rogers, W.S. and Head, G.C. (1966). The roots of fruit plants. *The Journal of the Royal Horticultural Society* **41**, 198–205.

Rogers, W.S. and Head, G.C. (1969). Factors affecting the distribution and growth of roots of perennial woody species. In *Root Growth*, ed. W.J. Whittington, pp. 280–95. London: Butterworths.

Rogers, W.S. and Parry, M.S. (1968). Effects of deep planting on anchorage and performance of apple trees. *Journal of Horticultural Science* **43**, 103–6.

Rogers, W.S. and Vyvyan, M.C. (1934). Root studies. V. Rootstock and soil effect on apple root systems. *Journal of Pomology and Horticultural Science* **12**, 110–50.

Rom, R.C. (1987). Roots. In *Rootstocks for Fruit Crops*, ed. R.C. Rom and R.F. Carlson, pp. 5–28. New York: John Wiley and Sons.

Rom, R.C. and Brown, S.A. (1979). Factors affecting burr-knot formation on clonal *Malus* rootstocks. *HortScience* **14**, 231–2.

Russell, R.S. (1972). Transport of nutrients from soils to crops. In *Crop Processes in Controlled Environments*, ed. A.R. Rees, K.E. Cockshull, D.W. Hand and R.G. Hurd, pp. 111–26. London: Academic Press.

Russell, R.S. (1977). *Plant Root Systems: Their Function and Interaction with the Soil.* London: McGraw Hill.

Sewell, G.W.F. and Roberts, A.L. (1985). Vesicular-arbuscular mycorrhizae and apple growth responses to soil sterilization. *Report of the East Malling Research Station for 1984*, 149–50.

Sewell, G.W.F., Roberts, A.L. and Samuelson, T.J. (1985). Interaction effects between apple, grass and a vesicular-arbuscular mycorrhizal fungus on nutrient uptake by apple. *Report of the East Malling Research Station for 1984*, 125.

Sriskandarajah, S., Mullins, M.G. and Nair, Y. (1982). Induction of adventitious rooting *in vitro* in difficult to propagate cultivars of apple. *Plant Science Letters* **24**, 1–9.

Stolzy, L.H. and Letey, J. (1964). Measurement of oxygen diffusion rates with the platinum microelectrode. 3. Correlation of plant response to soil oxygen diffusion rates. *Hilgardia* **35**, 567–76.

Stoutemyer, V.T. (1937). Regeneration in various types of apple wood. *Iowa Agricultural Experiment Station Research Bulletin* 220, pp. 308–52.

Takeno, K.J., Taylor, S., Sriskandarajah, S., Pharis, R.P. and Mullins, M.G. (1983). Endogenous gibberellin and cytokinin-like substances in cultured shoot tissues of apple, *Malus pumila* cv Jonathan, in relation to adventitious root formation. *Plant Growth Regulation* **1**, 261–8.

Tamasi, J. (1964). Effect of soil water level on root development of apple trees. *MTA Agartud Osztaly Kozl Budapest* XIII (1–2), 43–63.

Tamasi, J. (1986). *Root Location of Fruit Trees and its Agrotechnical Consequences.* Akademiai Kaido Budapest.

Terblanche, J.H., Wooldridge, L.G., Hesebek, I. and Joubert, M. (1979). The redistribution and immobilization of calcium in apple trees with special reference to bitter pit. *Communications in Soil Science and Plant Analysis* **10**, 195–215.

Tromp, J. (1980). Mineral absorption and distribution in young apple trees under various environmental conditions. In *Mineral Nutrition of Fruit Trees,* ed. D. Atkinson, J.E. Jackson, R.O. Sharples and W.M. Waller, pp. 173–82. London: Butterworths.

Tromp, J. (1983). Nutrient reserves in roots of fruit trees, in particular carbohydrates and nitrogen. *Plant and Soil* **71**, 401–13.

Tukey, L.D. (1959). Periodicity in growth of fruits of apples, peaches and sour cherries with some factors influencing this development. *Bulletin Pennsylvania Agricultural Experiment Station. 661*.

Tukey, L.D. (1963). Factors affecting rhythmic diurnal enlargement and contraction in fruits of the apple (*Malus domestica* Bork). *Proceedings 16th International Horticultural Congress Brussels 1962*, **3**, 528–536.

Van den Honert, T.H. (1948). Water transport in plants as a catenary process. *Discussions of the Faraday Society* **3**, 146–53.

Visser, T. Verhaegh, J.J. and de Vries, D.P. (1976). A comparison of apple and pear seedlings with reference to the juvenile period. 1. Seedling growth and yield. *Acta Horticulturae* **56**, 205–14.

Vyvyan, M.C. (1934). The distribution of growth between roots, stems and leaves in a young apple tree and its possible bearing on the problem of stock effect on tree vigour. *Report of the East Malling Research Station for 1933*, 122–31.

Webster, A.D., Atkinson, C.J., Lucas, A.S., Vaughan, S.J. and Taylor, L. (2000). Interactions of root restriction, irrigation and rootstock treatments on the growth and cropping of Queen Cox apple trees. 1. Effects on orchard growth and cropping. *Journal of Horticultural Science and Biotechnology* **75**, 181–9.

Webster, C.A. and Jones, O.P. (1992). Performance of field hedge and stoolbed plants of micropropagated dwarfing apple rootstock clones with different degrees of apparent rejuvenation. *Journal of Horticultural Science* **67**, 521–8.

Webster, D.H. (1978). Soil conditions associated with absence or sparse development of apple roots. *Canadian Journal of Plant Science* **58**, 961–9.

Welander, M. and Huntrieser, I. (1981). The rooting ability of shoots raised '*in vitro*' from the apple rootstock A2 in juvenile and in adult growth phase. *Physiologia Plantarum* **53**, 301–6.

Wellensiek, S.K. (1952). Rejuvenation of woody plants by formation of sphaeroblasts. *Proc. K. Ned. Akad. Wet. Amst. Ser. C* **55**, 567–73.

Weller, F. (1966). Horizontal distribution of absorbing roots and the utilization of fertilizers in apple orchards [in German]. *Erwerbsobstbau* **8**, 181–4.

Weller, F. (1971). A method for studying the distribution of absorbing roots of fruit trees. *Experimental Agriculture* **7**, 351–61.

Westwood, M.N. (1995). *Temperate Zone Pomology.* Portland, OR: Timber Press.

Westwood M.N. and Bjornstad, H.O. (1968). Chilling requirements of dormant seeds of 14 pear species as related to their climatic adaptation. *Proceedings of the American Society for Horticultural Science* **92**, 141–9.

White, G.C. and Holloway, R.I.C. (1967). The influence of simazine or a straw mulch on the establishment of apple trees in grassed down or cultivated soil. *Journal of Horticultural Science* **42**, 377–89.

Wiersum, L.K. (1980). The effect of soil physical conditions on roots and uptake. In *Mineral Nutrition of Fruit Trees*, ed. D. Atkinson, J.E. Jackson, R.O. Sharples and W.M. Waller, pp. 111–21. London: Butterworths.

Wilson, S.A. and Atkinson, D. (1979). Water and mineral uptake by fruit tree roots. *Proceedings of IUFRO Meeting, Nancy, France 1978.*

Young, E. and Werner, D.J. (1982). Early season root and shoot growth of 'Golden Delicious' apple on four rootstocks as affected by pruning at planting. *Journal of the American Society for Horticultural Science* **107**, 822–6.

4

The graft union, grafting and budding

Introduction

Grafting is the art of connecting two pieces of living tissue together in such a way that they unite and grow as one. In apples and pears it is generally used to combine a scion (fruiting) cultivar with a rootstock. Budding is a special form of grafting in which the initial scionwood component is reduced to a single bud.

This art has been practised for thousands of years, Garner (1988) noting that grafting with detached scions was used by the Chinese before 2000 BC. It was described by writers in ancient Greece and Rome and very widely employed in western Europe in the Renaissance period and subsequently.

The main purposes of grafting are to assist in the propagation and perpetuation of clones that cannot readily be propagated by other asexual means, and to enable the production of composite trees from rootstocks and scions each of which possesses specific and distinct desirable attributes. It is also used to change scion cultivars in established orchards, to hasten the fruiting of seedling selections in breeding programmes and as a research tool in the study of physiological processes and viruses.

Formation of the graft union

The formation of the graft union can be considered as resulting from the wound-healing processes which take place on the cut surfaces of the rootstock and scion, operating in the context of close contact between surfaces. The union is accomplished entirely by cells that develop after the grafting operation. The sequence of events is as follows.

1 The placement of freshly cut scion tissue in intimate contact with similar freshly cut stock tissue in such a way that their cambial regions, capable of

meristematic activity, are in close proximity. The two components must have the correct polarity, the morphologically proximal end of the scionwood piece, i.e. its basal end, being grafted on to the distal end of the rootstock stem, or, if it is grafted on to a root, to the end of this furthest from the root tip. The graft has to be secured by wrapping, tying, nailing or wedging. This is to ensure that the scion and stock cannot move relative to one another and dislodge the interlocking parenchyma cells that develop from the cut surfaces. The positioning of the rootstock and scionwood in some of the most widely used techniques of grafting and budding are shown in Figures 4.1 and 4.2.

It is essential that the cut surfaces of stock and scion which are placed together do not dry out. Exposure to air at below the saturation point inhibits callus formation (Shippy, 1930) because the parenchyma cells of the callus tissue are thin-walled and cannot resist desiccation. Speedy grafting or budding reduces the time over which the moist surfaces are exposed to air. Skilled grafters can average 70 whip-and-tongue grafts per hour and budders 100 buddings per hour when working in a team with a tyer (Garner, 1988). Shippy concluded that liquid moisture present as a film enclosing the cutting, such as is supplied by surrounding it with moderately moist peat, sphagnum or sand, provides the most favourable conditions for bringing about uniform callusing. Garner (1988), however, warns that deliberate wetting of the cut surfaces of either stock or scion in the budding or grafting process may prove detrimental. Where the plants have been bench-grafted, i.e. dormant scionwood grafted in winter on to rootstocks which have been lifted and cold-stored, adequate humidity is maintained by storing the grafted plants in callusing beds or boxes containing moist peat, sphagnum moss or sand. In countries with mild winters grafting is done out of doors, as is budding later in the year. Desiccation of these unions is prevented by sealing over grafts with wax, by smearing heavy-grade petroleum jelly over buds or by using sticky, watertight plastic tapes to seal as well as tie.

2 The cutting of the stock and scion results in the death of at least one cell layer at each surface. This necrotic material disappears or remains in pockets between newly formed living parenchyma cells.

3 Parenchymatous callus cells proliferate, mainly from the primary cortex of the scion and from secondary phloem of the stock and scion. The spongy callus cells mingle and growth pressure compresses them until cells from stock and scion become indistinguishable. The actual cambial layer takes little part at this stage (Sass, 1932).

4 Parenchyma cells at the edges of the newly formed callus mass which are touching the cambial cells of the stock and scion differentiate into new cambium cells within 2–3 weeks after grafting. This cambial formation

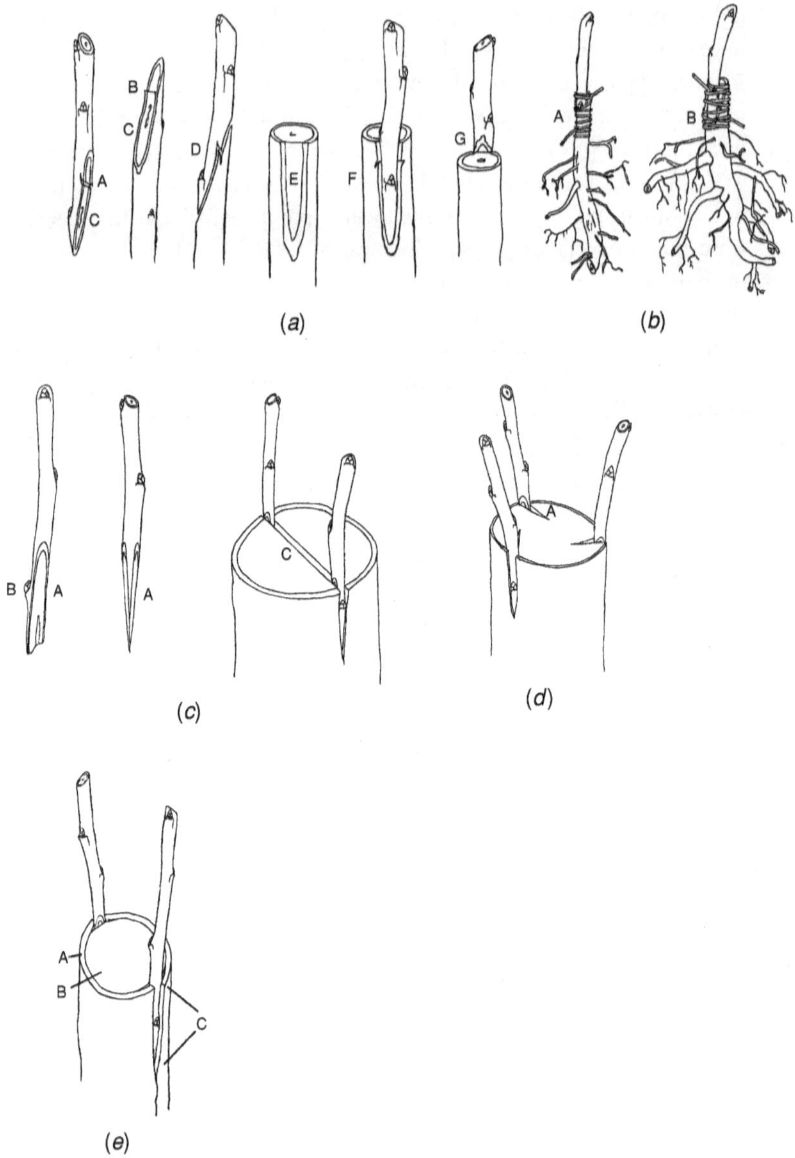

Figure 4.1 Methods of grafting.
(*a*) The whip-and-tongue graft is most suitable when the stock and scion are less than 25 mm in diameter. A flat slanting cut is made at the basal end of the scion and a downward pointing tongue (A) made in the apical part of this slanting surface. An upward slanting cut of corresponding length is made through the stock and an upward pointing tongue (B) made in this. The cut surfaces of stock and scion are placed together so that the tongues interlock (D) and the cambial regions (C) are in contact over as great a length as possible. If the

in the callus mass proceeds inwards from the original stock and scion cambium and through the callus bridge until there is a continuous cambial connection between stock and scion.

5 The newly formed cambial layer in the callus bridge lays down new xylem towards the inside and new phloem towards the outside. The new xylem tissue originates from the activities of the scion tissues rather than those of the stock (Yeager, 1944). Induction of vascular tissue in the callus appears to be under the control of material originating in shoot apices and can be artificially induced by the supply of auxins, cytokinins, gibberellins and sugars (Torrey *et al.*, 1971; Wetmore and Rier, 1963).

Union formation in T-budding and chip budding

In T-budding of apple, when the bark of the rootstock is lifted for bud insertion, the separation occurs in the young, undifferentiated, xylem and the cambial zone remains attached to the inside of the barkflaps. A necrotic plate develops from the cut cells shortly after the bud shield is inserted. After about two days callus parenchyma cells start to develop from the rootstock xylem

Figure 4.1 (*cont.*)
scion is much smaller in diameter than the stock the slice taken from the stock should be shallow (E) so that the cambia of stock and scion are well matched at the sides and top (F and G).
(*b*) Bench grafting on roots is usually done using whip-and-tongue grafts and can be used for stocks and scions of either similar (A) or dissimilar (B) size.
(*c*) The cleft graft is used in top-working trees with new scion cultivars. Scion bud sticks of 3 or 4 buds are inserted in splits made in the cut ends of tree trunks or large limbs. The scions are prepared with two slanting cuts at their base giving a long tapering wedge (A), preferably with a bud (B) between the cut surfaces as shown. The cleft (C) in the limb is opened by insertion of a wedge or grafting tool, the scions are inserted so that their cambial regions are in contact with those of the trunk or limb. The wedge or tool is removed and the scions are held fast by the pressure of the cleft moving to close.
In oblique cleft grafting (*d*) the clefts do not extend right across the branch. All exposed cut surfaces are sealed with grafting wax after first rubbing some clay into the clefts to prevent the wax running down inside.
(*e*) Rind or crown grafting involves inserting a scion, prepared as for whip-and-tongue grafting but without the tongue, between the rind (A) and the wood (B) of the tree to be top-worked. This can only be done when the bark will separate readily from the wood and the graft must be tied firmly in place before waxing. After Garner (1988).
Reproduced with permission of Mrs I.L. Garner.

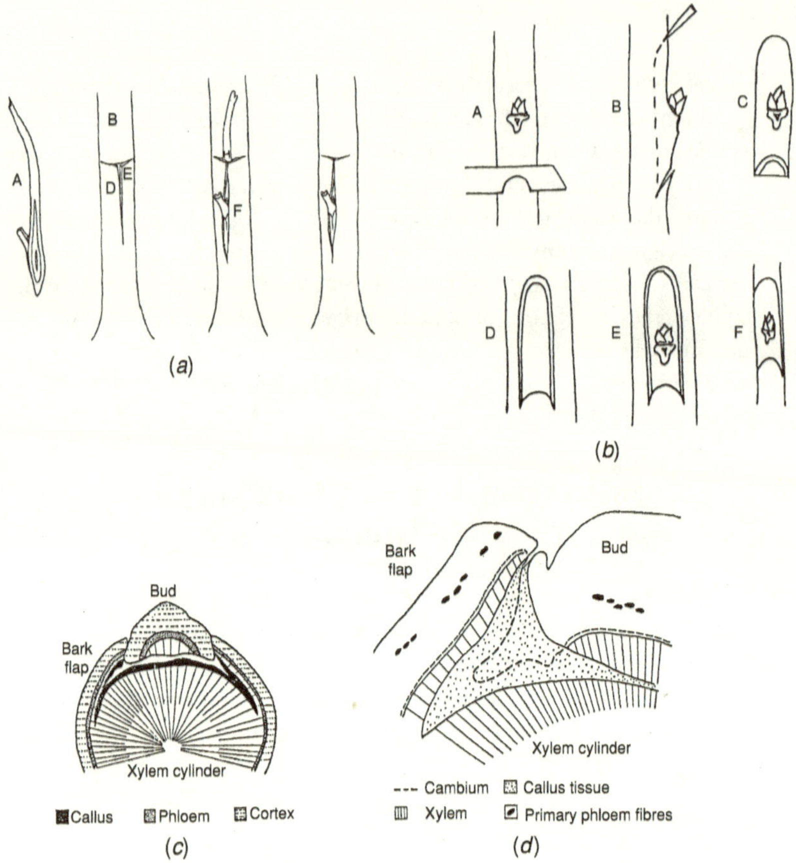

Figure 4.2 Shield budding (T budding) and chip budding.
(*a*) The technique of shield budding: (A) The bud nearest to the upper end of the budstick is removed by a shallow slicing cut. (B) The stock is prepared by making a T-shaped incision (D) through the rind down to the wood. The rind (E) is lifted and the bud inserted (F). The tail or handle (H) is cut off at the horizontal incision and the bud firmly tied. After Garner (1988). Reproduced with permission of Mrs I.L. Garner.
(*b*) The technique of chip budding. (A) The first cut is made downwards in the side of the stock. (B) A second cut, 20–55 mm long is made downwards to meet the first. (C) A bud chip is taken from the scion bud stick by making a horizontal cut 12 mm below the bud and a second cut starting 25 mm above the bud to join the first. This bud chip (C) is then fitted to the prepared stock (D) and temporarily held by the stock flap (E). The cambia of stock and scion should be matched, which usually leaves a layer of the outer rind of the stock visible (F). The bud is then very firmly tied and sealed in place. After Garner (1988). Reproduced with permission of Mrs I.L. Garner.
(*c*) The distribution of tissues in a 3-week-old apple bud-union (shield-budded). From Mosse and Labern (1960). Reproduced with permission of Oxford University Press and Dr B. Mosse.
(*d*) The location of connecting cambium in a shield-budded apple bud-union at 8 weeks old. From Mosse and Labern (1960). Reproduced with permission of Oxford University Press and Dr B. Mosse.

rays and, to a lesser extent, from the bud shield, and rupture through the necrotic plate. Callus originating almost entirely from rootstock tissue, mainly from the exposed surface of the xylem cylinder, surrounds the bud shield and holds it in place and, over a period of two or three weeks, fills all internal air pockets. A continuous cambium is then established, isolated tracheary elements appear and the callus lignifies. This lignification is completed by about 12 weeks after budding (Mosse and Labern, 1960; Wagner, 1969). In chip budding, in which the bud is removed from the scion as part of a wedge of tissue which is then seated in a corresponding notch on the rootstock stem, the juxtaposition of the xylem and cambial tissues of stock and scion results in the formation of a much more rapid and complete union (Figure 4.3).

Incompatibility

When two different plants can be grafted together to produce a long-lived, functional graft union and to develop successfully into one composite plant, they are said to be compatible. When this is not the case they are said to be incompatible. In general the more closely related the plants the higher is the chance of compatibility although permanent unions between one genus and another may occur. Pear (*Pyrus communis*) will form a lasting union with hawthorn (*Crataegus oxyacantha*), medlar (*Mespilus germanica*) and quince (*Cydonia oblonga*). Although apple and pear are usually incompatible the pear cultivar 'Fertility' has exceptionally high compatibility with apple and 'M.16' apple rootstock clone with pear (Garner, 1988). However, some economically-important pear cultivars are incompatible with otherwise desirable quince rootstocks and some apple cultivars with otherwise desirable apple rootstocks, while incompatibility can be induced between otherwise compatible rootstocks and scions by viruses and phytoplasmas.

Incompatibility has the following symptoms:

1 Initial failure to form graft or bud unions.
2 Poor growth of the scion often followed by premature death in the nursery: premature yellowing of foliage.
3 Breaking off of trees at the graft union, especially when they have been growing for many years and the break is clean and smooth.

The incompatibilities of the greatest commercial importance in apples and pears are:

1. Pear–quince incompatibility. When some pear cultivars are grafted on quince rootstocks a cyanogenic glycoside, prunasin, which is normally found in quince but not in pear is translocated into the pear phloem. The pear tissues

(*a*)

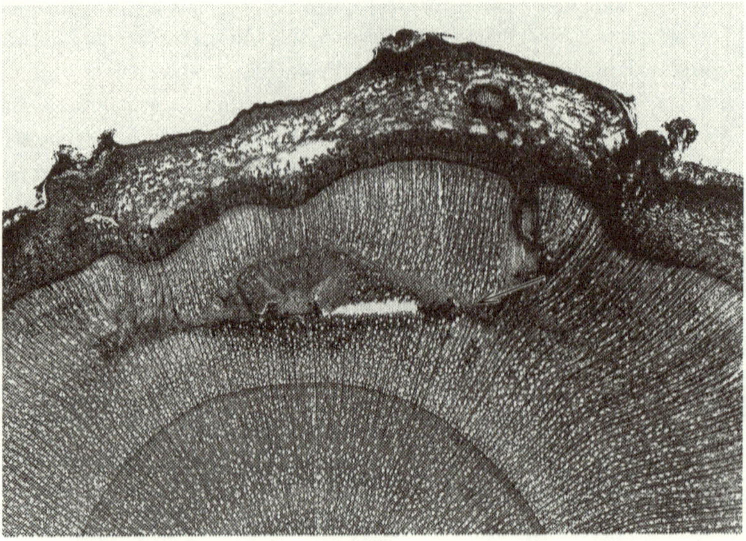

(*b*)

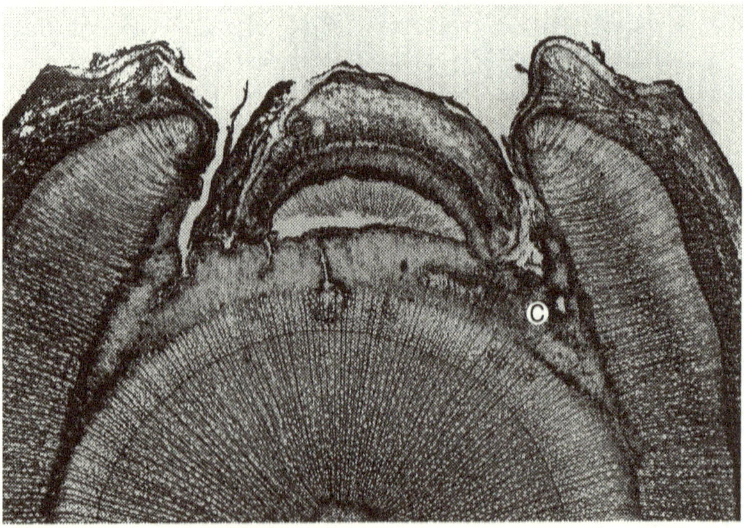

Figure 4.3 Transverse sections through 'Lord Lambourne' apple scion unions with 'M.26' apple rootstocks c. 0.5 cm below the bud (*a*) Chip budded, (*b*) T-budded, sampled in November. The junction between the scionwood chip and the rootstock is arrowed, callus formed after T-budding is indicated by C. From Howard *et al.* (1974). Reproduced with permission.

break down the prunasin in the region of the graft union to give hydrocyanic acid as one of the decomposition products. Different pear cultivars vary in their content of a water-soluble inhibitor of the action of the enzyme which catalyses the breakdown of prunasin. The presence of hydrocyanic acid at the graft union checks cambial activity there and also destroys phloem tissues at and above the union. Conduction of water and materials through the union is reduced. The reduction in the amount of sugars reaching the quince roots leads to further decomposition of prunasin, liberating hydrocyanic acid and killing quince phloem (Gur, 1957; Gur and Samish, 1965; Gur *et al.*, 1968). Further evidence that graft incompatibility between pear and quince involves the influence of cyanogenic glycosides produced by the quince has come from suspension culture studies. These showed that pear callus growth was inhibited in media in which quince cultures had been grown, and addition of a cyanogenic glycoside killed pear cultures (Moore, 1986). Moore concluded that, given this mechanism, incompatibility between pear and quince need not be associated with any particular stage of graft development. Other studies (Buchloh, 1960) showed failure of lignification in incompatible pear–quince unions.

Garner (1988) lists 26 cultivars of pear which show some degree of incompatibility with quince but will form more reliable unions if they are worked on an intermediate which is worked on the quince. The most widely used intermediates are 'Beurré Hardy' and 'Old Home'. The most important cultivars that are not directly compatible with quince are 'Williams'' ('Bartlett'), most Asian pears, 'Forelle', 'Bosc' and 'Jules Guyot'. Important compatible cultivars include 'Comice', 'Passe Crassane', 'Beurré Hardy', 'Clapps Favourite' and 'Conference'. Some selections of 'Williams'' are compatible with quince.

'M.26' is the only important apple rootstock to have shown incompatibility with a number of scions.

2. So-called virus-induced incompatibility affects both apples and pears. Union necrosis and decline of apple is caused by tomato ringspot virus which is transmitted by the soil-borne nematode *Xiphinema americanum*. Trees infected with this break at the union, with a clean break at least on one side. The rootstocks 'M.26' and 'MM.106' and the scion cv. 'Delicious' seem susceptible, while 'M.4', 'M.7', 'Ottawa 3', 'Robusta 5', 'Empire' and 'Golden Delicious' appear to be tolerant (Ferree and Carlson, 1987). Pear decline is caused by a phytoplasma (mycoplasma-like organism, MLO; see Chapter 13) which is spread by the insect *Cacopsylla pyricola* (Hibino and Schneider, 1970; Hibino *et al.*, 1971). This may cause little direct damage to a scion cultivar but as the phytoplasma moves across the graft union into the tissues of a susceptible rootstock (especially *P. ussuriensis* or *P. pyrifolia*) the rootstock phloem is killed, the

rootstock stem girdled and the tree dies. A rather different manifestation of its effects is shown by 'Conference' on 'Quince A' or on 'Quince C'. Young trees of these combinations show premature leaf red colour followed by early leaf fall. In the following spring the leaves remain small and pale, there being little or no shoot growth and no fruit production. A necrotic line is frequently visible in the bark at the stock/scion union. The disease is sometimes fatal but more commonly just reduces yield in the early orchard years. The quince rootstocks are rarely infected by the causal phytoplasma, such infection as there is being mainly in suckers. Quince should be regarded as being partially resistant or hypersensitive, rather than tolerant (D.L. Davies, personal communication). 'Conference' is readily infected by the phytoplasma, but, when grown on its own roots, is vigorous and, although infected, shows no evidence of pear decline (Davies *et al.*, 1992). Interestingly, although the cv. 'Comice' is readily infected it does not show symptoms when worked on quince, but a 'Comice' interstock between quince and 'Conference' does not prevent the latter suffering from pear decline.

Double working

A double-worked plant has three parts, the rootstock, the interstock and the scion; and two unions, one between the rootstock and interstock and one between the interstock and the scion. Double working is used.

1. In cases where the rootstock and the scion are mutually incompatible but each will form a good union with the interstock. A typical example is the propagation of 'Bartlett' pear on Quince rootstock by using a 'Beurré Hardy' or 'Old Home' interstock.
2. Where a rootstock with desirable properties, e.g. dwarfing, is in itself difficult to root but can give similarly desirable effects if used as an interstock. This is typified by the use of cold-tolerant, dwarfing, interstocks of the 'B.9' and 'P-series' rootstock clones inserted between a vigorous, cold-tolerant seedling rootstock and the scion cultivar in climates with severe winters.

Methods of double working

There are a number of different methods (Hartmann *et al.*, 1990). In England it is done as a combined operation using dormant scionwood of both the interstock material and the upper scion. The wood of the interstock cultivar is cut into 125 mm lengths in March and immediately grafted with three-bud lengths of the upper scion cultivar and the grafts sealed and stored upright in cool moist conditions. The interstock, with its scion attached, is then grafted on to an established rootstock in the nursery row so as to obtain a 50–75 mm interstock. The first graft may be made some weeks before the second. An even simpler way is double shield budding. A small budless shield of the interstock cultivar is slipped into the T-cut on the rootstock and a bud of the upper scion cultivar put on the upper and outer cut surface of the budless shield and both are slid home (Garner, 1953, 1988).

Effects of temperature on grafting and budding

The temperatures used in controlled conditions, e.g. in bench grafting, and the choice of season for budding and grafting in the nursery or orchard reflect the temperature requirements for three separate processes.

1 Bark-lifting (slipping): this is a requirement for those grafting and budding techniques in which scion material is inserted under bark which is parted from the wood. This is the case in crown grafting (rind grafting) in which pieces of scion graftwood with long sloping cut surfaces are pushed down between the rind and the wood around the perimeter of the decapitated trunk of a tree which it is desired to 'top-work' to a new cultivar. It is also the case for T-budding in which a scion bud is added to a complete rootstock stem by sliding it under the bark through an incision. In England the bark does not readily part from the wood until April and the best time for T-budding is from the end of June to mid-August.

2 Callus growth and union formation: the complete range of temperatures permitting the formation of callus from apple scion and rootstock cuttings and grafts was found by Shippy (1930) to lie between 0 °C and 40 °C. At 3–5 °C only a small amount of callus developed over several months and between 5 °C and 32 °C the rate of callus formation increased with rise in temperature. At temperatures above 32 °C injury usually resulted and at 40 °C death of tissues occurred within a few days. In general, temperatures below 20 °C are most satisfactory. Depending on temperature, apple grafts may therefore be callused over a period of several months or within a few days. As a consequence, even when there is no need for bark-lifting, as in chip budding where a piece of rootstock tissue including outer bark, phloem, cambium and some xylem is removed and a similar shaped piece of scion stem carrying a bud substituted, temperatures have to be high enough for active rootstock growth. In England this is from April to early September (Howard, 1977). Under subtropical conditions in Turkey, chip

budding as late as 10 November is successful (Kuden and Kaska, 1995). Grafting rootstocks with dormant scionwood in England is usually carried out in March and cleft-grafting of established trees from mid-February to June. Bench-grafting, carried out indoors, may be followed by two months of callusing at approximately 7 °C or 30 days at 21 °C (Hartmann *et al.*, 1990).

3 Budbreak: the buds of apples and pears enter into dormancy in autumn and emerge from this when their winter-chilling requirement has been satisfied and when subsequent ambient temperatures are high enough to permit growth (cf. Chapter 6). Attempting to use scionwood in which the buds are starting active growth usually results in failure: the buds leaf-out before the graft union is healed and water losses from these leaves cannot be replaced so the scions dry out and die. Scionwood should therefore be collected when dormant and, if it is not used immediately, stored in a cool chamber at 0–2 °C or, if external conditions are cold enough, bundled and heeled in on the shady side of a building.

When a high temperature is used for callusing bench grafts the material should be collected in autumn and the grafts made before any cold weather has overcome the rest (dormancy) requirement of the buds. After the union is well healed the grafts must be stored at 2–4 °C to meet the winter-chilling requirement of the buds and hold them dormant until planting (Howard and Hildreth, 1963).

Effects of height of budding or grafting on scion vigour and cropping

High budding, so as to give a greater length of rootstock stem, reduces the vigour of the scion. This effect persists indefinitely on dwarfing rootstocks but tends to diminish after a few years on vigorous rootstocks (Parry 1986; van Oosten, 1978). With 'M.9' as a rootstock (Table 4.1) the total extension shoot growth over 8 years of 'Cox' budded at 750 mm was only 59% of that of 'Cox' budded at 150 cm and that of 'Golden Delicious' was 66%. There was only a slight and non-significant reduction in cropping, so cropping efficiency was greatly increased. The relative effect of high budding on vigour of growth is much greater on 'M.26' than on 'M.9' but this is accompanied by undesirable fluting and burr-knot formation on the rootstock stem (Parry, 1986). Blanco (1989) showed that high budding of a vigorous pear scion on 'Quince A' rootstock reduced growth and increased precocity of cropping compared with low budding.

Table 4.1 *Effect of height of budding of 'Cox' and 'Golden Delicious' on 'M.9', 'M.26' and 'MM.111' on total extension growth (m/tree) over 8 years*

Rootstock	Height of budding (mm)	'Cox' extension growth (m/tree)	'Golden Delicious' extension growth (m/tree)
'M.9'	150	291	276
	300	254	225
	450	190	242
	750	172	181
'M.26'	150	418	325
	300	258	247
	450	177	152
	750	108	126
'MM.111'	150	583	523
	300	506	512
	450	518	419
	750	397	399
SED rootstocks		50.6	48.1
SED budding height		48.6	35.4

From Parry (1986). Reproduced with permission.

Effects of interstocks on root and shoot growth and cropping

It has been known for centuries that a small stem-piece or interstock of a dwarfing clone inserted between the scion and the rootstock of an apple tree can dwarf it. In general interstocks of stem material from the 'Malling' series dwarf 'Cox' and 'Worcester Pearmain' apple trees in rough proportion to their dwarfing effects as rootstocks (Parry and Rogers, 1968). 'M.9' was as dwarfing when used as an interstock on 'MM.104' as when used as a rootstock (Parry and Rogers, 1968; Preston, 1974), whereas other 'Malling' clones were less dwarfing as interstocks than as rootstocks. Moreover, in studies with 'M.9' as an interstock above rootstocks other than 'MM.104', it had a less dwarfing effect than when used as rootstock (Vyvyan, 1938; Tukey and Brase, 1943; Czynczyk, 1986), and Czynczyk (1986) also showed 'B.9' to be more dwarfing as a rootstock than as an interstock. In general, therefore, the effect of a dwarfing interstock on vigour is less than that of the same cultivar as rootstock. An interstock of 'M.9' inserted between a 'Crab' rootstock and a 'Bramley's Seedling' scion had a greater effect on root than on shoot growth, reducing it to half, whereas interstocks of 'M.2' or 'M.13' had no effect on either root or shoot growth (Swarbrick *et al.*, 1946).

The effect of a dwarfing interstock is proportional to the length of the interstock (Parry and Rogers, 1972; Lockard, 1974). Use of long interstocks of vigorous clones can result in increased tree size of both apple (Grubb, 1939) and pear (Wertheim and Toorenaar, 1966). Dwarfing interstocks induce precocity in cropping, the effect being greater the longer the interstock, while vigorous interstocks reduce precocity with this effect tending to be greater the greater the length of interstock (Parry and Rogers, 1972). Rauch (1968) found that increasing the length of interstock of 'Clark Dwarf' did not significantly reduce the size of the trees until after the onset of flowering, i.e. the effect on flowering was not a consequence of the effect on vigour. Long interstocks of 'M.26' are even more dwarfing than those of 'M.9' although the reverse is true with short interstocks (Parry and Rogers, 1972).

In general the relationship between yield and tree size (cropping efficiency) is similar whether tree size is controlled by rootstock or by interstock (cf. Chapter 2, especially Figures 2.3 and 2.4).

References

Blanco, A. (1989). First results on growth and cropping of pear trees budded at different heights. *Acta Horticulturae* **256**, 93–7.

Buchloh, G. (1960). The lignification in stock–scion junctions and its relation to compatibility. In *Phenolics in Plants in Health and Disease*, ed. J.B. Pridham, p. 67. Long Island City, NY: Pergamon Press.

Czynczyk, A. (1986). Eighteen years results with dwarf interstocks 'M.9' and 'B.9' on growth and cropping of two apple cultivars. *Acta Horticulturae* **160**, 49–55.

Davies, D.L., Guise, C.M., Clark, M.F. and Adams, A.N. (1992). Parry's disease of pears is similar to pear decline and is associated with mycoplasma-like organisms transmitted by *Cacopsylla pyricola*. *Plant Pathology* **41**, 195–203.

Ferree, D.C. and Carlson, R.F. (1987). Apple rootstocks. In *Rootstocks for Fruit Crops*, ed. R.C. Rom and R.F. Carlson, pp. 107–43. New York: John Wiley and Sons.

Garner, R.J. (1953). Double-working pears at budding time. *Report of the East Malling Research Station for 1952*, 174–5.

Garner, R.J. (1988). *The Grafter's Handbook*. London: Cassell.

Grubb, N.H. (1939). The influence of the intermediate in double-worked apple trees: nursery trials of the 'stembuilder' process at East Malling. *Journal of Pomology and Horticultural Science* **17**, 1–19.

Gur, A. (1957). The compatibility of the pear with quince rootstock. *Special Bulletin 10, Agricultural Research Station, Rehovot, Israel.*

Gur, A. and Samish, R.M. (1965). The relation between growth curves, carbohydrate distribution, and compatibility of pear trees grafted on quince rootstocks. *Horticultural Research* **5**, 81–100.

Gur, A., Samish, R.M. and Lifshitz, E. (1968). The role of the cyanogenic glycoside of the quince in the incompatibility between pear cultivars and quince rootstocks. *Horticultural Research* **8**, 113–34.

Hartmann, H.T., Kester, D.E. and Davies, F.T. (1990). *Plant Propagation, Principles and Practice.* New Jersey: Prentice-Hall.

Hibino, H. and Schneider, H. (1970). Mycoplasma-like bodies in sieve tubes of pear trees affected with pear decline. *Phytopathology* **60**, 499–501.

Hibino, H., Kaloostian, G.H. and Schneider, H. (1971). Mycoplasma like bodies in the pear psylla vector of pear decline. *Virology* **43**, 34–40.

Howard, B.H. (1977). Chip budding fruit and ornamental trees. *International Plant Propagators Society Combined Proceedings for 1977*, **27**, 357–64.

Howard, G.S. and Hildreth, A.C. (1963). Induction of callus tissue on apple grafts prior to field planting and its growth effects. *Proceedings of the American Society for Horticultural Science* **82**, 11–15.

Howard, B.H., Skene, D.S. and Coles, J.S. (1974). The effects of different grafting methods upon the development of one-year-old nursery apple trees. *Journal of Horticultural Science* **49**, 287–95.

Kuden, A. and Kaska, N. (1995). Extension of budding period of apples by using chip-budding, T-budding and English grafting. *Acta Horticulturae* **409**, 125–31.

Lockard, R.G. (1974). Effects of rootstocks and length and type of interstock on growth of apple trees in sand culture. *Journal of the American Society for Horticultural Science* **99**, 321–5.

Moore, R. (1986). Graft incompatibility between pear and quince: the influence of metabolites of *Cydonia oblonga* on suspension cultures of *Pyrus communis. American Journal of Botany* **73**, 1–4.

Mosse, B. and Labern, M.V. (1960). The structure and development of vascular nodules in apple bud-unions *Annals of Botany* **24**, 500–7.

Parry, M.S. (1986). The effects of budding height on the field performance of two apple cultivars on three rootstocks. *Journal of Horticultural Science* **61**, 1–7.

Parry, M.S. and Rogers, W.S. (1968). Dwarfing interstocks: their effect on the field performance and anchorage of apple trees. *Journal of Horticultural Science* **43**, 133–46.

Parry, M.S. and Rogers, W.S. (1972). Effects of interstock length and vigour on the field performance of Cox's Orange Pippin apples. *Journal of Horticultural Science* **47**, 97–105.

Preston, A.P. (1974). Apple rootstock studies: some rootstock and interstock comparisons. *Horticultural Research* **14**, 47–53.

Rauch, F.D. (1968). Floral induction and growth of apple. *Dissertation Abstracts Section B.* 1968, **28**, 4373. Iowa State University.

Sass, J.E. (1932). Formation of callus knots on apple grafts as related to the histology of the graft union. *Botanical Gazette* **94**, 364–80.

Shippy, W.B. (1930). Influence of environment on the callusing of apple cuttings and grafts. *American Journal of Botany* **17**, 290–327.

Swarbrick, T., Blair, D. and Singh, S. (1946). Studies in the physiology of rootstock and scion relationships. *Journal of Pomology and Horticultural Science* **22**, 51–61.

Torrey, J.G., Fosket, D.E. and Hepler, P.K. (1971). Xylem formation: a paradigm of cytodifferentiation in higher plants. *American Science* **59**, 338–52.

Tukey, H.B. and Brase, K.D. (1943). The dwarfing effects of an intermediate stem-piece of Malling IX apple. *Proceedings of the American Society for Horticultural Science* **42**, 357–64.

van Oosten, H.J. (1978). High working for control of vigour of apple trees. *Acta Horticulturae* **65**, 157–65.

Vyvyan, M.C. (1938). The relative influence of rootstock and of an intermediate piece of stock stem in some double-grafted apple trees. *Journal of Pomology and Horticultural Science* **16**, 251–73.

Wagner, D.F. (1969). Ultrastructure of the bud graft union in *Malus*. PhD dissertation, Iowa State University, Ames, Iowa.

Wertheim, S.J. and Toorenaar, G. (1966). Onderstammen en tussen stammenproefveld met peer. *Jaarverslag 1966 Proefstation voor de Fruitteelt, Wilhelminadorp*, 21–2.

Wetmore, R.H. and Rier, J.P. (1963). Experimental induction of vascular tissues in callus of angiosperms. *American Journal of Botany* **50**, 418–30.

Yeager, A.F. (1944). Xylem formation from ring grafts. *Proceedings of the American Society for Horticultural Science* **44**, 221–2.

5

Mechanisms of rootstock and interstock effects on scion vigour

Introduction

The use of dwarfing rootstocks for the control of tree vigour is a dominant feature of much of modern apple production and, to a lesser extent, pear production. The mechanism of vigour control by rootstocks has been studied for many years, partly in the hope that understanding the process will lead to its more effective utilization.

The effects of the rootstocks and interstocks are manifold, interactive and cumulative over years. Moreover, different mechanisms appear to be dominant in different species and even in different stock/scion cultivar combinations within a species. In practical terms the dwarfing effect of 'M.9' and its derivatives on apple scions and of quince rootstocks on pear scions are probably of the greatest importance so emphasis is given to these within the broader context.

Mechanisms of rootstock and interstock effects on vigour

The mechanisms of rootstock influence on tree vigour are best considered within the concept that the vigour of the composite tree reflects in an additive way the vigour of its components and that these interact. There is, as discussed earlier, a tendency to attainment of a functional equilibrium between roots and shoots. Roots supply shoots with nutrients and water, shoots supply roots with assimilates, and the roots and shoots appear to have specific roles in the production of the plant growth substances that control and coordinate activities in the plant. The graft union and the conducting tissues of the rootstock may also influence growth through their effects on translocation from root to shoot and from shoot to root.

Within this framework the following appear to make major contributions to the dwarfing effect of dwarfing rootstocks.

1. *The limited size of dwarfing rootstock root systems.* In general, the more dwarfing an apple rootstock the less vigorous it is as an 'unworked' tree and the smaller its root system. This characteristic is retained when the rootstock is grafted with a scion. Bane *et al.* (1935) and Beakbane and de Wet (1935) excavated the root systems of mature trees of 'Bramley' and 'Worcester' on various rootstocks and found that with the more vigorous scion the root systems were larger but the rank order of root system size of the different rootstocks was the same. This was also shown, much later, in trials in which rootstock clones ('M.9' crosses) of differing vigour were grafted on each other in all factorial combinations (Table 5.1). The limited size of the root system of 'M.9' and some other dwarfing rootstocks may be associated with their low capacity to produce roots. This is shown even *in vitro* when auxin and carbohydrate are supplied (Webster and Jones, 1989), after layering in the nursery or after auxin treatment of hardwood cuttings (Doud and Carlson, 1977). Grafted trees on 'M.9' show limited initial root growth following planting compared with those on some more vigorous rootstocks (Young and Werner, 1982). The small root systems of orchard trees on 'M.9' and other dwarfing rootstocks (Rogers and Vyvyan, 1934; Coker, 1958; Fernandez *et al.*, 1995) reflect this inherently limited growth potential and control scion growth. If a vigorous scion grafted on a dwarfing rootstock is allowed to develop scion roots into the soil the tree then grows much more rapidly, showing the predominance of root influence (Rogers and Beakbane, 1957).

2. *The special anatomical features of dwarfing rootstock roots.* The roots of dwarfing rootstocks have smaller xylem vessels and less than half as many xylem fibres as those of more vigorous rootstocks, and also a much higher percentage of bark and of wood ray tissue per unit of root cross-sectional area (Beakbane and Thompson 1939, 1947; McKenzie, 1961). Although grafting 'M.9' with more vigorous scions increased its mean vessel diameter this remained consistently smaller than that of other, more vigorous, rootstocks grafted with the same scions.

3. *Root system size is reduced by the presence of a dwarfing interstock.* The root systems of vigorous rootstocks are greatly reduced in size if they are separated from the scion by a piece of stem, an interstock, of a dwarfing rootstock such as 'M.9', 'M.26' or 'Clark Dwarf' (Swarbrick *et al.*, 1946; Dana *et al.*, 1962; Parry and Rogers, 1968). The interstock, in these studies, reduced the growth of the root system more than that of the scion.

4. *The size of the root system directly affects scion growth.* Rootstocks can be invigorating as well as dwarfing, as shown by stock-on-stock trials (Table 5.1) and by comparing trees of cultivars on their own roots (with no graft unions) with grafted trees. Trees of 'Starking Delicious' on 'M.4' rootstock are about

Table 5.1 *Growth of rootstocks when grafted in reciprocal stock × scion combinations*

Dry weight (kg) at 6 years from grafting

	Scion cv.									
	3431[a]		3436[a]		3428		3438		3430	
Rootstock cv.	Stock	Scion	Stock	Scion	Stock	Scion	Stock	Scion	Stock	Scion
3431[a]	0.51	1.56	0.28	1.54	0.92	3.16	0.76	2.60	1.36	4.87
3436[a]	1.94	6.81	0.63	3.30	3.06	7.75	2.44	6.91	3.93	9.29
3428	1.71	7.42	0.66	3.51	2.52	9.41	2.07	7.92	3.63	12.92
3438	2.82	9.70	0.95	4.18	3.70	10.20	2.66	8.39	4.94	13.41
3430	2.87	12.59	1.11	7.27	3.59	11.81	2.93	9.54	5.08	16.72

[a] 3431 is now known as 'M.27', 3436 as 'M.26'.

The rootstocks are arranged from top to bottom in order of increasing vigour: all are 'M.9' crosses.

From Tubbs, (1980). Reproduced with permission.

Table 5.2 *Growth of 'Lane's Prince Albert' trees grafted directly on rootstock roots or on rootstocks with both root and stem*

	Girth of trunk at 19 years old (cm) Rootstock			
Method of working	'M.9'	'M.2'	Seedling	'M.12'
On root	16.6	30.9	36.3	46.8
On stem at 2″	11.5	28.2	37.2	49.0
On stem at 7″	12.9	26.3	36.3	50.1

Significant ratio for rootstocks with the same method of working = 1.24 ($p = 0.05$).
Effect of method of working non-significant.
'Lane's Prince Albert' is rather more vigorous than 'M.2'.
Adapted from Beakbane and Rogers (1956), with permission.

twice as large as those of 'Starking Delicious' on its own roots (Gyuro and Gondor-Pinter, 1986) and those of 'Greensleeves' on 'MM.106' about twice as large as those of 'Greensleeves' on its own roots although the latter trees are larger than those of 'Greensleeves' on 'M.9' (Webster *et al.*, 1985). Limiting the size of apple root systems by root pruning (Geisler and Ferree, 1984) or by growing them within a root restriction membrane (Atkinson *et al.*, 1997) also reduces shoot growth.

5. *The main effect of a rootstock can be induced by the root system itself without involvement of the rootstock stem.* When scion trees of 'Lane's Prince Albert' were grafted on a range of rootstocks either directly on to root or on to rootstock stem, most of the rootstock effect on scion vigour was shown when the rootstock root alone was used (Beakbane and Rogers, 1956), although the presence of a piece of rootstock stem intensified this (Table 5.2). The relative importance of rootstock stem may, however, be greater in the case of 'M.26' in view of the effects of budding height on 'M.26' and of length of 'M.26' interstock noted in Chapter 4.

6. *Dwarfing rootstocks may be less effective than vigorous ones in uptake of nutrients and their supply to the shoot.* When uniform rooted stems (layers) of 'M.9', 'M.7' and 'M.16' were given ^{32}P a month after being established in solution culture the uptake within 96 hours, expressed as counts per minute per g shoot per mg root was four times as high with 'M.16' (vigorous) as with 'M.9' (dwarfing), with 'M.7' being intermediate (Bukovac *et al.*, 1958). A similar pattern was found for the uptake of ^{32}P and ^{45}Ca into scions of 'McIntosh' grafted on 'M.16', 'M.7' and 'M.9'.

Under orchard conditions the supply of nutrients is not usually limiting for growth and is controlled by scion demand, and although trees on different rootstocks may show differences in leaf nutrient content these are not systematically related to rootstock vigour. It is therefore unlikely that differences in nutrient uptake are generally responsible for rootstock effects on scion growth. There may be exceptions under conditions of very low fertility: Ruck and Bolas (1956) found the difference in growth between 'Crab C' (vigorous) and 'M.9' (dwarfing) to be much greater when in sand culture with 5 ppm N than with 200 ppm N.

7. *Apple trees on dwarfing rootstocks may suffer more from water stress.* Olien and Lakso (1984) found that the mid-day stem water potential of 'Empire' apple trees on 'M.9', 'M.26', 'M.7', 'MM.106' and 'M.4' was significantly correlated with tree size. They attributed the greater water stress in the trees on dwarfing rootstocks to a greater root resistance at the root surface and in the xylem. Atkinson *et al.* (2001), working with 'Queen Cox' on 'M.27', 'M.9' and 'MM.106', found that the hydraulic resistances of the rootstock shanks, rootstock graft unions and the scions worked on the different rootstocks were positively related to the rootstock's capacity to dwarf. Trees on dwarfing rootstocks also crop more heavily in relation to their size and, in general, heavy cropping is accompanied by higher stomatal conductance and greater water use (Jones *et al.*, 1985). These effects on water use and water stress could contribute to the effects of some dwarfing rootstocks on scion growth. Also the apple rootstock 'MM.106', generally classed as semi-dwarfing, gives much less vigorous scion trees on dry than on humid sites in comparison with other rootstocks (Parry, 1965).

8. *Cytokinin supply from roots to shoots is a function of rootstock vigour.* The xylem sap from decapitated apple root systems contains cytokinins which promote stem elongation and leaf production in isolated apple shoots (Jones, 1973) and are thought largely to control the development of the shoot system in intact trees (Lockard and Schneider, 1981). Kamboj (1996) and Kamboj *et al.* (1997a, 1999b) showed that the levels of cytokinin (zeatin + zeatin riboside) in the shoot xylem sap of both unworked rootstock trees and of 'Fiesta' scions grafted on rootstocks increased with the vigour of the rootstock (Table 5.3). It seems likely that differences in cytokinin production play a major role in causing the effects of rootstock root systems on scion growth. Cytokinin in the presence of auxin stimulates early stages of vascular differentiation and increases the sensitivity of tissues to auxin stimulation (Aloni, 1995).

9. *Graft unions between dwarfing rootstocks and scions frequently show unusual features of both xylem and phloem.* It is usual for apple graft unions on dwarfing rootstocks to be swollen (Mosse, 1962; Jones, 1974). Simons (1987) gives many examples of poor unions, with restricted vascular connections, which he attributes to bud quality, stage of development or particular tissue development at the time the rootstock was budded, usually involving dwarfing rootstocks with much slower growth rates than the scions budded or grafted on them.

Table 5.3 *Cytokinin (zeatin + zeatin riboside) total content and concentration in xylem sap from unworked rootstocks or 'Fiesta' apple scions worked on these*

(Back-transformed values, ng shoot^{-1} and ng ml^{-1} sap).

	Shoot			
	Unworked rootstock		'Fiesta' scion	
Rootstock	Content	Concentration	Content	Concentration
'M.27'	7.9	3.11	7.3	2.96
'M.9'	6.6	4.29	14.3	5.39
'MM.106'	25.6	6.77	26.3	5.59

Rootstock effects significant at $p = 0.05$.
Data from Kamboj *et al.* (1999b). Reproduced with permission.

Warmund *et al.* (1993) attributed vascular discontinuities at the union between 'Jonagold' and 'Mark' rootstock, followed by weak scion growth, to inadequate growth of the rootstock and/or scion tissues in the autumn. This problem was found in trees produced in some nurseries but not in others. Although Soumelidou *et al.* (1994a) found the stages in union formation between 'Bramley' on 'M.9' and 'Bramley' on 'MM.106' to be similar, the xylem linking the scion bud to the 'M.9' rootstock contained fewer and smaller vessels than did the xylem in the union of the scion bud to the 'MM.106' rootstock.

Mosse and Scaramuzzi (1956) found anatomical symptoms of incompatibility in all pear/quince unions studied. They were much more severe in the 'horticulturally-incompatible' unions between quince and 'Williams'' but also showed, to a lesser extent, at the unions of quince with 'horticulturally compatible' cultivars such as 'Hardy' and 'Fertility'. The symptoms consisted of necrotic cells in the phloem at the union, proliferation of phloem rays and, to a lesser extent, breaks in the xylem. The damage appeared to originate in the older part of the one-year-old phloem and spread to the rays. It was not shown in the pear-to-pear unions between pear interstocks and 'Williams''. These unions were so good that the line of union could not be detected.

The graft unions between dwarfing apple rootstocks, especially 'M.9', and the scions worked on them generally become very swollen with age, attaining several times the diameter of the rootstock stem below and the scion stem above (Jones, 1974; Simons, 1987). This swelling is progressive over time (Kamboj *et al.*, 1999b) and the associated abnormalities in conducting tissues may have increasing effects.

Table 5.4 *Effect of rootstock on solute concentration and nutrient content of the xylem sap above and below the graft union*

	Below graft union				Above graft union			
Rootstock	Solute (mg ml^{-1})	N ppm	P ppm	K ppm	Solute (mg ml^{-1})	N ppm	P ppm	K ppm
'M.9'	3.4	230	33	174	3.0	203	31	147
'M.27'	3.0	219	26	180	2.1	167	14	123
3426[a]	2.6				1.6			

[a] 3426 is more dwarfing than 'M.27'.
After Jones (1984). Reproduced with permission.

10. *With apple, the graft unions of dwarfing rootstocks or interstocks with the scions deplete solutes in the xylem sap and this effect is greater the more dwarfing the rootstock or interstock.* The depletion of the xylem sap (Table 5.4) is found with respect to N, P, K, Ca and Mg (Jones, 1976). The interstock effect appears to be produced in the upper union region between interstock and scion.

11. *Trees on dwarfing rootstocks show restricted canopy development.* Trees on vigorous rootstocks produce more shoots and greater shoot extension late in the season than those on 'M.9', even when all are deblossomed (Avery, 1969). The rootstock effects on shoot growth rate are the result of several factors: (1) Dwarfing rootstocks induce wider crotch angles to their lateral branches and a more spreading tree habit (Hatton, 1930, 1935; Crabbé, 1984). This is in keeping with the habits of the rootstocks growing on their own roots and is accentuated by the precocious and heavy cropping of scions worked on dwarfing rootstocks. Horizontal branches show both decreased vegetative growth and enhanced flowering and fruiting (Tromp, 1968, 1970, 1972), these effects being at least partially independent (Robbie *et al.*, 1993). (2) Dwarfing rootstocks check the growth rate of both vertical and horizontal shoots, the effect being greater on the former (Webster, 1995). (3) The heavy cropping induced by dwarfing rootstocks checks shoot growth in the following year. (4) Trees on dwarfing rootstocks show earlier termination of leaf production and earlier leaf senescence and spur leaf abscission (Webster, 1995).

The net effect is that leaf area per tree, especially towards the end of the season, is less the more dwarfing the rootstock (Webster, 1995).

12. *Trees on dwarfing rootstocks have lower net assimilation rates per unit leaf area than those on vigorous ones late in the season.* Net assimilation rates for whole trees, which represent the balance between photosynthetic production and respiratory losses were found to be higher for 'M.16' than for 'M.9' and for 'Cox'/'M.16' than for 'Cox'/'M.9' late in the season but not in the early part (Gregory, 1957). This may reflect earlier leaf senescence on dwarfing rootstocks

Table 5.5 *Effects of rootstock and deblossoming on the vegetative growth and fruit yield of 'Laxton's Superb' at 13 years*

	Accumulated weight (kg/tree)								
	Deblossomed				Cropping				
Rootstock	Leaf	Wood	Root	Total	Leaf	Wood	Root	Total vegetative	Fruit
Light pruned									
'M.9'	5	20	2	27	1	4	0.5	5.5	14.5
'M.16'	24	90	25	139	9	34	8.0	51.0	55.0
Hard pruned									
'M.9'	3	17	1	21	0.5	3	0.5	4.0	11.0
'M.16'	12	61	11	84	7.0	35	6.0	48.0	33.0

Data from Barlow and Smith (1971). Reproduced with permission.

and a lower proportion of young leaves with high photosynthetic rates late in the season. Measured effects of rootstock on photosynthesis per unit leaf area have been contradictory and usually small (Barden, 1978; Wunsche *et al.*, 1996). The results may be partly confounded by effects of crop load on photosynthesis and the comparisons of similar leaves may not give a relevant picture of total photosynthesis by leaf populations which differ in age structure. It does, however, seem that particularly in trees which have already begun to differ in size, the main effect of rootstock on photosynthetic potential is through its effect on leaf area.

13. *Dwarfing rootstocks induce precocious and heavy fruiting at the expense of vegetative growth.* Rootstocks have a major effect on the partitioning of assimilates. 'M.9' and its derivatives are precocious in flowering and induce early, heavy cropping in any scion cultivar grafted on them. This is not a response to their dwarfing effect but is shown before the latter becomes marked. At maturity trees on dwarfing rootstocks in general show a much higher ratio of fruit yield to vegetative growth increment than do trees on the more vigorous rootstocks. Heavy fruiting not only checks growth in the year of cropping, especially root growth, but also growth in the following year (Jackson, 1984). This indirect effect on scion vigour via effects on fruiting is responsible for an appreciable component of apple rootstock effects on vigour. Apple trees of 'Laxton's Superb' on 'M.9' which had been deblossomed from planting produced 22% as much accumulated dry matter (growth) over 13 years as corresponding trees on 'M.16'. When each had been allowed to crop, vegetative growth on 'M.9' was only 10% of that on 'M.16' (Table 5.5). The trees on 'M.9' that had been allowed to crop made only 20% of the growth of those that had been deblossomed.

Table 5.6 *Effect of rootstock on the transport of auxin ([^{3}H]-IAA) and carbohydrate ([^{14}C]-sorbitol) from mature leaves to roots*

	Total activity in roots after 24 h (dpm)		Specific activity in roots after 24 h (dpm mg^{-1} dry wt)	
	^{3}H	^{14}C	^{3}H	^{14}C
'M.27'	5140	73 080	2.84	40.73
'M.9'	9483	84 950	4.64	40.42
'MM.106'	18 888	119 423	12.12	72.72
'MM.111'	20 372	108 419	12.63	67.11
LSD 0.05	8183	31 948	5.81	23.27

From Kamboj (1996). Reproduced with permission.

Interstocks of dwarfing apple rootstocks, e.g. 'M.9', similarly induce precocious flowering and heavy cropping, the effects being proportional to the length of interstock (Roberts and Blaney, 1967; Parry and Rogers, 1972).

'Quince C', the dwarfing rootstock for pear, appears to exert its dwarfing effect primarily through its effect on cropping. It roots readily from cuttings and gives strong maiden growth in the nursery. Trees on it grow vigorously for the first few years in the orchard before their growth is checked by heavy, precocious cropping (Parry, 1972).

The size of apple trees on some relatively vigorous rootstocks may also be reduced to below their potential because they induce precocious cropping. This is particularly so for 'MM.106' and 'M.25'.

Although in general the root/shoot ratio is relatively constant for any given set of environmental conditions, irrespective of rootstock (Rogers and Beakbane, 1957), the root system of 'M.9' appears to be a particularly weak sink for assimilates in competition with shoot growth. Young and Werner (1982) found that shoot pruning of trees of 'Golden Delicious'/'M.9' at planting led to much poorer root growth, in the face of competing shoot growth, than was the case with 'Golden Delicious'/'MM.106'. The root/shoot ratio of trees on 'M.9' is usually lower than that on more vigorous rootstocks (Avery, 1969; Young and Werner, 1982) and is also reduced by an 'M.9' interstock (Swarbrick *et al.*, 1946), which may lead to root-supplied growth factors being limiting.

14. *Dwarfing rootstocks show limited ability to transport auxin from shoots to roots and have high ABA levels.* Basipetal transport of auxin from the region of the shoot tips towards the root system (Table 5.6) is slower in dwarfing than in more vigorous rootstocks (Soumelidou *et al.*, 1994b; Kamboj *et al.*, 1997b). Gur and Samish (1968) found that the amount of IAA destroyed by root and shoot bark of various apple rootstocks was negatively correlated with their effects

Table 5.7 *ABA contents (ng g⁻¹ fresh weight) in unworked apple rootstock shoot bark*

Rootstock	Month				Mean	Log transformed mean
	May	June	July	August		
'M.27'	97	188	285	172	186	5.09
'M.9'	58	210	154	116	135	4.73
'MM.106'	19	94	134	93	85	4.20
'MM.111'	19	92	134	85	83	4.17
LSD 0.05						0.30

From Kamboj *et al.* (1999a). Reproduced with permission.

on scion vigour. Limitations of transport of IAA to the roots should have an adverse effect on vascular development, since auxin controls cambial activity (Digby and Wareing, 1966; Lockard and Schneider, 1981), and therefore on transport of material to and from the roots. Auxin is the limiting and controlling factor for both phloem and xylem differentiation and induces phloem with no xylem at low auxin concentrations. It controls the size and density of vascular elements along the plant axis (Aloni, 1995). Limiting auxin flux could also have a directly negative effect on rooting in the context of high root turnover and the need to produce new roots from buried rootstock stems in the early orchard years. The very high ratio of phloem to xylem (bark to wood) in dwarfing rootstock roots is characteristic of tissues produced in a low IAA/high GA environment (Digby and Wareing, 1966). Also, dwarfing interstocks often lead to profuse suckering (shoot emergence) from the rootstock stem below them and many dwarfing rootstocks sucker profusely. This would be expected if there is inadequate auxin to maintain apical dominance and bud dormancy.

Yadava and Dayton (1972) found the ABA content of roots, shoots and leaves of 'M.9', 'M.7' and 'M.1' was negatively related to the invigorating effects of these rootstocks, and Kamboj *et al.* (1999a) found higher ABA concentrations in the shoot bark of 'M.27' and 'M.9' than of 'MM.106' and 'MM.111' (Table 5.7). ABA is involved in growth inhibition and dormancy and also is known to have an adverse effect on auxin transport.

15. *Phloem transport from leaves to roots is reduced when the scion is on a dwarfing rootstock or interstock.* Transport of ^{32}P applied to scion leaves down into the root system was greater per gram of root dry weight of invigorating than of dwarfing rootstocks (Bukovac *et al.*, 1958). There was no evidence of a build-up immediately above the graft union. Dickson and Samuels (1956) supplied ^{32}P to cut leaf petioles of scions on dwarfing interstocks of 'M.9' and 'Clark Dwarf'

above vigorous rootstocks. They found accumulation in the interstock especially just above the rootstock. Both studies could indicate reduced phloem transport, that of Dickson and Samuels either greater ^{32}P absorption in the swollen and actively growing interstock tissues or a specific phloem block at the lower graft union.

16. *Transport of carbohydrates from leaves to roots is reduced when the scion is on a dwarfing rootstock or interstock.* Movement of [^{14}C]-sorbitol (the predominant transport photoassimilate in apple) from mature leaves of cv. 'Fiesta' to the root system was greater when the trees were on 'MM.106' or 'MM.111' than when the root systems were on 'M.9' or 'M.27' (Kamboj, 1996). This was so even when activity was expressed per unit root weight and so was not just a consequence of the size of the root system (Table 5.6). This supports the concept that dwarfing rootstock roots are a less strong sink for assimilates than those of invigorating rootstocks or that transport to them is restricted by poorer transport systems or abstraction *en route*. Kamboj's data show the total ^{14}C activity to be higher in the graft union area than above or below it but this was not the case for specific activity. There was a tendency for the difference between the above- and below-union levels to be greater with the dwarfing than with the vigorous rootstocks but not enough to prove a role for the unions *per se* in the rootstock effect on sorbitol transport. On the other hand, when $^{14}CO_2$ was fed to scion leaves above different interstocks on 'MM.111' rootstock, trees with 'M.9' and 'M.26' dwarfing interstocks had six to nine times higher specific activity above the interstock than within or below it, while trees with the more vigorous 'MM.111' and '3430' interstocks had similar specific activity above, within and below these (Young *et al.*, 1986). This indicates a restriction of assimilate movement downwards at the upper graft union of a dwarfing interstock. Dana *et al.* (1963) found that scion leaves had higher reducing sugar content when on trees with dwarfing interstocks of 'Clark Dwarf' than when with 'own-stem' interstocks of 'Golden Delicious'. The concept that the root systems of dwarfing rootstocks have a limited capacity to make use of assimilates, or that the graft union limits transport to them, is supported by White (1970) who described a physiological disorder, Cox disease, accompanied by high leaf dry matter and sugar content and low concentrations of major nutrients, which occurs when trees on dwarfing rootstocks bear too little crop for this to be an effective sink for carbohydrates.

17. *Experimental treatments which interrupt phloem transport can partially mimic dwarfing rootstock effects.* Girdling the bark of apple trees or inverting a ring of bark has a dwarfing effect until new phloem develops (Sax, 1957). Even removing a ring of bark then replacing it in the normal position allows the growth of dormant buds below the disturbed bark, presumably as a result of interference with auxin transport. These effects are compatible with the finding that as much

as 60% of the bark of the graft union with very dwarfing rootstocks can be non-conducting phloem (Simons, 1987).

Conclusion and comments

It is clear that the vigour of a compound tree reflects the vigour of its component rootstock and scion. The use of a dwarfing rootstock as a management tool to control scion vigour is primarily dependent on the reduced vigour of the root system. This may be inherent, as when the rootstock provides the root system, or may be induced by limitation of root growth as a result of interstock, rootstock stem or graft union effects on the supply of auxins, carbohydrates and other metabolites. The greater check to growth as the length of rootstock stem or of dwarfing interstock is increased is compatible with abstraction from the downward fluxes to the root. The main effect of root system size on shoot growth is probably via supply of cytokinins but may also involve nutrient and water supply. The graft union *per se* may also restrict upward movement of water and nutrients.

Points of practical importance arising from other aspects of rootstock control of tree growth include the following.

1. When a vigorous scion is grafted on a dwarfing rootstock it must not be planted with the graft union below soil level. If this is done a vigorous scion root system may develop and the tree as a whole becomes vigorous.
2. The limited competitiveness of dwarfing rootstock root systems as sinks and their independent effect in stimulating precocious fruiting brings the risk of excessive dwarfing. This makes it important to plant trees which are already large and well branched and to manage their early fruiting so as to achieve the best balance between early economic returns and the rapid growth needed to attain the desired orchard canopy dimensions as soon as possible.
3. The effect of a dwarfing interstock on the size of the root system renders invalid the general concept of the use of a vigorous rootstock to give good anchorage coupled with a dwarfing interstock to control vigour. It does not rule out the possibility of such a combination being effective, but only if the rootstock root system has specific characteristics which improve anchorage, including mechanical toughness, not just vigour.
4. The potential to control tree growth by the use of dwarfing rootstocks, management of cropping level and appropriate tree training and pruning has greatly reduced the risks of excess vigour and consequently of excessive shade in orchards planted at high tree densities to achieve early yields and economic returns.

Recommended reading

Lockard, R.G. and Schneider, G.W. (1981). Stock and scion growth relationships and the dwarfing mechanism in apple. *Horticultural Reviews* **3**, 315–75.

Rogers, W.S. and Beakbane, A.B. (1957). Stock and scion relations. *Annual Review of Plant Physiology* **8**, 217–36.

Tubbs, F.R. (1973). Research fields in the interaction of rootstocks and scions in woody perennials. I and II. *Horticultural Abstracts* **43**, 247–53 and 325–35.

References

Aloni, R. (1995). The induction of vascular tissues by auxin and cytokinin. In *Plant Hormones*, ed. P.J. Davies, pp. 531–46. Dordrecht: Kluwer.

Atkinson, C.J., Else, M.A., Taylor, L. and Webster, A.D. (2001). The rootstock graft union: a contribution to the hydraulics of the worked fruit tree. *Acta Horticulturae* **557**, 117–22.

Atkinson, C.J., Webster, A.D., Vaughan, S. and Lucas, A.S. (1997). Effects of root restriction on the physiology of apple tree growth. *Acta Horticulturae* **451**, 587–95.

Avery, D.J. (1969). Comparisons of fruiting and deblossomed maiden apple trees and of non-fruiting trees on a dwarfing and an invigorating rootstock. *New Phytologist* **68**, 323–36.

Bane, W.A., Beard, F.H. and de Wet, A.F. (1935). A comparison of the root-systems of mature trees of Bramley's Seedling and Worcester Pearmain on various rootstocks. *Report of the East Malling Research Station for 1934*, 90–9.

Barden, J.A. (1978). Apple leaves, their morphology and photosynthetic potential. *HortScience* **13**, 644–6.

Barlow, H.W.B. and Smith, J.G. (1971). Effect of cropping on growth of the apple tree. *Report of the East Malling Research Station for 1970*, 52–3.

Beakbane, A.B. and de Wet, A.F. (1935). The root-systems of some Bramley's Seedling apple trees. *Report of the East Malling Research Station for 1934*, 87–9.

Beakbane, A.B. and Rogers, W.S. (1956). The relative importance of stem and root in determining rootstock influence in apples. *Journal of Horticultural Science* **31**, 99–110.

Beakbane, A.B. and Thompson, E. (1939). Anatomical studies of stems and roots of hardy fruit trees. II. The internal structure of the roots of some vigorous and some dwarfing apple rootstocks, and the correlation of structure with vigour. *Journal of Pomology* **17**, 141–9.

Beakbane, A.B. and Thompson, E.C. (1947). Anatomical studies of stems and roots of hardy fruit trees. IV. The root structure of some new clonal apple rootstocks budded with Cox's Orange Pippin. *Journal of Pomology* **23**, 206–11.

Bukovac, M.J., Wittwer, S.H. and Tukey, H.B. (1958). Effect of stock–scion interrelationships on the transport of ^{32}P and ^{45}Ca in the apple. *Journal of Horticultural Science* **33**, 145–52.

Coker, E.G. (1958). Root studies. XII. Root systems of apple on Malling rootstocks on five soil series. *Journal of Horticultural Science* **33**, 71–9.

Crabbé, J.J. (1984). Morphogenetic ways towards vigour restriction in spontaneous and man-made dwarf trees. *Acta Horticulturae* **146**, 113–20.

Dana, M.N., Lantz, H.L. and Loomis, W.E. (1962). Effects of interstock grafts on growth of 'Golden Delicious' apples. *Proceedings of the American Society for Horticultural Science* **81**, 1–11.

Dana, M.N., Lantz, H.L. and Loomis, W.E. (1963). Studies on translocation across dwarf interstocks. *Proceedings of the American Society for Horticultural Science* **82**, 16–24.

Dickson, A.G. and Samuels, E.W. (1956). The mechanism of controlled growth of dwarf apple trees. *Journal of the Arnold Arboretum* **37**, 307–13.

Digby, J. and Wareing, P.F. (1966). The relationship between endogenous hormone levels in the plant and seasonal aspects of cambial activity. *Annals of Botany*, N.S. **30**, 607–22.

Doud, S.L. and Carlson, R.F. (1977). Effects of etiolation, stem anatomy and starch reserves on root initiation in layered *Malus* clones. *Journal of the American Society for Horticultural Science* **102**, 487–91.

Fernandez, R.T., Perry, R.L. and Ferree, D.C. (1995). Root distribution patterns of nine apple rootstocks in two contrasting soil types. *Journal of the American Society for Horticultural Science* **120**, 6–13.

Geisler, D. and Ferree, D.C. (1984). Response of plants to root pruning. *Horticultural Reviews* **6**, 155–88.

Gregory, F.G. (1957). The place of plant physiology in fruit growing. *Report of the East Malling Research Station for 1956*, 41–8.

Gur, A. and Samish, R.M. (1968). The role of auxins and auxin destruction in the vigour effect induced by various apple rootstocks. *Beitr. Biol. Pflanzen.* **45**, 91–111.

Gyuro, F. and Gondor-Pinter, M. (1986). The effect of rootstock and variety on the productive surface of apple trees. *Acta Horticulturae* **160**, 83–96.

Hatton, R.G. (1930). Stock:scion relationships. *Journal of the Royal Horticultural Society* **LV, II**, 169–211.

Hatton, R.G. (1935). Apple rootstock studies. Effect of layered stocks upon the vigour and cropping of certain scions. *Journal of Pomology and Horticultural Science* **13**, 293–353.

Jackson, J.E. (1984). Effects of cropping on tree vigour. *Acta Horticulturae* **146**, 83–8.

Jones, H.G., Lakso, A.N. and Syvertsen, J.P. (1985). Physiological control of water status in temperate and subtropical fruit trees. *Horticultural Reviews* **7**, 301–44.

Jones, O.P. (1973). Effects of cytokinins in xylem sap from apple trees on apple shoot growth. *Journal of Horticultural Science* **48**, 181–8.

Jones, O.P. (1974). Xylem sap composition in apple trees. Effects of the graft union. *Annals of Botany* **38**, 463–7.

Jones, O.P. (1976). Effect of dwarfing interstocks on xylem sap composition in apple trees: effect on nitrogen, potassium, phosphorus, calcium and magnesium content. *Annals of Botany* **40**, 1231–5.

Jones, O.P. (1984). Mode-of-action of rootstock/scion interactions in apple and cherry trees. *Acta Horticulturae* **146**, 175–82.

Kamboj, J.S. (1996). Studies of the dwarfing mechanism of apple rootstocks. PhD dissertation, University of London.

Kamboj, J.S., Blake, P.S., Quinlan, J.D. and Baker, D.A. (1999b). Identification and quantitation by GC-MS of zeatin and zeatin riboside in xylem sap from rootstock and scion of grafted apple trees. *Plant Growth Regulation* **28**, 199–205.

Kamboj, J.S., Blake, P.S., Quinlan, J.D., Webster, A.D. and Baker, D.A. (1997a). Recent advances in studies on the dwarfing mechanism of apple rootstocks. *Acta Horticulturae* **451**, 75–82.

Kamboj, J.S., Browning, G., Blake, P.S., Quinlan, J.D. and Baker, D.A. (1999a). GC-MS-SIM analysis of abscisic acid and indole-3-acetic acid in shoot bark of apple rootstocks. *Plant Growth Regulation* **28**, 21–7.

Kamboj, J.S., Browning, G., Quinlan, J.D., Blake, P.S. and Baker, D.A. (1997b). Polar transport of [^{3}H]-IAA in apical shoot segments of different apple rootstocks. *Journal of Horticultural Science* **72**, 773–80.

Lockard, R.G. and Schneider, G.W. (1981). Stock and scion growth relationships and the dwarfing mechanism in apple. *Horticultural Reviews* **3**, 315–75.

McKenzie, D.W. (1961). Rootstock–scion interaction in apples with special reference to root anatomy. *Journal of Horticultural Science* **36**, 40–7.

Mosse, B. (1962). Graft incompatibility in fruit trees. *Technical communication No. 28. East Malling, Maidstone, Kent, UK.* Commonwealth Bureau of Horticulture and Plantation Crops.

Mosse, B. and Scaramuzzi, F. (1956). Observations on the nature and development of structural defects in the unions between pear and quince. *Journal of Horticultural Science* **31**, 47–54.

Olien, W.C. and Lakso, A.N. (1984). A comparison of the dwarfing character and water relations of five apple rootstocks. *Acta Horticulturae* **146**, 151–8.

Parry, M.S. (1965). Field trials of Malling–Merton rootstocks. *Report of the East Malling Research Station for 1964*, 91–6.

Parry, M.S. (1972). Dwarfing quince rootstocks for pear. *Proceedings of the Symposium on Pear Growing, Angers, France.* 199–209.

Parry, M.S. and Rogers, W.S. (1968). Dwarfing interstocks: their effect on the field performance and anchorage of apple trees. *Journal of Horticultural Science* **43**, 133–46.

Parry, M.S. and Rogers, W.S. (1972). Effects of interstock length and vigour on the field performance of 'Cox's Orange Pippin' apples. *Journal of Horticultural Science* **47**, 97–105.

Robbie, F.A., Atkinson, C.J., Knight, J.N. and Moore, K.G. (1993). Branch orientation as a factor determining fruit set in apple trees. *Journal of Horticultural Science* **68**, 317–25.

Roberts, A.N. and Blaney, L.T. (1967). Qualitative, quantitative and positional effects of interstock influence on growth and flowering of the apple. *Proceedings of the American Society for Horticultural Science* **91**, 39–50.

Rogers, W.S. and Beakbane, A.B. (1957). Stock and scion relations. *Annual Review of Plant Physiology* **8**, 217–36.

Rogers, W.S. and Vyvyan, M.C. (1934). Root studies. V. Rootstock and soil effects on apple root systems. *Journal of Pomology* **12**, 110–50.

Ruck, H.C. and Bolas, B.D. (1956). Studies in the comparative physiology of apple rootstocks. I. The effect of nitrogen on the growth and assimilation of Malling apple rootstocks. *Annals of Botany*, N.S. **20**, 57–68.

Sax, K. (1957). The control of vegetative growth and the induction of early fruiting of apple trees. *Proceedings of the American Society for Horticultural Science* **69**, 68–74.

Simons, R.K. (1987). Compatibility and stock–scion interactions as related to dwarfing. In *Rootstocks for Fruit Crops*, ed. R.C. Rom and R.F. Carlson, pp. 79–106. New York: John Wiley and Sons.

Soumelidou, K., Battey, N.H., John, P. and Barnett, J.R. (1994a). The anatomy of the developing bud union and its relationship to dwarfing in apple. *Annals of Botany* **74**, 605–11.

Soumelidou, K., Morris, D.A., Battey, N.H., Barnett, J.R. and John, P. (1994b). Auxin transport capacity in relation to the dwarfing effect of apple rootstocks. *Journal of Horticultural Science* **69**, 719–25.

Swarbrick, T., Blair, D. and Sham Singh (1946). Studies in the physiology of rootstock and scion relationships. *Journal of Pomology and Horticultural Science* **22**, 51–61.

Tromp, J. (1968). Flower-bud formation and shoot growth in apple as affected by shoot orientation. *Acta Botanica Neerlandica* **17**, 212–20.

Tromp, J. (1970). Shoot orientation effects on growth and flower-bud formation in apple. *Acta Botanica Neerlandica* **19**, 535–8.

Tromp, J. (1972). Effects of growth regulating substances and tree orientation on growth and flower-bud formation in apple. *Journal of Horticultural Science* **47**, 525–33.

Tubbs, F.R. (1980). Growth relations of rootstock and scion in apples. *Journal of Horticultural Science* **55**, 181–9.

Warmund, M.R., Barritt, B.H., Brown, J.M., Schaffer, K.L. and Jeong, B.R. (1993). Detection of vascular discontinuity in bud unions of 'Jonagold' apple on 'Mark' rootstock with magnetic resonance imaging. *Journal of the American Society for Horticultural Science* **118**, 92–6.

Webster, A.D. (1995). Rootstock and interstock effects on deciduous fruit tree vigour, precocity and yield productivity. *New Zealand Journal of Crop and Horticultural Science* **23**, 373–82.

Webster, A.D., Oehl, V.H., Jackson, J.E. and Jones, O.P. (1985). The orchard establishment, growth and precocity of four micropropagated apple scion cultivars. *Journal of Horticultural Science* **60**, 169–80.

Webster, C.A. and Jones, O.P. (1989). Micropropagation of the apple rootstock M9: effect of sustained subculture on an apparent rejuvenation *in vitro*. *Journal of Horticultural Science* **64**, 421–8.

White, G.C. (1970). 'Cox disease': a foliar disorder of apple trees. *Report of the East Malling Research Station for 1969*, 187–8.

Wunsche, J.N., Lakso, A.N., Robinson, T.L., Lenz, F. and Denning, S.S. (1996). The bases of productivity in apple production systems: the role of light interception by different shoot types. *Journal of the American Society for Horticultural Science* **121**, 886–93.

Yadava, U.L. and Dayton, D.F. (1972). The relation of endogenous abscisic acid to the dwarfing capability of East Malling apple rootstocks. *Journal of the American Society for Horticultural Science* **97**, 701–5.

Young, E. and Werner, D.J. (1982). Early season root and shoot growth of 'Golden Delicious' apple on four rootstocks as affected by pruning at planting. *Journal of the American Society for Horticultural Science* **107**, 822–6.

Young, E., Jones, O.P. and Bhat, K.K.S. (1986). Impedance of carbon movement downwards through dwarfing apple interstocks. *HortScience* **21**, Abstract 1639.

6

The shoot system

Introduction

The configuration and productivity of the individual apple or pear tree is determined by its height, the number and length of its branches and the angle of these to the vertical.

The size, density and arrangement of the branch and shoot framework determine the leaf area and light capture, and hence potential photosynthesis. They also determine the number of fruit buds and fruits.

Manipulation of shoot growth begins in the nursery, with the objective of producing trees with numerous lateral branches capable of bearing fruits in their early years in the orchard. It continues throughout the life of the tree with emphasis in the early years on branch initiation, development and training, followed by emphasis on the renewal of fruiting wood and ensuring adequate penetration of light into the canopy.

The stages in the development of an apple tree and the key elements of its above-ground structure are shown in Figure 6.1. Other tree forms may be used but the essentials are the same for both apples and pears.

Buds

All shoots of apple and pear scions arise from buds. The first in the life of the tree is the bud which is inserted into the rootstock stem by budding or is present on scionwood grafted on the rootstock. The buds on the orchard tree can be on the long (extension) shoots or the short (spur) shoots and may be terminal, i.e. at the end of the shoot, or lateral, i.e. in the axil of a leaf. The bourse bud which develops at the base of a flower cluster is considered as a distinct category. Terminal bud formation can be considered as a continuation of the axis after extension growth has ceased. Lateral buds develop from small sections of the apical meristem remaining in the axils of leaves. In the early

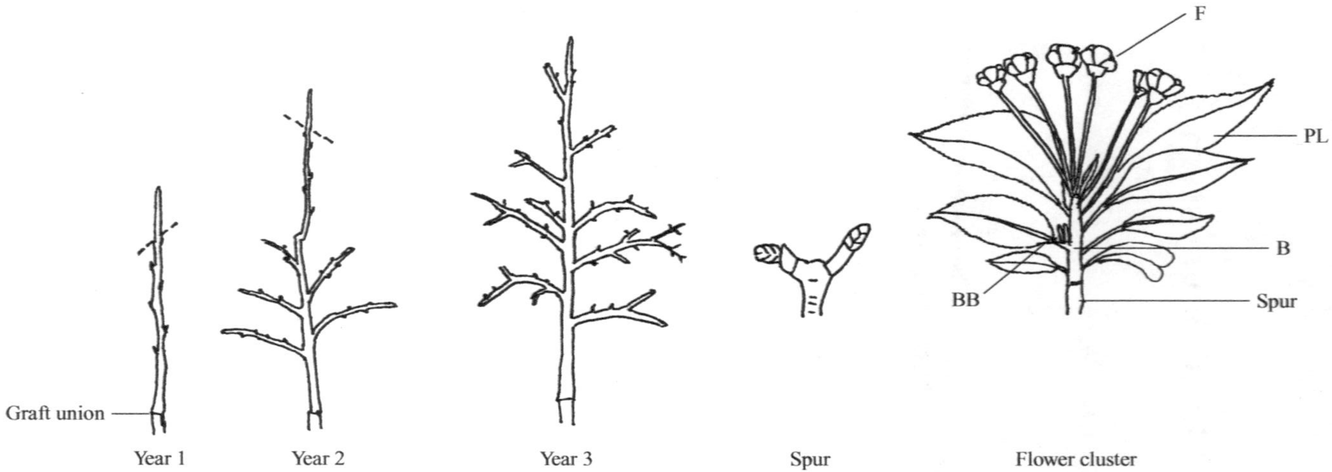

Figure 6.1 Development of an apple tree and diagrammatic representation of a spur and a blossom cluster. Year 1 shows a newly planted unbranched 'whip' with lateral buds on the shoot above the graft union. Pruning the top of this results in lateral branch development in year 2 and year 3. Planting a well-branched tree in year 1 shortens the timescale to produce a cropping tree. 'Spur' shows a short shoot or spur in winter. 'Flower cluster,' a flower cluster developed on a spur. BB, a bourse bud; B, a bourse; PL, a primary leaf; F, a flower.

Figure 6.2 Bud structure showing, in ascending order, nine bud scales, three transition leaves, six true leaves and three bracts. The axis is terminated by a flower primordium (the king flower) and lateral primordia are formed in the axils of the three bracts and three distal leaves. After Abbott (1970). Reproduced with permission.

part of the growing season both extension shoots and spurs are terminated by a naked bud but as the season progresses bud scales instead of leaves develop and enclose the terminal bud. At this stage it is commonly referred to as a resting bud. The time when terminal and axillary buds form on extension shoots varies with the length of the shoot.

Each bud can be considered as a very compressed incipient or unelongated shoot. It has a very short axis on which are borne, in spiral sequence, 21 leaf formations made up of nine bud scales, three transition leaves, six true leaves and three bracts (Figure 6.2). All of the buds are potentially mixed buds, commonly referred to as fruit buds in which the axis is terminated by a flower primordium (the 'king flower' primordium), and lateral flower primordia are formed in the axils of the bracts and three upper true leaves. Vegetative primordia form in the axils of the lower leaves. Flower induction does not, however, begin until a critical node (leaf formation) number has been reached. In apple the critical node number is about 20 for 'Cox's Orange Pippin' and 16 for 'Golden Delicious' (Luckwill, 1974). If this critical node number is not reached the bud remains smaller and more pointed, and is capable of only vegetative growth. Such 'leaf buds' or 'wood buds' are the ones used in budding and grafting, and pruning to just above them is a standard technique for inducing new shoots.

Whether a bud becomes a fruit bud or remains at the solely vegetative stage depends on the type of bud (development of buds on pre-existent spurs starts earlier than that of terminal buds on long shoots or lateral buds on these), the length of the growing season (the shorter the season the greater the probability that buds, especially buds on long shoots, will fail to produce flower primordia) and the presence of fruits (heavily cropping trees have a lower proportion of their buds producing flower primordia). These and other factors determining the type of bud that forms are discussed in detail in the section on flowering, but it is important at this stage to recognize that the balance between fruit buds and vegetative buds is a variable and is controlled by cultivar, climate and tree management.

Bud dormancy

Dormancy is used as a general term to indicate a period of temporary suspension of visible growth of a plant structure containing a meristem (Lang, 1987). This is a practical definition which in the case of buds includes those that are growing very slowly, such as fruit buds in winter (Abbott, 1970), and axillary 'trace buds' which may persist for years under the bark while growing enough each year for the tip to keep pace with cambial growth (Esau, 1965).

Bud dormancy enables plants to survive adverse events and environments. Entry into dormancy and emergence from it are therefore likely to involve mechanisms relevant to the conditions under which the particular plant genotype evolved (Vegis, 1964).

The traditional view of apple and pear bud dormancy in temperate regions is as follows.

1. In summer and early autumn the primary mechanism controlling bud dormancy is correlative inhibition and the buds are classified as paradormant (Lang, 1987). At this time they can be stimulated to grow out quickly if the source of correlative inhibition is removed, either in the field when temperatures are suitable for growth or under so-called forcing conditions when cut shoots are kept at adequate temperatures with their bases in water.
2. In late autumn and early winter (September, October, November and December in western Europe) the buds on intact shoots require a much longer period under forcing conditions to break dormancy. They are said to be in a state of endodormancy, deep dormancy or rest.
3. Following exposure to a period of low temperatures (winter chilling) the buds lose their endodormancy and can once more be induced into rapid budbreak under forcing conditions. Dormancy in the field is maintained by low temperature. This period of ecodormancy lasts until the buds have been exposed to enough high temperature to attain budbreak, which in western Europe is usually in April or May.

Detailed studies have shown that these three types of dormancy interact, overlap in time, and may involve mechanisms in common. In particular, lateral buds can be induced to break under forcing conditions at any time of year and chilling influences the effect of post-chilling temperatures on the rate of bud development.

In relation to the practical management of bud dormancy it is most relevant to consider it under two headings. The first is correlative inhibition, understanding of which is fundamental to tree management by pruning and branch training. The second is seasonal bud dormancy. This is of critical importance with respect to adaptation to different environments and the overcoming of some major environmental constraints. It involves correlative inhibition as well as endodormancy and ecodormancy effects of temperature and, in some circumstances, water stress.

Dormancy through correlative inhibition

Apical dominance

Dormant lateral buds represent a very small commitment of resources both in terms of tree constituents and of maintenance respiration. They are maintained in a state of dormancy by the shoot distal to them. This apical dominance is shown throughout the tree structure. If the apical portions of a shoot are eaten by browsing animals, or if a branch is broken or pruned back, the lateral buds are released from dormancy and grow as new replacement shoots. The uppermost bud nearest to the previous source of dominance breaks out first, followed by the few just below it, and, as it grows, re-imposes the dormancy influence on the buds lower down. This process is relatively conservative in the use of resources and tends to ensure effective utilization of available light by provision of replacement shoots where they can best exploit this.

Axillary buds on shoots sprout to give lateral branches when the growing shoot tips or young growing leaves are removed (Barlow and Hancock, 1960, 1962). Mika (1986) confirmed this and also found that removal of fully expanded leaves does not release the axillary buds from dormancy although there is some evidence of inconsistency in this respect. Removal of a segment of stem tissue including bark and cambium above a lateral bud (notching) will also release the bud from dormancy (Greene and Autio, 1994). Bending branches towards the horizontal has a similar effect, the lateral buds on the upper side of the branch being released from dormancy to give either long or short shoots (spurs) while those on the lower side stay dormant (Faust, 1989).

The apical dominance exerted by a shoot tip over lateral buds is characteristically much more pronounced in warm-winter areas with low levels of

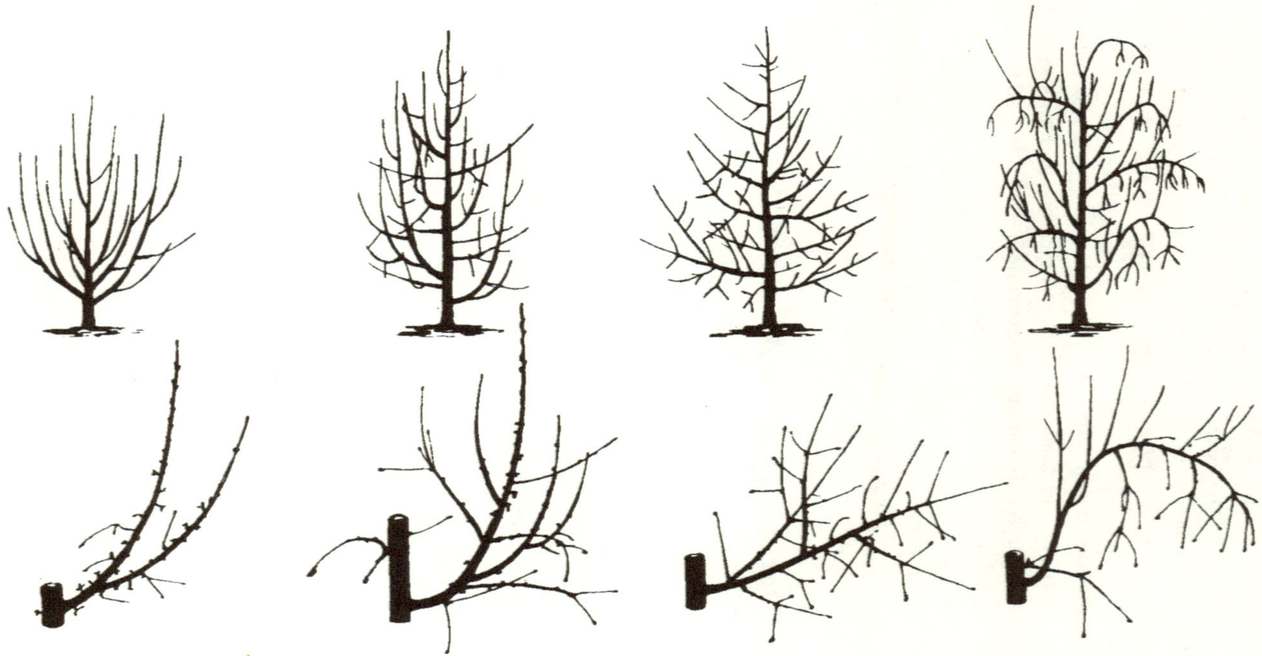

(*a*) (*b*) (*c*) (*d*)

Figure 6.3 Differences in growth habit between apple cultivars with different degrees of apical dominance. (*a*) Spur type; (*b*) 'Reine des Reinettes' type; (*c*) 'Golden Delicious' type; (*d*) 'Rome Beauty' type. After Lespinasse and Delort (1986). Reproduced with permission.

winter chilling. The failure of lateral buds to break under these conditions leads to bare unbranched shoots with just a tuft of leaves and fruits at the tips. This condition may be, in part, a consequence of the terminal buds breaking long before lateral buds in warm-winter areas and therefore establishing greater dominance (Saure, 1985).

There are large differences between cultivars of apple, and pear, in the degree of apical dominance. This is shown by differences in the proportion of the lateral buds which 'break' to give 'feathers', i.e. side branches, on nursery trees. Volz *et al.* (1994) showed that trees of 'Fuji', if not treated, produced fewer lateral branches than trees of 'Braeburn' or 'Gala', although all cultivars had similar numbers of side branches following bud-breaking hormonal treatments. Similarly, Jaumien *et al.* (1993) found that untreated trees of 'Cortland' and 'Gloster' developed fewer lateral branches than those of 'Jonagold' and 'Melrose'. Mature trees of different cultivars show large differences in growth habit attributable to differences in the degree of apical dominance and control exerted by the leading shoot (Crabbé, 1984a; Lespinasse and Delort, 1986) as shown in Figure 6.3

There is direct experimental evidence (Faust *et al.*, 1995) that cultivars with a low requirement for winter chilling, e.g. 'Anna', have weaker apical dominance than those with a high chilling requirement.

Spur-type cultivars are sometimes said to show strong apical dominance because, especially in the case of the 'Wijcik' types, they exhibit strongly vertical growth with few if any side branches. They do not, however, show inhibited lateral budbreak: the buds break but give rise to spurs instead of long shoots. They should therefore be considered as a special case.

Trace buds, buried under the bark in secondarily thickened branches including the central trunk, are held dormant by correlative inhibition and are released from dormancy when the branch above them is pruned off.

Other types of correlative inhibition

Champagnat and Côme (1986) concluded that, from spring to autumn, typical apical dominance is succeeded by a more diffuse inhibition from the large leaves followed by an inhibition located in the stem and then, in November, an inhibition located in the bud itself. Abbott (1970) found that removal of bud scales after incomplete winter chilling hastened and increased budburst. Swartz *et al.* (1984) showed that removal of bud scales can stimulate budbreak in winter, especially from December onwards, in typical cultivars including 'Idared' and 'Millerspur Delicious'. Removal of apical bud scales also increased percentage budbreak of the first lateral bud of 'Millerspur Delicious' throughout the winter. Notodimedjo *et al.* (1981) found that bud slicing, which had the effect of removing bud scales, had a dramatic effect on lateral budburst

Figure 6.4 Effect of NAA applied in a post-pruning paint on the emergence of shoots from previously dormant buds: left: whole cut surface treated with NAA; centre: control; right: half cut surface treated with NAA. After Blanco-Braña (1976). Reproduced with permission.

under tropical conditions in Indonesia. Removal of mature leaves under tropical conditions is followed by budburst of terminal buds on spurs (Janick, 1974; Edwards and Notodimedjo, 1987).

Mechanisms of correlative inhibition

There is no single theory that provides a fully satisfactory explanation of the mechanism of correlative inhibition (Hillman, 1984). There is, however, considerable evidence that apically produced auxin indirectly suppresses axillary bud outgrowth that is promoted by cytokinin originating from roots or shoots (Cline, 1991; Tamas, 1995). Other hormones may also be involved and there may be a critical role for nutrients and for water as a possible inducing signal for bud outgrowth.

AUXINS AND CORRELATIVE INHIBITION

Thimann and Skoog (1933, 1934) originally demonstrated the presence of auxin in apical buds of *Vicia faba*. They showed that the amount of inhibition of budbreak of laterals was related to the auxin content of the apical bud and that indoleacetic acid (IAA) applied to a decapitated stump can to some extent substitute for the apex. Wang *et al.* (1994) quantified the concentration of IAA needed to substitute for apple shoot apices. They found that exogenous IAA inhibited the increase of free water in lateral buds that is an initial step in their release from dormancy. Abbas (1978) showed that the shoot tips of apple cultivars that are freely branching have lower concentrations of auxin-like substances than those of cultivars showing little branching. The application of an auxin transport inhibitor below the apical zones of intact apple shoots induces budbreak of the hitherto dormant lateral buds below the treated zone (Duckworth *et al.*, 1979). The emergence of shoots from dormant trace buds in the tissues of secondarily thickened branches following pruning can be prevented by painting IAA or the synthetic auxins IBA or NAA on the pruning cuts (Blanco-Braña and Jackson, 1982a), as shown in Figure 6.4.

The most rapid synthesis of auxin occurs in expanding leaves (Sachs, 1993), but it also occurs even in mature leaves and other parts of shoots. Dominant organs, which are usually the first to develop, have a high IAA export rate while reducing the auxin export by inhibited organs.

Studies on auxin movement tend to emphasize polar transport. In shoots this is basipetal, with auxin moving preferentially from morphologically apical to more basal regions. In roots it then continues to move in the same physical direction as in the shoots, but this is now called acropetal movement because it is in the direction of the root tip. In young roots, however, there may also be movement away from root tips. This polar transport is specific for IAA and

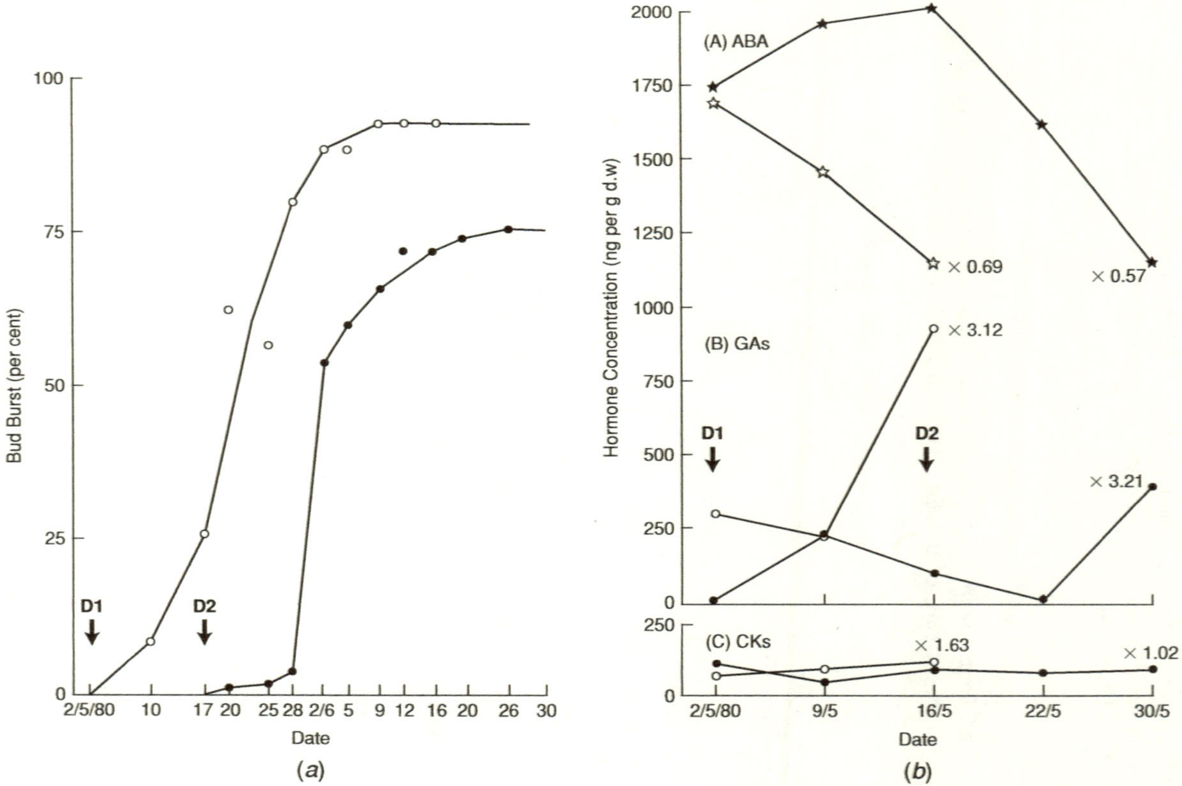

Figure 6.5 Effects of defoliation at times D1 and D2 (*a*) on budburst, and (*b*) on bud hormone status, of 'Rome Beauty' apple in Indonesia. Open symbols relate to early defoliation, closed symbols to defoliation 2 weeks later. ×0.69, ×3.12, *etc.* show changes in hormone concentrations in the 2 weeks following defoliation. Modified from Taylor *et al.* (1984). Reproduced with permission.

synthetic auxins. It appears to be through parenchymatous cells, particularly those associated with or differentiating into vascular tissues, but not through the vascular elements themselves (Rubery, 1987). It requires energy and is inhibited by anaerobiosis and metabolic poisons such as cyanide and dinitrophenol. It is also inhibited by specific auxin transport inhibitors such as 2,3,5-tri-iodo-benzoic acid (TIBA).

Both endogenous and applied auxins can also move in the plants' vascular system. Auxin applied to mature leaves is translocated in the phloem like a photoassimilate and may also move in the xylem transpiration stream. This transport in the vascular system is by mass flow down gradients of osmotic potential or water potential. Inactive auxin conjugates may move in the vascular system and be activated by enzymatic hydrolysis before being distributed by polar transport.

In apple, Soumelidou *et al.* (1994) found predominantly basipetal but also some acropetal movement in shoots. Blanco-Braña and Jackson (1982b) found that labelled NAA applied to the cut surface of a short shoot (spur) growing vertically from a horizontal apple branch moved into the latter and along it, mainly towards the trunk but with some acropetal movement as well.

CYTOKININS AND CORRELATIVE INHIBITION

In many species, including apple, cytokinin treatment of buds releases them from correlative inhibition (Sachs and Thimann, 1967; Tamas, 1995), and a relationship between endogenous cytokinin and axillary bud growth has been demonstrated by comparing two tomato lines with different degrees of apical dominance. Williams and Stahly (1968), Williams and Billingsley (1970) and many other authors have shown that the cytokinin 6-benzyladenine (BA) can be used to induce breaking of lateral buds of apple otherwise kept dormant by apical dominance. Thidiazuron (TDZ), which has cytokinin-like properties, is even more effective (Wang *et al.*, 1986).

Although cytokinins are thought to be mainly produced in the roots, there is no need for contemporaneous root production of cytokinins to stimulate budburst when a shoot apex is removed. This budburst happens even when isolated shoots without roots are decapitated.

GIBBERELLINS AND CORRELATIVE INHIBITION

Gibberellins do not appear to be involved directly in apical bud dominance (Tamas, 1995), but the outgrowth of axillary buds can be enhanced by the application of GA_3. The terminal budbreak of apples in the tropics which is induced by leaf removal is preceded by a large increase in both concentration and total amount of gibberellins in the buds (Taylor *et al.*, 1984; Edwards, 1985), as shown in Figure 6.5.

ABSCISIC ACID AND CORRELATIVE INHIBITION

The breaking of isolated apple buds in tissue culture is inhibited by abscisic acid (ABA) (Borkowska, 1981; Borkowska and Powell, 1982/83), even if the chilling requirement of the buds is fully satisfied. The effect of ABA is, however, only at a late stage of the bursting process, i.e. it inhibits extension of the bud axis. Edwards (1985) found that defoliation of apples in the tropics, triggering budburst, was accompanied by a decline in bud ABA content which he attributed to the removal of the source of their ABA, i.e. the mature leaves.

MINERAL NUTRITION AND CORRELATIVE INHIBITION

Early investigators of correlative inhibition postulated that the apical bud monopolized nutrients to the detriment of lateral buds. This is no longer considered to be the prime mechanism of apical dominance although nutrient and water status may influence the process. Phillips (1975) noted that inorganic nutrient deficiency, particularly of nitrogen, can increase the strength of apical dominance and depress cytokinin levels.

ENVIRONMENTAL EFFECTS ON APICAL DOMINANCE

Tromp (1992a, b) found that apical dominance in young apple trees is greater at low soil temperatures and low atmospheric humidity. He attributed this to more adverse water status and possibly reduced supplies of root-produced cytokinins.

Correlative inhibition of buds is generally enhanced at low light intensity, perhaps as a result of lower photoassimilate and higher auxin levels (Phillips, 1975).

Gravity regulates bud growth and lateral organ orientation. Placement of a normally negatively geotropic (i.e. upward-growing) shoot at or near to the horizontal usually reduces correlative inhibition of buds by the main apex. This may relate to the higher auxin content of terminal buds of vigorous upright shoots than of terminal buds of horizontal or weeping shoots in apple (Kato and Ito, 1962). In general, the physically highest and upwardly directed bud on a shoot achieves dominance. When all buds are at equal height as on a horizontal branch, the upwardly directed bud nearest to the root system is the most likely to become dominant. Under tropical and subtropical conditions all the buds on the upper side of a horizontal branch often break to give short fruiting spurs. In general, under the influence of gravity, the IAA levels are higher in the lower than the upper halves of horizontal plant organs (Reinhold, 1978). Blanco-Braña and Jackson (1982b) found that NAA applied to the cut end of a horizontal apple branch moved mainly to the lower half of this, as shown by development of characteristic auxin-induced xylem.

Releasing buds from correlative inhibition

NURSERY TREE TREATMENTS

The economic success of a modern orchard depends on the ability of the fruit grower to bring his trees into cropping at an early age (Jackson *et al.*, 1981). The potential of a young tree to produce adequate crops in its early years is related to the number and length of the lateral branches present when the tree, whether apple or pear, is transplanted from the nursery (van Oosten, 1978; Lawes *et al.*, 1997).

Removal of the shoot tips of trees in the nursery releases some lateral buds from dormancy, but, in the initial absence of auxin flow from above, these lateral shoots tend to form very narrow angles with the main stem. Such 'narrow crotch angles' are mechanically weak and the branches readily break off at the junction with the trunk when bearing heavy crops. Removal of the young, not fully expanded, leaves while leaving the growing point intact can lead to lateral budbreak giving wide-angled side shoots (Wertheim, 1978), presumably because the interruption of auxin supply is less prolonged.

Treatment with an auxin transport inhibitor (M & B 25105, which is *n*-propyl-3-*t*-butylphenoxyacetate) can increase lateral budbreak to give wide-angled branches of apple and pear (Wertheim, 1978; Quinlan, 1981; Cody *et al.*, 1985).

Application of BA to lateral buds of nursery trees leads to an increase in budbreak (Figure 6.6) and to branches with wide crotch angles (Williams and Billingsley, 1970). This is commonly applied in combination with GA_{4+7} which, although it does not increase budbreak, induces greater extension growth. The combination of BA and GA_{4+7} (Promalin) therefore increases both the number and the length of lateral branches.

Lateral budbreak and the development of a large number of 'feathers' can be greatly increased by growing nursery plants at a wide rather than close within-row spacing (Wilson and Jarassamrit, 1994). This was considered to be a response to reducing interplant competition for light. Jaumien *et al.* (1993) reported adverse effects of drought on branching of nursery trees in the absence of irrigation and that, in general, the better quality (size) of the rootstock the more lateral branches were produced by scions grafted on them. Chip budding, which gives a more rapid graft union than shield budding, hence a longer effective growing season and, presumably, less risk of water stress in the scion, induces more laterals per scion tree (Howard *et al.*, 1975).

TREATMENT OF ORCHARD TREES

Failure of lateral budbreak in orchard trees results in problems of bare wood, i.e. failure to occupy the canopy volume with an adequate number of fruit-bearing long shoots and spurs.

(*a*) (*b*)

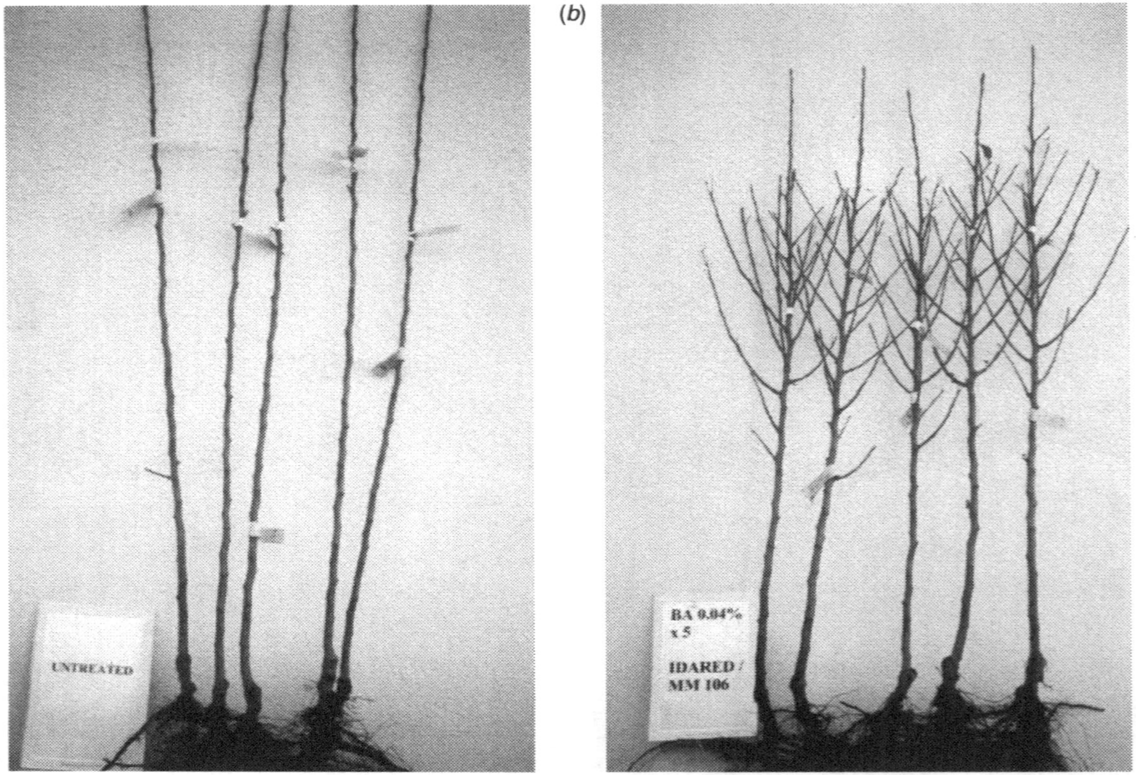

Figure 6.6 The effect of five applications of 0.04% benzyladenine in the nursery on lateral branch formation, 'Idared' on 'MM.106'. (*a*) Control, (*b*) treated. From Hrotkó *et al.* (1997). Reproduced with permission.

If the trees when planted do not have enough lateral branches, ideally more than ten, additional branches can be induced in the orchard by:

1 Heading-back, i.e. removing the apical part of the tree to stimulate budbreak below the pruning cut.
2 Notching, i.e. removal of a thin band of bark above each lateral bud.
3 Periodic removal of immature apical leaves.
4 Tying the leading shoot down, first in one direction then in the other.
5 Spraying with Promalin (BA + GA_{4+7}).

Each of these can be effective but their individual utility varies with the initial level of branching, with cultivar and with growing conditions (Volz *et al.*, 1994; Parker and Young, 1995; Ouellette *et al.*, 1996).

These techniques can also induce budbreak on older trees, but reducing apical dominance by training branches to a horizontal or a nearly horizontal orientation assumes greater importance. In many systems of tree management the angle made by the lateral branches to the main stem is increased by tying them down to pegs in the ground, weighting them with transferable concrete weights, inserting 'spreaders' between the main stem and the branch, or training the branches along wires.

Under subtropical conditions with limited winter chilling, branching in the nursery and orchard can be induced by cold storage of nursery trees or chemical treatments (pp. 185–7).

Seasonal bud dormancy

Introduction

Seasonal plant dormancy is, in general, a mechanism for surviving regularly recurring periods (seasons) of drought or low temperature.

Seasonal dormancy in apples and pears can be expected to be adapted to conditions at their centres of origin and also to conditions in the areas in which the cultivated varieties have been selected. In some of these climates temperatures become so low in winter that shoots, stems and roots which have not 'hardened' are killed, and cessation of growth is a precondition for such hardening (Westwood, 1993). The capacity to initiate physiological changes in advance of potentially lethal conditions, and for buds (and seeds) to remain dormant in winter until there is little risk of lethal frost, must have been essential features of adaptation in the wild. The adverse economic effects of even intermittent frost damage to buds and blossom have also resulted in selection pressure for appropriate timing of budbreak of cultivated apples and pears.

In trees, in general, the mechanisms regulating the time of budburst are usually highly heritable and finely tuned, e.g. the temperature requirements for budburst of Douglas-fir (*Pseudotsuga menziesii*) can differ inherently between natural populations only a few kilometres apart, as discussed in a review by Cannell (1989). In apples and pears the mechanisms of adaptation to winter conditions may result in very negative effects if the cultivar is grown beyond its zone of adaptation. Thus most cultivars selected under conditions of prolonged winters show inability of the buds to break dormancy when grown in climates with short mild winters, even if temperatures are adequate for growth and there is no correlative inhibition (Samish, 1954; Saure, 1985). Conversely, cultivars selected in short-winter areas emerge from dormancy too soon and suffer frost damage if grown in less mild regions (Bernardi, 1988). The rapid increase in apple and pear production in warm-temperate, subtropical and even tropical regions, through the use of adapted cultivars and of technologies for dormancy breaking, has increased the economic importance of understanding the nature and mechanisms of seasonal dormancy.

Cessation of growth and bud formation

Lateral buds which are dormant because of correlative inhibition remain dormant throughout the winter. They show no visible change of state as shoot growth slows and ceases and the trees lose their leaves with the onset of winter.

Terminal buds form as shoot growth ceases. This takes place very early in the summer on short shoots (spurs), Fulford (cf. Abbott, 1970), noting that the formation of bud scales under English conditions begins in early May. The buds so formed are initially held dormant by correlative inhibition; this is shown by the fact that spur buds can be stimulated into growth by pruning away the distal part of the branch (Abbott, 1970). Formation of terminal buds on long shoots can occur at widely varying times. Vigorous young shoots can continue to grow until at least mid-October in southern England if amply supplied with nutrients and water (Hancock and Barlow, 1958). Commonly, however, growth ceases and a terminal bud forms much earlier in the season. Cessation of growth is earlier in trees on dwarfing rootstocks than in those on invigorating ones, in heavily cropping trees than in vegetative ones and in water-stressed trees than in those well supplied with water. Webster (1995) showed that extension growth of horizontal branches of apple trees on some dwarfing rootstocks in England can cease as early as 5 July. It is obvious, from the wide range of dates of cessation of growth and formation of buds, that this event is not triggered by a single clear environmental signal such as the change in photoperiod which is associated with the formation of resting buds in other tree species. Instead, the formation of terminal buds in apple appears to be a response to a number of different factors that check growth

(Powell, 1987) and to dominance effects of young expanding leaves (Abbott, 1970).

Initially, the development of dormant buds is reversible if growing conditions improve and a new flush of growth can occur following heavy summer rain, the application of nitrogenous fertilizer in summer or early summer pruning. Abbott (1970) concluded that effects of stress leading to shoot growth termination and the temporary reversal of these are mediated by effects on root growth and the supply of growth substances from the roots. Alternatively, auxin or inhibitors may be involved.

Development of deep dormancy, endodormancy or rest

As summer, autumn and early winter progress in the temperate zone, apple and pear terminal buds enter a state of endodormancy (Lang, 1987). In this state they cannot be stimulated into rapid budbreak by transfer into conditions of temperature and water supply suitable for normal growth, even in the absence of external sources of correlative inhibition. The 'depth' of such endodormancy, otherwise referred to as true dormancy, deep dormancy or rest, is usually measured in an arbitrary way as the number of days at a temperature in the range 15–25 °C, that are needed to induce a given stage of bud development (Spiegel-Roy and Alston, 1979; Saure, 1985).

At Geneva, New York State, some cultivars show appreciable endodormancy even on 14 July and most cultivars show significant levels of such dormancy before any winter chilling or winter frost occurs (Hauagge and Cummins, 1991 b). Cultivars can be divided into three categories based on the number of days of 'forcing' required to induce 50% of shoots to show terminal budbreak. Cultivars in the first category, which include 'Anna', 'Dorsett Golden' and 'Ein Shemer', show only shallow dormancy throughout. They give 50% budbreak with fewer than 35 days 'forcing' at 19 °C at any time in the season. These cultivars show excellent adaptation to subtropical climates with mild winters (Miller and Baker, 1982; Bepete and Jackson, 1995). Jackson (1990) reported that in tropical Zimbabwe 'Anna' can produce two crops a year without any defoliation and without any prior exposure of the flowers of the second crop to chilling. This is compatible with the conclusion of Hauagge and Cummins that these cultivars never exhibit deep dormancy. The majority of standard, temperate-zone cultivars show a very different pattern, characterized by abrupt changes in the intensity of dormancy and very deep dormancy (Figure 6.7). Some cultivars reached deep dormancy by 1 September, before the onset of cold weather, others not until after the first frosts and the beginning of accumulation of chilling units (see p. 175). Maximum dormancy for most cultivars occurs at the time of complete leaf senescence, as based on leaf

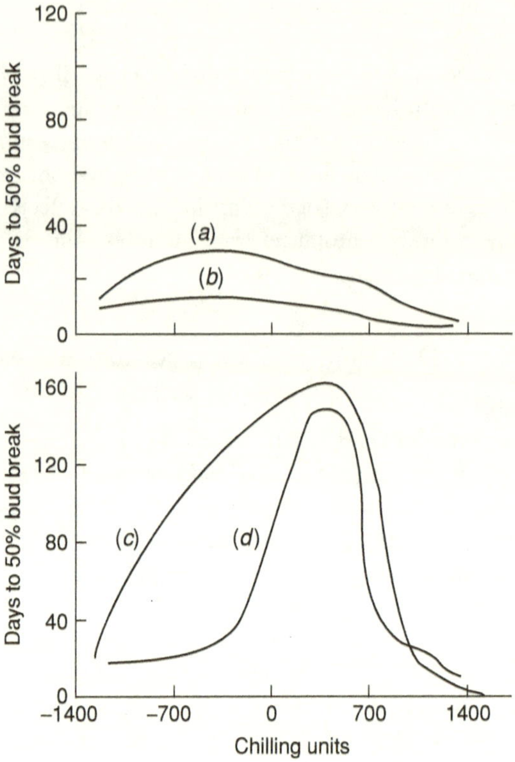

Figure 6.7 Seasonal changes in the intensity of bud dormancy in 'low-chilling-requirement' and 'high-chilling-requirement apple cultivars. Days to 50% budbreak under 'forcing' conditions of (*a*) 'Dorsett Golden', (*b*) 'Anna', (*c*) 'McIntosh' and (*d*) 'Delicious'. Negative chilling units have no physiological meaning but are used as a measure of time. Sampling on 14 July and 1 Sept corresponded to −1276 CU and −238 CU. Data from Hauagge and Cummins (1991b). Reproduced with permission.

colour and advanced leaf abscission. A third group of cultivars are intermediate in their behaviour. In general the cultivars which develop only shallow dormancy reach their peak dormancy earlier in the season than those which develop deep dormancy. Under subtropical conditions, however, 'Gala' may attain much deeper dormancy than 'Golden Delicious' and 'Fuji', even though it enters into dormancy much earlier (Herter *et al.*, 1988). Mauget and Rageau (1988) found terminal buds of orchard trees of 'Golden Delicious' in central France to reach deepest dormancy by 24 September, well before winter.

Westwood (1993) concluded that at low latitudes the cues for endodormancy induction are weak and some cultivars do not attain full endodormancy. They do not, therefore, require as much chilling to break dormancy. Sherman and

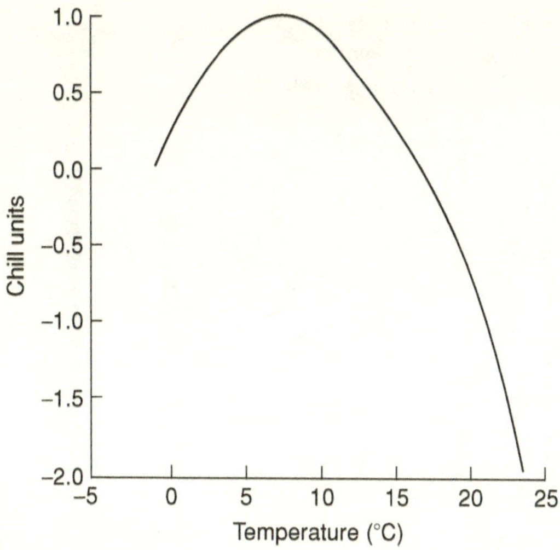

Figure 6.8 The relative efficiencies of different temperatures in meeting the chilling requirements of 'Starkrimson Delicious' apple. From Shaltout and Unrath (1983). Reproduced with permission.

Crocker (1982) found that in Florida, at latitude 30° N, the Asian pears 'Tsu-li' and 'Ya-li' had chilling requirements of 365–480 hours whereas in Oregon at 45° N 'Tsu-li' needs at least 600 h and 'Ya-li' 900–1400 h. Hauagge and Cummins (1991 a) noted that some apple cultivars which show good adaptation and a short chilling requirement under subtropical conditions had a longer chilling requirement, much more similar to that of conventional, high-chilling requirement cultivars, under the conditions of Geneva, NY.

As has been known for many years, the depth of dormancy begins to decline after exposure to a period of cold weather. Initially the most relevant measurement of chilling for deciduous tree fruits was taken as hours below 45 °F (7.2 °C). Further studies showed that temperatures below freezing did not contribute, that some temperatures within the chilling range were more effective than others, and that high temperatures within the chilling period could have a negative effect. 'Starkrimson Delicious' apple trees showed a maximum chilling response at 7.2 °C, with declining responses as temperatures increased to 16.5 °C and fell to −1.1 °C and progressively more negative effects of temperatures from 19 °C to 23.3 °C (Figure 6.8). This response curve is used to calculate 'chilling units' (CU). However, Thompson *et al.* (1975) found that budbreak of 'Jonathan' was stimulated to a greater extent by chilling at 2 °C than at 6 °C and exposure to 23.9 °C had a greater negative (de-chilling) effect than exposure to 18.3 °C when the chilling periods were interrupted by treatment at these higher temperatures. A chilling temperature of 2 °C was

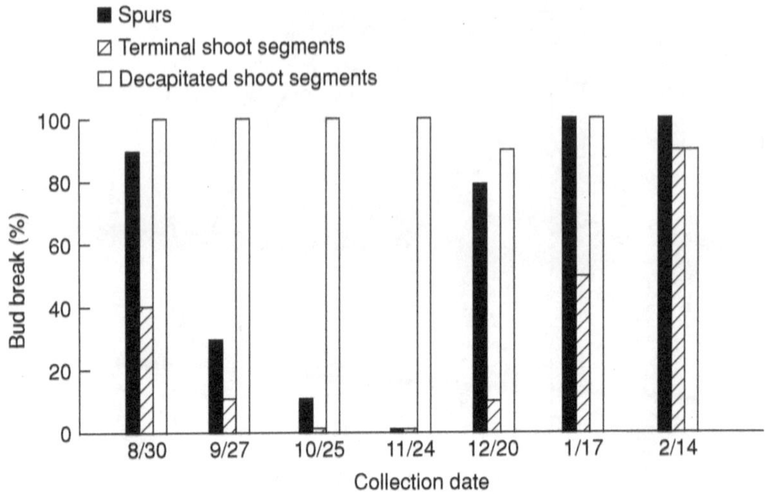

Figure 6.9 Seasonal changes in depth of endodormancy as measured by budbreak after 28 days of forcing of spur buds, shoot terminal buds and uppermost lateral buds on decapitated shoot segments. After Latimer and Robitaille (1981). Reproduced with permission.

also found to be more effective than 6 °C for 'Gala' and 'Fuji' (Petri and Stuker, 1988).

Latimer and Robitaille (1981) showed that spur terminal buds enter into rest later, as measured by budbreak under forcing conditions, and emerge earlier than shoot terminal buds (Figure 6.9).

Temperature effects on emergence from dormancy

Two aspects of temperature determine the time at which buds emerge from dormancy. These are low temperatures, to meet the chilling requirement as described above, and accumulated temperatures above the threshold for bud growth calculated as growing-degree-hours (GDH).

These effects were originally considered to be sequential and additive; for example, Shaltout and Unrath (1983) found that 'Starkrimson Delicious' apple trees in North Carolina attained budbreak after about 1200 CU followed by 7082 GDH.

There is, however, increasing evidence of inter-relationships between chilling and the thermal time (GDH) requirement. Increased chilling can be accompanied by a decrease in the number of days at 'forcing' temperatures needed to achieve budbreak for high-chilling-requirement cultivars. This effect is also shown for low-chilling-requirement cultivars but is much less pronounced and ceases to be evident once a fairly modest amount of chilling

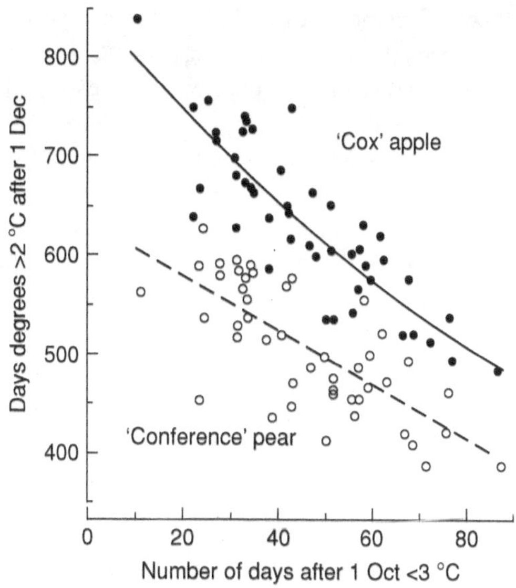

Figure 6.10 The inverse relationship between thermal time to full bloom of 'Cox' apple and 'Conference' pear and the accumulated number of 'chill days' in winter at East Malling, Kent, England. From Cannell (1989), from *Manipulation of Fruiting* by C.J. Wright. Reprinted by permission of Elsevier Science Ltd.

has accumulated (Figure 6.7). Statistical studies on the factors controlling the date of budburst over many years and sites of 'Cox' apples and 'Conference' pears in England showed that the greater the number of days from 1 October in which temperatures were lower than 3 °C, the fewer day-degrees above 2 °C after 1 December were needed to attain full bloom, i.e. 50% budburst (Figure 6.10). There have also been numerous reports that chilling over and above the accepted chilling requirement advances budbreak. This is explained by the concept that at any time during dormancy a bud (vegetative or mixed) has the potential to respond to warm temperatures with morphological development, extension growth or a change in physiological activity. The effect of chilling is to change the response of the bud to temperature, i.e. increase the potential rate of bud development (Vegis, 1973; Campbell, 1978; Cannell, 1989). Young (1990) and Young *et al.* (1995) showed that chilling dormant apple trees increases the effect of temperature on subsequent shoot respiration (the Q_{10}).

In general those cultivars with little requirement for winter chilling also have low heat requirements to induce budbreak. This is true for both apple (Petropoulou, 1985; Hauagge and Cummins, 1991b) and pear (Spiegel-Roy and Alston, 1979). Both chilling requirement and heat requirement are closely correlated with the time of budburst. There are exceptions to these relationships, however. Some cultivars have a lower heat requirement than might have been expected from their chilling classification, and some experience budbreak either earlier or later than might be expected from their apparent chilling and heat requirements (Spiegel-Roy and Alston, 1979). These discrepancies may reflect the weakness of applying standard equations to cultivars which differ in their underlying sensitivities to temperature. For example, Shaltout and Unrath (1983) used 4.4 °C as the base temperature for calculating growing degree-hours for 'Starkrimson Delicious', but Gianfagna and Mehlenbacher (1985) found that the buds of some cultivars, even after their chilling requirement had been met, did not grow at all at 10 °C. Hauagge and Cummins (1991b) also reported that budbreak of 'Elstar' appeared to be dependent on accumulated temperatures above 10 °C and that of other *Malus* on accumulated temperatures above 0 °C.

EFFECTS OF ROOT CHILLING ON BUDBREAK

Chilling of both roots and shoots is essential to achieve maximum budbreak of dormant apple trees (Young and Werner, 1984, 1985b; Young, 1987). Chilling only the roots or the shoots during rest elicits about 50% budbreak while chilling both gives 100%.

With pear, fully chilled 'Bartlett' buds grew less when budded on to inadequately chilled host trees than when budded on fully chilled hosts (Westwood and Chestnut, 1964).

EFFECTS OF ROOTSTOCK ON SCION CHILLING AND GDH REQUIREMENTS

Apple rootstock cultivars have a wide range of chilling and GDH requirements for budbreak (Young and Werner, 1985a). Petropoulou (1985) found that, in general, the more dwarfing the rootstock the lower its chilling and heat requirements, but Young and Werner found 'M.7a' to have lower chilling and GDH requirements than more dwarfing as well as more invigorating rootstocks.

In Brazil, in a subtropical climate, Couvillon *et al.* (1984) found that 'Rome Beauty' apple trees on 'MM.106' and 'MM.104' rootstocks showed classic symptoms of insufficient chilling whereas those on 'M.7' and especially 'M.26' had a greater degree of lateral vegetative budbreak.

Different pear species have very different chilling requirements, reflecting their areas of origin. *Pyrus communis* has a fairly high chilling requirement, *P. calleryana* a low chilling requirement. Both are used as rootstocks. Westwood and Chestnut (1964) found that 'Bartlett' (*P. communis*) on a *P. calleryana* rootstock had a much lower chilling requirement than when on its own (*P. communis*) roots. When trees of 'Bartlett' on *P. communis* and *P. calleryana* rootstocks were each exposed to 480 hours at 4.4 °C and buds from them were budded on a fully chilled host plant, the growth of the buds that had been on *P. calleryana* was much greater than that of buds that had been on *P. communis*. This indicates that

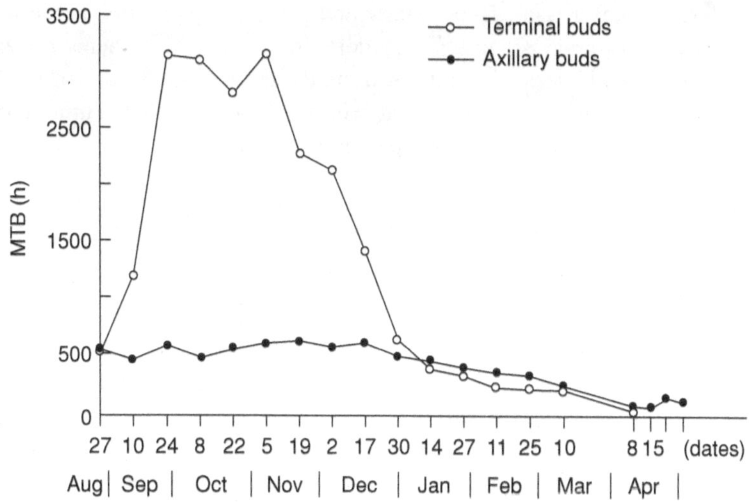

Figure 6.11 Bud dormancy of 'Golden Delicious', measured as mean time to budburst (MTB) of single-node cuttings under forcing conditions, throughout the winter in central France. From Mauget and Rageau (1988). Reproduced with permission.

P. communis buds on *P. calleryana* could lose their dormancy through exposure to a chilling regime inadequate to satisfy the chilling requirement of 'Bartlett' on *P. communis* and that the dormancy level was then intrinsic to the bud (Westwood, 1970).

Chilling requirements of lateral buds

Under conditions of limited chilling, terminal buds on intact shoots may break long before lateral buds (Saure, 1985). From this it has been deduced that lateral buds have a much greater chilling requirement than terminal buds. However, in numerous studies (Latimer and Robitaille, 1981; Williams *et al.*, 1979; Skene, 1980; Mauget and Rageau, 1988; Hauagge and Cummins, 1991 b) it has been shown that the uppermost lateral bud on a decapitated shoot, or a lateral bud on a single-node cutting, can be induced to break out quickly at any time during the winter under forcing conditions (Figure 6.11). Subsequent to the period over which terminal buds show endodormancy, terminal and lateral buds are influenced by growing temperature in an identical way (Mauget and Rageau, 1988).

Paiva and Robitaille (1978) also found that notching above a lateral bud could induce it to grow out readily even during the period of terminal bud endodormancy. This strongly suggests that the major cause of lateral bud dormancy in winter is continued correlative inhibition combined with ecodormancy as a result of low winter and spring temperatures. There was, however, also some evidence of endodormancy in the buds themselves because the rate of lateral bud outgrowth on decapitated shoots under forcing conditions still showed a seasonal pattern. It was slowest in late autumn and throughout the 'deep rest' period up to 20 December.

Positional and shoot bending effects on winter dormancy of buds

Crabbé (1984b) showed that on vertically-growing apple shoots the lateral buds nearest to the apex showed their deepest dormancy in December and at that time were more deeply dormant than those further from the apex. Subsequently, in February and March, their intensity of dormancy declined. On shoots which had been arched over in September so that the tip was near the ground, budbreak from the zone near the tip was achievable fairly quickly throughout the winter and buds on the basal part of the shoot broke readily in December but were deeply dormant in February. Thus the maximum depth of dormancy can be altered and the time at which it occurs can be shifted by branch training. This again suggests a major influence of correlative inhibition on the winter dormancy of lateral buds.

Mechanisms involved in seasonal dormancy

This is still very much a 'black box' area of knowledge, with highly heritable controlling factors that cannot be defined in physiological or biochemical terms. The following elements appear the most relevant to placing apple and pear bud winter dormancy in the context of overall theories of bud dormancy and to its practical management.

DEVELOPMENT OF ENDODORMANCY

There are very large variations in time of entry into endodormancy and in patterns of depth of endodormancy between cultivars, types of buds and buds on upright and bent-over shoots. These make it difficult to believe that endodormancy is triggered and developed in response to a single dominant environmental signal operating directly on the buds. It is possible that the concurrent development of the deepest dormancy with full leaf senescence (Hauagge and Cummins, 1991b) indicates a causal link: the timing of leaf senescence can vary with all the above factors and leaf removal before senescence can prevent the onset of endodormancy (Notodimedjo *et al.*, 1981). Some of the variability could also be explained if the level of endodormancy is the consequence of a balance of bud growth promoters and inhibitors, i.e. a balance of hormonal factors.

AUXINS AND SEASONAL BUD DORMANCY

The obvious effects of the terminal bud and of branch bending on lateral bud endodormancy indicate that auxin can influence chilling requirements and increase the depth of endodormancy. Auxin (indolebutyric acid, IBA) applied to apple roots increased budbreak (Young, 1987), presumably as a consequence of its effect in increasing new root growth which is likely to have led to increased cytokinin production. Tying down to give horizontal branches as part of the technique for dormancy control in the tropics presumably involves auxin-mediated effects.

ABSCISIC ACID AND SEASONAL BUD DORMANCY

Abscisic acid (ABA) was originally thought to be the most important inhibitor preventing growth and to be a causal factor of dormancy, whether drought-induced or developed in anticipation of winter (Faust *et al.*, 1997). ABA is effective in delaying budbreak in tissue-cultured shoots of apple (Dutcher and Powell, 1972) and the ABA content of apple buds declines through the winter (Borkowska and Powell, 1982/83), reaching a minimum level at the time of budbreak (Seeley and Powell, 1981). There is no clear evidence of a link between ABA and chilling effects (Powell, 1987). However, Swartz *et al.* (1984) concluded that ABA may be responsible for all or part of the effect of bud scales in inhibiting budbreak, and Taylor *et al.* (1984) and Edwards (1985) found that defoliation under tropical conditions was followed by a decrease in bud ABA content followed by budburst. Hydrogen cyanamide treatment, which can overcome winter dormancy in the tropics, induces a reduction in ABA levels in both apical and lateral buds of apple, followed by budburst (Subhadrabandhu, 1995). It is possible that ABA is involved in dormancy through its effects on dehydrins and changes in membrane permeability (Faust *et al.*, 1997).

CYTOKININS AND SEASONAL BUD DORMANCY

Broome and Zimmerman (1976) showed that externally applied cytokinins were at least as effective as chilling in inducing budbreak of otherwise dormant lateral buds of *Malus hupahensis*. Their plants were seedlings grown in continuous long days in a greenhouse so the dormancy, though responsive to chilling, may have been atypical. When Thidiazuron (TDZ), a potent and stable compound with cytokinin-like activity, was applied prior to winter chilling of shoots from 'Delicious' and 'Northern Spy' it markedly reduced their chilling requirement for budbreak (Steffens and Stutte, 1989). It had a similar but smaller effect on 'Anna'.

The concentration of cytokinins in the xylem sap of orchard trees of 'Granny Smith' increases steadily during the six weeks prior to budbreak, followed by a rapid increase immediately before and at budbreak, followed by a decline (Cutting *et al.*, 1991). Treatment with the rest-breaking agents dinitro-*o*-cresol

(DNOC) oil and hydrogen cyanamide resulted in earlier and more rapid increases in sap cytokinins and concomitant advancement and intensification of budbreak. Tromp and Ovaa (1990) similarly showed a sharp increase in the concentration of cytokinins prior to budbreak of 'Cox' and 'Discovery.' Young (1989) found that increases in xylem cytokinins followed forcing, i.e. exposure to high growing temperatures, irrespective of winter chilling. Cytokinin levels then decreased significantly as budbreak occurred in fully chilled trees but they did not in unchilled trees, which showed very little budbreak. When 6-BA was applied after artificial root and shoot chilling it increased budbreak by a fairly constant amount whether the roots, or the shoots, or both, had been chilled but did not induce budbreak of unchilled trees (Young and Werner, 1986). The evidence therefore points to cytokinins having a major controlling role in budbreak, adding to or amplifying the effects of chilling but not being able to totally replace this. It should be noted, however, that the normal requirement by apple seeds for a long period at chilling temperatures before they will germinate can be totally replaced by soaking embryos in 6-benzylaminopurine (Zhang and Lespinasse, 1991). This may be significant in view of the many parallels between seed and bud dormancy.

GIBBERELLINS AND SEASONAL BUD DORMANCY

El-Banna *et al.* (1995) found that 200 ppm GA_3 applied during winter dormancy in an environment with sub-optimal winter chilling in Egypt led to advancement of vegetative budbreak and of flowering of 'LeConte' pears by about two weeks. It increased vegetative budbreak from 22% to 78%, flower budbreak from 21% to 56% and yield from 50 to 227 fruits per tree. The effects were very similar to, although rather more pronounced than, those of a standard 'dormancy-breaking' treatment with DNOC.

Taylor *et al.* (1984) and Edwards (1985) found that defoliation of apple trees in Indonesia was followed by a three-fold increase in gibberellin-like substances and, subsequently, by budburst. Exogenous application of GA_3 increased budburst of both defoliated and undefoliated trees.

Although GA_3 sometimes stimulates germination of dormant seed embryos the effects are less consistent than those of 6-benzylaminopurine (Zhang and Lespinasse, 1991). Chilling appears to enhance the gibberellin content of seeds and may induce a shift in free and bound forms of GA_4 (Powell, 1987).

CHANGES IN STATE OF WATER DURING DORMANCY

During dormancy water is closely associated with macromolecules and is in a bound state (Faust *et al.*, 1997). It is gradually freed during the winter dormant period and rapidly converted to free water when resumption of growth is triggered by TDZ or by forcing conditions. TDZ also induces a change from bound water to free water in buds subjected to correlative inhibition.

MEMBRANE CHANGES DURING DORMANCY

With resumption of bud growth there is a change in membrane composition allowing increased permeability of solutes and water to the cytoplasm. Wang and Faust (1990) found that the seasonal changes in membrane lipids in apple buds from dormancy in winter to budbreak, and bud growth in spring, were similar to the general pattern of lipid metabolism induced by TDZ. Apple buds exposed to low winter temperatures responded with an increase in the degree of unsaturation of the fatty acids of their membrane lipids, changes in polar head group composition, increases in membrane phospholipid content and changes in sterol level and composition. The ratio of sterols to phospholipids decreased during budbreak and bud growth in spring.

Controlling seasonal budbreak

The major problem is that of poor budbreak in climates with mild winters and little winter chilling. A second problem is that of early budbreak and risk of damage from spring frosts in temperate-zone fruit-growing areas.

SELECTION AND BREEDING OF ADAPTED CULTIVARS

In general the date of budbreak reflects both the chilling and the heat requirements of the cultivar in a specific environment (Spiegel-Roy and Alston, 1979; Petropoulou 1985; Hauagge and Cummins, 1991 a). There is considerable variation between existing cultivars in these respects. At the National Fruit Trials in Kent, England the range of budbreak (flowering) dates of apple cultivars was from 30 April for 'Nico' to 11 June for 'Spätblühender Taffetapfel' (Morgan and Richards, 1993). In Zimbabwe the date of first budbreak of 'Anna' was about 3 months earlier than that of 'Starking' (Bepete and Jackson, 1995) and in Brazil Bernardi (1988) found 'Anna' to be more than 3 months earlier than 'Red Delicious'.

The most widely used apple cultivars in warm-winter areas are of two types.

The first are those which show only shallow depth of dormancy as well as limited requirements for chilling. 'Anna' and 'Dorsett Golden' are the most widely grown and adapted to the mildest winters (Miller and Baker, 1982). 'Ein Shemer', 'Elah', 'Maayan' and 'Michal' have also performed well in tropical and subtropical areas with inadequate winter chilling for mainstream cultivars.

The second type are exemplified by 'Rome Beauty' which, although classified as having a high chilling requirement, can be prevented from entering deep dormancy by leaf stripping following branch bending (Janick, 1974). Its tip-bearing habit may be important in this respect.

In intermediate areas, still too warm in winter for traditional cultivars, 'Mutsu' and 'Braeburn' have both shown excellent cropping and leafing out in Zimbabwe, with fewer than 300 chilling hours and without any dormancy-breaking sprays. 'Mollies Delicious' and 'Drakenstein' are also reasonably adapted to these conditions (Jackson and Bepete, 1995).

In breeding programmes it has been found that some crosses among 'Mollies Delicious', 'Gala', 'Golden Delicious' and 'Fuji' segregate out for lower chilling requirements than their parents (Denardi *et al.*, 1988). There are also 'Golden Delicious' mutations, e.g. 'Panorama Golden' and 'Davilla Spur Golden Delicious' with a lower chilling requirement and some lower chilling requirement mutants of 'Delicious' and 'Fuji' (Hauagge and Cummins, 2000). A number of *Malus* species with very low chilling requirements are being evaluated as parents but the greatest chance of producing cultivars combining high commercial quality with a minimal chilling requirement may lie in programmes including 'Anna' as a parent. The low-chilling-requirement character present in 'Anna' is thought to be controlled by at least one major dominant gene with minor genes interacting to modulate its effects (Hauagge and Cummins, 1991c). Usually 50% or more of 'Anna' descendants have a low chilling requirement and fail to develop a deep dormancy state.

Within *Pyrus*, *P. pashia*, *P. calleryana* and *P. amygdaliformis* have lower chilling requirements than *P. pyrifolia* and *P. communis* and grow in areas with warmer winter climates. Westwood and Bjornstad (1968) found that the seeds of species from warm-winter climates required less chilling for germination than those from colder climates and the former species had higher optimum chilling temperatures (7–10 °C) than the latter (3–5 °C). Temperatures below freezing were relatively ineffective in breaking rest of all species. Seeds within a species showed considerable variation in their chilling requirement, e.g. 90% of *P. calleryana* seeds germinated after 28 days of chilling but a few took three times as long. Trees grown from these latter seeds also had longer than ordinary bud chilling requirements. This is in accordance with the conclusion by Vegis (1963) that seeds of many species and varieties in the *Rosaceae* behave, in respect of their dormancy, in the same way as their buds. Seeds from interspecific crosses showed chilling requirements between those of their parents. It seems probable that natural selection for seed chilling requirements has played a major role in the evolution of highly adapted ecotypes with bud chilling requirements parallel to those of the seeds. Selection for seed chilling requirement might also provide a rapid preliminary screen for bud chilling requirement.

Within a north European environment Spiegel-Roy and Alston (1979) found 'Packham's Triumph' and 'Conference' to have much lower chilling and post-dormant heat requirements than 'Comice' or, especially, 'Williams''. The perry pear 'Fleurissant Tard' had very high chilling and heat requirements, which makes it a useful parent in breeding for avoidance of spring frost.

'Africana' is a lower chilling-requirement mutant of 'Packham's Triumph' (Hauagge and Cummins, 2000).

The effects of rootstocks on the chilling requirements for scion budbreak are of practical importance. Use of 'M.26' and 'M.7' apple rootstocks, which induce a lower chilling requirement in the scions than do more vigorous rootstocks, is advantageous in subtropical conditions. The lower chilling requirement of 'Bartlett' trees on *P. calleryana* than on *P. communis* rootstocks is disadvantageous in areas prone to spring frost (Westwood, 1970).

MODIFYING THE THERMAL ENVIRONMENT

Several technologies have been developed to facilitate the production of cultivars of high perceived quality in areas of unsuitable winter climates.

In southern Brazil young orchard trees of 'Fuji' and 'Gala' produce only two or three shoots from below the initial 'heading' pruning cut, as a result of inadequate winter chilling in a climate with fewer than 600 hours below 7.2 °C. Cold storage of young 'Gala' and 'Fuji' apple trees after lifting from the nursery and before planting in the orchard increased subsequent budbreak and production of lateral branches (Petri and Stuker, 1988). Storage at 2 °C gave better results than storage at 6 °C. Increasing the duration of cold storage from 15 days, in steps of 15 days to 60 days, gave increasing budbreak, the 45- and 60-day treatments resulting in formation of a good tree framework. Application of 4% mineral oil plus 0.16% DNPB (dinitro-butyl-phenol) immediately after planting in the orchard resulted in additional budbreak especially of unchilled trees or those given only short periods of chilling, but much of the budbreak induced by chemical treatment gave only short spurs unsuitable for branch formation.

In Thailand, shoots of high quality Asian pears which had been exposed to winter chilling in Taiwan were imported annually for grafting on low-chilling, poor quality cultivars. When this procedure became too expensive, locally produced budwood of the high-chilling-requirement cultivars was stored at 11 °C for two weeks prior to grafting. Budbreak was increased by this procedure (Krisanapook and Subhadrabandhu, 1995).

Water sprinkling to give evaporative cooling can be used to minimize the de-chilling effects of high daytime temperatures in either the nursery or the orchard (Erez, 1995, 2000).

Water sprinkling has also been tested as a method of delaying apple blossoming by evaporative cooling of buds, i.e. by effectively delaying the accumulation of growing-degree-hours. Budbreak was delayed by 14 days in accordance with expectation from heat balance calculations, but the buds from sprinkled trees had a higher water content than those of unwatered trees at any given stage of development and were more damaged by frost at any given time (Hamer, 1981).

TREE MANAGEMENT TO MODIFY BUD DORMANCY

Defoliation is used to modify or manipulate bud dormancy, both in tropical conditions without strong seasonal patterns of temperature and in the subtropics where chilling is inadequate. In Indonesia, leaf stripping one month after harvest results in budbreak, mainly of terminal buds on long shoots and spurs, within 20 to 30 days. If the stripping is carried out too soon after harvest the buds have not yet developed to give flower buds, so the growth is purely vegetative. Carried out at the right time, the buds are mixed buds, i.e. produce flowers as well as leaves. Two crops a year can be obtained in this way (Janick, 1974; Edwards and Notodimedjo, 1987; Notodimedjo *et al.*, 1981). This technique is effective with cultivars normally considered to have a high chilling requirement, e.g. 'Rome Beauty', and is essentially based on prevention of entry into endodormancy.

In Mexico defoliation is carried out in circumstances where normal leaf drop does not occur. Chemical defoliation induces more budbreak than manual defoliation and is particularly effective if carried out in early January after some chilling has occurred. Diaz *et al.* (1987) reported very positive effects of defoliation by use of copper sulphate or urea applied to 'Anna' apple trees.

Bending branches towards the horizontal is widely practised in the tropics. This may have a direct effect on the depth of bud dormancy, as shown by Crabbé (1984b). It also has the very important effect of releasing lateral buds from apical dominance. These grow out to give numerous short shoots or spurs along the upper surface of the branches. Such spurs bear terminal buds which break dormancy with less chilling than lateral buds on intact shoots, and these spurs give most of the crop in the Indonesian production system (Janick, 1974; Erez and Lavi, 1985).

Induction of water stress by stopping irrigation in a dry environment is commonly used in warm-winter conditions together with the above tree management practices to provide an effective substitute for winter chilling.

CHEMICAL INDUCTION OF BUDBREAK

Mineral oils in conjunction with chemicals such as dinitro-ortho-cresol (DNOC) were widely used from the late 1940s as dormancy-breaking agents in countries such as South Africa with sub-optimal winter chilling. Their use has been largely discontinued because of toxicity to humans (Erez, 1995).

Hydrogen cyanamide is very effective as a bud-breaking agent for a wide range of apple cultivars (Jackson and Bepete, 1995) and for pears including *P. pyrifolia* (Krisanapook and Subhadrabandhu, 1995). It can be used in the early orchard years to stimulate vegetative budbreak and lateral branch production. When used in cropping orchards it induces budbreak, compresses the period of budbreak within a cultivar and synchronizes the flowering, i.e. breaking of mixed buds, of different cultivars. As with a number of other dormancy-breaking treatments its effect is cumulative over years, stimulation of vegetative budbreak resulting in the production of both lateral branches and spurs. This results in the tree having many more buds which, in turn, can be induced to break.

Hydrogen cyanamide can induce the budbreak of single buds when painted on these at a concentration of 1.5%, and may be used in this way in the early years of the trees to economize on material. Lower concentrations, e.g. 0.5–1.25%, showed promise if combined with 2–5% winter oil (North, 1995).

The dinitro compounds only induced budbreak if applied when there were already signs of bud growth, but hydrogen cyanamide can be effective on endodormant buds and is damaging if applied after the buds have started to grow (Erez, 1995). Subhadrabandhu (1995) found that cyanamide had a smaller effect if applied to buds early in December than in the months before and after this when endodormancy could be expected to be less deep.

Hydrogen cyanamide application synchronizes budbreak both within and between cultivars. The dates of full bloom of 12 apple cultivars in Zimbabwe ranged from 19/09 to 27/09 when treated with hydrogen cyanamide, from 25/09 to 22/10 if untreated (Jackson and Bepete, 1995). This is in conformity with the results of Subhadrabandhu (1995) from which it could be inferred that cyanamide can induce budbreak largely irrespective of unfilled CU and GDH requirements. In Zimbabwe, with fewer than 300 chilling hours below 7.2 °C, the cropping response to cyanamide was positively correlated with the chilling requirement of the cultivar as shown by the date of budbreak on untreated trees, i.e. it overcame the shortfalls in cropping due to inadequate chilling (Table 6.1).

A number of other chemicals, e.g. potassium nitrate, thiourea and gibberellic acid, have been found to have some effects on budbreak under subtropical conditions (Erez, 2000), gibberellic acid being the most effective (El-Banna *et al.*, 1995).

Shoot extension growth

The total shoot growth on a tree in one year depends on the number of buds that break, as has already been discussed, and the growth of these individual shoots. This growth is controlled by a large number of separate, though often interacting, factors.

Use of reserves in shoot growth

The growth of shoots in spring starts with cambial activity in the stem before budbreak. Evert (1960, 1963), cited by Oliveira and Priestley (1988)

Table 6.1 *Effect of application of 1.5% hydrogen cyanamide in 1988, 1989, 1990 and 1991 on cropping efficiency (accumulated yield per cross-section of trunk), the calculated reduction in cropping efficiency as a result of inadequate winter chilling, and the averaged date of full bloom 1989–1992*

Cultivar	Accumulated yield (kg) per trunk cross-sectional area (cm^2): Cyanamide treated (T)	Accumulated yield (kg) per trunk cross-sectional area (cm^2): Unsprayed (U)	% cropping efficiency deficit (T – U)/T	Average date of full bloom (days from 31 August)
'Goldjon'	1.996	1.975	1	25
'Mutsu'	1.945	1.919	1	25
'Mollies Delicious'	2.425	1.997	18	45
'Canvada'	2.591	1.836	29	39
'Drakenstein'	3.365	2.052	40	37
'Golden Delicious'	3.586	2.014	44	39
'NJ46'	1.645	0.907	45	44
'Marjorie Pye'	2.901	1.348	54	36
'Spartan'	1.450	0.610	58	42
'Starking'	3.086	1.265	59	45
'Granny Smith'	1.629	0.565	65	49
'Ohinemuri'	1.698	0.463	73	50

Correlation between increases in accumulated yield in response to cyanamide (%) and date of full bloom $r = 0.82^{***}$ $R^2 = 0.67$
Correlation between cropping efficiency deficit and date of full bloom $r = 0.80^{***}$ $R^2 = 0.65$
From Jackson and Bepete (1995). Reproduced with permission.

reported that cambial activity in apples and pears resumes early in April when cells at the outer margin of the cambial zone begin to differentiate into phloem sieve elements. Later, in mid-May, xylem and sclereid differentiation begins. The differentiation of new conducting tissue below the buds prior to budburst is obviously dependent on the use of carbohydrate reserves.

The initial stages of new shoot extension growth after budbreak are also dependent on the use of reserves of carbohydrates, nitrogen and other mineral elements. Hansen (1971) estimated that half to two thirds of the building materials used in spurs until the flowers show colour, and in extension shoots up to the five- or six-leaf stage, come from reserves. The main source of the reserves of carbohydrates appears to be the root system (Quinlan, 1969; Hansen and Grauslund, 1973), but Kandiah (1979), while confirming the importance of carbohydrates from the root, considered that the whole perennial structure

Table 6.2 *Number of leaves produced (nodes) and length of shoot (cm) made during a 6-week period at the temperature shown (means of six trees)*

Temp of CE rooms		Pruning treatments, shoots per tree						
		1	2		3			
			lower	upper	lower	middle	upper	Mean
25 °C	Leaves	27	23	25	20	21	22	23
	Shoot	64	54	57	45	52	55	54
17.5 °C	Leaves	15	14	13	12	13	12	13
	Shoot	35	30	25	23	27	22	27
10 °C	Leaves	4	3	2	3	4	3	3
	Shoot	4	3	2	2	3	3	3

After Abbott (1984). Reproduced with permission.

of an apple tree can behave as a storage organ for carbohydrates. Carbohydrate reserves are primarily used up in respiration, supplying energy for growth, rather than in provision of structural material. Most of the nitrogen for early growth comes from the bark of branches and shoots near the expanding buds (Mason and Whitfield, 1960; Tromp and Ovaa, 1971; Titus and Kang, 1982).

Effects of temperature

Abbott (1984), using controlled environment rooms, showed that the rate of extension growth of shoots arising following decapitation of potted 'Cox' trees increased progressively over the range of 10–25 °C (Table 6.2). This was so even when only a single shoot was allowed to grow as a replacement leader. These results provide experimental support for the field observations of Barlow (1975a) and Johnson and Lakso (1985), who showed shoot growth to be a function of temperature. Tromp (1992a) found a relatively small effect of increasing soil temperature from 12 °C to 22 °C on the terminal shoot growth of newly budded trees but a large effect on total lateral shoot growth. He also (Tromp, 1992b) showed greatly reduced lateral shoot numbers and consequently total shoot growth when root temperatures were at 7 °C compared with 14 °C or higher.

Effects of light

Artificial shading of whole trees results in a reduction in total shoot growth (Priestley, 1969; Jackson and Palmer, 1977a). This effect is primarily through a reduction in the number of shoots that grow and, to a lesser extent, to a reduction in weight per unit length of shoot, rather than in the length of the individual shoots (Jackson and Palmer, 1977a).

Effects of water stress

Tromp (1992a) found that reducing water stress by growing apple trees at 90% Relative Humidity (RH), as contrasted with 50% RH, increased total shoot growth by more than 40%. Part of this effect was a result of increase in individual shoot length but most of it was a result of the emergence of more lateral shoots. Maggs (1961), working with pot-grown trees disbudded so as to give only one shoot per plant, found that growth of this shoot under a high water supply regime, expressed as dry weight increment, was more than twice as great as under a low water supply regime. The shoots grown under conditions of ample water supply were also about 50% longer (taller) than those with limited water supply. These results are compatible with results from field trials that showed the main effect of irrigation to a minimal soil water deficit (25 mm) in England, as compared with no irrigation, was to increase the number of shoots, but that in a dry season mean shoot length was increased by 50% also (Goode and Ingram, 1971).

Effects of nitrogen and other nutrients

The nitrogen reserves available at the beginning of the season determine the length of time for which the shoots will grow without further supplies. Such supplies delay the formation of terminal buds and prolong the season of growth (Hill-Cottingham, 1963).

Deficiencies of other mineral nutrients, or toxic levels of these, can also reduce shoot growth as part of their general effects of inducing stunting, often associated with poor leaf development, as in the little-leaf syndrome induced by zinc deficiency, or with leaf shed.

Effects of plant hormones

Shoot extension is generally considered to be controlled by both growth inhibitors and growth promoters and the rate of extension to reflect the balance between these.

Auxins stimulate cell enlargement, cell division and stem growth (Davies, 1987). They also have a major effect on root system development and, through this, on the growth of shoots. There is, however, also evidence that the apical dominance effect of shoot-tip-derived auxin reduces the growth of lateral shoots as well as influencing lateral bud dormancy.

Abscisic acid (ABA), which is generally regarded as a growth inhibitor, stopped the elongation of apple shoots and induced terminal bud formation in single stem apple trees into which it was injected (Robitaille and Carlson, 1971). There is also evidence of growth inhibitors, which have not been characterized chemically, in apple rootstocks and in dwarf scions (Grochowska *et al.*, 1984).

There are a large number of plant gibberellins (GAs) and some of these are active in promoting apple shoot growth.

GAs applied to a wide range of intact plants induce elongation of stem tissue, this effect being more pronounced in rosette or in dwarf species (Jones, 1973; Métraux, 1987). The increase in length of the extending shoot results from increase in length of existing as well as newly divided cells and may also be accompanied by an increase in cell number. There appear to be at least three sites of GA biosynthesis in higher plants, in developing fruits and seeds, in elongating shoot apical regions, and in roots (Graebe and Ropers, 1978; Sponsel, 1987). They seem to be transported both in the xylem and in the phloem. In apple and pear the application of inhibitors of GA biosynthesis such as cyclocel and paclobutrazol results in a reduction in shoot elongation, with short internodes. Such application has been, and is, of considerable commercial importance and has also provided a tool for studying control of growth by endogenous gibberellins. Paclobutrazol (2RS,3RS)-1-(4-chlorophenyl)-4,4-dimethyl-2-(1H-1,2,4-triazol-1-yl) pentan-3-ol inhibits three steps in the oxidation of the GA precursor *ent*-kaurene to *ent*-kaureneoic acid (Hedden and Graebe, 1985) and its effects on apple can be partially reversed by GA_3 application (Steffens *et al.*, 1985). It is transported acropetally in the xylem when applied to young stem internodes and, to a lesser extent, to the youngest unrolled leaf (Richardson and Quinlan, 1986). When applied to the soil it is taken up by the roots and transported upwards in the xylem stream. When applied in this way it can be relatively slow to act and may be reversibly bound to vascular tissues, to be released and to check extension growth in the following year. Richardson and Quinlan (1986) applied paclobutrazol to different regions of apple rootstock shoots growing in a heated greenhouse in late spring, with the effects shown in Table 6.3. These results show that paclobutrazol applied directly to, or translocated to, the apical region of the shoot was much more effective in retarding shoot extension growth than the chemical localized in other parts of the shoot. Similarly, growth of internodes and leaves produced from a treated shoot tip was reduced. This indicates that the shoot tip, and possibly the young leaves, are important sites for the synthesis of gibberellins influencing shoot development.

Repeated use of paclobutrazol results in the development of a spur-type habit of growth with lateral buds giving rise to short fruiting spurs rather than long shoots (Elfving and Proctor, 1986; Tukey, 1986), which suggests that whether a shoot becomes a long shoot or a short shoot is at least partly controlled by gibberellin biosynthesis.

Table 6.3 *Effect of site of application of paclobutrazol on shoot extension and leaf production of 'M.26' rootstock shoots during 30 days following treatment*

Site of application	Number of new unrolled leaves	Shoot extension (cm)
One-year-old wood	11.8	28.2
Laminae	11.5	23.4
Petioles	10.9	23.1
Shoot tip	10.3	17.3
New stem	10.4	15.5
Shoot tip + new stem	8.1	7.8
Complete shoot	10.0	9.5
Untreated	12.4	28.1
LSD 0.05 between any two values	1.6	3.6

From Richardson and Quinlan (1986). Reproduced with permission.

Cytokinins are generally considered to regulate cell division, and also play an important role in budbreak and therefore help determine the number of shoots (see pp. 167–70). Jones (1967) found a marked response of apple shoot growth *in vitro* to benzyladenine. Applications of this in the field have sometimes increased and sometimes reduced mean shoot length.

Effects of other plant parts on shoot extension

Leaves, as the source of carbohydrates produced by photosynthesis, play a major role in all plant growth. Rapidly growing shoot tips are a strong sink for assimilates, and after the first five or six shoot leaves have been produced current photosynthesis becomes a more important carbon source than reserves (Hansen, 1971). Photosynthates from the upper leaves of the shoot are in general exported upwards, those from the lower part being exported to other parts of the tree. This pattern can, however, be influenced by factors which affect the balance of assimilate supply and demand, e.g. by fruiting and by partial defoliation of the shoot (Hansen, 1967, 1969; Quinlan, 1965, 1966): it is therefore not simply a function of pre-existing vascular connections.

Leaves also have effects on shoot extension growth in ways other than through their supply of assimilates. Kato and Ito (1962) found that apple shoot growth was promoted by the removal of expanded leaves. This effect was a consequence of increased internode length and was associated with an increase in shoot auxin content. Abbott (1984) found that removal of young expanding leaves was followed by a virtual cessation of stem elongation and then by an increase in this to result in longer shoots and a later termination of growth. Barlow and Hancock (1956) and Kato and Ito (1962) also found that

Table 6.4 *Effects of treatments which induced large differences in fruit yields in 1973 on cropping and growth in 1973 and 1974. 'Cox's Orange Pippin' apples with results averaged over 4 rootstocks*

	Treatments in 1973				
	P	C	t	T	Significance
Fruits per tree 1973	333	221	177	114	***
Fruit yield, kg/tree, 1973	20.9	19.5	17.8	14.6	***
Fruit yield, kg/tree, 1974	1.1	7.1	7.1	12.1	***
Mean shoot length cm, 1973	26.6	30.1	30.6	31.7	*
Mean shoot length cm, 1974	9.1	12.6	13.0	14.8	***
Relative girth increment 1973[a]	0.12	0.19	0.19	0.23	***
Relative girth increment 1974[a]	0.21	0.23	0.23	0.22	NS

[a] Calculated as $\log_e$ girth$^{8/3}$ end of year $-\log_e$ girth$^{8/3}$ beginning of year, which also gives a measure of relative growth rate.

P, Hand-pollinated, C, control; t, lightly thinned; T, heavily thinned.

Data from Blasco (1976). Reproduced with permission.

removal of unexpanded tip leaves caused an immediate reduction in length of internodes produced and the latter workers related this to the high GA content of the leaves removed.

Root system size and function has a dominant effect on shoot growth. This is found, as discussed earlier, when the shoot system is grafted on to an inherently dwarfed root system, when root system size is reduced by root pruning or constrained by growing it in a root-restriction nylon container. Although the response to root pruning may involve checks to shoot growth as a result of competition from compensatory root growth, the effects of the genetically-limited rootstocks and of root restriction membranes under conditions of ample water and nutrient supply suggest an intrinsic hormonal control mechanism (cf. pp. 105–6).

Fruiting results in diversion of metabolites from vegetative growth, including shoot growth, to fruits (Maggs, 1963). Although the effects can be shown within a single season (Avery, 1969), the effect of crop on shoot numbers and on average shoot length is usually greater in the following season. This has been shown by experiments in which different levels of cropping were induced (Table 6.4). Regression studies of the relationship between year-to-year variations in shoot growth and those in fruit yield have shown either no correlation in the same season but a major adverse effect of fruit yield on shoot growth in the following year (Rogers and Booth, 1964), or approximately equal negative effects of cropping on shoot growth in the same and the following year (Barlow, 1975b). The direct (same season) effect may simply reflect competition for currently available resources, which may be influenced by the lower

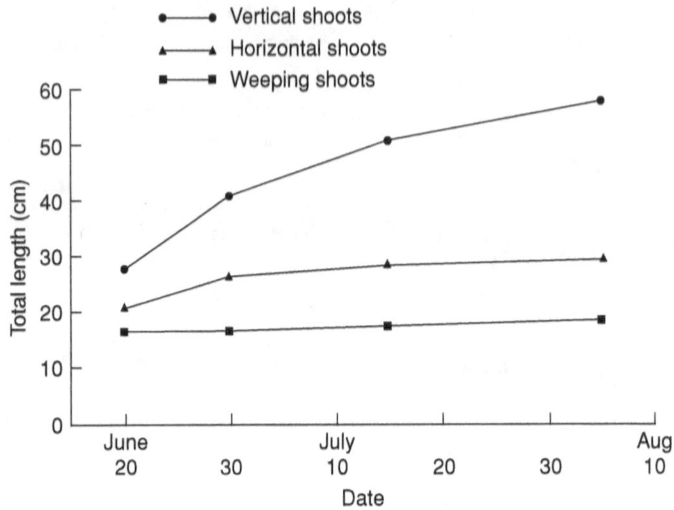

Figure 6.12 Growth of vertical, horizontal and weeping apple shoots. From Kato and Ito (1962). Reproduced with permission.

auxin content of shoots of fruiting as compared with defruited trees (Kato and Ito, 1962). The residual effect may be a consequence of effects on bud development, which are also shown as effects on the number and 'strength' of flower buds, and the very severe effect of heavy cropping on root growth. The adverse effect of cropping on total shoot growth over a number of years is primarily by a reduction in the number of long shoots and their proportion of all shoots, although there is some effect on their average length (Barlow, 1966). Cumulative effects of fruiting have been shown in Table 5.5.

Effects of shoot angle to the horizontal

Vertically-growing shoots continue to grow later in the season than do horizontal shoots, and become longer. Weeping shoots, bending below the horizontal, grow least vigorously and become the shortest (Figure 6.12). Kato and Ito (1962) found that the concentration of endogenous auxin and gibberellins was highest in the vertically-growing shoots and lowest in the weeping ones. Vertical shoots also had the lowest carbohydrate and the highest nitrogen content.

Effects of bark ringing or girdling

Removal of a ring of bark (ringing or girdling) greatly reduces the extension growth of the shoot above the ring although it increases the carbohydrate content of this shoot. Girdling results in a decline in auxin content of the terminal bud of shoots (Kato and Ito, 1962).

Effects of pruning

Pruning is carried out either in the dormant, winter, season or in the season of active growth, when it is referred to as summer pruning. It is carried out to develop a desirable tree framework and to maintain tree size to that which can be easily sprayed, thinned and harvested: also to maintain fruit quality by reducing the number of fruits, so increasing the size of those remaining, and by facilitating light penetration, important for fruit colour development, into the tree. In dormant-season pruning, shoots or older branches may be shortened or removed entirely. Shoot shortening is called tipping or heading, branch shortening is called stubbing or heading. Removal of complete shoots or branches is referred to as thinning. In summer pruning leafy shoots may be tipped or pinched, or removed entirely, i.e. thinned. The effects of these treatments on shoot growth, reviewed by Mika (1986), are as follows.

- The individual shoots arising from a pruned branch are larger than those on an unpruned branch.
- The new terminal shoot following shortening of a branch is longest if 70% of the branch is pruned away.
- For any given degree of pruning the size of the shoot growing from a pruned branch is correlated with the length of the latter before pruning.
- If the same amount of wood is removed, heading cuts induce more new shoot growth than thinning cuts, largely because they release more buds from correlative inhibition.
- Despite the faster growth of individual shoots on pruned trees, the total length of old branch and new terminal shoot is less than if no pruning has taken place: pruning reduces the total size of the shoot system although it stimulates the growth of new shoots.
- Dormant pruning results in major increases in the levels of cytokinins, auxins and, especially, gibberellins in the above-ground parts of pruned compared with unpruned trees. These may be involved in the shoot growth responses noted.
- Summer pruning early in the season tends to result in the production of a greater number of shoots than dormant-season pruning (Maggs, 1965). The earlier in the season that summer pruning is carried out, the greater the amount of regrowth: possibly as a result of the higher levels of cytokinins and other growth promoters early in the season and the build-up of inhibitors late in the season (Ferree *et al.*, 1984). Late summer pruning results in shorter shoots than on controls or winter pruned trees. Summer pruning has a minimal effect on shoot growth in subsequent years (Marini and Barden, 1987).
- Summer pruning tends to lead to the production of spurs rather than long shoots.

- Responses to summer pruning are very variable, being influenced by time of pruning, the type of buds immediately below the pruning cut, rootstock and scion cultivar.
- Pinching out of shoot tips or removal of shoot apices or new leaves greatly reduces growth in shoot length (Table 6.5).

Cultivar differences in shoot growth

Scion cultivars may show extreme contrasts in their vigour of shoot growth. Barlow (1964) reported cumulative extension growth over 10 years of only 31 m per tree of 'Miller's Seedling'/'M.9' under conditions where 'Laxton's Superb'/'M.9' produced 219 m per tree. Such differences may reflect intrinsic differences in shoot extension growth potential but also result from cultivar differences in cropping level, in lateral budbreak, and in the angle of lateral branches to the horizontal.

The physiological basis for intrinsic differences between cultivars with respect to extension growth has not been studied to any great extent. This partly reflects the difficulty and complexity of such studies. It also reflects the fact that scion cultivars are selected primarily for the characteristics of their fruits. Control of tree vigour is achieved primarily by the use of size-controlling rootstocks, by control of the level of fruiting, by branch training and pruning and by the use of plant growth regulators. The exception to this generalization is the selection and utilization of spur-types.

Spur-type cultivars have been selected and propagated vegetatively following observation of branches with short internodes and a higher than average number of fruiting spurs per unit length of branch. These have arisen spontaneously by bud mutation within many of the leading cultivars, most commonly within 'Delicious' ('Red Delicious'). Others have been selected following induction of mutation by irradiation (Campbell and Lacey, 1973). They are considered advantageous because they offer a measure of vigour control with less restriction to a specific rootstock than the standard cultivars while retaining the fruit characteristics of these.

Spur-type trees typically have about the same number of nodes per shoot as their 'parent' cultivar but the internodes are 10–15% shorter. There may be no difference between the 'normal' cultivars and the spur-types in the number of laterals produced in the first year, but subsequently a greater proportion of the axillary buds develop into short fruiting spurs, instead of long vegetative shoots, in the spur-types (Arasu, 1968). These differences have cumulative effects so that in mature trees spur-types may show only half to one third as much total annual shoot growth as their standard counterparts (Curry and Looney, 1986).

Since most spur-types have arisen by mutation from different 'parents' it is not surprising that they show variation in the factors which contribute to

Table 6.5 *Effect of pruning treatments on mean length of annual shoot (cm) of apple trees: average for the years 1976–79*

Treatment	'McIntosh'		'Melba'	
	'M.26' rootstock	'A.2' rootstock	'M.26' rootstock	'A.2' rootstock
1 1/3 of dormant shoots removed	40.4 e[a]	42.9 fg	36.2 gh	39.2 g
2 Dormant shoots headed by 1/3	38.8 e	45.6 g[a]	39.2 i	47.1 h[b]
3 Dormant shoots headed by 2/3	47.9 g	56.6[b]	43.7 j	50.8 i[b]
4 Pinching in June to 4–5 leaves	29.2 c	33.0 d	29.1 e	31.5 cd
5 Pinching in July to 7 leaves	22.2 b	26.6 b	21.8 b	26.8 b
6 1/3 of shoots removed in August	39.2 e	40.9 ef	32.6 f	33.9 def
7 All dormant lateral buds removed	44.2 f	46.8 h	37.6 hi	41.2
8 1/2 of dormant lateral buds removed	34.7 d	42.6 fg[b]	32.0 f	35.8 f
9 Shoot apexes removed every 2 weeks	15.8 a	16.4 a	14.7 a	15.4 a
10 New leaves removed every week	27.7 c	30.3 cd	23.0 bc	32.0 cde[b]
11 Control, untreated	38.7 e	38.6 c	33.7 fg	34.9 ef

[a] Mean separation within columns by Duncan's multiple range test, 5% level.
[b] Significant difference between the rootstocks at 5% level.
From Mika *et al.* (1983). Reproduced with permission.

their relative growth and vigour. Walsh and Miller (1984) identified variability between spur-types of different cultivars in their lateral budbreak on both current season's and second-year wood, in their internode length and in the formation of 'long spurs' between 5 and 20 cm in length. Faust and Zagaja (1984) reported that although short internodes were more frequent in the low-vigour than the high-vigour seedlings arising from crosses between short-internode parents of apple, there were some long-internode, low-vigour and some short-internode, high-vigour progeny. The same was true for pears. Wide crotch angles tended to be associated with low vigour. Grochowska *et al.* (1984), contrasting low- and high-vigour seedlings from crosses between spur-types, found much higher auxin-like and gibberellin-like activity in the phloem of low-vigour seedlings and concluded that their lack of growth was a consequence of production of a growth inhibitor. Faust and Zagaja (1984) concluded that expansion growth is generally under polygenic control.

The situation appears much simpler in the 'compact columnar' apples now being bred using the 'Wijcik McIntosh' mutant as a parent. This arose as a natural mutation in the orchard and is an extreme spur-type which seldom gives lateral long shoots. This growth habit is governed by the compact (*Co*) gene. It is highly heritable, with 40–45% of the progeny of crosses involving 'Wijcik' being compact and columnar (Lapins, 1969; Lapins and Watkins, 1973). The shoot tips of 'Wijcik McIntosh' are low in polar gibberellins, have high endogenous cytokinin levels and exhibit very high tolerance to benzyladenine in tissue culture (Looney and Lane, 1984).

Effects of *rol* genes on shoot growth

Holefors *et al.* (1998) have shown that infection of 'M.26' apple with *Agrobacterium tumefasciens* containing the *rol*A gene, which may operate through a decrease in gibberellin or conjugated polyamine content and increased sensitivity to auxin, led to transformed plants with short internodes. Bell *et al.* (1999) found that the *rol*C gene reduced tree height and number of nodes of transgenic 'Beurré Bosc' pear trees (cf. pp. 476–7).

Effects of rootstocks on shoot growth

Even non-cropping scions on dwarfing rootstocks show reduced total shoot growth, both because fewer of their buds give rise to shoots and because their shoots extend more slowly in the later part of the season and cease to grow sooner than those on more vigorous rootstocks (Avery, 1969). Rootstock effects are also discussed in detail in Chapters 2 and 5.

Water shoot growth

Water shoots arise from trace buds buried in older stem tissues after heavy pruning. They grow very strongly, with all their assimilates directed to the rebuilding of new shoots, limbs and trunks (Mika, 1986). They never supply assimilates to fruits even if these are nearby. This appears to illustrate the general principle that fruit trees react to severe pruning in such a way as to restore the functional equilibrium between shoots and roots.

Secondary thickening

In the stems of woody plants the cambium begins to differentiate from the procambium in a given region just before that region ceases to elongate (Eames and MacDaniels, 1947). A continuous cambial cylinder forms and produces annual increments of secondary vascular tissues for the life of that part of the tree. Evert (1963) found this production of new secondary vascular tissues in pear to start in April in Montana, USA, reaching a peak in June and early July. Knight (1961) found new secondary xylem in apple stems in England by 25 April. Gross measurements of trunk girth as an indicator of cambial activity show this to occur from June to September, or from May to August, in different studies (Pratt, 1990). Girth often continues to increase after extension growth has ceased.

The increments in trunk girth each year as a result of secondary thickening reflect factors influencing growth in general. Nutrient deficiencies and water stress reduce trunk growth, as do dormant-season pruning, fruiting, root pruning and the use of dwarfing rootstocks.

Water stress, however, reduces trunk growth less than it reduces new stem growth (Maggs, 1961) but more than it reduces fruit enlargement (Iancu, 1985). The negative effect of heavy fruiting on trunk growth is greater in the year of fruiting than in the following year. This has been demonstrated experimentally by Blasco (1976), as shown in Table 6.4 and in regression studies by Barlow (1975b).

Secondary thickening in relation to trace buds and to branches

Secondary xylem produced by cambium at the nodes surrounds and buries leaf and branch traces.

The base of a side branch becomes buried in growth rings of secondary tissues of the trunk. This produces a mechanically strong crotch, in contrast to the weak crotch between approximately equal leading branches, as shown in Figure 6.13. Secondary xylem in general has high mechanical strength. Apple

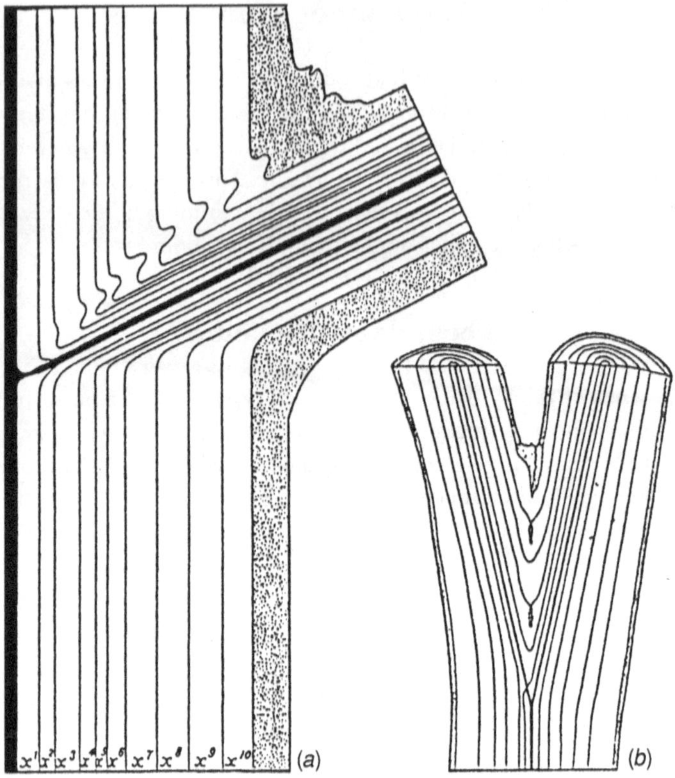

Figure 6.13 Tree crotches in diagrammatic longitudinal sections. (*a*) Burial of the base of a living branch by 10 growth rings (x^1 to x^{10}). On the upper side of the branch secondary xylem, cambium, secondary phloem and cortex are folded as a result of confined growth in a narrow angle. Only the pith is unaffected. (*b*) Weak crotch between approximately equal branches showing 5 growth rings and occluded bark. Cortex and phloem stippled; pith solid black. From Eames and MacDaniels, *An Introduction to Plant Anatomy* (1947), McGraw-Hill. Reproduced with permission of the McGraw-Hill Companies.

wood is diffuse-porous and heavy with numerous fibres and with thick walls to the vessels, fibre tracheids and parenchyma.

References

Abbas, M.F. (1978). Association between branching in maiden apple trees and levels of endogenous auxins. *Acta Horticulturae* **80**, 59–62.

Abbott, D.L. (1970). The role of budscales in the morphogenesis and dormancy of the apple fruit bud. In *Physiology of Tree Crops*, ed. L.C. Luckwill and C.V. Cutting, pp. 65–82. London: Academic-Press.

Abbott, D.L. (1984). *The Apple Tree: Physiology and Management.* London: Grower Books.

Arasu, N.T. (1968). Spur-type sports in apple. *Report of the East Malling Research Station for 1967*, 113–19.

Avery, D.J. (1969). Comparison of fruiting and deblossomed maiden apple trees, and of non-fruiting trees on a dwarfing and an invigorating rootstock. *New Phytologist* **68**, 323–336.

Barlow, H.W.B. (1964). An interim report on a long-term experiment to assess the effect of cropping on apple tree growth. *Report of the East Malling Research Station for 1963*, 84–93.

Barlow, H.W.B. (1966). The effect of cropping on the number and kind of shoots on four apple varieties. *Report of the East Malling Research Station for 1965*, 120–4.

Barlow, H.W.B. (1975a). Effects of temperature on apple shoot growth. In *Climate and the Orchard*, ed. H.C. Pereira, pp. 41–3. Farnham Royal, UK: Commonwealth Agricultural Bureaux.

Barlow, H.W.B. (1975b). Effects of cropping on the growth of orchard trees. In *Climate and the Orchard*, ed. H.C. Pereira, pp. 98–102. Farnham Royal, UK: C.A.B.

Barlow, H.W.B. and Hancock, C.R. (1956). Internode growth in stoolbed shoots of the apple rootstock Crab C. *Report of the East Malling Research Station for 1955*, 89–100.

Barlow, H.W.B. and Hancock, C.R. (1960). The experimental modification of branch form in an apple rootstock. *Botanical Gazette* **121,** 208–15.

Barlow, H.W.B. and Hancock, C.R. (1962). The influence of the leaf on the development of its axillary meristem. *Report of the East Malling Research Station for 1961*, 71–6.

Bell, R.L., Scorza, R., Srinivasan, C. and Webb, K. (1999). Transformation of 'Beurré Bosc' pear with the *rol C* gene. *Journal of the American Society for Horticultural Science* **124**, 570–4.

Bepete, M. and Jackson, J.E. (1995). Apple cultivar performance and response to a chemical dormancy breaking spray under 'marginal' winter-chilling conditions in Zimbabwe. *Acta Horticulturae* **409,** 121–4.

Bernardi, J. (1988). Behaviour of some apple cultivars in the subtropical region of Santa Catarina, Brazil. *Acta Horticulturae* **232**, 46–50.

Blanco-Braña, A. (1976). Effects of growth regulating hormones on the emergence of shoots following pruning of fruit trees. PhD thesis, London University (East Malling).

Blanco-Braña, A. and Jackson, J.E. (1982a). Effects of applying growth regulating hormones following fruit tree pruning. I. Effects of different types and concentrations of auxin and of GA_{4+7} and 6-benzylaminopurine on shoot emergence and wound healing in apple trees. *Journal of Horticultural Science* **57**, 17–30.

Blanco-Braña, A. and Jackson, J.E. (1982b). Transport of NAA applied to the cut surfaces of pruned apple branches. *Journal of Horticultural Science* **57**, 31–44.

Blasco, A.B. (1976). Rootstock effects on growth and cropping of apples with special reference to fruit quality. PhD thesis, London University (East Malling).

Borkowska, B. (1981). Dormancy and development of apple axillary buds investigated *in vitro*. *Acta Horticulturae* **120**, 161–4.

Borkowska, B. and Powell, L.E. (1982/83). Abscisic acid relationships in dormancy of apple buds. *Scientia Horticulturae* **18**, 111–17.

Broome, O.C. and Zimmerman, R.H. (1976). Breaking bud dormancy in tea crabapple (*Malus hupehensis* (Pamp) Rehd) with cytokinins. *Journal of the American Society for Horticultural Science* **101**, 28–30.

Campbell, A.I. and Lacey, C.N.D. (1973). Compact mutants of Bramley's Seedling apple induced by gamma radiation. *Journal of Horticultural Science* **48**, 397–402.

Campbell, R.K. (1978). Regulation of bud burst timing by temperature and photoperiod during dormancy. In *Proceedings of the Fifth North American Forest Biology Workshop*, ed. C.A. Hollis and A.E. Squillace, pp. 19–33. Forestry Department, University of Florida, Gainesville.

Cannell, M.G.R. (1989). Chilling, thermal time and the date of flowering of trees. In *Manipulation of Fruiting*, ed. C.J. Wright, pp. 99–113. London: Butterworths.

Champagnat, P. and Côme, D. (1986). Some thoughts on hormonal regulation of bud and seed dormancies. *Acta Horticulturae* **179**, 117–27.

Cline, M.G. (1991). Apical dominance. *The Botanical Review* **57**, 318–58.

Cody, C., Larson, F.E. and Fritts, R. (1985). Induction of lateral branches in tree fruit nursery stock with propyl 3-*t*-butylphenoxy acetate (MB25, 105) and promalin (GA_{4+7} + 6-benzyladenine). *Scientia Horticulturae* **26**, 111–18.

Couvillon, G.A., Finardi, N., Magnani, M. and Freire, C. (1984). Rootstock influences the chilling requirement of 'Rome Beauty' apple in Brazil. *HortScience* **19**, 255–6.

Crabbé, J.J. (1984a). Morphogenetical ways towards vigour restriction in spontaneous and man-made dwarf trees. *Acta Horticulturae* **146**, 113–20.

Crabbé, J.J. (1984b). Correlative effects modifying the course of bud dormancy in woody plants. *Zeitschrift für Pflanzenphysiologie* **113 S**, 465–9.

Curry, E.A. and Looney, N.E. (1986). Effect of cropping on shoot growth of spur-type 'Golden Delicious' trees. *HortScience* **21**, 1015–17.

Cutting, J.G.M., Strydom, D.K., Jacobs, G., Bellstedt, D.U., Van Der Merwe, K.J. and Weiler, E.W. (1991). Changes in xylem constituents in response to rest-breaking agents applied to apple before budbreak. *Journal of the American Society for Horticultural Science* **116**, 680–3.

Davies, P.J. (1987). The plant hormones: their nature, occurrence and function. In *Plant Hormones and their Role in Plant Growth and Development*, ed. P.J. Davies, pp. 1–11. Dordrecht: Martinus Nijhoff.

Denardi, F., Hough, L.F. and Bonetti, J.I. (1988). Low chilling and disease resistance as main objectives of apple breeding in Santa Catarina, Brazil. *Acta Horticulturae* **232**, 15–25.

Diaz, D.H., Alvarez, A. and Sandoval, J. (1987). Cultural and chemical practices to induce uniform bud break of peach and apple under warm climates in Mexico. *Acta Horticulturae* **199**, 129–36.

Duckworth, S.J., Abbas, M.F. and Quinlan, J.D. (1979). Influence of endogenous growth regulators on branching. *Report of the East Malling Research Station for 1978*, 39.

Dutcher, R.D. and Powell, L.E. (1972). Culture of apple shoots from buds *in vitro*. *Journal of the American Society for Horticultural Science* **97**, 511–14.

Eames, A.J. and MacDaniels, L.H. (1947). *An Introduction to Plant Anatomy*, 2nd edn. New York: McGraw-Hill.

Edwards, G.R. (1985). Changes in endogenous hormones in apple during bud burst induced by defoliation. *Acta Horticulturae* **158**, 203–10.

Edwards, G.R. and Notodimedjo, S. (1987). Defoliation, bending and tip pruning of apple under tropical conditions. *Acta Horticulturae* **199**, 125–8.

El-Banna, S., El-Shall, S. and Shahin, B. (1995). Effect of dormant season spray applications on growth and fruiting of Le-Conte pear trees. *Acta Horticulturae* **409**, 207–13.

Elfving, D.C. and Proctor, J.T.A. (1986). Long term effects of paclobutrazol (Cultar) on apple shoot growth, cropping and fruit–leaf relations. *Acta Horticulturae* **179**, 473–80.
Erez, A. (1995). Means to compensate for insufficient chilling to improve bloom and leafing. *Acta Horticulturae* **395**, 81–95.
Erez, A. (2000). Bud dormancy: phenomenon, problems and solutions in the tropics and subtropics. In *Temperate Fruit Crops in Warm Climates*, ed. A. Erez, pp. 17–48. Dordrecht: Kluwer.
Erez, A. and Lavi, B. (1985). Breaking bud rest of several deciduous fruit tree species in the Kenyan highlands. *Acta Horticulturae* **158**, 239–48.
Esau, K. (1965). *Plant Anatomy*, 2nd edn. New York: John Wiley.
Evert, R.F. (1960). Phloem structure in *Pyrus communis*, L. and its seasonal changes. *University of California Berkeley Publications Bot.* **32**, 127–94.
Evert, R.F. (1963). The cambium and seasonal development of the phloem in *Pyrus malus*. *American Journal of Botany* **50**, 149–59.
Faust, M. (1989). *Physiology of Temperate Zone Fruit Trees*. New York: Wiley.
Faust, M. and Zagaja, S.W. (1984). Prospects for developing low vigour fruit tree cultivars. *Acta Horticulturae* **146**, 21–7.
Faust, M., Erez, A., Rowland, L.J., Wang, S.Y. and Norman, H.A. (1997). Bud dormancy in perennial fruit trees: physiological basis for dormancy induction, maintenance and release. *HortScience* **32**, 623–9.
Faust, M., Liu, D., Wang, S.Y. and Stutte, G.W. (1995). Involvement of apical dominance in winter dormancy of apple buds. *Acta Horticulturae* **395**, 47–56.
Ferree, D.C., Myers, S.C., Rom, C.R. and Taylor, B.H. (1984). Physiological aspects of summer pruning. *Acta Horticulturae* **146**, 243–52.
Gianfagna, T.J. and Mehlenbacher, S.A. (1985). Importance of heat requirement for bud break and time of flowering in apple. *HortScience* **20**, 909–11.
Goode, J.E. and Ingram, J. (1971). The effect of irrigation on the growth, cropping and nutrition of Cox's Orange Pippin apple trees. *Journal of Horticultural Science* **46**, 195–208.
Graebe, J.E. and Ropers, H.J. (1978). In *Phytohormones and Related Compounds a Comprehensive Treatise*, Vol. I, ed. D.S. Letham, P.B. Goodwin and T.J.V. Higgins, pp. 107–204. Amsterdam: Elsevier/North Holland.
Greene, D.W. and Autio, W.R. (1994). Notching techniques increase branching of young apple trees. *Journal of the American Society for Horticultural Science* **119**, 678–82.
Grochowska, M.J., Buta, G.J., Steffens, G.L. and Faust, M. (1984). Endogenous auxin and gibberellin levels in low and high vigour apple seedlings. *Acta Horticulturae* **146**, 125–34.
Hamer, P.J.C. (1981). The effects of evaporative cooling on apple bud development and frost resistance. *Journal of Horticultural Science* **56**, 107–12.
Hancock, C.R. and Barlow, H.W.B. (1958). Seasonal fluctuations in stem growth in the apple rootstock Crab C. *Report of the East Malling Research Station for 1957*, 91–5.
Hansen, P. (1967). ^{14}C-Studies on apple trees. II. Distribution of photosynthates from top and base leaves from extension shoots. *Physiologia Plantarum* **20**, 720–5.
Hansen, P. (1969). ^{14}C-Studies on apple trees. IV. Photosynthate consumption in fruits in relation to leaf–fruit ratio and to the leaf–fruit position. *Physiologia Plantarum* **22**, 186–98.

Hansen, P. (1971). ^{14}C-Studies in apple trees. VII. The early seasonal growth in leaves, flowers and shoots as dependent upon current photosynthates and existing reserves. *Physiologia Plantarum* **25**, 469–73.

Hansen, P. and Grauslund, J. (1973). ^{14}C-Studies on apple trees. VIII. The seasonal variation and nature of reserves. *Physiologia Plantarum* **28**, 24–32.

Hauagge, R. and Cummins, J.N. (1991 a). Phenotypic variation in length of bud dormancy in apple cultivars and related *Malus* species. *Journal of the American Society for Horticultural Science* **116**, 100–6.

Hauagge, R. and Cummins, J.N. (1991 b). Seasonal variation in intensity of bud dormancy in apple cultivars and related *Malus* species. *Journal of the American Society for Horticultural Science* **116**, 107–15.

Hauagge, R. and Cummins, J.N. (1991 c). Genetics of length of dormancy period in *Malus* vegetative buds. *Journal of the American Society for Horticultural Science* **116**, 121–6.

Hauagge, R. and Cummins, J.N. (2000). Pome fruit genetic pool for production in warm climates. In *Temperate Fruit Crops in Warm Climates*, ed. A. Erez, pp. 267–303. Dordrecht: Kluwer.

Hedden, P. and Graebe, J.E. (1985). Inhibition of gibberellin biosynthesis by paclobutrazol in cell-free homogenates of *Cucurbita maxima* endosperm and *Malus pumila* embryos. *Journal of Plant Growth Regulation* **4**, 111–22.

Herter, F.G., Finardi, N.L. and Mauget, J.C. (1988). Dormancy development in apple trees cvs Gala, Golden and Fuji in Pelotas RS. *Acta Horticulturae* **232**, 109–15.

Hill-Cottingham, D.G. (1963). Effect of the time of application of fertilizer nitrogen on the growth, flowering and fruiting of maiden apple trees grown in sand culture. *Journal of Horticultural Science* **38**, 242–51.

Hillman, J.R. (1984). Apical dominance. In *Advanced Plant Physiology*, ed. M.B. Wilkins, pp. 127–48. London: Pitman.

Holefors, A., Xue, Z-T., and Welander, M. (1998). Transformation of the apple rootstock M26 with the *rol* A gene and its influence on growth. *Plant Science* **136**, 69–78.

Howard, B.H., Preston, A.P. and Quinlan, J.D. (1975). Large, uniform and well-feathered fruit trees should be aimed for. *The Grower*, July 19, 118–20.

Hrotkó, K., Magyar, L., Yao, C. and Rónay, Z. (1997). Effect of repeated B.A. (benzyladenine) application on feathering of 'Idared' apple nursery trees. *Horticultural Science – Kertészeti Tudomány* **29**, 46–51.

Iancu, M. (1985). Growth rate of apple trunk and fruit – additional indicators for water needs of fruit trees. *Acta Horticulturae* **171**, 417–25.

Jackson, J.E. (1990). Cropping and multiple cropping of apples in medium altitude tropical Zimbabwe. In *Abstracts of the 23rd International Horticultural Congress, International Society for Horticultural Science, Florence, 1990*, Abst. #4325.

Jackson, J.E. and Bepete, M. (1995). The effect of hydrogen cyanamide (Dormex) on flowering and cropping of different apple cultivars under tropical conditions of sub-optimal winter chilling. *Scientia Horticulturae* **60**, 293–304.

Jackson, J.E. and Palmer, J.W. (1977 a). Effects of shade on the growth and cropping of apple trees. I. Experimental details and effects on vegetative growth. *Journal of Horticultural Science* **52**, 245–52.

Jackson, J.E., Parry, M.S. and Stephens, C.P. (1981). Intensification of tree fruit production: current constraints, relevant research and an alternative system strategy for the 1980s. *Acta Horticulturae* **114**, 399–406.

Janick, J. (1974). The apple in Java. *HortScience* **9**, 13–15.

Jaumien, F., Czarnecki, B., Poniedzialek, W. and Mitrut, T. (1993). Very similar effects of a mixture of GA_3 and BA (6-benzylaminopurine) and of GA_{4+7} and

BA on branching of some apple cultivars in nursery. *Acta Horticulturae* **329**, 35–42.

Johnson, R.S. and Lakso, A.N. (1985). Relationships between stem length, leaf area, stem weight and accumulated growing degree-days in apple shoots. *Journal of the American Society for Horticultural Science* **110**, 586–90.

Jones, O.P. (1967). Effect of benzyladenine on isolated apple shoots. *Nature* **215**, 1514–15.

Jones, R.L. (1973). Gibberellins: their physiological role. *Annual Review of Plant Physiology* **24**, 571–98.

Kandiah, S. (1979). Turnover of carbohydrates in relation to growth in apple trees. II. Distribution of ^{14}C assimilates labelled in autumn, spring and summer. *Annals of Botany* **44**, 185–95.

Kato, T. and Ito, H. (1962). Physiological factors associated with the shoot growth of apple trees. *Tohuku Journal of Agricultural Research* **13**, 1–21.

Knight, R.C. (1961). Taper and secondary thickening in stems and roots of apple. *Report of the East Malling Research Station for 1960*, 65–71.

Krisanapook, K. and Subhadrabandhu, S. (1995). Effect of some practices and hydrogen cyanamide on budbreak of 'Shinseiki' pear. *Acta Horticulturae* **395**, 149–51.

Lang, G.A. (1987). Dormancy, a new universal terminology. *HortScience* **22**, 817–20.

Lapins, K.O. (1969). Segregation of compact growth types in certain apple seedling progenies. *Canadian Journal of Plant Science* **49**, 765–8.

Lapins, K.O. and Watkins, R. (1973). Genetics of compact growth habit. *Report of the East Malling Research Station for 1972*, 136.

Latimer, J.G. and Robitaille, H.A. (1981). Sources of variability in apple shoot selection and handling for bud rest determinations. *Journal of the American Society for Horticultural Science* **106**, 794–8.

Lawes, G.S., Spence, C.B., Tustin, D.S. and Max, S.M. (1997). Tree quality and canopy management effects on the growth and floral precocity of young 'Doyenne du Comice' pear trees. *New Zealand Journal of Crop and Horticultural Science* **25**, 177–84.

Lespinasse, J.M. and Delort, J.F. (1986). Apple tree management in vertical axis: appraisal after ten years of experiments. *Acta Horticulturae* **160**, 139–55.

Looney, N.E. and Lane, W.D. (1984). Spur-type growth mutants of 'McIntosh' apple: a review of their genetics, physiology and field performance. *Acta Horticulturae* **146**, 31–46.

Luckwill, L.C. (1974). A new look at the process of fruit bud formation in apple. *Proceedings of 19th International Horticultural Congress*, Warsaw, **3**, 237–45.

Maggs, D.H. (1961). Changes in the amount and distribution of increment induced by contrasting watering, nitrogen and environmental regimes. *Annals of Botany*, N.S. **25**, 353–61.

Maggs, D.H. (1963). The reduction in growth of apple trees brought about by fruiting. *Journal of Horticultural Science* **38**, 119–28.

Maggs, D.H. (1965). Dormant and summer pruning compared by pruning young apple trees once on a succession of dates. *Journal of Horticultural Science* **40**, 249–65.

Marini, R.P. and Barden, J.A. (1987). Summer pruning of apple and peach trees. *Horticultural Reviews* **9**, 351–75.

Mason, A.C. and Whitfield, A.B. (1960). Seasonal changes in the uptake and distribution of mineral elements in apple trees. *Journal of Horticultural Science* **35**, 34–55.

Mauget, J.C. and Rageau, R. (1988). Bud dormancy and adaptation of apple tree to mild winter climates. *Acta Horticulturae* **232**, 101–8.

Métraux, J.-P. (1987). Gibberellins and plant cell elongation. In *Plant Hormones and their Role in Plant Growth and Development*, ed. P.J. Davies, pp. 296–317. Dordrecht: Martinus Nijhoff.

Mika, A. (1986). Physiological response of fruit trees to pruning. *Horticultural Reviews* **8**, 339–78.

Mika, A., Grochowska, M.J., Karaszewska, A. and Williams, M.W. (1983). Effects of dormant and summer pruning, disbudding and growth retardants on growth, flower bud formation and fruiting of young apple trees. *Journal of the American Society for Horticultural Science* **108**, 655–60.

Miller, E.P. and Baker, L.H. (1982). An evaluation of apple cultivars for central and north Florida. *Proceedings of the Florida State Horticultural Society* **95**, 88–90.

Morgan, J. and Richards, A. (1993). *The Book of Apples.* London: Ebury Press.

North, M.S. (1995). New rest-breaking agents for the control of delayed foliation of apples. *Acta Horticulturae* **409**, 151–4.

Notodimedjo, S., Danoesastro, H., Sastrosumarto, S. and Edwards, G.R. (1981). Shoot growth, flower initiation and dormancy of apple in the tropics. *Acta Horticulturae* **120**, 179–86.

Oliveira, C.M. and Priestley, C.A. (1988). Carbohydrate reserves in deciduous fruit trees. *Horticultural Reviews* **10**, 403–30.

Ouellette, D.R., Unrath, C.R. and Young, E. (1996). Manual and chemical branch inducement in fall- and spring-planted 'Empire' apple on two rootstocks. *HortScience* **31**, 82–8.

Paiva, E. and Robitaille, H.A. (1978). Breaking bud rest on detached apple shoots: effects of wounding and ethylene. *Journal of the American Society for Horticultural Science* **103**, 101–4.

Parker, M.L. and Young, E. (1995). Leader management and soil fumigation affect branching and precocity of young apple trees. *HortScience* **30**, 1193–5.

Petri, J.L. and Stuker, H. (1988). Effects of temperature and length of cold storage of nursery plants on the subsequent development of young apple trees. *Acta Horticulturae* **232**, 117–26.

Petropoulou, S.P. (1985). Temperature related factors as selection criteria in apple breeding. PhD thesis, University of London.

Phillips, I.D.J. (1975). Apical dominance. *Annual Review of Plant Physiology* **26**, 341–67.

Powell, L.E. (1987). The hormonal control of bud and seed dormancy in woody plants. In *Plant Hormones and Their Role in Plant Growth and Development*, ed. P.J. Davies, pp. 539–52. Dordrecht: Martinus Nijhoff.

Pratt, C. (1990). Apple trees: morphology and anatomy. *Horticultural Reviews* **12**, 265–305.

Priestley, C.A. (1969). Some aspects of the physiology of apple rootstock varieties under reduced illumination. *Annals of Botany* **33**, 967–80.

Quinlan, J.D. (1965). The pattern of distribution of ^{14}Carbon in a potted apple rootstock following assimilation of ^{14}Carbon dioxide by a single leaf. *Report of the East Malling Research Station for 1964*, 117–18.

Quinlan, J.D. (1966). The effects of partial defoliation on the pattern of assimilate movement in an apple rootstock. *Report of the East Malling Research Station for 1965*, 128–9.

Quinlan, J.D. (1969). Mobilisation of ^{14}C in the spring following autumn assimilation of $^{14}CO_2$ by an apple rootstock. *Journal of Horticultural Science* **44**, 107–10.

Quinlan, J.D. (1981). Recent developments in the chemical control of tree growth. *Acta Horticulturae* **114**, 144–51.
Reinhold, L. (1978). Phytohormones and orientation of growth. In *Phytohormones and Related Compounds – a Comprehensive Treatise*, Vol. 2, ed. D.S. Letham, P.B. Goodwin, and T.J.V. Higgins, pp. 251–89. Amsterdam: Elsevier.
Richardson, P.J. and Quinlan, J.D. (1986). Uptake and translocation of paclobutrazol by shoots of M26 apple rootstock. *Plant Growth Regulation* **4**, 347–56.
Robitaille, H.A., and Carlson, R.F. (1971). Response of dwarfed apple trees to stem injections of gibberellic and abscisic acids. *HortScience* **6**, 539–40.
Rogers, W.S. and Booth, G.A. (1964). Relationship of crop and shoot growth in apple. *Journal of Horticultural Science* **39**, 61–5.
Rubery, P.R. (1987). Auxin transport. In *Plant Hormones and their Role in Plant Growth and Development*, ed. P.J. Davies, pp. 341–62. Dordrecht: Martinus Nijhoff.
Sachs, T. (1993). The role of auxin in plant organization. *Acta Horticulturae* **329**, 162–8.
Sachs, T. and Thimann, K.V. (1967). The role of auxins and cytokinins in the release of buds from dominance. *American Journal of Botany* **54**, 136–44.
Samish, R.M. (1954). Dormancy in woody plants. *Annual Review of Plant Physiology* **5**, 183–203.
Saure, M.C. (1985). Dormancy release in deciduous fruit trees. *Horticultural Reviews* **7**, 239–300.
Seeley, S.D. and Powell, L.E. (1981). Seasonal changes of free and hydrolyzable abscisic acid in vegetative apple buds. *Journal of the American Society for Horticultural Science* **106**, 405–9.
Shaltout, A.D. and Unrath, C.R. (1983). Rest completion prediction model for 'Starkrimson Delicious' apples. *Journal of the American Society for Horticultural Science* **108**, 957–61.
Sherman, W.B. and Crocker, T.E. (1982). Low chilling pears. In *The Pear*, ed. T. van der Zwet and N.F. Childers. Gainesville, FL: Horticultural Publications.
Skene, D.S. (1980). Estimating potential blossom on 'Cox's Orange Pippin' apple shoots by forcing isolated buds. *Journal of Horticultural Science* **55**, 145–8.
Soumelidou, K., Morris, D.A., Battey, N.H., Barnett, J.R. and John, P. (1994). Auxin transport capacity in relation to the dwarfing effects of apple rootstocks. *Journal of Horticultural Science* **69**, 719–25.
Spiegel-Roy, P. and Alston, F.H. (1979). Chilling and post-dormant heat requirement as selection criteria for late-flowering pears. *Journal of Horticultural Science* **54**, 115–20.
Sponsel, V.M. (1987). Gibberellin biosynthesis and metabolism. In *Plant Hormones and their Role in Plant Growth and Development*, ed. P.J. Davies, pp. 43–71. Dordrecht: Martinus Nijhoff.
Steffens, G.L. and Stutte, G.W. (1989). Thidiazuron substitution for chilling requirement in three apple cultivars. *Journal of Plant Growth Regulation* **8**, 301–7.
Steffens, G.L., Byun, J.K. and Wang, S.Y. (1985). Controlling plant growth via the gibberellin biosynthesis system – I. Growth parameter alterations in apple seedlings. *Physiologia Plantarum* **63**, 163–8.
Subhadrabandhu, S. (1995). Induction of budbreak in apple trees that received insufficient chilling by hydrogen cyanamide. *Acta Horticulturae* **409**, 171–8.
Swartz, H.J., Geyer, A.S., Powell, L.E. and Lin, H.C. (1984). The role of bud scales in the dormancy of apples. *Journal of the American Society for Horticultural Science* **109**, 745–9.

Tamas, I.A. (1995). Hormonal regulation of apical dominance. In *Plant Hormones and their Role in Plant Growth and Development*, 2nd edn, ed. P.J. Davies, pp. 572–97. Dordrecht: Kluwer.

Taylor, J.S., Pharis, R.P., Loveys, B., Notodimedjo, S. and Edwards, G.R. (1984). Changes in endogenous hormones in apple during bud burst induced by defoliation. *Plant Growth Regulation* **2**, 117–34.

Thimann, K.V. and Skoog, F. (1933). Studies on the growth hormone of plants. III. The inhibiting action of the growth substance on bud development. *Proceedings of the National Academy of Science* (USA) **19**, 714–16.

Thimann, K.V. and Skoog, F. (1934). On the inhibition of bud development and other functions of growth substance in *Vicia faba*. *Proceedings of the Royal Society B* **114**, 317–39.

Thompson, W.K., Jones, D.L. and Nichols, D.G. (1975). Effects of dormancy factors on the growth of vegetative buds in young apple trees. *Australian Journal of Agricultural Research* **26**, 989–96.

Titus, J.S. and Kang, S.M. (1982). Nitrogen metabolism, translocation and recycling in apple trees. *Horticultural Reviews* **4**, 204–46.

Tromp, J. (1992a). Lateral shoot formation in apple in the first year after budding as affected by air humidity and soil temperature. *Acta Horticulturae* **322**, 141–51.

Tromp, J. (1992b). The effect of soil temperature on lateral shoot formation and flower bud formation in apple in the first year after budding. *Journal of Horticultural Science* **67**, 787–93.

Tromp, J. and Ovaa, J.C. (1971). Spring mobilization of storage nitrogen in isolated shoot sections of apple. *Physiologia Plantarum* **25**, 16–22.

Tromp, J. and Ovaa, J.C. (1990). Seasonal changes in the cytokinin composition of xylem sap of apple. *Journal of Plant Physiology* **136**, 606–10.

Tukey, L.D. (1986). Cropping characteristics of bearing apple trees annually sprayed with paclobutrazol (PP 333). *Acta Horticulturae* **179**, 481–8.

van Oosten, H.J. (1978). Effect of initial tree quality on yield. *Acta Horticulturae* **65**, 123–7.

Vegis, A. (1963). Climatic control of germination, budbreak, and dormancy. In *Environmental Control of Plant Growth*, ed. L.T. Evans, p. 273. London: Academic Press.

Vegis, A. (1964). Dormancy in higher plants. *Annual Review of Plant Physiology* **15**, 185–224.

Vegis, A. (1973). Dependence of the growth processes on temperature. In *Temperature and Life*, ed. H. Precht, C. Christophersen, H. Hensel and W. Larcher, pp. 145–70. Berlin: Springer-Verlag.

Volz, R.K., Gibbs, H.M. and Popenoe, J. (1994). Branch induction on apple nursery trees: effects of growth regulators and defoliation. *New Zealand Journal of Crop and Horticultural Science* **22**, 277–83.

Walsh, C.S. and Miller, A.N. (1984). Observations on the growth and vigour of spur and non-spur-type apple trees. *Acta Horticulturae* **146**, 211–14.

Wang, S.Y. and Faust, M. (1990). Changes in membrane lipids in apple buds during dormancy and budbreak. *Journal of the American Society for Horticultural Science* **115**, 803–8.

Wang, S.Y., Faust, M. and Line, M.J. (1994). Apical dominance in apple (*Malus domestica* Borkh): the possible role of indole-3-acetic acid (IAA). *Journal of the American Society for Horticultural Science* **119**, 1215–21.

Wang, S.Y., Steffens, G.L. and Faust, M. (1986). Breaking bud dormancy in apple with a plant bioregulator, Thidiazuron. *Phytochemistry* **25**, 311–17.

Webster, A.D. (1995). Rootstock and interstock effects on deciduous fruit tree vigour, precocity, and yield productivity. *New Zealand Journal of Crop and Horticultural Science* **23**, 373–82.

Wertheim, S.J. (1978). Manual and chemical induction of side-shoot formation in apple trees in the nursery. *Scientia Horticulturae* **9**, 337–45.

Westwood, M.N. (1970). Rootstock–scion relationships in hardiness of deciduous fruit trees. *HortScience* **5**, 418–21.

Westwood, M.N. (1993). *Temperate Zone Pomology, Physiology and Culture.* Portland, OR: Timber Press.

Westwood, M.N. and Bjornstad, H.O. (1968). Chilling requirement of dormant seeds of 14 pear species as related to their climatic adaptation. *Proceedings of the American Society for Horticultural Science* **92**, 141–9.

Westwood, M.N. and Chestnut, N.E. (1964). Rest period chilling requirement of 'Bartlett' pear as related to *Pyrus calleryana* and *P. communis* rootstocks. *Proceedings of the American Society for Horticultural Science* **84**, 82–7.

Williams, M.W. and Billingsley, H.D. (1970). Increasing the number and crotch angles of primary branches of apple trees with cytokinins and gibberellic acid. *Journal of the American Society for Horticultural Science* **95**, 649–51.

Williams, M.W. and Stahly, E.A. (1968). Effect of cytokinins on apple shoot development from axillary buds. *HortScience* **3**, 68–9.

Williams, R.R., Edwards, G.R. and Coombe, B.G. (1979). Determination of the pattern of winter dormancy in lateral buds of apples. *Annals of Botany* **44**, 575–81.

Wilson, S.J. and Jarassamrit, N. (1994). Nursery factors influencing lateral shoot development in a spur type apple cultivar. *Scientia Horticulturae* **56**, 207–15.

Young, E. (1987). Effects of 6-BA, GA_{4+7} and IBA on growth resumption of chilled apple roots and shoots. *HortScience* **22**, 212–13.

Young, E. (1989). Cytokinins and soluble carbohydrate concentrations in xylem sap of apple during dormancy and budbreak. *Journal of the American Society for Horticultural Science* **114**, 297–300.

Young, E. (1990). Changes in respiration rate and energy of activation after chilling and forcing dormant apple trees. *Journal of the American Society for Horticultural Science* **115**, 809–14.

Young, E. and Werner, D.J. (1984). Effects of rootstock and scion chilling during rest on resumption of growth in apple and peach. *Journal of the American Society for Horticultural Science* **109**, 548–51.

Young, E. and Werner, D.J. (1985a). Chill unit and growing degree hour requirements for vegetative bud break in six apple rootstocks. *Journal of the American Society for Horticultural Science* **110**, 411–13.

Young, E. and Werner, D.J. (1985b). Effects of shoot, root and shank chilling during rest in apple and peach on growth resumption and carbohydrates. *Journal of the American Society for Horticultural Science* **110**, 769–74.

Young, E. and Werner, D.W. (1986). 6-BA applied after shoot and/or root chilling and its effect on growth resumption in apple and peach. *HortScience* **21**, 280–81.

Young, E., Dautlick, T.K. and Belding, R.D. (1995). Respiratory changes during dormancy breaking of apple trees. *Acta Horticulturae* **395**, 21–33.

Zhang, Y.X. and Lespinasse, Y. (1991). Removal of embryonic dormancy in apple (*Malus* × *domestica* Borkh) by 6-benzylaminopurine. *Scientia Horticulturae* **46**, 215–23.

7

Leaves, canopies and light interception

Leaf anatomy and morphology

Leaf primordia are initiated by periclinal divisions in layers 2 or 3 of the tunica on the flanks of apical meristems (Pratt, 1990). They develop into protuberances flattened on the adaxial side. The base of the leaf primordium is an intercalary meristem and forms the petiole. The leaf blade develops concomitantly from two layers of cells derived from marginal and sub-marginal initials which produce a marginal meristem on the lateral flanks of the midrib. The superficial cells of the marginal meristems divide anticlinally to form epidermal cells. The sub-epidermal cells on the adaxial (upper) side become the first row of palisade cells and those of the abaxial side form the spongy mesophyll. Cells between these two layers form the central mesophyll and the smaller vascular bundles. On the adaxial side sub-epidermal and central meristematic cells differentiate into one to three or more layers of palisade cells depending on leaf type and environmental conditions. Each mature palisade cell is surrounded by air space continuous with that in the spongy mesophyll, except in the vicinity of a vein.

At maturity (Figure 7.1) the adaxial cuticle consists of a layer of wax and cutin. The abaxial epidermal cells have thinner cuticles than the adaxial ones. They are variable in thickness and shape except for the paired, kidney-shaped guard cells which are nearly constant in size within a cultivar and surround the pores or stomates through which gas exchange with the external air takes place. Initially the stomata are immature and non-functional with no evidence of a pore, but even under English conditions leaves on the flower clusters can have a high proportion of functional stomata in April and May, and by June all stomata are functional (Slack, 1974). Increase in turgor causes the thin guard cell walls adjacent to the epidermal cells to stretch and the elliptical pore opens. Stomata in apple and pear are found only on the lower, abaxial, leaf surface and, contrary to earlier studies, Slack (1974) found no systematic distribution pattern over this surface and a mean

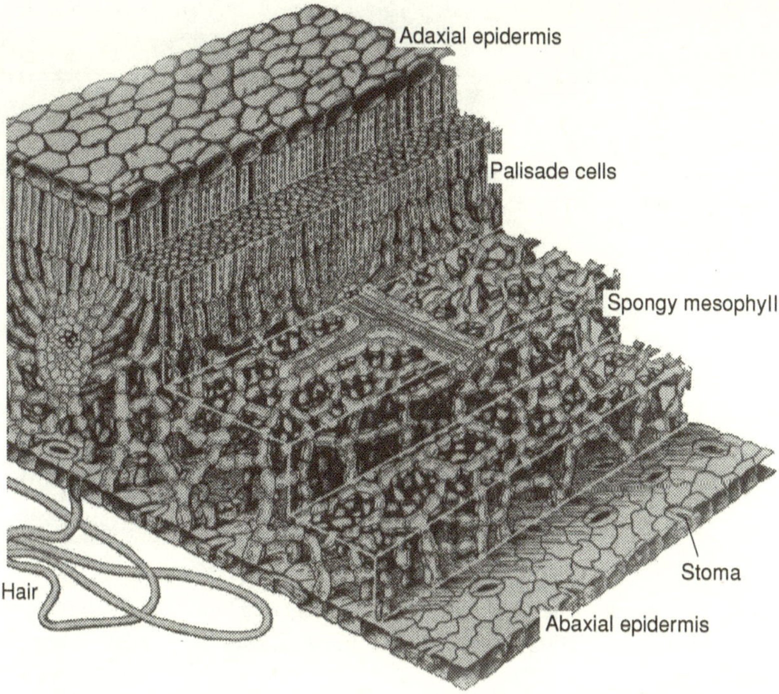

Figure 7.1 Apple leaf in 3-dimensional diagram showing epidermis, palisade cells, spongy mesophyll, minor veins, stomata and a hair. From MacDaniels and Cowart (1944). Originally drawn by Rita B. Eames.

stomatal pore length of 19–21 μm with the distance between the pore centres of 50–57 μm depending on cultivar.

The apple epidermis has unicellular hairs, originating from single epidermal cells. They are abundant on both surfaces of young leaves but occur primarily on the abaxial surface of lower leaves. These may have an adaptive function in reducing water loss by maintaining humid air near the leaf surface. The epidermis also has simple hydathodes associated with vascular tissue on the adaxial surface, at the margins and tips of leaves and at the end of minor veinlets. There are also glandular hairs with stalks and multicellular heads which contain secretory cells. The secretions may favour the growth of parasitic organisms.

This basic leaf structure varies with a number of factors. At least some spur-type cultivars have thicker leaves and a greater number of palisade layers than the corresponding standard cultivars (Liu and Eaton, 1970). The number of palisade mesophyll cells per unit area of leaf lamina, which depends on cell diameter, the number of palisade layers and the amount of intercellular space, is greater in vigorous than in dwarf apple rootstock cultivars (Beakbane, 1967),

Table 7.1 *Effects of shading, from 2 June 1970, on the structure of leaves from the middle regions of extension shoots on 22 Sept 1970*

	% full daylight					
	100	37	25	11	SED	
Leaf dry wt (mg cm^{-2})	11.65	8.63***	7.87***	6.33***	0.375	15 df
Leaf thickness (μm)	266	212***	201 ***	182***	9.46	
Depth of palisade layer (μm)	145	105***	98***	86***	4.08	
Number of cells in palisade						
Layer 1	30.2	28.3*	29.0	28.5*	0.72	12 df
Layer 2	26.7	24.2*	23.9	22.2*	1.06	
Layer 3	18.8	11.6***	10.1 ***	7.9***	0.96	
Layer 4	3.7	0.7***	0.2***	0.0***	0.41	

*, **, *** denote differences from the unshaded control at the 5%, 1% and 0.1% levels of significance.
From Jackson and Palmer (1977). Reproduced with permission.

but the rootstock influence on scion leaf structure is inconsistent. Hard-pruned young trees of dwarfing rootstocks have much thicker leaves than unpruned older trees.

Spur leaves are thinner than extension shoot leaves, with fewer layers of palisade parenchyma, shorter palisade parenchyma cells, less chlorophyll and lower specific leaf weight (Ghosh, 1973). Basal leaves on shoots tend to have thinner layers of palisade cells than apical leaves (Cowart, 1935). These two effects of leaf position may be associated with differences in exposure to light, the effects of which are discussed below.

Cultivars also differ in the number of stomata per unit leaf area, ranging from around 200 to around 450 mm^{-2} (Cowart, 1935; Slack, 1974). The leaves of vigorous apple rootstock cultivars have a much greater stomatal density than those of dwarfing ones (Beakbane and Majumder, 1975).

Basal leaves have many fewer stomata per unit area than apical leaves, the increasing frequency being associated with increasing structural density (Cowart, 1935).

Perhaps the most important variations in leaf structure are those attributable to variations in light intensity within the tree canopy. Leaf thickness and specific leaf weight (mg cm^{-2}) are positively correlated with incident radiation (Barden, 1974, 1978; Barritt *et al.*, 1987; Jackson and Beakbane, 1970; Tustin *et al.*, 1992). The extreme cases are obvious 'sun' and 'shade' leaves. Jackson and Beakbane (1970) showed that the percentage of air space in the spongy mesophyll decreased and the thickness of the palisade layer increased with light intensity. Artificial shading experiments confirmed these effects (Table 7.1) and also reductions in stomatal density (number cm^{-2}) and chlorophyll ($\mu g\ cm^{-2}$) with increasing shade (Asada and Ogasawara, 1998).

Barden (1974) showed that the specific leaf weight was increased by exposure to full greenhouse sun in comparison to artificial shade both during and after the period of leaf expansion. Shading of spurs may result in the spur and bourse shoot leaves (see Figure 6.1, p. 158) emerging from them in the following year having lower specific leaf weights (Tustin *et al.*, 1992), although Jackson and Palmer (1977) found that shading whole trees did not result in lower specific leaf weight in the following year, possibly because the previously shaded trees had less vigorous shoot growth and hence better current-season light exposure.

Shade also influences leaf chloroplast structure. The thickness of grana within chloroplasts of mature apple leaves increases with depth in the palisade layer and with external shade. Transferring trees from light to shade and *vice versa* indicated that grana thickness increased with increasing shade but did not decrease with increased light intensity (Skene, 1974). In general chloroplasts with thick grana have low photosynthetic electron transport capacity, and light saturation of photosynthesis occurs at relatively low light intensities. They also, however, have low respiration rates and so enable efficient use of low levels of light.

The mechanism by which 'sun' or 'shade' leaf characteristics are induced by relative light exposure during leaf expansion is not clear. One hypothesis is that the leaves which transpire most receive a greater proportion of the nutrients and hormones that are transported upwards in the transpiration stream in the xylem (Flore and Lakso, 1989). Evidence to support this comes from the facts that leaf nutrient concentration per unit area is generally much higher the greater the light intensity (Table 7.2), that many of the characteristics of 'sun' leaves can be induced by cytokinin treatment of low-light leaves, and that shading mature leaves on apple shoots resulted in the developing leaves attaining more pronounced 'sun-type' characteristics (Flore and Lakso, 1989). Beakbane (1965) found that treatment with gibberellic acid resulted in changes in the structure of the palisade mesophyll and the spongy mesophyll of apple leaf disks to give a 'shade-type' leaf structure.

Leaf production and growth

The first leaves to emerge are the primary leaves on spurs (Forshey *et al.*, 1987; Lakso, 1984). Hansen (1971) reported usually 5 or 6 leaves per bud in Denmark, Barritt *et al.* (1991) an average of 8.6 primary leaves per fruiting spur of 'Oregon Spur Delicious' in Washington State, USA, by late April. Most

Table 7.2 *Effects of artificial shading throughout the growing seasons in 1970 and 1971 on the dry matter and nutrient element content (mg cm^{-2}) in leaves from the mid position of extension shoots in mid-August 1971*

	% daylight in 1970 and 1971									
1970	100	100	100	100	37	25	11	37	25	11
1971	100	37	25	11	37	25	11	100	100	100
Dry matter	10.8	7.8	6.6	5.0	8.0	6.7	5.3	11.6	11.6	13.1
N	0.293	0.211	0.177	0.135	0.210	0.178	0.131	0.285	0.252	0.255
P	0.025	0.018	0.015	0.012	0.017	0.015	0.012	0.026	0.027	0.027
K	0.162	0.149	0.135	0.132	0.154	0.171	0.135	0.209	0.206	0.237
Ca	0.105	0.071	0.066	0.043	0.071	0.054	0.041	0.096	0.102	0.072
Mg	0.022	0.016	0.016	0.012	0.016	0.014	0.012	0.022	0.019	0.018

Calculated from data presented by Jackson and Palmer (1977).

Table 7.3 *Leaf area on short shoots as a percentage of the total leaf area of mature apple trees of 'Cox's Orange Pippin'/'M.9'*

Pruning	Cropping trees	Non-cropping trees
Light	86.4	69.1
Heavy	55.4	35.2

From Barlow (1969). Reproduced with permission.

spur leaves emerge before full bloom. After flowering some secondary leaves may develop on the spurs and their development is usually complete about a month after full bloom.

The other leaves develop on the new shoots, i.e. on terminal, lateral and bourse shoots. Proctor and Palmer (1991) found that at full bloom true spur leaves made up 88% of the leaf area of the spurs of 'Cox' but only 56% of that of 'Golden Delicious', the balance being bourse shoot leaves. At maturity the ratio of spur leaves to new shoot leaves is very variable. Barlow (1969) found that the proportion of the leaf area borne on short shoots (spurs) was reduced by heavy pruning and the absence of crop (Table 7.3), i.e. by treatments which promote extension shoot growth. The treatments in this trial were extreme, giving something near to the maximum range of spur to shoot leaf ratios for mature trees. Palmer and Jackson (1977) found 63–80% of the leaf area of 'Golden Delicious'/'M.9' pruned as slender spindle bushes to be spur leaves. Wünsche *et al.* (1996) found the percentage of spur leaf area to range from 48% to 62% of the total in a trial of different rootstocks and training systems with 'Empire'/'M.9'. The relative pattern of spur and shoot leaf development over the season was well illustrated by Forshey *et al.* (1987) working with 'MacSpur McIntosh' (Figure 7.2), who found that at the time of full canopy development spur leaves made up more than half of the total leaf area, after having made up a much larger proportion early in the season.

The number of leaves which develop on new shoots is a function of temperature (Abbott, 1984; Johnson and Lakso, 1985; Lindhagen, 1996): effects of temperature under controlled environment conditions are shown in Table 7.4. Leaf production is also influenced by many of the factors discussed in relation to shoot growth. The number of leaves produced is reduced by water stress and nutrient deficiencies and is under hormonal control. Tustin *et al.* (1992) showed that shading of spurs reduced the number of leaves produced on their bourse shoots in the subsequent year as a result of the earlier cessation of growth of these bourse shoots.

As discussed under shoot growth, the removal of expanding leaves can result in a check to shoot growth followed by its prolongation and production of more leaves than might otherwise have occurred (Abbott, 1984).

Table 7.4 *Number of leaves produced during a six week period at different temperatures: 'Cox's Orange Pippin' apple trees*

	Temperature of C.E. rooms		
	25.0 °C	17.5 °C	10.0 °C
Leaves per shoot	23	13	3

From Abbott (1984). Reproduced with permission.

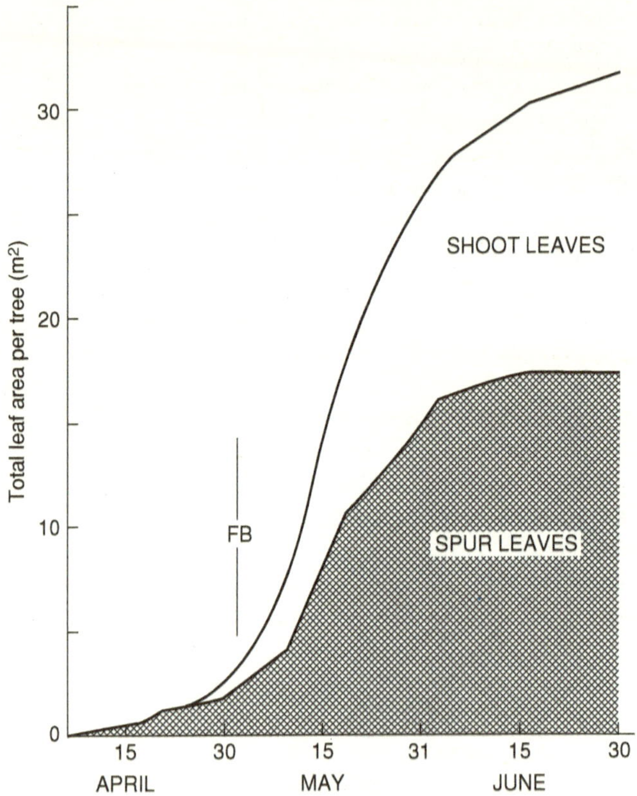

Figure 7.2 Seasonal development of the leaf canopy of a 7-year-old 'MacSpur McIntosh' apple tree. FB, Full Bloom. From Forshey *et al.* (1987). Reproduced with permission.

Hancock and Barlow (1960) showed that individual apple leaves had a typically sigmoid growth curve and, in England, took about 25 days to grow from 10 cm in length to a maximum of 78 cm (Figure 7.3). Apple leaves are simple in shape and their areas have a consistent relationship to that of the rectangle

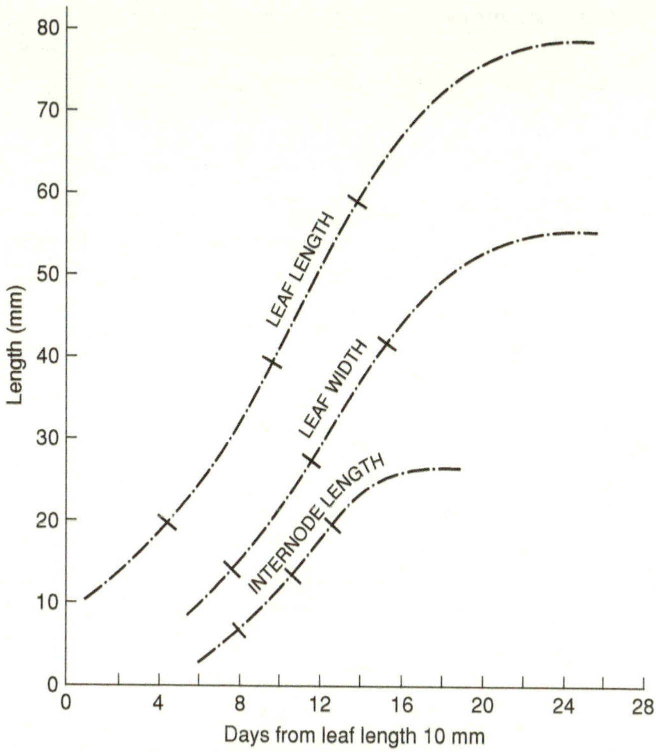

Figure 7.3 Curves representing the 'normal growth' of 'Crab C' leaves and internodes during mid-summer 1957 on a common scale of days from the time when the leaf reaches a length of 10 mm. Bars across the curves indicate the growth stages of 25%, 50% and 75% of final size. From Hancock and Barlow (1960). Reproduced with permission.

which would contain them. Their leaf areas can therefore be calculated as length × breadth × a factor which, in different studies, has been found to range from 0.68 to 0.74 (cf. references reviewed by Jackson, 1980). There is therefore a considerable increase in total leaf area well after the cessation of shoot growth and the production of new leaves. This was also illustrated by Abbott (1984). Hancock and Barlow (1960) found that the rate of increase in leaf length during the most rapid period of leaf expansion was positively linked to temperature, but final length was not. Barritt *et al.* (1987) and others found little or no effect of position in the canopy on individual leaf area but Jackson and Palmer (1977) found increased area per shoot leaf with increasing imposed shade. These apparently conflicting results may reflect the counteracting effects of reduction in the potential for leaf growth and increase in the leaf area per unit weight. Leaf expansion is checked by water stress and by nutrient deficiencies, an extreme case of the latter being the 'little-leaf' disease induced by zinc deficiency. Spur leaves are smaller than extension shoot leaves. On extension shoots the leaves are larger the more vigorous the shoot (Barlow, 1980).

Sources of carbon and nitrogen for leaf growth

Early in spring carbohydrates move upwards in the tree. Both Hansen (1967b) and Quinlan (1969) found the new leaves to contain ^{14}C that had been supplied as $^{14}CO_2$ to the trees in the previous autumn. Hansen (1971) found that a third to a half of all building materials used in spurs up to the time of flower colour development, and in shoots during the development of the first 5 or 6 leaves, originated from reserves. These reserves also provide the energy for early leaf growth. Young, growing leaves use the majority of the photosynthates that they produce. Hansen (1967a) found that 30–35% of the ^{14}C supplied to young leaves was incorporated into the structural material of the leaf (methanol-insoluble substances) whereas fully developed leaves incorporated only 4–9%. Quinlan (1965) found that growing leaves consumed most of the $^{14}CO_2$ supplied to them but usually exported some to younger leaves and the shoot tip. They also received some ^{14}C when $^{14}CO_2$ was supplied to mature leaves in the apical part of the same shoot. Jankiewicz *et al.* (1967) also found that young leaves retained most of their assimilates and received assimilated ^{14}C from mature leaves on the same shoot; moreover, the youngest leaves on the leading shoot receive ^{14}C when $^{14}CO_2$ is supplied to lateral shoots.

At the beginning of leaf development the supplies of nitrogen, phosphorus and potassium used in the formation of new tissue are derived mainly from tree reserves, principally the bark of branches and stem (Mason and Whitfield, 1960). Subsequently N is translocated upwards from the roots in the form of amino acids in the xylem (Titus and Kang, 1982), together with other nutrient elements.

Leaf senescence and shed

Premature leaf shed can be induced by pest and disease attack and nutrient deficiencies. Leaf retention is favoured by the presence of fruits.

The general process of leaf senescence and abscission in autumn is primarily controlled by temperature. Jonkers (1980) showed that apple and pear trees grown under controlled temperature conditions as well as outdoors had accelerated leaf drop when kept at 9 °C or 13 °C but greatly retarded drop when maintained at 17 °C, 21 °C or 25 °C. He also found that at 13 °C all apple leaves had yellowed by end-November whereas in trees kept at 17 °C, 21 °C or 25 °C this did not occur until late January. Lakso and Lenz (1986) showed that

Table 7.5 *Effects of rootstock on tree size and leaf area per tree (m^2) and per unit of orchard land (LAI, m^2 leaf m^{-2} land), cv. 'Red Delicious' apple, Penticton, BC, Canada*

Rootstock	Tree height (m)	Number of trees ha^{-1}	Leaf area (m^2 $tree^{-1}$)	LAI
Seedling	6.1	119	317	3.78
'M.4'	4.9	239	160	3.82
'M.7'	3.7	331	129	4.27
'M.9'	2.4	1075	43	4.58

Calculated from data given by Heinicke (1964).

autumnal senescence could be essentially stopped by transferring apple trees to an 18 °C day, 10 °C night temperature regime. This effect of temperature provides an explanation for the failure of apple leaves to shed synchronously in the tropics and the need for chemical or manual defoliation to release fruit buds from correlative inhibition by leaves. In temperate climates, light frost hastens the low-temperature induced abscission process (Larsen, 1973).

Neither light intensity nor daylength appears to have any effect on the abscission process (Jonkers, 1980).

Cultivars differ in their time of leaf shed within the overall constraints imposed by temperature. This characteristic is highly heritable, with some evidence for major gene effects (Tydeman, 1963).

The rootstock can have a marked effect on the time of onset of leaf senescence and this can be delayed by nitrogen fertilizer application (Heinicke, 1934). Senescing leaves export carbohydrates and also N, P, and K, but not Ca, to the rest of the plant prior to shed (Mason and Whitfield, 1960).

Individual tree and orchard leaf area

The canopy size of the individual tree reflects the accumulated effects of its genetic make-up, in terms of both scion and rootstock, of tree management practices and of the external environment.

The rootstock has a very large effect on leaf area. Heinicke (1964) found the leaf area per tree of 'Red Delicious' apple on seedling rootstock to be 317 m^2 and on the dwarfing 'M.9' rootstock to be only 43 m^2 when the trees were planted at the then typical spacings (Table 7.5). The trees described in Table 7.5 had much denser leaf canopies than those typical of more modern orchards.

Giuliani *et al.* (1997) cited a leaf area of 14.5 m^2 per tree (LAI, 1.61) for heavily cropping 'Smoothee Golden Delicious'/'Pajam 2' planted at 1111 trees ha^{-1}, and Palmer and Jackson (1977) a leaf area of 3.9 m^2 per tree (LAI, *c*. 1.5) for 'Golden Delicious'/'M.9' planted at 3831 trees ha^{-1}.

Cropping and pruning also influence tree leaf area. Barlow and Smith (1971) found that consistently deblossomed trees had up to 6 times the leaf weight of cropping trees. Trees that had consistently been lightly pruned had up to twice the leaf weight of heavily pruned trees. Lightly pruned or unpruned trees also show a much more rapid development of their leaf area early in the season than heavily pruned trees because a higher proportion of their leaves are borne on spurs (Lakso, 1984).

Temperature has major effects that result in differences in tree leaf area at different latitudes. Annual mean daily air temperature is a linear function of latitude. Leafing out and blossoming is earlier at low latitudes, leading to an increase in leaf area duration, i.e. leaf area integrated over the season (Heim *et al.*, 1979). Apple flower development starts 2.5 days earlier per degree lower latitude between 43° N and 65° N and apple trees on 'M.9' have 25% and 40% more leaf area at 51° than at 55° and at 43° than at 51°, respectively (Wagenmakers, 1991a). Areas with a late onset of winter, especially if late-fruiting cultivars are used, may have an increase in leaf area duration because of delayed senescence.

Orchard leaf area, or leaf area index in m^2 leaf m^{-2} ground (LAI) depends, as shown above, on individual tree leaf area and plant population. These are not independent. Wagenmakers (1989) found that leaf area per tree of 'Comice'/'Quince C' pears decreased linearly with planting density over the range of 2000–5000 trees ha^{-1} although total orchard LAI still increased with planting density. Verheij (1972) showed that with apple the leaf area per tree declined with increasing planting density even with unpruned trees, and was accompanied by a relative suppression of lateral growth in the lower parts of the trees.

It is obvious from the magnitude of the differences in leaf area per tree and per orchard, and from the trend towards much higher densities of planting but with lower leaf area indices, that tree and orchard leaf canopy development are best regarded as variables under management control. The relevant choices of rootstock, plant population and tree management systems are largely in order to optimize light interception and distribution.

Although in detailed studies tree leaf area may be assessed by different sampling and measuring procedures an approximate technique is often used. In general a given leaf area is associated with a certain cross-sectional area of subtending stem, both with respect to leaf area per branch and to leaf area per tree (Holland, 1968; Barlow, 1969; Verheij, 1972; Jackson, 1980). The basic linear regression equation is

$$\log A = \log K + b \log G \tag{7.1}$$

where A is leaf area, K is a constant and G is branch or trunk girth. This corresponds to the relationship

$$A = KG^{b} \tag{7.2}$$

The regression coefficient b, which determines the slope of the relationship, was found by Holland (1968) to range from 2.1 to 2.3. Other studies gave similar results.

Equations 7.1 or 7.2 can be used to estimate tree or branch leaf area from girth once K and b have been determined. Equation 7.2 also provides the rationale for using girth^2 (which is widely used as a measure of tree weight because of its close correlation with this, cf. Pearce (1952), Moore (1978) and Chapter 2, p. 56) as an estimate of tree size in relation to potential assimilation. The limitations to the use of equation 7.1 to estimate leaf area are that K, and to a much lesser extent b, vary with factors such as pruning, cropping level, time of season and tree age, and so need to be established from samples of the population being studied. For example, leaf area per tree will level off at maturity, when the trees are pruned to contain them to the chosen size even though trunk girth and cross-sectional area (TCA) continue increasing. The limitations to the use of girth^2 or TCA as a basis for estimating light interception and potential assimilation are discussed on p. 232.

Effects of light interception and of within-tree shade

Total dry matter production, hence the upper limit to potential yield, is usually directly proportional to light interception by fruit tree orchards as well as other crops (Monteith, 1977; Jackson, 1980; Palmer, 1989a; cf. Figure 7.4). Light that is not intercepted by the crop canopy is not used by it!

The slope of the regression for dry matter production on intercepted photosynthetically active radiation (PAR) gives 1.95 G MJ^{-1} for the apple orchards studied by Palmer (1989a). The efficiency with which intercepted radiation is used in dry matter production is affected by a number of factors, discussed in Chapter 8 on photosynthesis and respiration. Factors controlling the proportion of assimilated carbon which is harvested as fruit are also discussed in Chapters 2, 8 and 9. Some of these factors are very important determinants of orchard fruit yield, especially the differences between cultivars in partitioning between fruiting and vegetative growth and the differences between rootstocks in their effects on this partitioning (cf. Table 2.10; Figure 2.3). However, with productive cultivars on efficient rootstocks the yields of well-managed orchards are largely dependent on their light interception.

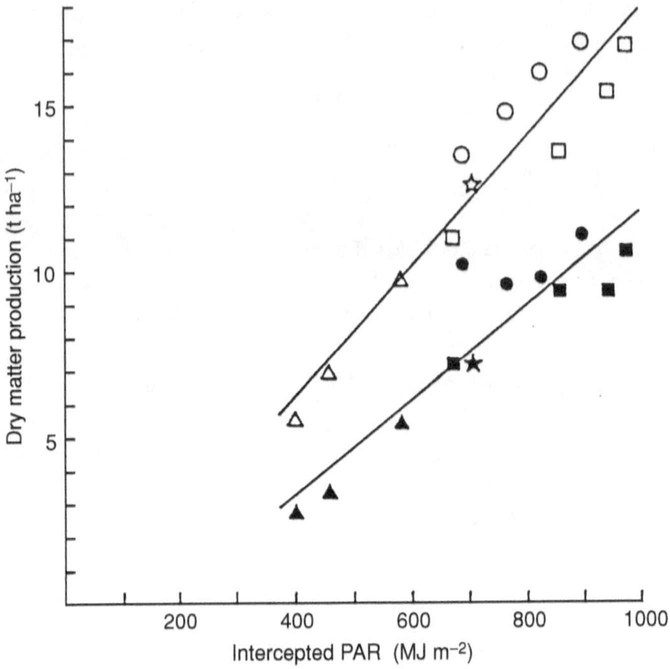

Figure 7.4 The effect of seasonal intercepted PAR on total (open symbols) and fruit (filled symbols) dry matter production from a range of intensive apple orchards. 'Golden Delicious'/'M.9': spindlebushes: △, 1974; ☆, 1975; 'Crispin'/'M.27' bed systems: O, 1982; □, 1983. From Palmer (1989a), from *Manipulation of Fruiting* by C. J. Wright. Reprinted by permission of Elsevier Science Ltd.

Robinson and Lakso (1991) found that 86% of the differences (variance) in yield, in a trial involving two scion cultivars on four rootstocks and with three pruning systems, was accounted for by a positive linear regression of yield on intercepted PAR. Similar results have been found in numerous studies, especially on planting density in modern orchard systems (Jackson, 1978; Robinson, 1997).

The speed with which orchard canopies obtain high levels of light interception is a dominant factor in determining the precocity of orchard yield, which in turn is a key determinant of profitability (Jackson *et al.*, 1981; Jackson, 1985). Where, however, orchards of large trees had formed 'fully closed stands' orchard yields were increased by thinning out the trees (Verheij, 1968). This presumably resulted from removing the adverse effects of excess shade on fruit bud formation and fruit set (cf. Chapter 9), and therefore on dry matter partitioning.

Shade in the immediate environment of apple fruits has pronounced effects on their size, red colour development and eating quality. There are critical levels of within-canopy light intensity below which it is not possible to produce high-value fruits (Jackson, 1980; see Figure 2.1; also Chapters 9 and 10). Since production costs generally rise with increase of tree size (Jackson, 1985; Wagenmakers, 1991a), it is not desirable to have large volumes of excessively shaded, unproductive tree canopy.

There are, consequently, major questions of optimizing orchard canopy design so as to attain high light interception, including in the early years, while maintaining adequate radiation levels in the fruiting zone. Unlike many annual crops there is much scope for canopy manipulation by tree training, pruning, control by growth regulators and vigour-controlling rootstocks as well as planting density.

Understanding the light relationships of three-dimensional tree canopies is essential to their effective manipulation.

Shading by stems, fruits and leaves

Some shading is caused by stems and fruits. Wagenmakers (1991b) found that the projected branch cast-shadow area for 'Comice' pear was about 5% of the leaf area. The relative effect of branches on shade is likely to be less than this because of the way in which the leaves are positioned on the branches. The projected area of the fruits can be up 15% of the leaf area in some segments of heavily cropping 'Cox' trees (Jackson, 1970), but much lower in other segments. The 6.5–6.6% found by Wagenmakers for 'Golden Delicious' apple trees grown in high-density planting systems is probably the best estimate.

The leaves are therefore the most important organs for shade creation as well as for interception of light and its use in photosynthesis. The relationship between leaf area and light interception and distribution depends on leaf transmission and reflection of light, leaf poise in relation to the sources of light, and the degree of leaf folding. It also depends on the extent to which leaves are uniformly distributed within the trees, and the size and shape of the trees and their arrangement within the orchard.

Leaf transmittance and reflectance

Palmer (1977a) found that leaf reflectance and transmittance of PAR in the wavelength range 400–700 nm were both highest early in the season when the leaves were shiny and relatively thin (Table 7.6). Absorptance ranged from 82% to 90%. Both reflectance and transmittance showed a very pronounced peak for green light at around 550 nm. There is high transmission in the range 750–1350 nm.

Table 7.6 *Mean percentage leaf reflectance, transmittance and absorptance (over the 400–700 nm waveband) and dry weight per unit area for sun and shade leaves of 'Golden Delicious' at four occasions during the season*

	23 May 1975	14 June 1974	19 July 1974	12 Sept. 1974
Sun leaves				
Reflectance	10.4	7.9	8.2	8.5
Transmittance	6.8	3.1	1.9	1.7
Absorptance	82.8	89.0	89.9	89.8
Leaf dry wt/unit area (mg cm^{-2})	6.2	9.7	13.6	11.5
Shade leaves				
Reflectance	11.1	8.6	8.5	8.9
Transmittance	7.1	4.7	4.7	4.5
Absorptance	81.8	86.7	86.8	86.6
Leaf dry wt/unit area (mg cm^{-2})	4.2	5.0	5.4	6.0

From Palmer (1977a). Reproduced with permission.

Leaf angles and leaf folding

Jackson (1970) found that most of the leaves of 'Laxton's Superb' apple trees grown as hedgerows had the plane of their laminas at less than 45° to the horizontal with 11–35° as the most common angle. The leaf petiole was commonly bent so that the upper surface was at right angles to the incident light, noticeably in the lower parts of the trees with shading from above but exposure to light from the side. In 'Golden Delicious' and 'Goldspur' trees more than 80% of leaves inclined less than 50° above the horizontal (Wagenmakers, 1991b), although Proctor *et al.* (1975) found little preferred inclination or orientation on bush trees of 'McIntosh' apple.

Johnson and Lakso (1986) reported that 'Jonamac' apple leaves are usually appreciably folded: the leaf folding angle between each leaf half and the horizontal position averaged 45°.

Foliage clumping

Apple and pear leaves are obviously not uniformly or randomly distributed within the orchard space. Extension shoot leaves are systematically arranged around the stems that bear them and short-shoot or spur leaves are in compact clusters. In general, the greater the non-uniformity of distribution the less light will be intercepted per unit leaf area. The same concept of 'clumping' can be applied at the orchard level: dense individual trees or hedgerows separated by wide gaps being much more 'clumped' than, for example, orchards of V- or Y-trellised trees especially if these almost meet above the alleyways.

Canopy light interception and distribution

The light intensity I_L at any given depth in a continuous leafy canopy, like a field of grass, is related to the vertically-summed leaf area index L (LAI measured as m^2 leaf m^{-2} ground) above that plane according to equation 7.3 in which I_0 is light intensity above the canopy and K the light extinction coefficient.

$$\frac{I_L}{I_0} = e^{-KL} \tag{7.3}$$

The total percentage interception can be calculated from this with LAI summed down to ground level.

The light extinction coefficient depends on the average spectral properties of the leaves and their orientation in relation to the spatial distribution of solar radiation. In theory, stands with randomly spaced opaque horizontal leaves have $K = 1$; in practice K ranges from about 0.9 for planophile canopies to 0.3 for erectophile canopies. Clumping of leaves decreases the interception per unit leaf area and so reduces the value of K. In practice K is usually calculated using equation 7.3 after measurement of I_L, I_0 and the intervening leaf area index L. This derivation ignores the fact that some of the reduction in light intensity from I_0 to I_L may be a result of intervening stems, etc. but has the practical advantage of relating the photosynthesizing leaf surface to irradiance levels. For apple, K measured in this way ranges from 0.44 to 0.77 for PAR, with a mean value of 0.6 (Jackson, 1978, 1980).

In terms of light climate a continuous canopy can be considered as a series of strata separated by horizontal contours of irradiance as measured on a horizontal plane. The leaf area, in LAI terms, above the irradiance contour at which the irradiance is I_L can be defined as L_I and is determined by K (equation 7.4).

$$L_I = (\ln I_L)/(-K) \tag{7.4}$$

or L_I is the total LAI if this is less.

Some apple orchards, e.g. mature orchards of 'open centre' bush trees, horizontally trellised systems and very closely planted multirow bed systems (Wagenmakers, 1994) may give virtually continuous canopies. In these cases, and assuming an extinction coefficient of 0.6, light levels of more than 40% of those above the canopy can only be achieved in the strata above a vertically-summed LAI, L_{40}, of 1.5, the 30% irradiance contour would correspond with an L_{30} of 2 and the 20% irradiance contour with an L_{20} of 2.7.

Much more commonly, however, orchards consist of individual hedgerows or trellised systems separated by alleyways and, in their early years, almost all

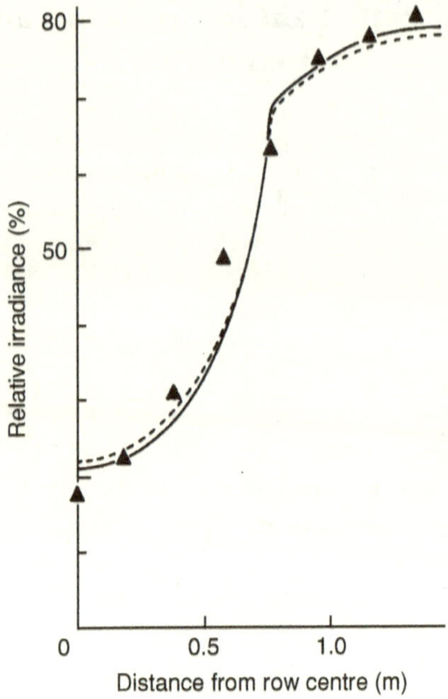

Figure 7.5 Comparison of observed (▲) values of relative irradiance across a row of hedgerow orchards at 2.9 × 0.9 m spacing under diffuse light conditions with those predicted using a 45° leaf angle (solid line) and a spherical leaf angle distribution (broken line). From Palmer (1977b). Reproduced with permission.

orchards consist of trees that are spatially separated within as well as between rows. In this situation light may be received on the lower, but outer, portions of the tree canopy without passing through foliage vertically above. It may, however, have been attenuated by passage, at an angle, through the upper part of the foliage of an adjacent tree or row of trees. It may also be intercepted by stems or fruits.

The light transmission to the ground, or to any particular point in the canopy, can be calculated as a process in which beams of light pass through the canopy in the manner of a point quadrat. Penetration depends on the altitude and azimuth of the source of light during the day, the areas and arrangement of leaves, stems and fruits per unit canopy volume and separate light extinction coefficients for each of these shading structures. Computer models have been developed, both for direct light penetration and interception, and for that of diffuse light. The models of Palmer (1977b) and Wagenmakers (1991b) give quite close correspondence with measured values of light transmission to the orchard floor (Figure 7.5). Wagenmakers' simulations showed that correct assessment of crown size and leaf area were very important to accurate estimation of light interception and she concluded that fruit contributed little. Charles-Edwards and Thorpe (1976) found that a model using only leaf areas and tree dimensions gave simulated transmission values very close to experimental measurements.

Solar altitude and azimuth vary throughout the day, throughout the year, and with latitude. Light interception and penetration vary not only with tree or hedgerow dimensions but also with canopy density and hedgerow orientation. The range of potential orchard designs made possible by the use of rootstocks giving different degrees of vigour control and of training and pruning systems to control tree shape, together with the change of tree dimensions over the years, limits relevant field investigations. The computer modelling approach enables light interception and light levels within canopies to be calculated for orchards with different configurations and leaf area densities at any latitude over the chosen period (Figures 7.6, 7.7). Results are conventionally expressed as percentage interception and percentage of above-canopy irradiance but can readily be converted to absolute values if irradiance at the site is known.

An alternative modelling approach is to consider light transmission to the orchard floor (T) to be made up of that which misses the trees altogether (T_f) and that which penetrates through the tree canopy (T_c) including sunflecks on the ground within the cast shadow area, i.e.

$$T = T_f + T_c \tag{7.5}$$

In this T_f depends on tree form (size and shape) and between-tree spacing, and T_c on leaf density within the tree outline expressed as leaf area per unit of shaded area, i.e. the cast-shadow area including sunflecks (Jackson 1981). The fractional interception of light (F) can then be calculated from the fractional interception which would occur if the trees were opaque (F_{max} which equals $1 - T_f$) minus the transmission through the trees. This transmission, from equation 7.3, will be $F_{max}e^{-KL'}$ where F_{max} is the ground area shaded and L' is the projected leaf area on the cast shadow area, i.e. LAI divided by F_{max}. Thus, from Jackson and Palmer (1979, 1981),

$$F = F_{max} - F_{max}e^{-KL'} \tag{7.6}$$

F_{max} can be determined using relatively simple computer models assuming opaque 'trees' and calculating their interception of direct and diffuse light throughout the day and season at different latitudes (Jackson and Palmer, 1972), or by computer graphics (Johnson and Lakso, 1991). It can also be measured using opaque scale models of orchards placed on surfaces such as solar panels or arrays of cosine-corrected sensors giving electrical output proportional to incoming radiation (Jackson, 1981; Middleton and Jackson,

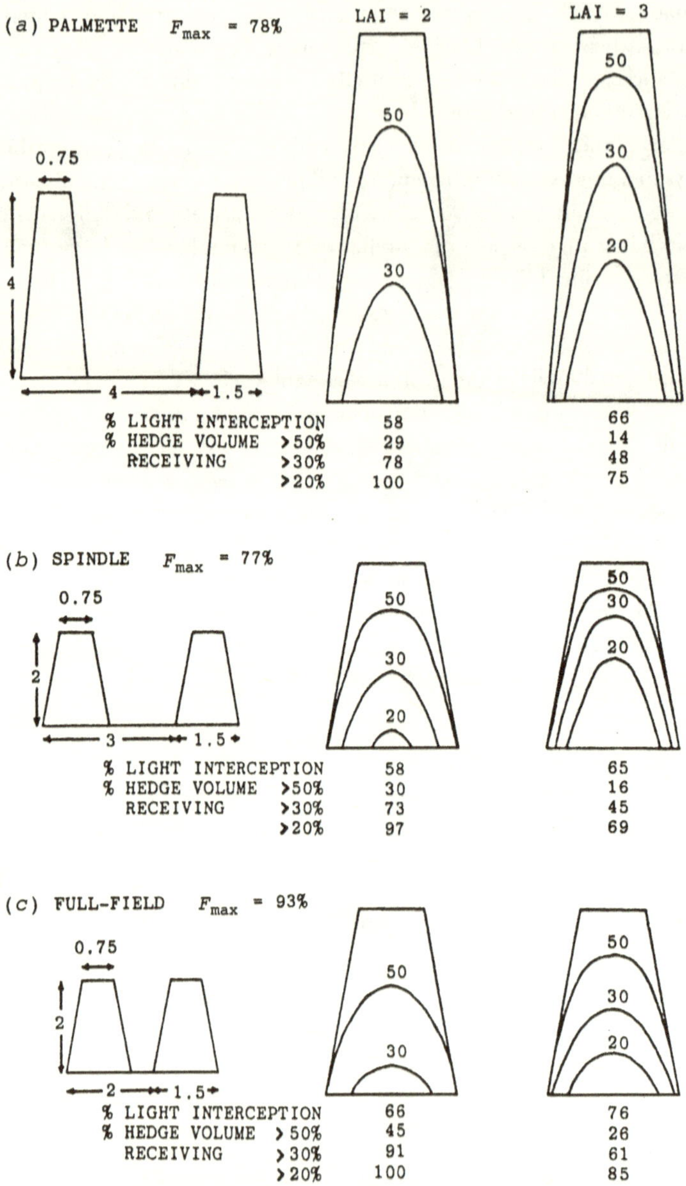

Figure 7.6 Calculated light profiles within individual sections of hedge of given dimensions (m) and two LAIs for three orchard systems. Hedgerow orientation N–S. From Palmer (1981). Reproduced with permission.

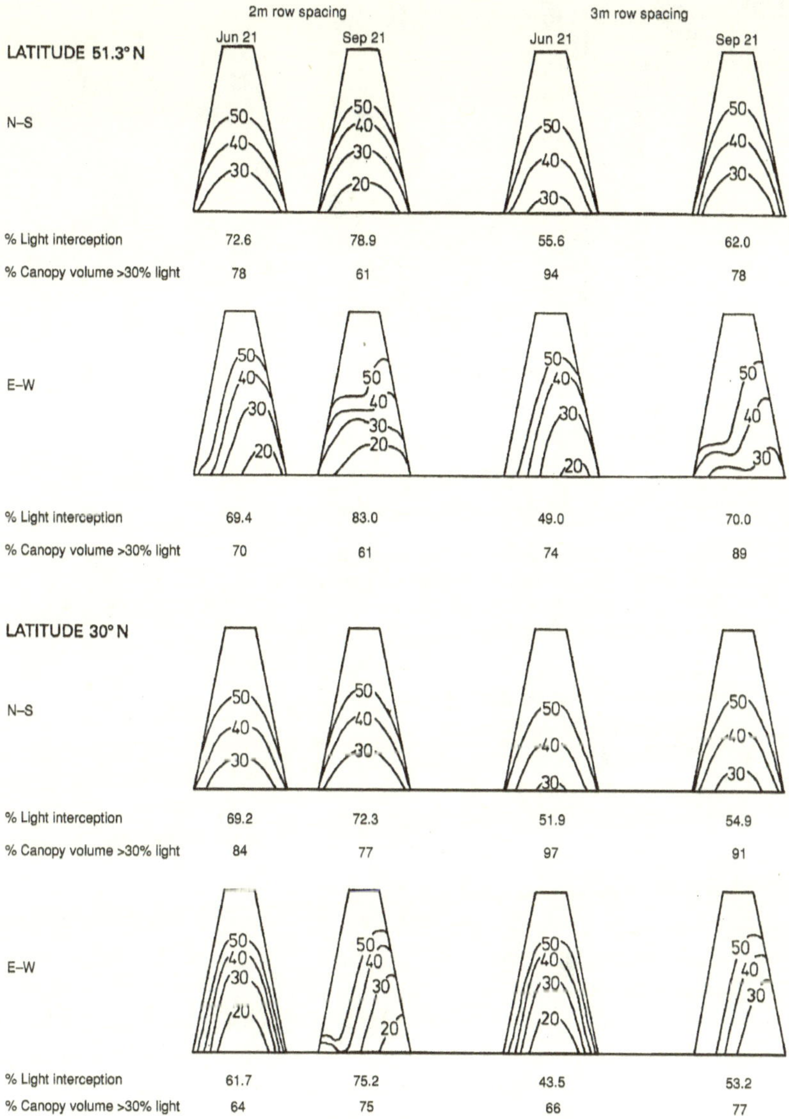

Figure 7.7 Calculated daily light distributions under sunny conditions within hedgerows at two latitudes, two dates, two row spacings and two row orientations. Each tree outline shows contours of percentage irradiance. Hedgerows are 2.5 m tall, 1.5 m thick at the base and 0.5 m thick at the top, with a leaf area density of 2.6 $m^2\ m^{-3}$. From Palmer (1989b). Reproduced with permission.

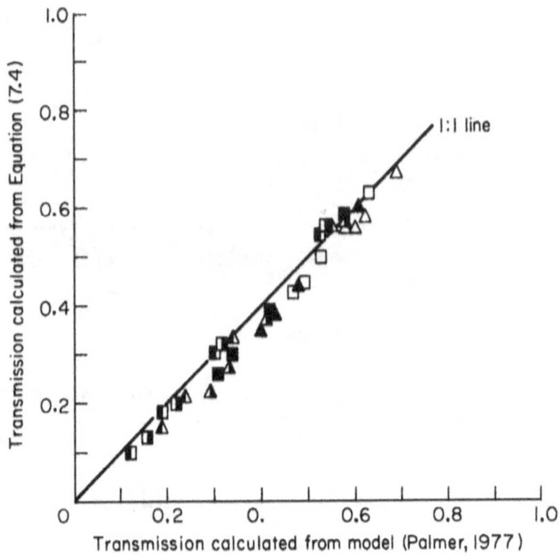

Figure 7.8 Comparison of calculated daily transmission from the Palmer (1977b) computer model with 50% direct and 50% diffuse light to that estimated from Equation (7.4) with $K = 0.6$. Comparisons for the range of hedgerow orchards as shown:

Triangular cross-section	△	▲	◮	◭
Rectangular cross-section	□	■	◨	◧
Hedge height (m)	1	3	1	3
Row spacing (m)	4	4	2	2

For LAIs of 1, 2, 3, 4.
All hedges of 1.5 m spread at ground level.
From Jackson and Palmer (1977). Reproduced with permission.

1989). This latter method copes with natural patterns of direct and diffuse irradiation which may differ from 'normal'.

Results obtained by the use of equation 7.6 agree closely with those calculated from more complex models (Figure 7.8). This approach can also be extended to calculate the leaf area, L_I, which will be in receipt of irradiance at above any given light intensity, I_L expressed as a decimal fraction of above canopy irradiance (Jackson and Palmer, 1981).

$$L_I = F_{\max}[(\ln I_L)/(-K)] \tag{7.7}$$

or LAI if this is less.

Results obtained by the use of equation 7.7 agree closely with those calculated from more complex models (Figure 7.9).

$F_{\max}$, LAI and K are therefore the determinants both of total orchard canopy light interception and of the leaf area or canopy volume (CV_I) external

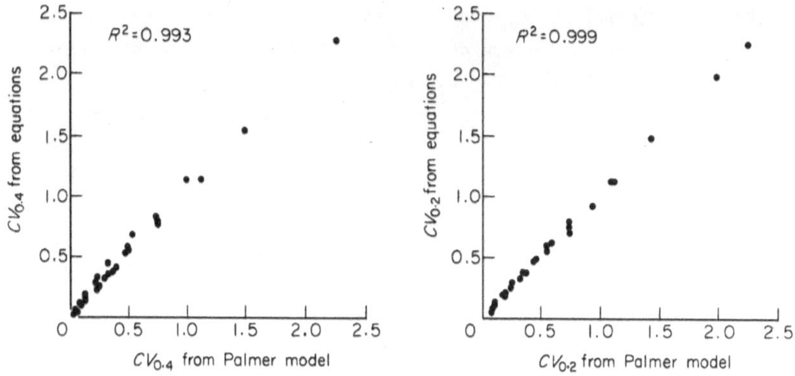

Figure 7.9 Comparison of estimates of canopy volumes (m^3 m^{-2} orchard) receiving more than 40% ($CV_{0.4}$) or more than 20% ($CV_{0.2}$) of full daylight (daily integrals) for N–S oriented hedgerows 1.5 m wide at ground level giving all combinations of LAI 1, 2, 3 and 4, between row spacings of 2 m and 4 m, heights of 1 m and 3 m and rectangular and triangular profiles in section. 'Palmer model' is the computer model of Palmer (1977b). The equations are equation 7.8 and those preceding it in this text. From Jackson and Palmer (1981) with permission.

to any chosen contour of mean irradiance where

$$CV_I = L_I/(\text{leaf area density m}^2\ \text{m}^{-3}) \qquad (7.8)$$

It follows from this analysis:

1 That potential orchard light interception and yield cannot be predicted from such factors as tree height, surface area, LAI, etc. alone; virtually identical levels of light interception can be achieved with very different tree forms and LAI values (Table 7.7).

2 That, especially for tall hedgerow orchards, light interception and distribution are partly controlled by orchard geometry in relation to direct-beam solar radiation. North–south hedgerows are generally preferable to east–west ones and latitude influences the effect of orchard geometry.

3 That light interception per unit leaf area is higher the greater the value of F_{max}. This is because, at the extremes, increasing orchard leaf area by increasing F_{max} may have a linearly proportional effect on light interception (e.g. with higher plant populations in the first years after planting). On the other hand, increasing orchard leaf area by increasing leaf density within a constant canopy volume will increase light interception in proportion to the increase in the logarithm of leaf area only.

4 The actual light intensities within the canopy are a function of above canopy irradiance. This has implications for optimal canopy dimensions

Table 7.7 *Examples of similar light interceptions achieved with truncated triangular cross-section hedgerows of different dimensions, spacings, leaf areas, canopy volumes and canopy surface areas*

% light interception	Canopy height (m)	Clear alley width (m)	Basal thickness (m)	LAI	Canopy volume ($m^3\ m^{-2}$)	Canopy surface area ($m^2\ m^{-2}$)
62.4	3.0	2.5	1.2	4.3	0.73	1.79
62.1	2.5	1.5	0.6	3.1	0.54	2.53
62.2	3.0	1.0	1.5	2.4	1.35	2.72
62.3	1.5	1.0	0.9	3.1	0.53	1.83
62.2	1.5	0.5	0.3	2.5	0.42	3.94

From Jackson and Palmer (1980). Reproduced with permission.

at different latitudes. For example, Davis, California has a high-light climate with 4.13 GJ m^{-2} over a 5-month growing season whereas Wilhelminadorp, Netherlands has an average of only 2.5 GJ m^{-2} over a similar period. If 30% of above-canopy radiation at Wilhelminadorp is taken to be the critical level for 'good' apple fruit bud formation and fruit size development there, the same light intensity will be attained within the canopy at Davis with only 18% of above-canopy irradiance. With F_{max} at 0.7 and K at 0.6, the LAI receiving this critical light level in absolute terms will be 1.4 at Wilhelminadorp and 2.0 at Davis, because at Wilhelminadorp I_L/I_0 needs to be 0.30 but at Davis only 0.18 (Jackson, 1997).

Trunk cross-sectional area and light interception

Trunk cross-sectional area, or girth^2, is related to leaf area by equations 7.1 and 7.2, although both b and K may vary. Individual tree leaf area can thus be estimated from trunk girth^2 or TCA; orchard LAI from the summation of these.

The relationship between summed orchard TCA and light interception will therefore follow the relationship between orchard leaf area and light interception. It will be a combination of linear effects (e.g. increase in planting density in the early years influencing F_{max} proportionately) and effects based on the exponential relationship between light interception and leaf area, as discussed earlier. The latter will become increasingly important as the trees grow to their final size with increasing within-canopy shading. This provides the basis for the good fit of a quadratic curve to the relationship between light interception and TCA over a number of orchard systems (Robinson and Lakso, 1991) and also, at least in part, to the curvilinear relationship between yield and girth^2.

References

Abbott, D.L. (1984). *The Apple Tree: Physiology and Management.* London: Grower Books.

Asada, T. and Ogasawara, M. (1998). The effect of shading on the growth of young 'Fuji' apple trees. *Journal of the Japanese Society for Horticultural Science* **67**, 655–9.

Barden, J.A. (1974). Net photosynthesis, dark respiration, specific leaf weight, and growth of young apple trees as influenced by light regime. *Journal of the American Society for Horticultural Science* **99**, 547–51.

Barden, J.A. (1978). Apple leaves, their morphology and photosynthetic potential. *HortScience* **13**, 644–6.

Barlow, H.W.B. (1969). The relation of leaf area to stem cross section. *Report of the East Malling Research Station for 1968*, 117–19.

Barlow, H.W. B. (1980). The relationship between leaf size and shoot length in apple. *Journal of Horticultural Science* **55**, 279–83.

Barlow, H.W.B. and Smith, J.G. (1971). Effect of cropping on growth of the apple tree. *Report of the East Malling Research Station for 1970*, 52.

Barritt, B.H., Rom, C.R., Guelich, K.R., Drake, S.R. and Dilley, M.A. (1987). Canopy position and light effects on spur leaf and fruit characteristics of 'Delicious' apple. *HortScience* **22**, 402–5.

Barritt, B.H., Rom, C.R. Konishi, B.J. and Dilley, M.A. (1991). Light level influences spur quality and canopy development and light interception influence fruit production in apple. *HortScience 26*, 993–9.

Beakbane, A.B. (1965). Some effects of gibberellin on the anatomy of apple leaf disks. *Report of the East Malling Research Station for 1964*, 102–3.

Beakbane, A.B. (1967). A relationship between leaf structure and growth potential in apple. *Annals of Applied Biology* **60**, 67–76.

Beakbane, A.B. and Majumder, P.K. (1975). A relationship between stomatal density and growth potential in apple rootstocks. *Journal of Horticultural Science* **50**, 285–9.

Charles-Edwards, D.A. and Thorpe, M.R. (1976). Interception of diffuse and direct-beam radiation by a hedgerow apple orchard. *Annals of Botany* **40**, 603–13.

Cowart, F.F. (1935). Apple leaf structure as related to position of the leaf upon the shoot and to type of growth. *Proceedings of the American Society for Horticultural Science* **33**, 145–8.

Flore, J.A. and Lakso, A.N. (1989). Environmental and physiological regulation of photosynthesis in fruit crops. *Horticultural Reviews* **11**, 111–57.

Forshey, C.G., Wieres, R.W. and Van Kirk, J.R. (1987). Seasonal development of the leaf canopy of 'MacSpur McIntosh' apple trees. *HortScience* **20**, 881–3.

Ghosh, S.P. (1973). Internal structure and photosynthetic activity of different leaves of apple. *Journal of Horticultural Science* **48**, 1–9.

Giuliani, R., Corelli-Grappadelli, L. and Magnanini, E. (1997). Effects of crop load on apple photosynthetic responses and yield. *Acta Horticulturae* **451**, 303–11.

Hancock, C.R. and Barlow, H.W.B. (1960). Short term fluctuations in leaf and internode growth in the apple rootstock Crab C. *New Phytologist* **59**, 227–37.

Hansen, P. (1967a). ^{14}C-studies on apple trees. II. Distribution of photosynthates from top and base leaves from extension shoots. *Physiologia Plantarum* **20**, 720–5.

Hansen, P. (1967b) ^{14}C-studies on apple trees. III. The influence of season on storage and mobilization of labelled compounds. *Physiologia Plantarum* **20**, 1103–11.

Hansen, P. (1971). ^{14}C-studies on apple trees. VII. The early seasonal growth in leaves, flowers and shoots as dependent upon current photosynthates and existing reserves. *Physiologia Plantarum* **25**, 469–73.

Heim, G., Landsberg, J.J., Watson, R.L. and Brain, P. (1979). Eco-physiology of apple trees: dry matter partitioning by young 'Golden Delicious' trees in France and England. *Journal of Applied Ecology* **16**, 179–94.

Heinicke, A.J. (1934). Photosynthesis in apple leaves during late fall and its significance in annual bearing. *Proceedings of the American Society for Horticultural Science* **32**, 77–80.

Heinicke, D.R. (1964). The microclimate of fruit trees. III. The effect of tree size on light penetration and leaf area in 'Red Delicious' apple trees. *Proceedings of the American Society for Horticultural Science* **85**, 33–41.

Holland, D.A. (1968). The estimation of total leaf area on a tree. *Report of the East Malling Research Station for 1967*, 101–4.

Jackson, J.E. (1970). Aspects of light climate within apple orchards. *Journal of Applied Ecology* **7**, 207–16.

Jackson, J.E. (1978). Utilization of light resources by HDP systems. *Acta Horticulturae* **65**, 61–70.

Jackson, J.E. (1980). Light interception and utilization by orchard systems. *Horticultural Reviews* **2**, 208–67.

Jackson, J.E. (1981). Theory of light interception by orchards and a modelling approach to optimizing orchard design. *Acta Horticulturae*, **114**, 69–79.

Jackson, J.E. (1985). Future fruit orchard design: economics and biology. In *Attributes of Trees as Crop Plants*, ed. M.G.R. Cannell and J.E. Jackson, pp. 441–59. Monks Wood, UK: Institute of Terrestrial Ecology, Natural Environment Research Council.

Jackson, J.E. (1997). Light interception and canopy characteristics at low latitudes in relation to orchard system design. *Acta Horticulturae* **451**, 417–25.

Jackson, J.E. and Beakbane, A.B. (1970). Structure of leaves growing at different light intensities within mature apple trees. *Report of the East Malling Research Station for 1969*, 87–9.

Jackson, J.E. and Palmer, J.W. (1972). Interception of light by model hedgerow orchards in relation to latitude, time of year and hedgerow configuration and orientation. *Journal of Applied Ecology* **9**, 341–57.

Jackson, J.E. and Palmer, J.W. (1977). Effects of shade on the growth and cropping of apple trees. I. Experimental details and effects on vegetative growth. *Journal of Horticultural Science* **52**, 245–52.

Jackson, J.E. and Palmer, J.W. (1979). A simple model of light transmission and interception by discontinuous canopies. *Annals of Botany* **44**, 381–3.

Jackson, J.E. and Palmer, J.W. (1980). A computer model study of light interception by orchards in relation to mechanised harvesting and management. *Scientia Horticulturae* **13**, 1–7.

Jackson, J.E. and Palmer, J.W. (1981). Light distribution in discontinuous canopies: calculation of leaf areas and canopy volumes above defined 'irradiance contours' for use in productivity modelling. *Annals of Botany* **47**, 561–5.

Jackson, J.E., Parry, M.S. and Stephens, C.P. (1981). Intensification of tree fruit production: current constraints, relevant research and an alternative system strategy for the 1980's. *Acta Horticulturae* **114**, 399–406.

Jankiewicz, L.S., Antoszewski, R and Klimowicz, E. (1967). Distribution of labelled assimilates within a young apple tree after supplying $C^{14}O_2$ to a leaf or a shoot. *Biologia Plantarum* (Prague) **9**, 116–21.

Johnson, R.S. and Lakso, A.N. (1985). Relationships between stem length, leaf area, stem weight and accumulated growing degree-days in apple shoots. *Journal of the American Society for Horticultural Science* **110**, 586–90.

Johnson, R.S. and Lakso, A.N. (1986). Carbon balance model of a growing apple shoot. I. Development of the model. *Journal of the American Society for Horticultural Science* **111**, 160–4.

Johnson, R.S. and Lakso, A.N. (1991). Approaches to modelling light interception in orchards. *HortScience* **26**, 1002–4.

Jonkers, H. (1980). Autumnal leaf abscission in apple and pear. *Fruit Science Reports* **7**, 25–9.

Lakso, A.N. (1984). Leaf area development patterns in young pruned and unpruned apple trees. *Journal of the American Society for Horticultural Science* **109**, 861–5.

Lakso, A.N. and Lenz, F. (1986). Regulation of apple tree photosynthesis in the autumn by temperature. In *The Regulation of Photosynthesis in Fruit Trees*, ed. A.N. Lakso, and F. Lenz, pp. 34–37. Special Publication, New York State Agricultural Experiment Station, Geneva.

Larsen, F.E. (1973). Promotion of leaf abscission in fruit nursery stock. *Acta Horticulturae* **34**, 129–33.

Lindhagen, M. (1996). Model analysis of leaf unfolding rate in *Malus domestica* Borkh. *Scientia Horticulturae* **67**, 65–78.

Liu, A. and Eaton, G.W. (1970). Comparative leaf anatomy of two standard and two compact apple mutants. *Canadian Journal of Plant Science* **50**, 733–5.

MacDaniels, L.H. and Cowart, F.F. (1944). *The Development and Structure of the Apple Leaf.* Cornell University (Ithaca) Agricultural Experiment Station, Memoir 258.

Mason, A.C. and Whitfield, A.B. (1960). Seasonal changes in the uptake and distribution of mineral elements in apple trees. *Journal of Horticultural Science* **35**, 34–55.

Middleton, S. and Jackson, J.E. (1989). 'Solid model' estimation of light interception by apple orchard systems. *Acta Horticulturae* **240**, 83–6.

Monteith, J.L. (1977). Climate and the efficiency of crop production in Britain. *Philosophical Transactions of the Royal Society of London B*, **281**, 277–94.

Moore, C.S. (1978). Biometric relationship in apple trees. *Journal of Horticultural Science* **53**, 43–51.

Palmer, J.W. (1977a). Light transmittance by apple leaves and canopies. *Journal of Applied Ecology* **14**, 505–14.

Palmer, J.W. (1977*b*). Diurnal light interception and a computer model of light interception by hedgerow apple orchards. *Journal of Applied Ecology* **14**, 601–14.

Palmer, J.W. (1981). Computed effects of spacing on light interception and distribution within hedgerow trees in relation to productivity. *Acta Horticulturae* **114**, 80–8.

Palmer, J.W. (1989a). Canopy manipulation for optimum utilization of light. In *Manipulation of Fruiting*, ed. C.J. Wright, pp. 245–62. London: Butterworths.

Palmer, J.W. (1989b). The effects of row orientation, tree height, time of year and latitude on light interception and distribution in model apple hedgerow canopies. *Journal of Horticultural Science* **64**, 137–45.

Palmer, J.W. and Jackson, J.E. (1977). Seasonal light interception and canopy development in hedgerow and bed system apple orchards. *Journal of Applied Ecology* **14**, 539–49.

Pearce, S.C. (1952). Studies in the measurement of apple trees. 1. The use of trunk girths to estimate tree size. *Report of the East Malling Research Station for 1951*, 101–4.

Pratt, C. (1990). Apple trees: morphology and anatomy *Horticultural Reviews* **12**, 265–305.

Proctor, J.T.A. and Palmer, J.W. (1991). The role of spur and bourse leaves of three apple cultivars on fruit set and growth and calcium content. *Journal of Horticultural Science* **66**, 275–82.

Proctor, J.T.A., Kyle, W.J. and Davies, J.A. (1975). The penetration of global solar radiation into apple trees. *Journal of the American Society for Horticultural Science* **100**, 40–4.

Quinlan, J.D. (1965). The pattern of distribution of ^{14}Carbon in a potted apple rootstock following assimilation of carbon dioxide by a single leaf. *Report of the East Malling Research Station for 1964*, 117–18.

Quinlan, J.D. (1969). Mobilization of ^{14}C in the spring following autumn assimilation of $^{14}CO_2$ by an apple rootstock. *Journal of Horticultural Science* **44**, 107–10.

Robinson, T.L. (1997). Interaction of tree form and rootstock on light interception, yield and efficiency of 'Empire', 'Delicious' and 'Jonagold' apple trees trained to different systems. *Acta Horticulturae* **451**, 427–36.

Robinson, T.L. and Lakso, A.N. (1991). Bases of yield and production efficiency in apple orchard systems. *Journal of the American Society for Horticultural Science* **116**, 188–94.

Skene, D.S. (1974). Chloroplast structure in mature apple leaves grown under different levels of illumination and their responses to changed illumination. *Proceedings of the Royal Society of London, B* **186**, 75–8.

Slack, E.M. (1974). Studies of stomatal distribution on the leaves of four apple varieties. *Journal of Horticultural Science* **49**, 95–103.

Titus, J.S. and Kang, S.M. (1982). Nitrogen metabolism, translocation and recycling in apple trees. *Horticultural Reviews* **4**, 204–46.

Tustin, S., Corelli-Grappadelli, L. and Ravaglia, G. (1992). Effect of previous-season and current light environments on early-season spur development and assimilate translocation in 'Golden Delicious' apple. *Journal of Horticultural Science* **67**, 351–60.

Tydeman, H.M. (1963). The inheritance of time of leaf break and leaf fall in seedling apples. *Report of the East Malling Research Station for 1962*, 58–63.

Verheij, E.W.M. (1968). *Yield-density Relations in Apple: Results of a Planting System Experiment in Hungary.* Publication **37**, Wageningen, ITT.

Verheij, E.W.M. (1972). *Competition in apple, as influenced by alar sprays, fruiting, pruning and tree spacing.* Mededelingen Landbouwhogeschool Wageningen. Publicatie 73, pp. 1–54.

Wagenmakers, P.S. (1989). High-density planting system trial with pear. *Acta Horticulturae* **243**, 303–7.

Wagenmakers, P.S. (1991 a). Planting systems for fruit trees in temperate climates. *Critical Reviews in Plant Sciences* **10**, 369–85.

Wagenmakers, P.S. (1991 b). Simulation of light distribution in dense orchard systems. *Agricultural and Forest Meteorology* **57**, 13–25.

Wagenmakers, P.S. (1994). Light relations in orchard systems. Thesis, Wageningen.

Wünsche, J.N., Lakso, A.N., Robinson, T.L., Lenz, F. and Denning, S.S. (1996). The bases of productivity in apple production systems: the role of light interception by different shoot types. *Journal of the American Society for Horticultural Science* **121**, 886–93.

8

Photosynthesis, respiration, and carbohydrate transport, partitioning and storage

Introduction

The greater part of the dry weight of apple and pear trees is derived from photosynthetically fixed carbon. Moreover, sugars comprise a critically important component of the economic product, the fruits, being a major factor in controlling their taste.

Early in the season, around budburst, the carbohydrate needed to provide energy for growth and to supply a basis for structural material comes from reserves stored in perennial tissues in the previous season or seasons. As the leaves expand their photosynthesis supplies growing tissues and replenishes reserves.

One very important aspect of apple and pear carbohydrate metabolism is that of the ripening and the post-harvest fruit. This is dealt with in later chapters.

Photosynthesis

In photosynthesis energy from solar radiation is converted into chemical energy, which enables the reduction of carbon dioxide to produce carbohydrates, according to equation (8.1).

$$n\mathrm{CO_2} + 2n\mathrm{H_2O} \xrightarrow[\text{chlorophyll}]{\text{light}} (\mathrm{CH_2O})_n + n\mathrm{O_2} + n\mathrm{H_2O} \qquad (8.1)$$

This process involves both 'light' and 'dark' reactions in very close conjunction. In the 'light' reaction light energy is converted into chemical energy in the form of adenosine triphosphate (ATP) and nicotinamide adenine dinucleotide phosphate (NADPH). In the 'dark' reaction, which is closely coupled to the 'light' reaction, CO_2 is incorporated by carboxylation of ribulose phosphate by ribulose 1,5-bisphosphate carboxylase–oxygenase (RuBPC/O). The first stable products of this reaction are two molecules of glycerate 3-phosphate (glycerate-3-P). Plants such as apple and pear in which the initial carboxylation results in the formation of 3-carbon acids are termed C_3 plants. Glycerate-3-P is phosphorylated when ATP is present to form glycerate 1,3-P which is reduced, when NADPH is present, to give glyceraldehyde-3-P. This triose phosphate, together with dihydroxyacetone phosphate (DHAP) can be combined to form fructose 1,6-bisphosphate (fructose-P_2) through the action of fructose bisphosphatase (FBPase). Fructose 6-phosphate can then be used to form starch.

The flow of carbon within this overall photosynthetic carbon reduction (PCR) cycle is largely controlled by regulation of the enzymes within it. The enzyme present in the highest concentration in the chloroplast stroma is RuBPC/O, which may constitute up to 50% of the total protein within the chloroplast. The activities of several of the enzymes are regulated by light and, in addition, changes in the stromal pH and magnesium concentration affect the activities of both RuBPC/O and FBPase.

Both O_2 and CO_2 compete for RuBP at the catalytic site on the enzyme so photosynthesis is directly inhibited by O_2. The oxygenase reaction results in the formation of phosphoglycolate at the start of the photosynthetic oxidative carbon cycle which is the metabolic pathway for photorespiration. The release of CO_2 from this pathway may be equivalent to 15–50% of the rate of net CO_2 assimilation (Sharkey, 1988).

Sorbitol, a 6-carbon alcohol which plays several key roles in woody *Rosaceae*, is formed by the reduction of D-fructose and D-glucose. Glucose-P is reduced to sorbitol-P by aldose 6-P reductase and the P split from sorbitol-P by a phosphatase. A glucitol 6-P phosphatase has been found in apple leaves (Grant and ap-Rees, 1981).

When $^{14}CO_2$ is fed to apple leaves 58–80% of the total activity in the sugar fraction is found in sorbitol immediately after supply (Hansen, 1967a, b). The rest is mainly in sucrose, with some in glucose and fructose. Similar results are found with pear leaves (Bieleski, 1977). Sorbitol is the major end product of apple and pear photosynthesis (Priestley, 1983; Oliveira and Priestley, 1988).

Transport of photosynthate from leaves to fruits appears to be mainly as sorbitol. After supply of $^{14}CO_2$ to apple leaves the concentration of ^{14}C-sorbitol in these falls quickly, labelled sorbitol is found in the conducting tissues between the leaves and adjacent fruits and sorbitol appears as the main labelled 'sugar' in the fruit within a few hours (Hansen, 1970). Sorbitol is the main carbohydrate found in apple phloem (Bieleski, 1969). It is particularly suited to translocation in the phloem because it is not metabolized there.

Sorbitol is the most abundant compound in the soluble fraction of carbohydrate reserves in woody *Rosaceae* (Oliveira and Priestley, 1988). However, sorbitol transported from leaves is quickly converted in the fruits into sucrose, glucose and, especially, fructose (Hansen, 1970). This maintains the concentration gradient which facilitates the leaf-to-fruit translocation. It is consistent

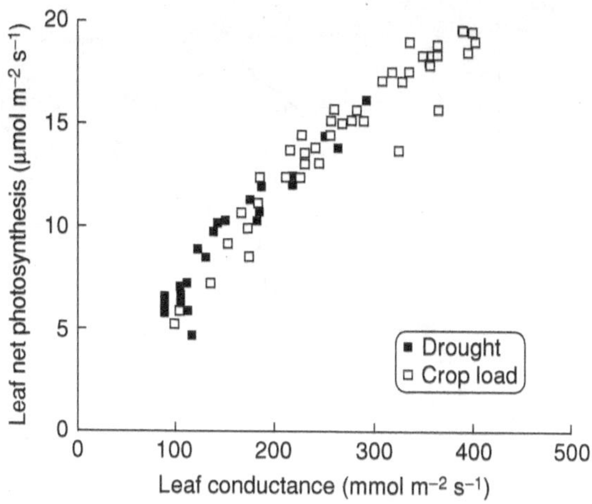

Figure 8.1 The relationship between photosynthesis and leaf conductance in separate studies of drought stress and of varying crop loads with 'Empire'/'M.9' apple trees in the field. From Lakso (1994). Reproduced with permission.

with the isolation from apple fruits of a sorbitol dehydrogenase that oxidizes sorbitol to fructose.

Traditionally, leaf photosynthetic rate (A) was expressed as the mass of CO_2 fixed per unit of leaf area per unit of time (mg CO_2 dm^{-2} h^{-1}). More recently it has been expressed in molar units (μmol m^{-2} s^{-1}) with an approximate conversion of $0.63 \times$ mg CO_2 dm^{-2} h^{-1} = μmoles m^{-2} s^{-1} (Flore and Lakso, 1989). Photosynthesis responds to light in the visible wavelengths (400–700 nm), which makes up about 47% of total solar radiation, with some variation depending on cloudiness and solar position relative to the earth. This is measured in W m^{-2} or μE m^{-2} s^{-1}. Commonly photosynthetically active radiation (PAR) or photosynthetic photon flux (PPF) within or below the canopy is expressed as a percentage of above-canopy irradiance, but it must be remembered that the latter varies from site to site and through the season.

Stomatal conductance and photosynthesis

Clearly the rate of entry of CO_2 into the leaf is potentially a major limiting factor to photosynthesis. Effects of both environmental and within-plant factors on photosynthesis may therefore be mediated by effects on stomatal conductance, measured as leaf conductance or canopy conductance. Stomata and stomatal aperture control are discussed in Chapter 7 and Chapter 12.

There is frequently a close relationship between stomatal conductance and net photosynthesis of apple leaves (Palmer, 1992; Lakso, 1994 (cf. Figure 8.1); Giuliani *et al.*, 1997b). The pattern of causality in this relationship may be

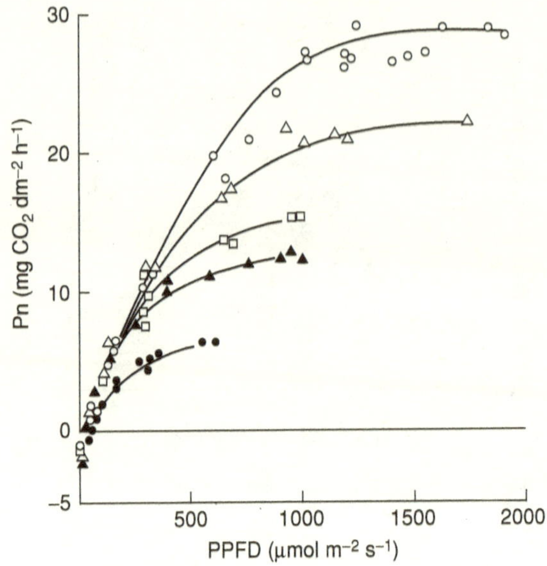

Figure 8.2 Light response curves of mid-shoot leaves from young 'Fuji' apple trees grown under full daylight (○—○), 52.9% daylight (△—△), 34.2% daylight (□—□), 15.8% daylight (▲—▲) and 5.0% daylight (●—●). From Asada and Ogasawara (1998). Reproduced with permission. Under natural light conditions 2000 μmol m^{-2} s^{-1} PAR corresponds to *c.* 1000 W m^{-2} of total radiation measured by radiometer or 100 Klux measured with a light meter.

complex. 'Sun' leaves in exposed (exterior) canopy positions have higher net photosynthesis per unit light and higher stomatal conductance (Campbell *et al.*, 1992), but also have numerous other adaptations leading to higher photosynthetic potential. Humidity may control stomatal conductance directly (see Chapter 11, pp. 430–1), hence photosynthesis, and in other cases photosynthesis appears to control stomatal behaviour rather than vice versa (Lakso, 1994).

Light response curves

Apple leaf photosynthesis is of the C_3 type with a hyperbolic light response that typically saturates at 500–1000 μmol quanta m^{-2} $^{-}s^{1}$ (Figure 8.2). The light compensation point, i.e. the light level below which net CO_2 exchange is negative, with respiration exceeding photosynthesis, is 20–60 μmol quanta m^{-2} s^{-1}. Good rates of photosynthesis per unit leaf area for healthy exposed leaves are around 15 μmol CO_2 m^{-2} s^{-1} (Lakso, 1994) although much higher rates have been recorded, e.g. over 35 mg CO_2 dm^{-2} h^{-1} by Avery (1977), about 34 mg CO_2 dm^{-2} h^{-1} by Bravdo (1986) and 25–35 mg CO_2 dm^{-2} h^{-1} by Looney (1968).

Although stomata are closed in the dark they open fully at light levels well below photosynthetic light saturation and the photosynthesis light response curve does not reflect changes in stomatal conductance (Kriedemann and Canterford, 1971; Lakso, 1994).

When solar irradiance and air temperatures are very high, shading may actually increase photosynthesis. Bravdo (1986), working in Israel, found that when the maximum radiation flux density at noon was 930 W m^{-2} or 2000 μE m^{-2} s^{-1} PAR shading by 25% reduced leaf temperatures from 36 °C to 31 °C, reduced stomatal resistance to about a third and increased photosynthesis by up to around 3.5 times. This shade effect should be regarded as overcoming temperature and water stress effects rather than indicating supraoptimal PAR levels.

The net photosynthesis of *Pyrus communis* cultivars at saturating light levels is similar to that of apples: Kriedemann and Canterford (1971) found 'Bartlett' leaves to be light saturated at a light intensity equivalent to 200 W m^{-2} of PAR under the conditions of the experiment, at which net photosynthesis of single leaves was approximately 21 mg CO_2 dm^{-2} h^{-1}. The maximum rate in the orchard was around 30–32 mg CO_2 dm^{-2} h^{-1}, achieved early in the day with solar radiation at approximately 300 W m^{-2} and PAR about half of this. The light compensation point was 100 fc (4 W m^{-2} of PAR).

Several studies have shown *P. serotina* to have relatively low maximum photosynthetic rates. Honjo *et al.* (1990) found net photosynthesis to increase slowly to 11.7 μmol m^{-2} s^{-1} at PPF = 1200 μmol m^{-2} s^{-1} under conditions where *P. communis* cv. 'Bartlett' reached a near maximum of 19.0 μmol m^{-2} s^{-1} at PPF = 400 μmol m^{-2} s^{-1}. Higgins *et al.* (1992) also found the light compensation point of the Asian pear *P. serotina* cv. '20th Century' was, at 28 μmol quanta m^{-2} s^{-1}, about half that of the apple cv. 'Granspur'.

The typical rates for apple and European pear at saturating light intensity (A_{max}) are comparable to those of *Prunus* species and grapes but higher than those of citrus (Flore and Lakso, 1989; Lakso, 1994). Comparisons of individual published rates can be difficult to interpret because of the effects of a large number of plant factors, e.g. fruiting and leaf type, on A_{max} and the differing effects of e.g. temperature on A_{max} and of leaf-to-air vapour pressure differences on stomatal conductance and net photosynthesis of different species (Higgins *et al.*, 1992).

Conversion of light energy to dry matter by apple trees, at 1.95 g MJ^{-1} of PAR, is slightly higher than that calculated for forest stands (Palmer, 1989).

Effects of temperature

The response curves of apple net photosynthesis to temperature show broad optima over the range 15–30 °C (Lakso *et al.*, 1999), 16–26 °C (Watson *et al.*,

1978) or 15–25 °C with a distinct peak at around 20 °C (Higgins *et al.*, 1992; some data of Watson *et al.*, 1978). These optima are lower than many cited in earlier literature. Apple photosynthesis at 10 °C is 70–80% of the maximum at optimal temperature and is very much higher at low temperatures than that of peach, grape and some other fruit trees. The decline with temperature from 30 °C to 45 °C, where it is near zero, is fairly consistent and is at least partly due to concomitant effects on stomatal aperture with change in leaf-to-air vapour pressure differences.

Kriedemann and Canterford (1971) found net photosynthesis of pear leaves to show a broad temperature optimum of 20–30 °C, at which it was about 24 mg CO_2 dm^{-2} h^{-1}, with lower rates at 35 °C (*c.* 21 mg CO_2 dm^{-2} h^{-1}) and at 10 °C and 40 °C (*c.* 16 mg CO_2 dm^{-2} h^{-1}) under laboratory conditions of saturating light and high relative humidity.

Effects of water stress and water potential

Water stress is used as the descriptive term for an imbalance between the supply of and the demand for water. It is accompanied by changes in plant water potential which may or may not have deleterious effects on plant processes. Lakso (1979) found that net photosynthesis can occur at very low apple leaf water potentials and that substantial reduction of photosynthesis may not occur until the water potential (ψ) falls below −30 bar. Kriedemann and Canterford (1971) found that pear leaf photosynthesis did not decline until leaf water potential had fallen to below −30 bar although a history of water shortage reduced the daily maximum rate of photosynthesis. Photosynthesis in apples and pears can withstand much lower water potentials than is the case in many annual crops and grapes. There is also some evidence of lower photosynthetic rates at very high as well as very low water potentials (Flore and Lakso, 1989). It may be that photosynthesis declines above an optimum cell volume.

Lakso (1979) showed a high correlation between net photosynthesis and stomatal conductance in leaves from field and potted apple trees with varying water stresses imposed. Lakso (1994) has suggested that the close coupling between photosynthesis and stomatal behaviour may indicate that photosynthesis controls stomatal behaviour as well as vice versa.

Effects of flooding

Flooding results in dramatic reductions in leaf photosynthesis of apple. Fernandez *et al.* (1997) found that flooding of 'Jonnee' apples on a range of rootstocks reduced net photosynthesis to less than 20% of that of control trees, some trees showing deleterious effects within 3 days.

Effects of atmospheric CO_2

Minchin *et al.* (1997) found that lowering CO_2 in a leaf chamber from a typical ambient level of 360 ppm to 160 ppm reduced the leaf photosynthetic rate from 0.8 to 0.3 μmol min^{-1}. Carbon dioxide depletion is not usually considered as a major factor operating in the orchard.

Corelli-Grappadelli and Magnanini (1993) found a doubling of net photosynthesis at high (>1500 μmol m^{-2} s^{-1}) PAR levels when enclosed mature cropping trees of 'Golden Delicious'/'M.27' were fed with 900 ppm instead of ambient CO_2. The effect of higher CO_2 levels was smaller at lower light intensities, the net result being that the light response curve was always higher. The photosynthesis rates at high CO_2 levels continued to rise with increasing light intensities well beyond the level at which the photosynthesis/light response curves at ambient CO_2 had reached light saturation. This has implications for potential photosynthesis if atmospheric CO_2 concentrations rise.

Effects of leaf nitrogen

Kriedemann and Canterford (1971) showed that the photosynthesis rates of nitrogen-deficient pear leaves were only about a quarter as high as those with high nitrogen levels (Figure 8.3). This effect was associated with lower chlorophyll levels but the latter explained only about half of the reduction in photosynthesis. Effects on enzyme production may also be implicated.

Effects of leaf type, history and structure

There is ample evidence that leaves exposed to high light levels (sun leaves) have a higher photosynthetic capacity per unit leaf area than leaves developed in the shade (Figure 8.2). They have a higher specific leaf weight (mg cm^{-2}) as discussed earlier and Barden (1978) found a close linear relationship (r = 0.926) between net photosynthesis (P_n) and the specific leaf weight of detached leaves taken from interior (shaded) and peripheral (exposed) positions within mature apple trees. Campbell *et al.* (1992) found that spur leaves from the outer, high-light zone of apple trees had higher specific leaf weight than interior and intermediate leaves from full bloom onwards. They consistently throughout the season had higher net photosynthesis at all light levels and had higher light saturation levels. These effects may result from the greater number of palisade layers, and higher chlorophyll and nitrogen content per unit leaf area commonly found in 'sun' leaves. Asada and Ogasawara (1998) found that leaves of young trees of 'Fuji' grown in the open had maximum photosynthesis rates of about 29 mg CO_2 dm^{-2} h^{-1} at light saturation of 800 μmol m^{-2} s^{-1} PPFD. Those from trees grown at increasing levels of shade had corresponding

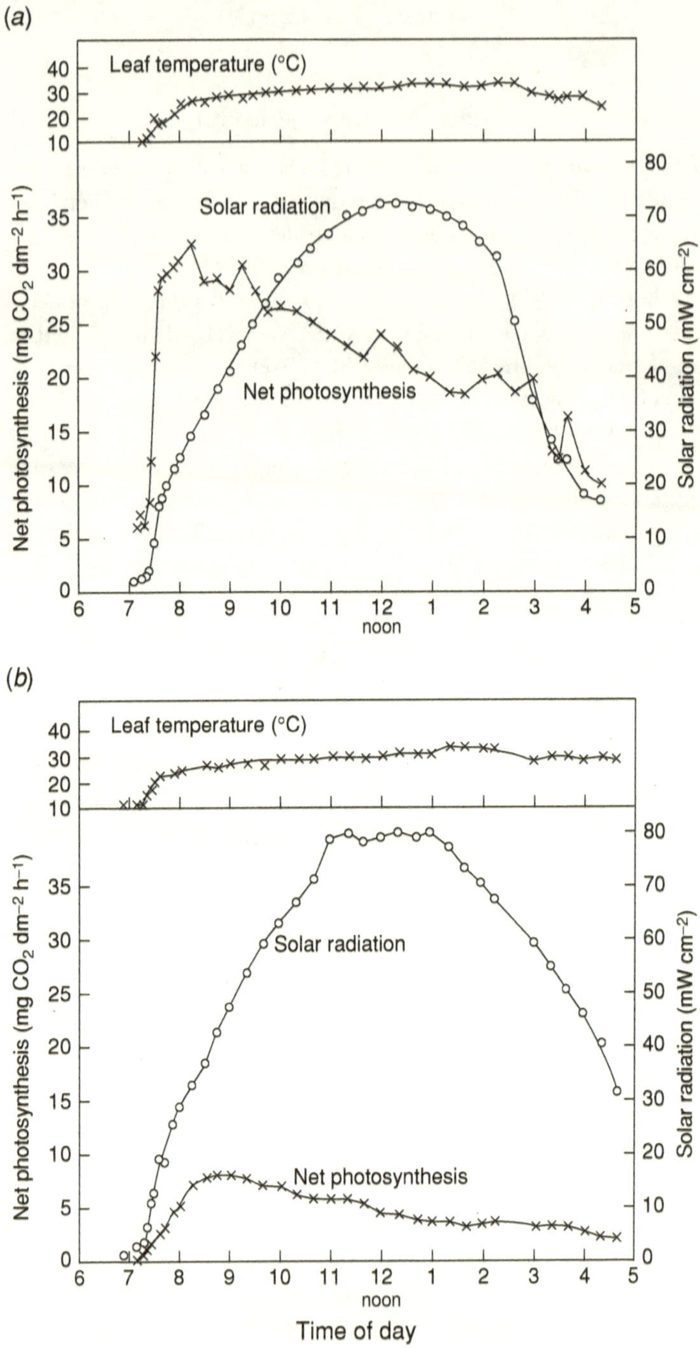

Figure 8.3 Diurnal changes in photosynthesis, leaf temperature and solar radiation: *Pyrus communis* (L) cv. 'Williams' Bon Chrétien' ('Bartlett') trees growing at a high (a) and a low (b) level of nitrogen nutrition. From Kriedemann and Canterford (1971). Reproduced with permission.

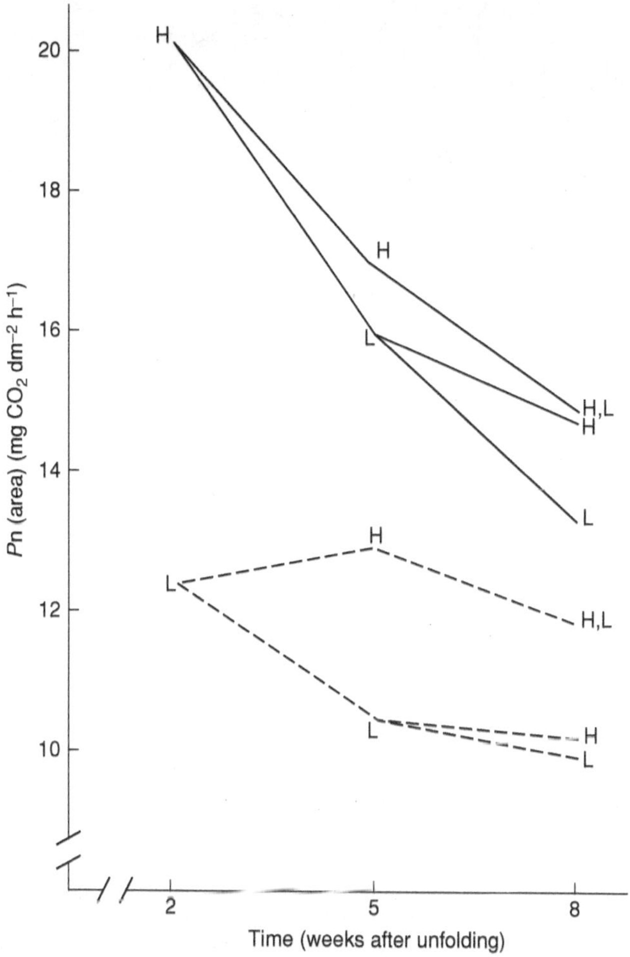

Figure 8.4 Net photosynthesis rates at light saturation of apple leaves as affected by light regime during the preceding weeks. H, full greenhouse sunlight; L, 20% of full greenhouse sunlight. From Barden (1974). Reproduced with permission.

changes in leaf structure and lower maximum rates, e.g. 15 mg CO_2 dm^{-2} h^{-1} by leaves from trees grown at 34% full sunlight.

By comparing all sequences of sun versus heavy shade over three 3-week periods starting in April, Barden (1974) showed that net photosynthesis at saturating illumination was increased by exposure to high light even after leaf expansion had ceased, although the effect of the light climate in the earliest period of leaf development was greatest (Figure 8.4). This is in keeping with effects on specific leaf weight.

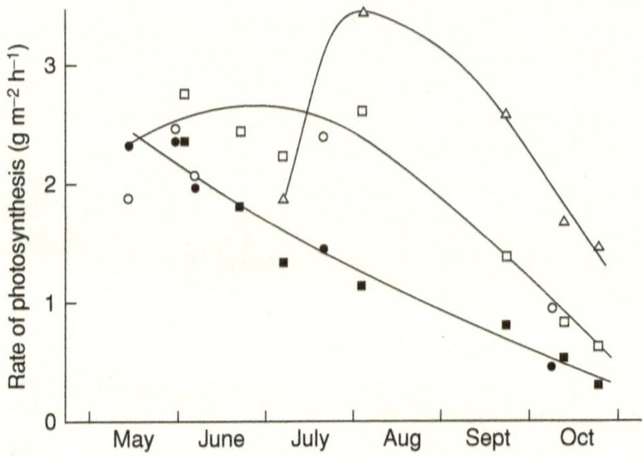

Figure 8.5 Changes in the rate of leaf photosynthesis of different leaf types. ●, spur leaves 1980; ■, spur leaves 1981; ○, vegetative spur leaves 1980; □, basal extension shoot leaves 1981; △, mid-position extension shoot leaves 1981. East Malling, UK. From Palmer (1986).

Extension shoot leaves show much higher leaf photosynthesis at saturating light intensity (>1000 μE m^{-2} s^{-1}) than spur leaves (Figure 8.5). This is in keeping with their greater thickness and chlorophyll content (Schechter *et al.*, 1992).

Effects of sink activity

Numerous studies, reviewed by Avery (1975, 1977) and Flore and Lakso (1989), have shown that there is greater production of dry matter per unit leaf area by fruiting than by defruited trees. Expressed as average net assimilation rates over the season, Avery (1969) calculated that the leaves on fruiting trees were 21 % more efficient. Some of the implicit enhancement of photosynthesis by fruiting may have been due to poorer leaf development, therefore less shaded canopies, in the cropping trees. In a number of cases, however, the presence of fruit caused an increase in total dry matter production (Chandler 1934; Maggs 1963; Avery 1969).

Direct measurements have confirmed that the presence of fruits can lead to substantial enhancement of photosynthesis (Avery 1977; Lenz 1986; Palmer 1992; Greer *et al.*, 1997). There have also been many experiments in which no such effects were demonstrated, possibly as a result of experimental conditions in which photosynthesis was source-limited or non-fruit sinks were of dominant importance (Flore and Lakso, 1989). Palmer (1992) compared heavily cropping trees of 'Crispin'/'M.27' with ones thinned to various levels, including complete deblossoming, under English conditions. Only in July

Table 8.1 *Mean rates of light-saturated net photosynthesis of leaves of 6-year-old orchard-grown trees of 'Braeburn'/'M.26' apple in relation to crop load during the growing season in New Zealand*

Number of fruits per tree	Net photosynthesis (μmol m^{-2} s^{-1})				
	23 Nov	20 Dec	25 Jan	14 Mar	24 April
330	15.7	15.3	16.1	13.8	9.4
250	15.1	14.3	15.6	13.9	9.8
140	15.7	14.1	14.7	12.7	10.3
0	13.9	12.2	10.8	9.8	10.3
$LSD_{0.05}$	1.59	1.16	1.65	1.81	1.63

Based on data presented by Greer *et al.* (1997). Reproduced with permission.

and August, the period of maximum fruit dry matter increment, was light-saturated photosynthesis significantly higher (+47%) in un-thinned than in totally deblossomed trees, with intermediate photosynthesis rates at intermediate cropping levels. Greer *et al.* (1997) subsequently obtained very similar results in New Zealand (Table 8.1). Chlorophyll fluorescence studies indicated that the non-cropping trees had a higher proportion of the absorbed energy dissipated as heat rather than by the photochemical process. The leaves of these non-cropping trees had increased starch concentration possibly leading to feadback inhibition of photosynthesis (Wünsche and Palmer, 1997b).

Giuliani *et al.* (1997a, b) found that whole canopy photosynthesis was higher for fruiting than non-fruiting apple trees of cv. 'Smoothee'/'Pajam 2', particularly in the afternoon. This was associated with increased canopy conductance (see also Chapter 12).

Daily pattern of photosynthesis

The daily pattern of pear leaf photosynthesis under conditions of ample solar radiation shown in Figure 8.3 has also been noted for apple. Most commonly maximum photosynthesis occurs before solar noon (Cheng and Luo, 1997) and rates are lower in the afternoon at similar levels of irradiance and optimal temperatures. Part of the decline in the afternoon may be due to accumulation of assimilates and feedback inhibition: Giuliani *et al.* (1997a, b) found that at saturating light intensity fruiting trees showed only a slight decline in photosynthesis from morning to afternoon whereas this was more pronounced in non-fruiting trees. On the other hand, where irradiance levels are limiting, the daily course of photosynthesis reflects these. Mika and

Antoszewski (1974) found that photosynthesis rates on the east sides of north–south oriented hedgerows peaked in mid-morning, and those on the west sides in the early afternoon, reflecting the daily course of illumination on the two sides.

Effects of leaf age

Kennedy and Fuji (1986) found that, following apple leaf emergence in spring, net photosynthesis at saturating irradiance increased rapidly as the leaves expanded, reaching a peak of 40 to 43 mg CO_2 dm^{-2} h^{-1} at about 60–75% of full leaf expansion. Rates then dropped by about 25% to a plateau which could be maintained for more than 5 months, to mid-October, before beginning a rapid decline. Porpiglia and Barden (1980) found that well-exposed apple spur leaves showed little change in photosynthesis for about 4 months. In contrast, Palmer (1986) and others have found ageing effects from quite early in the season (Figure 8.5).

Seasonal patterns

Seasonal patterns reflect the ageing of individual leaves, which has been slow in most studies, and a very rapid decline due to, or triggered by, environmental factors at the end of the season. Cheng and Luo (1997), measuring photosynthesis in the field, found very little decline in the daily maximum photosynthesis rate of mid-position shoot leaves in Shandong, China, over the months of May to October inclusive but a sharp decline in daily total CO_2 gain in September and, especially, October. This appears to relate to the lower irradiance and leaf temperatures in the later part of the day in those months. Lakso and Lenz (1986) found that autumnal senescence and reduction in photosynthesis could be stopped by transferring apple trees to a regime with higher temperatures (18 °C day, 10 °C night) than those obtaining outdoors.

Photosynthesis by flowers, fruits and stems

These organs may contain chlorophyll and fix carbon dioxide. Vemmos and Goldwin (1993, 1994) found that before flower opening, sepals and receptacles had chlorophyll concentrations similar to those found in rosette leaves although they decreased subsequently. At the green-cluster stage flowers contributed 40% to the green surface area, 30% to total chlorophyll and 25% to the number of stomata of floral clusters. Flower gross photosynthesis was about a third of that of leaves at the balloon stage and at 15 days after full bloom, although much less at full bloom. Photosynthesis by flowers apparently contributed 15–33% of their own carbohydrate need.

Apple fruits have chloroplasts in the green tissue of the hypodermal and inner perivascular tissues (Blanke and Lenz, 1989). Hypodermal chloroplasts in the five to six layers below the epidermis are smaller than in the inner tissues, resemble those found in leaves and exhibit grana throughout fruit development (Phan, 1973). Inner central tissues in young fruits are more photosynthetically active than in mature fruits but still less active than the hypodermis. The fruit surface has stomata but these are only about 1–10% as frequent as those in leaves. The fruits also, however, possess a system which re-fixes CO_2 from the mitochondrial respiration of predominantly imported carbon. This pathway produces malate by the action of phosphoenolpyruvate carboxylase (PEPC). Fruit photosynthesis appears to be intermediate in status between C_3 non-autotropic and C_4/CAM photosynthesis. Fruit photosynthesis often compensates for respiratory CO_2 loss in the light but, due to respiration in the dark, there is a continual net loss of CO_2 from fruits throughout their development. Nevertheless, given the very high proportion of the tree mass which is made up of fruits, their ability to compensate for respiratory losses in the light is very important.

Respiration

Respiration is the process in which carbon substrates are utilized in the production of energy.

$$C_6H_{12}O_6 + 6O_2 \longrightarrow 6CO_2 + 6H_2O + \text{energy}$$

Low-energy bonds in the carbon substrates are converted to high-energy bonds in reduced nucleotides (NADH, NADPH and $FADH_2$) and ATP. The oxidation of one glucose molecule has the potential to yield 36 ATP molecules. Respiration is divided into that associated with the biosynthesis of new plant tissue (growth respiration) and that needed for organ maintenance (maintenance respiration). These definitions have to be used with care and are, to some extent, oversimplifications. Not all growth has the same energy requirement: in particular, the cell division phase may involve much more energy use than that of cell expansion. Furthermore, processes such as those involved in senescence, and translocation and storage of metabolites which cannot be construed as growth, are energy-intensive. Measurements of respiration are of particular value as an index of when energy-requiring processes are taking place or being inhibited. They are also used to help construct carbon-balance sheets and help in understanding the effects of environmental factors on these, and, consequently, on potential growth and cropping. Attention has been concentrated on effects of temperature and on seasonal patterns which may or may not be temperature-dependent. Another important aspect, the post-harvest respiration of fruits, is discussed in Chapter 10 also.

Root respiration

The root respiration of apple trees differs from their shoot respiration in that it responds to increases in temperature in winter even when very few chilling units have been accumulated and the overall chilling requirement is not satisfied. Under these circumstances, Young (1990) found the Q_{10} for respiration, i.e. the ratio between the rates of respiration at temperature $(T + 10)$ °C to that at T °C, to be 2.26. However, both respiration rates and the effects of temperature on these are greater when the chilling requirement is satisfied.

Later in the season, Buwalda *et al.* (1992) found that root respiration, under conditions of high irradiation and nitrogen supply, declined from about 5.3 to 2.8 nmol CO_2 g^{-1} dry wt s^{-1} between 46 and 138 days after budburst. Shading and low nitrogen also reduced root respiration. Ebert (1991) found that potassium and phosphorus deficiency, and reduction of the oxygen content in the root medium, particularly below 2% O_2, reduce root respiration. Andersen *et al.* (1985), however, found that the respiration rates of roots of *Pyrus* species and *Cydonia oblonga* (quince) were much less affected by anaerobiosis in solution culture than were those of *Prunus persica*, being reduced by no more than 50% after 21 days of anaerobiosis.

Ebert (1991) found that the respiration of 'M.9' roots increased linearly with rising temperature.

Respiration of the above-ground framework

Respiration by the above-ground tree framework (wood respiration) is low when the trees are dormant. Butler and Landsberg (1981) showed that the respiration of bare branches rises rapidly when physiological activity resumes.

In apple and pear this is dependent on the satisfaction of the winter-chilling requirement, or chemical treatments to substitute for this, as well as the rise in temperatures in spring. During endodormancy and the accumulation of their chilling requirement, apple and pear trees carry on metabolic functions through maintenance respiration. As trees enter ecodormancy and begin accumulating heat units respiration usually increases significantly. Young *et al.* (1995) concluded that the trees need to have received 80% of their chilling requirement before shoot respiration will increase with higher temperatures during ecodormancy. The Q_{10} for respiration increases from 1.5 to 2.5 as trees complete their chilling requirement and the respiratory energy of activation (EA) decreases from 18 to 12 kcal mol^{-1}. As trees are chilled up to 80% of their chilling requirement, their respiratory quotient (the ratio of CO_2 evolved to O_2 consumed in respiration) increases from 0.25 to 1. This may indicate a shift from the use of lipids as a significant substrate for respiration early in the dormancy period, to carbohydrates. The respiration of both bud and internode tissue increases under warm temperatures after adequate chilling. This indicates that physiological changes related to carbohydrate utilization occur throughout the shoot during budbreak.

Stem respiration peaks at about the time the leaves emerge from the buds, presumably at the time of maximum mobilization of reserves. In England the calculated peak of about 0.05 mg m^{-2} surface area s^{-1} (20 °C values) occurred in late April, and by late May had fallen to half of this (Butler and Landsberg, 1981).

Leaf respiration

Leaf dark respiration (R_d) per unit area is greatest at full bloom and immediately afterward, and then declines. R_d is greater in leaves from the exterior parts of apple trees, well exposed to light, than in leaves from inner or intermediate zones (Porpiglia and Barden, 1980; Campbell *et al.*, 1992). At full bloom and full bloom + 2 weeks, Campbell *et al.* (1992) found rates of 1.8 μmol m^{-2} s^{-1} for exterior leaves and 1.0–1.2 μmol m^{-2} s^{-1} for interior leaves, declining to 1.3 for exterior leaves and 0.2–0.4 for interior leaves during the period from full bloom + 6 weeks to full bloom + 22 weeks. Porpiglia and Barden (1980) found a decline from about 1.9 mg CO_2 dm^{-2} h^{-1} in early May to about 1 mg CO_2 dm^{-2} h^{-1} in August and September for exposed leaves and correspondingly lower values for interior leaves. The differences in R_d between sun and shade leaves is reduced if respiration is expressed per unit leaf weight instead of area. Barden (1974) showed that, as with photosynthetic rate and specific leaf weight, artificial shading could induce changes in R_d even after leaf expansion ceased. Leaf dark respiration is greater in fruiting than in non-fruiting trees. This may be partly due to stomata opening wider and staying open to later in the evening in the fruiting trees (Blanke, 1997). Temperature is the dominant environmental factor controlling leaf respiration. Watson *et al.* (1978) showed an excellent linear relationship between the logarithm of leaf dark respiration and temperature, R_d increasing ten-fold between 5 °C and 35 °C (Figure 8.6) according to the equation $R_d = 0.006e^{kT}$ where $k = 0.09$. Higgins *et al.* (1992) found a similar temperature effect for apple, with a k value of 0.079, and appreciably lower R_d values for Asian pear.

Proctor *et al.* (1976) found R_d values for the stems and leaves of leafy shoots of 'Golden Delicious' apple to be similar whether on a dry weight or a fresh weight basis (Table 8.2), but the dark respiration of leafy shoots little more than doubled between 14 °C and 30 °C.

Table 8.2 *Estimates of dark respiration (R_d) of parts of an apple tree expressed on various bases*

	R_d (μg CO_2 h^{-1})		
Tree part	g dry wt^{-1}	g fresh wt^{-1}	cm^{-2}
Leaf	930	366	12
Stem	910	335	50
Fruits	75	10	10
Roots	394	352	–

Values for leaf area for one side only. Values obtained with air temperatures from 20°C to 30°C and not corrected for temperature.
From Proctor *et al.* (1976). Reproduced with permission.

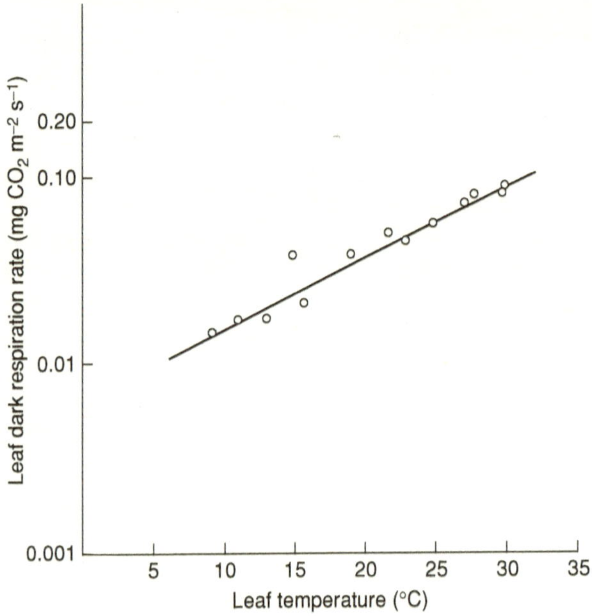

Figure 8.6 The response of dark respiration (R_d) of 'Golden Delicious' apple leaves to temperature. The fitted line is described by the equation $R_d = 0.006e^{kT}$ where $k = 0.09$. From Watson *et al.* (1978). Reproduced with permission.

Fruit respiration

Fruit respiration rates are high early in the season during the cell division phase of fruit growth when, for apple, they can be about 120 ng CO_2 g^{-1} fresh wt s^{-1}(= 432 mg CO_2 kg^{-1} h^{-1}) at four weeks after full bloom (Jones,

1981). They then decline, much more steeply than is the case for other plant parts, to less than 3 ng g^{-1} fresh wt s^{-1} by late September. This is followed by the climacteric rise at maturity discussed in a later chapter. Bepete and Lakso (1997) found some apples to have R_d values of more than 250 ng g^{-1} fresh weight s^{-1} when only about 10 mm diameter, falling to less than 10 when over 35 mm diameter. Indeed, maximum respiration per fruit occurred at only 15–20 mm diameter. Fruit respiration showed an exponential temperature response under controlled conditions, the equation for the response curve being $Y = 0.0695\ (10^{0.614X})$ where Y is net carbon dioxide evolution and X is the temperature of the fruit. Despite this, exposed fruits, with temperatures as much as 10 °C above shaded fruits, have much lower 'respiration' as measured by CO_2 output. In the morning they showed no output, the CO_2 presumably being re-fixed by fruit photosynthesis, and even later in the day gave out less than a quarter of the CO_2 given out by shaded fruits.

Net CO_2 exchange by canopies

Net carbon dioxide exchange (NCE) of canopies, measured by enclosing whole trees in plastic chambers (cuvettes) and recording the changes in CO_2 between the air pumped in and that coming out, integrates the effects of photosynthesis and respiration of leaves, stems, buds, flowers and fruits including the ways in which organs affect the CO_2 exchanges of other organs.

Canopy light response curves

Canopy net photosynthesis light response curves show lower maxima at saturating light intensity than do individual well-exposed leaves or shoots (Proctor *et al.*, 1976). This is because whole trees or branches have more respiratory losses from non-photosynthetic tissues and include shade-adapted leaves with low photosynthetic potential as well as having low light intensities in parts of the canopy.

Daily and seasonal patterns

The whole canopy carbon exchange follows the diurnal pattern of incoming radiation, with some evidence of higher rates of net carbon assimilation per unit of light in the morning than in the afternoon (Figure 8.7). It also follows a pattern that reflects changes in incident radiation and leaf area through the season (Figure 8.8). Net photosynthesis and carbon assimilation per unit leaf area peak very early in the season with maximum rates of 8.3 and 7.7 μmol CO_2 m^{-2} s^{-1} in April at Bonn in northern Europe (Wibbe *et al.*, 1993), but net photosynthesis per tree peaked in August at 3.6 g CO_2 tree^{-1} h^{-1} (4.2 μmol CO_2 m^{-1} s^{-1}), and dark respiration, although also high in mid-summer, peaked in October

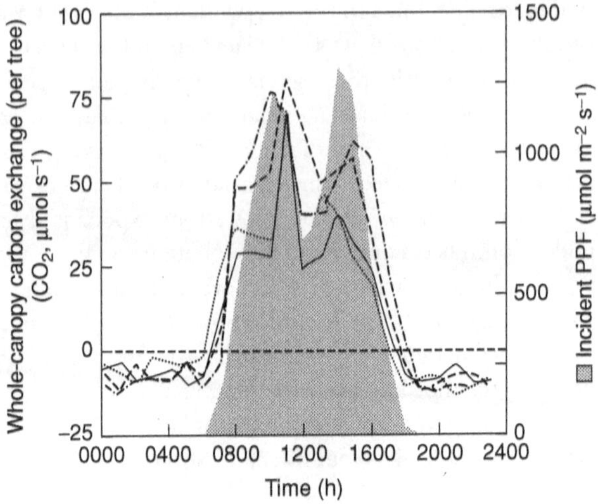

Figure 8.7 Diurnal patterns of whole canopy carbon exchange of 4 cropping 'Braeburn' apple trees *c.* 25 weeks after full bloom. The shaded area represents photosynthetic photon flux. From Wünsche and Palmer (1997a). Reproduced with permission.

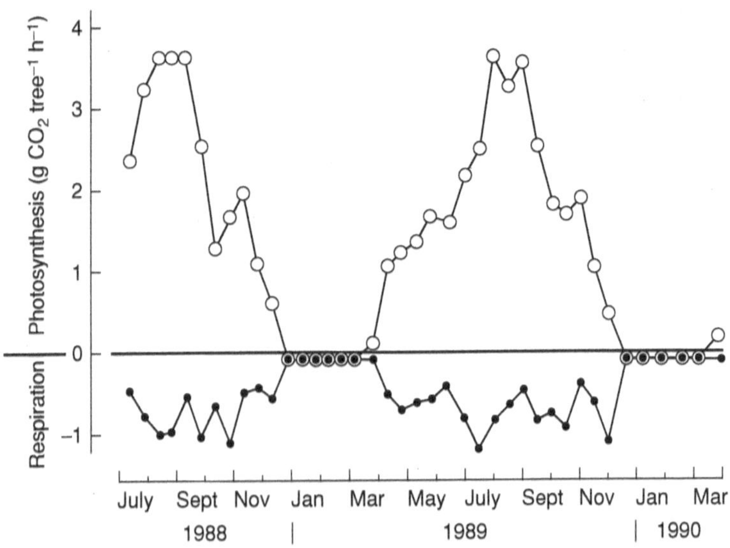

Figure 8.8 Seasonal course of net photosynthesis in the light and dark respiration of 'Golden Delicious' apple trees at Bonn, Germany. Values based on weekly 48-h measurements. From Wibbe *et al.* (1993). Reproduced with permission.

at 1.2 g CO_2 tree^{-1} h^{-1} (1.4 μmol CO_2 m^{-1} s^{-1}). The high level of respiration late in the season was attributed to the energy needs of translocation of carbohydrates from the leaves to the woody parts of the trees. Dark respiration levels are very low in the winter months, following leaf shed.

Measured effects of light and temperature

Over any short period it is expected that high light intensities will have a positive effect on net carbon exchange, through effects on photosynthesis, and high temperatures a negative effect, through effects on respiration.

Francesconi *et al.* (1997) found that shading to 23% of full sun reduced the net carbon exchange rate (NCER) of potted trees of 'Gala'/'M.9' by about 60%. Trees at 31.4 °C had approximately 60% lower NCER than those at 21.3 °C and trees at 15.3 °C about 20% higher NCER than the latter at around noon. The differences in temperature had substantial effects on vapour pressure deficit, which may have been involved in the effects on NCER. They were not accompanied by differences in irradiation such as would frequently, though not invariably, occur in nature in the daytime but may explain the negative relationship between night temperature and NCE found by Wünsche and Palmer (1997b). They were obtained over single days.

Over longer periods of time the effects of higher temperatures may be less negative. In a number of crops it has been found useful to divide total respiration, R_t, into two components, maintenance (R_m) and synthetic respiration (R_s). R_m is associated with the maintenance of established physiologically active tissue and increases in proportion to the amount of tissue to be maintained. As the metabolic rate of cells increases rapidly with temperature, R_m is strongly dependent on temperature. R_s is associated with the synthesis of new cellular material and on a daily basis is governed by two main factors, the level of respiratory substrate which is normally determined by the photosynthetic rate, and the temperature, which governs the maximum rate of synthesis that can occur. When photosynthesis is limited, e.g. by dull weather or drought, R_s is governed mainly by the rate of assimilate supply. Dewar *et al.* (1999) have concluded that in general over the long term, the respiration to photosynthesis ratio is relatively insensitive to temperature, but more information is needed for apples and pears. This is important in relation to understanding effects of latitude, and climate in general, on orchard productivity and also the use of shading and orchard misting treatments in areas of high insolation.

Higher temperatures in spring and autumn increase leaf area duration, with shoots becoming net exporters of photosynthates earlier in the season (Johnson and Lakso, 1986) and leaf senescence being delayed (Lakso and Lenz, 1986). Both of these effects can be expected to have positive impacts on net carbon exchange.

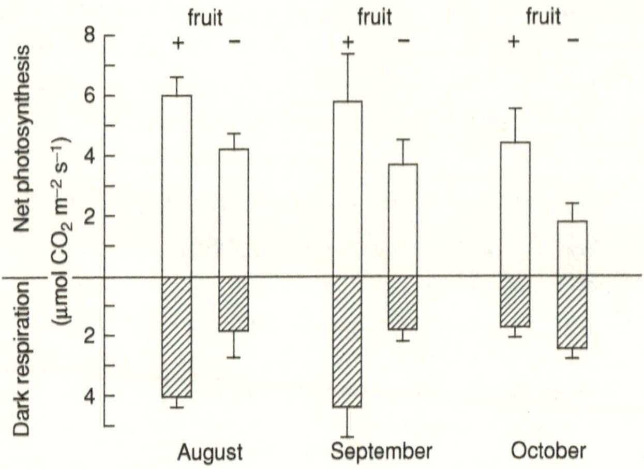

Figure 8.9 Effect of defruiting [either] in August, September or October on net photosynthesis and dark respiration of 'Golden Delicious' apple trees. Each column represents the mean of continuous measurements over 5 days prior to (left columns) or after fruit removal. From Wibbe and Blanke (1995). Reproduced with permission.

Effects of fruiting

Wibbe *et al.* (1993) and Wibbe and Blanke (1995) found that fruiting trees had much higher net photosynthesis from June to October inclusive and higher dark respiration from June to September inclusive, than trees without fruits (Figure 8.9). The parallel effects on net photosynthesis and on dark respiration may have been attributable to defruiting effects on stomatal behaviour, since dark respiration largely depends on stomatal aperture. Defruiting increased net respiration in October and had a very negative effect on NCER, which may be associated with earlier leaf senescence and translocation of carbohydrates for storage in de-fruited trees.

Wünsche and Palmer (1997b) found that fruiting may not affect NCE early in the season, presumably because of compensatory shoot and leaf growth on the trees with few or no fruits. As shoot and leaf growth ceased the sink strength effect of fruits was evidenced by higher CO_2 uptake by fruiting than non-fruiting trees.

Source–sink relationships and carbohydrate partitioning

In many annual crop plants the greatest gains in economic yield have been achieved by breeding and selection of cultivars which direct a greater proportion of their assimilates into the harvested product, i.e. have a higher 'harvest index' rather than a higher total biological yield (Gifford and Evans, 1981). Similarly for apples, the use of dwarfing rootstocks has resulted in a much increased ratio of crop yield to total dry matter production. Their perennial growth habit, however, introduces further complexities. In the early years after planting it is usually desirable that the trees grow rapidly, as well as bear crop, in order to attain high levels of light interception, hence potential yield, on an orchard basis. Moreover, excessive cropping can inhibit fruit bud formation, although not necessarily as a result of effects on carbohydrate partitioning, and reduce cropping in the following year. It can also reduce shoot growth in both the year of cropping and the subsequent year, so influencing future cropping potential. Knowledge of the relative 'sink strength', i.e. power to attract assimilates, of different tissues and organs throughout the year is therefore basic to effective tree management.

In general the underlying basis of sink strength (assuming a pressure-flow mechanism of phloem translocation) is thought to be the ability to lower the concentration of photosynthate in the sieve elements servicing the sinks and thus establish a favourable concentration gradient between source and sink (Wardlaw, 1990). This is achieved either by rapid metabolic use of the carbohydrate being transported or its conversion into a storage compound.

The supply of assimilate to any particular sink, either directly or after storage, depends on source limitations and on the competitive strength of the sink. There are many examples of source-limitation, e.g. photosynthesis is low in the shaded parts of trees and the supply of carbohydrates in spring, from storage, can be reduced by premature autumn defoliation (Abusrewil and Larsen, 1981).

An approximate measure of net carbon increment and its partitioning over the season is given by changes in dry matter, more than 90% of the dry matter produced by the plant originating from photosynthesis by the leaves (Hansen, 1977). Figure 8.10 shows the cumulative dry matter production of fruiting and non-fruiting apple trees. Up to approximately the end of August (Northern hemisphere) the total dry matter increment of the fruiting trees exceeded that of the non-fruiting trees, i.e. the latter were sink-strength limited. Throughout the season the dry matter increment of root and stem was much greater in defruited than fruiting trees, showing the effects of the fruits, as the dominant sink, on vegetative growth.

Relative sink strength is not, however, an intrinsic and unvarying property: it is a consequence, not a primary cause, and varies with the time of season and the physiological state of the tissues or organs. Shoots form terminal buds and cease growth early in the season. Fruits, especially those of early maturing cultivars, approach their final size well before cessation of leaf activity. Stem thickening and root growth may occur much later and different patterns of

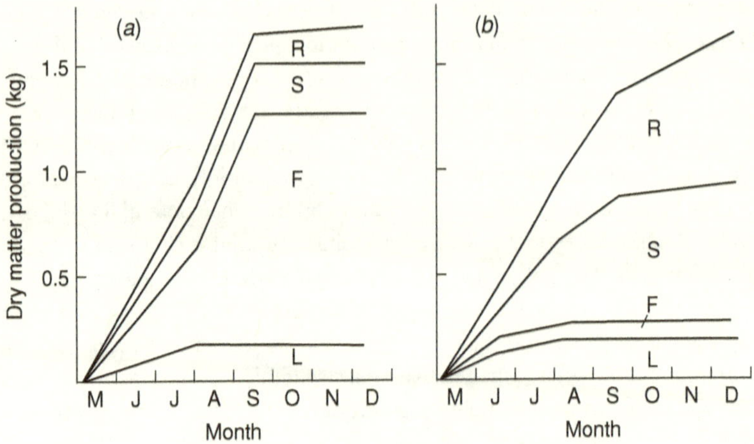

Figure 8.10 Cumulative dry matter production of fruiting (*a*) and defruited (*b*) trees of 'Starkspur Golden Delicious' at Montpellier, France. Fruiting trees include weight of shed fruits, de-fruited trees weight of fruitlets removed in May. R, root; S, shoot; F, fruit; L, leaf. From Heim *et al.* (1979). Reproduced with permission.

dormancy in different parts of the tree structure may also result in changes in competitive ability. For this reason it is important to look at competition through the year.

Kandiah (1979) found that leaves fed with $^{14}CO_2$ in October still contained 27% of the total plant extractable activity at leaf fall and the stem above the upper fed leaves had negligible amounts of the tracer. By December the label was primarily in the wood of the old stems and, especially, the roots. Forty-eight per cent of the extractable activity and 35% of the activity in the residues at that time was in the roots.

In the spring following autumn supply of $^{14}CO_2$ to the leaves, radioactivity is concentrated in the first-formed leaves although present in all new leaf and shoot growth (Quinlan, 1969). Hansen (1967c) found that over the period from leaf fall to spring after leafing-out there is a large reduction in the content of previously supplied ^{14}C in the root system and bark. This stored carbohydrate is very important as a source of respiratory energy as well as for shoot growth. Circumstances that reduce carbohydrate availability in the autumn, and stored carbohydrate levels in winter, have a number of adverse effects. Hand defoliation of single-shoot apple nursery trees (whips) at the beginning of October followed by storage over winter and planting in spring led to low carbohydrate content, and poor tree survival and growth, compared with controls or trees defoliated later (Abusrewil and Larsen, 1981). Such small trees would have both low reserves, even in the absence of defoliation, and very high demand for both root and shoot growth. Early harvesting of fruits, which should enable more assimilates to be stored over winter, increased flowering and fruit set in the following year (Williams *et al.*, 1980). This effect may have resulted from increased carbohydrate availability during fruit bud development in autumn or increased reserves available in spring. Defoliation of fruiting trees after harvest delays blossoming and reduces fruit set (Faby and Naumann 1986; Tustin *et al.*, 1997), spur leaf size and fruit size (Tustin *et al.*, 1997). There was no evidence as to whether the defoliation had direct (autumnal) or deferred (winter and spring) effects on the developmental processes. Faby and Naumann showed that defoliation prevented the translocation of nitrogen out of the leaves to the bark but N fertilization did not overcome the negative effects of defoliation.

Vigorous vegetative shoot growth in spring appears to compete successfully with fruitlet growth of apple and to lead to increased fruit drop. It can be prevented by shoot tip removal (Quinlan and Preston, 1971; Quinlan, 1975). Saunders *et al.* (1991) obtained similar results by pruning pear shoots before and at the time of anthesis. Pruning had a particularly large effect on the set of seedless fruits, which are normally considered to be weaker competitors than seeded fruits. Saunders *et al.* (1991) considered that the effect of pruning before any visible shoot growth indicated that the inhibition of fruitlet development was a result of correlative dominance rather than direct competition.

Later in the season vigorous shoot growth, e.g. that resulting from heavy pruning, leads to increased fruit shed during the 'June drop' period (Quinlan, 1975) but subsequently the fruits become increasingly dominant. The more fruits are carried by a tree the smaller the proportion of the total annual dry matter increment which goes to the stems and shoots and, especially, to the roots (Figure 8.10). Fruit–fruit competition is also very evident as crop load increases. Hansen (1969) found that the leaf area required to meet the assimilate needs of a fruit of 'Golden Delicious' apple in August was 230 cm^2 (17 leaves). The corresponding figures for 'Gravenstein', an earlier maturing but larger fruited cultivar were 670 cm^2 or 42 leaves. Other cultivars, measured in different environments, required up to 1175 cm^2 (50 leaves) to give maximum fruit size. At lower ratios of leaf area to fruit there is increasingly strong competition between fruits and reduced fruit size. The relative sink strength of different fruits is determined very early in their lives. The cells of incipient apple flesh double 21 times before anthesis and only four or five times thereafter (Coombe, 1976), and the potential number of cells may be already evident in differentiated flowers in the previous autumn (Bergh, 1985). Fruit sink-strength potential is greater on 'strong' spurs with good vascular connections and in seeded than in seedless fruits (Abbott, 1984). However, it may also be modified by within-tree conditions in the growing season. Thorpe (1974) showed that the temperature differences between fruits in shaded and exposed parts of apple trees could in large measure, through their effects on metabolism and sink strength, account for the larger size of exposed fruits.

Minchin *et al.* (1997) showed experimentally that apple fruitlets import photosynthate at a rate matching their utilization capacity, which in turn can adapt to supply. Increases in availability of photosynthate cause only a small increase in import at first but over several hours further and much larger increases occur, suggesting an increase in enzyme activity induced by increased availability. Short-term reduction in photosynthate supply resulted in an incapacity of the fruitlet to return to its initial import rate, indicative of the reduction in activity of enzymes. Where two fruits were supplied by a single leaf fed with $^{11}CO_2$ the partitioning between them responded to changing their relative temperatures.

Net carbon exchange and orchard productivity

Dry matter increment in an orchard over the season is dependent on:

1 The percentage of available sunlight intercepted (Chapter 7).
2 The irradiance per unit surface,
3 The efficiency with which light energy is used in dry matter production.
4 Carbon losses by respiration.

The yield of fruit depends on the total dry matter increment and the proportion of this directed into fruit.

Percentage light interception is the major factor determining differences in yield between different orchard systems, especially in the years before attainment of final canopy dimensions (Chapter 7). It also contributes to the higher yields at lower latitudes where growing seasons are longer and leaf area durations greater (Heim *et al.*, 1979).

Irradiance varies with latitude, cloud cover and altitude (Wagenmakers, 1991; Jackson, 1997, 2000). It generally increases with decreasing latitude down to around 30°, further increases as the equator is approached being confined to autumn and winter months. Growing-season irradiance is, however, higher over most of the summer months in fruit-growing areas of New Zealand (41° S) than those of South Africa (34° S) as well as being about 50% higher than those in Kent, England (51° N). It decreases with cloud cover, so is higher in arid regions, and increases with altitude. Light energy is the driving force of photosynthesis although not the sole controlling factor. The photosynthesis of a single, fully exposed, leaf is light-saturated at a relatively low irradiance level, but this saturation level is higher the greater the irradiance under which the leaf is grown. The leaf area receiving light energy above any specific level is, moreover, a function of above-canopy irradiance as well as the light transmitting characteristics of the canopy (Chapter 7). Assuming a closed canopy with a transmission coefficient (K) of 0.6, Monteith (1981) calculated that a doubling of above-canopy irradiance from 10 to 20 MJ m^{-2} day^{-1} could double canopy dry matter production. In fruit trees with low LAI values the effect is less but still appreciable. Wagenmakers (1991) estimated apple orchard fruit production potential from canopy photosynthesis assuming a 5-month growing season with respiration costing 40% of gross photosynthesis. With a constant LAI of 2.5, energy differences due to latitude alone would result in fruit yields being about 11 % higher at 35° than at 55°. Where cloudiness is also taken into account the potential yield difference becomes 37%, i.e. about 2 t ha^{-1} per degree of latitude. If a higher LAI value (4.5 instead of 2.5) is assumed for 35° compared with 55°, the relative potential yields become about 172 t ha^{-1} compared with about 98 t ha^{-1} i.e. a ratio of 1.75 to 1 essentially due to irradiance intensity. Higher yields at lower latitudes are further a result of higher temperature giving a longer growing season. These conclusions are borne out by growth and yield data. Cripps (1972) found that shading to give 70% of full sunlight in Western Australia, i.e. to levels still well above those of northern Europe in energy terms, reduced total dry matter production by 10%. Folley (1973) estimated that yields in southern France were about 40% higher than in southern England and, whereas absolute maximum experimental yields recorded in Denmark and England are 90 and 100 t ha^{-1} (Wagenmakers 1991), commercial orchards have attained yields of 130–170 t ha^{-1} in New Zealand (McKenzie, 1981), some yields having reached 180 t ha^{-1}. Experimental yields have reached 140 t ha^{-1} in Tasmania (Jotic, 1981) and at least 130 t ha^{-1} in Israel (Bravdo, 1986).

High irradiance levels are frequently, but not invariably, accompanied by high temperatures. Night-time temperatures are generally lower in relation to day-time temperatures and irradiation in arid areas. Day and, especially, night temperatures are lower in relation to irradiation at higher altitudes. This may influence the balance of respiratory losses to photosynthetic gain.

Many factors, e.g. leaf nitrogen status and water-stress effects on stomatal behaviour influence the efficiency with which light energy is used in dry matter production. By and large they are optimized by management practices developed to overcome obvious adverse effects on growth. Selection of cultivars, e.g. spur-types with thicker leaves and more palisade layers, giving higher photosynthesis per unit area, may achieve increased photosynthesis without a commensurate increase in within-canopy shade. This may also be achieved by the use of plant growth regulators.

The net carbon budget for apple trees, defined as the difference between the amount of CO_2 lost at night and gained during the day, follows the net photosynthesis curve over the season, with a fairly consistent ratio of dark respiration to net photosynthesis in June, July, August and September of 16–25% in western Europe. Dark respiration increases sharply in October, partly as a consequence of fruit reaching a respiratory climacteric, while net photosynthesis declines (Wibbe *et al.*, 1993).

It is not correct to regard the dry matter going to fruit as a proportion of an independently-controlled dry matter production. Net photosynthesis by a fruiting apple tree canopy can be 2–2.5 times as high as that of a corresponding de-fruited tree and dark respiration about 50% higher, resulting in a two-fold carbon gain of the fruiting compared with the non-fruiting tree (Wibbe *et al.*, 1993). The rise in net photosynthesis may be attributable to reduced photorespiration, to increased canopy conductance and to feedback control of assimilation (Wibbe *et al.*, 1993; Giuliani *et al.*, 1997a,b).

The effects of temperature on fruit set and fruit growth, discussed in the next chapter, are therefore major determinants of net photosynthesis per unit leaf or canopy. High levels of fruit production, however, reduce root and shoot growth through competition for resources and therefore check leaf canopy development. The overall effect of fruiting on orchard dry matter production thus depends on timescale and the extent to which canopy development has reached the level at which it will be controlled.

References

Abbott, D.L. (1984). *The Apple Tree: Physiology and Management.* London: Grower Books.

Abusrewil, G.S. and Larsen, F.E. (1981). Tree fruit nursery stock defoliation: carbohydrate levels pre- and post storage and shoot length of Delicious apple hand stripped or treated with 'Dupont WK Surfactant' and ethephon. *Acta Horticulturae* **120**, 83–8.

Andersen, P.C., Montano, J.M. and Lombard, P.B. (1985). Root anaerobiosis, root respiration, and leaf conductance of peach willow, quince and several pear species. *HortScience* **20**, 248–50.

Asada, T. and Ogasawara, M. (1998). The effect of shading on the growth of young 'Fuji' apple trees. *Journal of the Japanese Society for Horticultural Science* **67**, 655-9.

Avery, D.J. (1969). Comparisons of fruiting and deblossomed maiden apple trees and of non-fruiting trees on a dwarfing and an invigorating rootstock. *New Phytologist* **68**, 323–6.

Avery, D.J. (1975). Effects of fruits on photosynthetic efficiency. In *Climate and the Orchard*, ed. H.C. Pereira, pp. 110–12. Farnham Royal, UK: Commonwealth Agricultural Bureaux.

Avery, D.J. (1977). Maximum photosynthetic rate: a case study in apple. *New Phytologist* **78**, 55–63.

Barden, J.A. (1974). Net photosynthesis, dark respiration, specific leaf weight, and growth of young apple trees as influenced by light regime. *Journal of the American Society for Horticultural Science* **99**, 547–51.

Barden, J.A. (1978). Apple leaves, their morphology and photosynthetic potential. *HortScience* **13**, 644–6.

Bepete, M. and Lakso, A.N. (1997). Apple fruit respiration in the field: relationships to fruit growth rate, temperature and light exposure. *Acta Horticulturae* **451**, 319–26.

Bergh, O. (1985). Effect of the previous crop on cortical cell number of *Malus domestica* cv Starking Delicious apple flower primordia, flowers and fruit. *South African Journal of Plant and Soil* **2**, 191–6.

Bieleski, R.L. (1969). Accumulation and translocation of sorbitol in apple phloem. *Australian Journal of Biological Science* **22**, 611–20.

Bieleski, R.L. (1977). Accumulation of sorbitol and glucose by leaf slices of *Rosaceae*. *Australian Journal of Plant Physiology* **4**, 11–24.

Blanke, M.M. (1997). Effect of fruit load on whole tree carbon assimilation, dark respiration and water relations in apple. *Acta Horticulturae* **451**, 313–17.

Blanke, M.M. and Lenz, F. (1989). Fruit photosynthesis – a review. *Plant, Cell and Environment* **12**, 31–46.

Bravdo, B-A. (1986). Effect of environmental factors on leaf photosynthesis, leaf resistance and yield of apples and palm dates. In *The Regulation of Photosynthesis in Fruit Trees*, ed. A.N. Lakso and F. Lenz, pp. 75–9. Special Publication, New York State Agricultural Experiment Station, Geneva.

Butler, D.R. and Landsberg, J.J. (1981). Respiration rates of apple trees, estimated by CO_2-efflux measurements. *Plant, Cell and Environment* **4**, 153–9.

Buwalda, J.G., Fossen, M. and Lenz, F. (1992). Carbon dioxide efflux from roots of calamodin and apple. *Tree Physiology* **10**, 391–401.

Campbell, R.J., Marini, R.P. and Birch, J.B. (1992). Canopy position affects light response curves for gas exchange characteristics of apple spur leaves. *Journal of the American Society for Horticultural Science* **117**, 467–72.

Chandler, W.H. (1934). Dry matter residue of trees and their products in proportion to leaf area. *Proceedings of the American Society for Horticultural Science* **31**, 39–56.

Cheng, L. and Luo, X. (1997). Diurnal and seasonal stomatal regulation of water use efficiency in leaves of field-grown apple trees. *Acta Horticulturae* **451**, 375–82.

Coombe, B.G. (1976). The development of fleshy fruits. *Annual Review of Plant Physiology* **27**, 507–28.

Corelli-Grappadelli, L. and Magnanini, E. (1993). A whole-tree system for gas-exchange studies. *HortScience* **28**, 41–5.

Cripps, J.E.L. (1972). The effect of shading and Alar application on apple root:shoot ratios in Western Australia. *Journal of Horticultural Science* **47**, 291–9.

Dewar, R.C., Medlyn, B.E. and McMurtie, R.E. (1999). Acclimation of the respiration/photosynthesis ratio: insights from a model. *Global Change Biology* **5**, 615–22.

Ebert, G. (1991). Root respiration of apple trees. I. Influence of environmental factors. *Erwerbsobstbau* **33**, 227–30.

Faby, R. and Naumann, W.D. (1986). Effects of defoliation of apple trees after harvest. II. Mineral and carbohydrate contents in shoots, crop yield. *Gartenbauwissenschaft* **51**, 136–42.

Fernandez, R.T., Perry, R.L., Flore, J.A. and MacLean, R.M. (1997). Photosynthesis, ^{14}C-photosynthate distribution and shoot and root growth of young apple trees on 3 rootstocks exposed to flooding. *Acta Horticulturae* **451**, 351–9.

Flore, J.A. and Lakso, A.N. (1989). Environmental and physiological regulation of photosynthesis in fruit crops. *Horticultural Reviews* **11**, 111–57.

Folley, R.R.W. (1973). An estimate in the net difference in yield of apple orchards in northern and in southern Europe attributable to climate differences; with notes upon the economic significance of superiority in yield. *Scientia Horticulturae* **1**, 43–55.

Francesconi, A.H.D., Lakso, A.N. and Denning, S.S. (1997). Light and temperature effects on whole-canopy net carbon dioxide exchange rates of apple trees. *Acta Horticulturae* **451**, 287–94.

Gifford, R.M. and Evans, L.T. (1981). Photosynthesis, carbon partitioning and yield. *Annual Review of Plant Physiology* **32**, 485–509.

Giuliani, R., Corelli-Grappadelli, L. and Magnanini, E. (1997a). Effects of crop load on apple photosynthetic responses and yield. *Acta Horticulturae* **451**, 303–11.

Giuliani, R., Nerozzi, F., Magnanini, E. and Corelli-Grappadelli, L. (1997b). Influence of environmental and plant factors on canopy photosynthesis and transpiration of apple trees. *Tree Physiology* **17**, 637–45.

Grant, C.R. and ap-Rees, T. (1981). Sorbitol metabolism by apple seedling. *Phytochemistry* **20**, 1505–11.

Greer, D.H., Wünsche, J.N. and Palmer, J.W. (1997). Effect of fruiting on seasonal apple leaf chlorophyll fluorescence. *Acta Horticulturae* **451**, 345–50.

Hansen, P. (1967a). ^{14}C-studies on apple trees. I. The effect of the fruit on the transport and distribution of photosynthates. *Physiologia Plantarum* **20**, 382–91.

Hansen, P. (1967b). ^{14}C-studies on apple trees. II. Distribution of photosynthates from top and base leaves from extension shoots. *Physiologia Plantarum* **20**, 720–5.

Hansen, P. (1967c). ^{14}C-studies on apple trees. III. The influence of season on storage and mobilization of labelled compounds. *Physiologia Plantarum* **20**, 1103–11.

Hansen, P. (1969). ^{14}C-studies on apple trees. IV. Photosynthate consumption in fruits in relation to the leaf–fruit ratio and to the leaf–fruit position. *Physiologia Plantarum* **22**, 186–98.

Hansen, P. (1970). ^{14}C-studies on apple trees. V. Translocation of labelled compounds from leaves to fruit and their conversion within the fruit. *Physiologia Plantarum* **23**, 564–73.

Hansen, P. (1977). Carbohydrate allocation. In *Environmental Effects on Crop Physiology*, ed. J.J. Landsberg and C.V. Cutting, pp. 247–58. London: Academic Press.

Heim, G., Landsberg, J.J., Watson, R.L. and Brain, P. (1979). Ecophysiology of apple trees: dry matter production and partitioning by young Golden Delicious trees in France and England. *Journal of Applied Ecology* **16**, 179–94.

Higgins, S.S., Larsen, F.E., Bendel, R.B., Radamaker, G.K., Bassman, J.H., Bidlake, W.R. and Al Wir, A. (1992). Comparative gas exchange characteristics of potted, glasshouse-grown, almond, apple, fig, grape, olive, peach and Asian pear. *Scientia Horticulturae* **52**, 313–29.

Honjo, H., Kotobuki, K., Asakura, T., and Kamoto, F. (1990). Net photosynthesis and dark respiration of pear leaves infected by Japanese Pear Rust. *Journal of the Japanese Society for Horticultural Science* **59**, 99–105.

Jackson, J.E. (1997). Light interception and canopy characteristics at low latitudes in relation to orchard system design. *Acta Horticulturae* **451**, 417–25.

Jackson, J.E. (2000). Light regimes in temperate fruit-tree orchards grown at low latitudes. In *Temperate Fruit Crops in Warm Climates*, ed. A. Erez, pp. 1–15. Dordrecht: Kluwer.

Johnson, R.S. and Lakso, A.N. (1986). Carbon balance model of a growing apple shoot. II. Simulated effects of light and temperature on long and short-shoots. *Journal of the American Society for Horticultural Science* **111**, 164–9.

Jones, H.G. (1981). Carbon dioxide exchange of developing apple (*Malus pumila* Mill) fruits. *Journal of Experimental Botany* **32**, 1203–10.

Jotic, P. (1981). Performance of six apple orcharding systems in Tasmania. *Acta Horticulturae* **114**, 344–5.

Kandiah, S. (1979). Turnover of carbohydrates in relation to growth of apple trees. II. Distribution of ^{14}C assimilates labelled in autumn, spring and summer. *Annals of Botany* **44**, 185–95.

Kennedy, R.A. and Fuji, J. (1986). Seasonal and developmental changes in apple photosynthesis: enhancement effects due to flowering and fruit maturation. In *The Regulation of Photosynthesis in Fruit Trees*, ed. A.N. Lakso and F. Lenz, pp. 27–29. Special Publication, New York State Agricultural Experiment Station, Geneva.

Kriedemann, P.E. and Canterford, R.L. (1971). The photosynthetic activity of pear leaves (*Pyrus communis* L.). *Australian Journal of Biological Science* **24**, 197–205.

Lakso, A.N. (1979). Seasonal changes in stomatal response to leaf water potential in apple. *Journal of the American Society for Horticultural Science* **104**, 58–60.

Lakso, A.N. (1994). Apple. In *Handbook of Environmental Physiology of Fruit Crops*, Vol. 1. *Temperate Crops*, ed. B. Schaffer and P.C. Anderson, pp. 3–42. Boca Raton, FL: CRC Press.

Lakso, A.N. and Johnson, R.S. (1990). A simplified dry matter production model for apple using automatic programming simulation software. *Acta Horticulturae* **276**, 141–8.

Lakso, A.N. and Lenz, F. (1986). Regulation of apple tree photosynthesis in the autumn by temperature. In *The Regulation of Photosynthesis in Fruit Trees*, ed. A.N. Lakso and F. Lenz, pp. 34–7. Special Publication New York State Agricultural Experiment Station, Geneva.

Lakso, A.N., Piccioni, R.M., Denning, S.S., Sottile, F. and Costa Tura, J. (1999). Validating an apple dry matter production model with whole canopy gas exchange measurements in the field. *Acta Horticulturae* **499**, 115–22.

Lenz, F. (1986). Fruit effects on transpiration and dry matter production in apples. In *The Regulation of Photosynthesis in Fruit Trees*, ed. A.N. Lakso and F. Lenz, pp. 101–4. Special Publication, New York State Agricultural Experiment Station, Geneva.

Looney, N.E. (1968). Comparison of photosynthetic efficiency of two apple cultivars with their compact mutants. *Proceedings of the American Society for Horticutural Science* **92**, 34–6.

Maggs, D.H. (1963). The reduction in growth of apple trees brought about by fruiting. *Journal of Horticultural Science* **38**, 119–28.

McKenzie, D.W. (1981). The ideal apple tree unit in New Zealand. *Acta Horticulturae* **114**, 343.

Mika, A. and Antoszewski, R. (1974). Photosynthesis efficiency of apple trees trained as hedgerows. *Fruit Science Reports* **1**, 10–17.

Minchin, P.E.H., Thorpe, M.R., Wünsche, J.N., Palmer, J.W. and Picton, R.F. (1997). Carbon partitioning between apple fruits: short- and long-term response to availability of photosynthate. *Journal of Experimental Botany* **48**, 1401–6.

Monteith, J. (1981). Does light limit crop production? In *Physiological Processes Limiting Plant Productivity*, ed. C.B. Johnson, pp. 23–38. London: Butterworths.

Oliveira, C.M. and Priestley, C.A. (1988). Carbohydrate reserves in deciduous fruit trees. *Horticultural Reviews* **10**, 403–30.

Palmer, J.W. (1986). Seasonal variation of light saturated photosynthesis rate of Golden Delicious apple leaves as influenced by leaf type and crop load. In *The Regulation of Photosynthesis in Fruit Trees*, ed. A.N. Lakso and F. Lenz, pp. 30–3. Special Publication, New York State Agricultural Experiment Station, Geneva.

Palmer, J.W. (1989). Canopy manipulation for optimum utilization of light. In *Manipulation of Fruiting*, ed. C.J. Wright, pp. 245–62. London: Butterworths.

Palmer, J.W. (1992). Effects of varying crop load on photosynthesis, dry matter production and partitioning of Crispin/M27 apple trees. *Tree Physiology* **11**, 19–33.

Phan, E.T. (1973). Chloroplasts of the peel and the internal tissues of apple fruits. *Experimentia* **29**, 1555–7.

Porpiglia, P.J. and Barden, J.A. (1980). Seasonal trends in net photosynthetic potential, dark respiration and specific leaf weight in apple as affected by canopy position. *Journal of the American Society for Horticultural Science* **105**, 920–3.

Priestley, C.A. (1983). Interconversions of ^{14}C-labelled sugars in apple tree tissues. *Journal of Experimental Botany* **34**, 1740–7.

Proctor, J.T.A., Watson, R.L. and Landsberg, J.J. (1976). The carbon budget of a young apple tree. *Journal of the American Society for Horticultural Science* **101**, 579–82.

Quinlan, J.D. (1969). Mobilization of ^{14}C in spring following autumn assimilation of $^{14}CO_2$ by an apple rootstock. *Journal of Horticultural Science* **44**, 107–10.

Quinlan, J.D. (1975). Reduction of crop yield by growth competition. In *Climate and the Orchard*, ed. H.C. Pereira, pp. 106–9. Farnham Royal, UK: CAB.

Quinlan, J.D. and Preston, A.P. (1971). The influence of shoot competition on fruit retention and cropping of apple trees. *Journal of Horticultural Science* **46**, 525–34.

Saunders, R.C., Jacobs, G. and Strydom, D.K. (1991). Effect of pruning on fruitset and shoot growth of 'Packham's Triumph' pear trees. *Scientia Horticulturae* **47**, 239–45.

Schechter, I., Proctor, J.T.A. and Elfving, D.C. (1992). Morphological differences among apple leaf types. *HortScience* **27**, 101–3.

Sharkey, T.D. (1988). Estimating the rate of photorespiration in leaves. *Physiologia Plantarum* **73**, 147–53.

Thorpe, M.R. (1974). Radiant heating of apples. *Journal of Applied Ecology* **11**, 755–60.

Tustin, D.S., Stanley, C.J. and Adams, H.M. (1997). Physiological and phenological responses of apple trees to artificial reduction of the growth period from harvest to leaf fall. *Acta Horticulturae* **451**, 383–92.

Vemmos, S.N. and Goldwin, G.K. (1993). Stomatal and chlorophyll distribution of 'Cox's Orange Pippin' apple flowers relative to other cluster parts. *Annals of Botany* **71**, 245–50.

Vemmos, S.N. and Goldwin, G.K. (1994). The photosynthetic activity of 'Cox's Orange Pippin' apple flowers in relation to fruit setting. *Annals of Botany* **73**, 385–91.

Wagenmakers, P.S. (1991). Planting systems for fruit trees in temperate climates. *Critical Reviews in Plant Sciences* **10**, 369–85.

Wardlaw, I.F. (1990). The control of carbon partitioning in plants. *New Phytologist* **116**, 341–81.

Watson, R.L., Landsberg, J.J. and Thorpe, M.R. (1978). Photosynthetic characteristics of the leaves of 'Golden Delicious' apple trees. *Plant, Cell and Environment* **1**, 51–8.

Wibbe, M. and Blanke, M.M. (1995). Effects of defruiting on source–sink relationship, carbon budget, leaf carbohydrate content and water use efficiency of apple trees. *Physiologia Plantarum* **94**, 529–33.

Wibbe, M., Blanke, M.M. and Lenz, F. (1993). Effect of fruiting on carbon budgets of apple tree canopies. *Trees* **8**, 56–60.

Williams, R.R., Arnold, G.M., Flook, V.A. and Jefferies, C.J. (1980). The effect of picking date on blossoming and fruit set in the following year for the apple cv Bramley's Seedling. *Journal of Horticultural Science* **55**, 359–62.

Wünsche, J.N. and Palmer, J.W. (1997a). Portable through-flow cuvette system for measuring whole-canopy gas exchange of apple trees in the field. *HortScience* **32**, 653–8.

Wünsche, J.N. and Palmer, J.W. (1997b). Effects of fruiting on seasonal leaf and whole canopy carbon dioxide exchange of apple. *Acta Horticuturae* **451**, 295–301.

Young, E. (1990). Changes in respiration rate and energy of activation after chilling and forcing dormant apple trees. *Journal of the American Society for Horticultural Science* **115**, 809–14.

Young, E., Dautlick, K. and Belding, R.D. (1995). Respiratory changes during dormancy breaking of apple trees. *Acta Horticulturae* **395**, 21–33.

Young, E., Motomura, Y. and Unrath, C.R. (1987). Influence of root temperature during dormancy on respiration, carbohydrates and growth resumption in apple and peach. *Journal of the American Society for Horticultural Science* **112**, 514–19.

9

Flowers and fruits

The defining characteristics of apple and pear flowers and fruits were summarized in Chapter 2 and the morphology and anatomy of apple flowers and fruits comprehensively reviewed by Pratt (1988). Flowers are initiated and develop within the buds borne terminally on fruiting spurs (short shoots) and terminally or laterally on long shoots. These buds, as discussed in Chapter 6, consist essentially of a shortened shoot axis with a 'leaf formation' inserted at each node. These 'leaf formations' in apple commonly consist of about nine bud scales, three transition leaves, six true leaves and three bracts (cf. Figure 6.2, p. 159). Flower primordia may form at the apex and in the axils of the bracts and the three uppermost leaves (Abbott, 1970). Alternatively, the buds may fail to develop far enough to produce flowers and remain vegetative.

Juvenility

Seedling apple and pear trees usually remain vegetative for several years after seed germination. During this 'juvenile' phase they readily form adventitious roots, as discussed in Chapter 3, are often thorny and have a distinctive leaf shape and, especially, cannot be induced to flower (Zimmerman, 1972). They emerge from this juvenile state after reaching a minimum size, characteristic of the cultivar, measured as height or number of main stem nodes. The duration of the juvenile phase can be shortened by growing the seedling tree under conditions which speed up its growth or by specific horticultural practices such as grafting the seedling on to a dwarfing rootstock. Failure to flower as a result of juvenility is primarily of importance to fruit breeders. Apple and pear trees in the orchard are usually compound trees in which the scionwood grafted or budded on to the rootstock was taken from trees already in the mature, non-juvenile phase.

Flowering

Initiation of flowering

Flower bud development involves the transformation of the vegetative apex to a reproductive structure (Buban and Faust, 1982). This can occur in different buds on the same tree at very different times so is obviously not triggered by a specific stimulus of temperature or daylength. Following detailed studies of the morphogenesis of apple buds, Fulford (1966b) postulated that the bud meristems will always form flowers unless prevented from doing so. The transformation from a vegetative to a reproductive bud takes place only when the structure of the vegetative bud is complete. The effects of the large number of factors that influence the proportion of buds giving rise to flowers have generally been interpreted in terms of an in-built propensity to flowering and interference with the attainment of this. The controlling factors may involve hormonal balance (Luckwill, 1974; Hoad, 1984), availability of nutrients, especially carbohydrate (Sachs, 1977), and the interaction of these (Ryugo, 1986).

Most flowers and fruits are borne on spurs and Abbott (1977) described the sequence of events starting with a vegetative apple spur bud in winter. This breaks in spring to form either a leafy shoot or, more commonly, a rosette of leaves in the centre of which a new bud starts to form. At inception this consists of one immature scale and five or six leaf primordia. As the season advances it adds primordia at the apex while the older ones become transformed into bud scales or transition leaves. Activity may then slow down and the bud go into dormancy, or the bud may become floral. In the latter case the apex ceases to provide new leaf primordia, but instead produces bract primordia and flower initials.

Whether the bud becomes floral or not seems to depend mainly on the rate of production of new primordia, i.e. on the plastochron, or interval of time between production of successive primordia (Fulford, 1966b, c; Abbott, 1977). A 'strong' growing point does not form a bud but grows out as a leafy shoot. A weak growing point with a long plastochron fails to reach the node number, i.e. the number of primordia, at which flowers form. Fulford concluded that a plastochron of 18 days is associated with failure of the buds to form flowers, and Abbott (Figure 9.1) concluded that flower initiation takes place when about 20 nodes have developed in buds of 'Cox's Orange Pippin'. Failure to attain this number of nodes before the onset of dormancy is therefore considered to lead to failure to initiate flowers. In England the average plastochron leading to flowering of 'Cox's Orange Pippin' is about 7 days. An apex starting with six nodes at the beginning of the season must therefore grow for about 100 days to attain the 20-node stage. Flower initiation would not, therefore, be expected until about mid-August (Luckwill, 1974). Critical

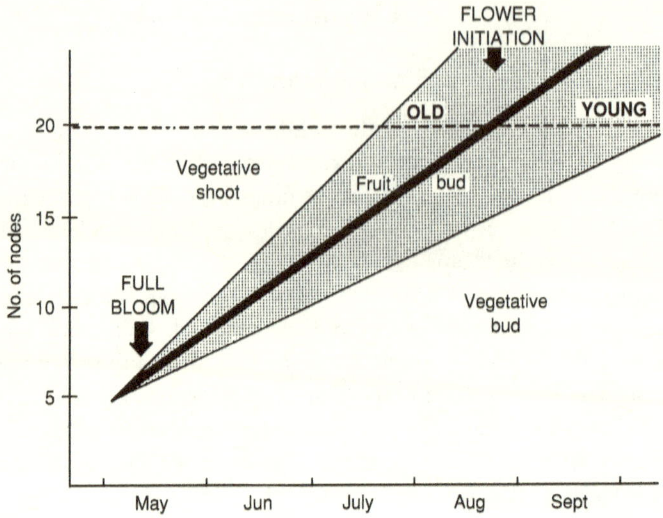

Figure 9.1 Flower initiation and fruit bud differentiation in relation to node production within the bud. From Abbott (1977). Reproduced with permission.

node numbers of 16 for 'Golden Delicious' apple, 20 for 'Comice' pear and 12 for Japanese pear have been reported by Luckwill and Silva (1979), Dheim and Browning (1988) and Banno *et al.* (1985a, b) respectively. The critical node number may be lower for buds on long shoots than for spur buds, as discussed later (p. 277).

The first doming of the bud apex (Figure 9.2), indicating the change from the vegetative to the floral condition, usually coincides with the cessation of shoot growth of apple (Fulford, 1966b; Luckwill and Silva, 1979; Dencker and Hansen, 1994) and of Japanese pear (Banno *et al.*, 1985a, b, 1986). This may be a causal relationship (Luckwill, 1974). Where it appears to be the case, the positive effect of checking shoot growth on flower development may be very localized and is probably due to weakening of correlative inhibition. Some treatments appear, however, to influence flower initiation and development and shoot growth independently (Tromp, 1970, 1973). Development of flower buds does not start uniformly throughout the tree. The terminal buds of short shoots (spurs) begin their transformation to flower buds 4–6 weeks earlier than do lateral buds (Buban, cited by Faust, 1989). Young spurs show earlier flower initiation than old spurs. In general, where flower induction occurs relatively late in the season the fruit buds are 'young' at the onset of winter, are relatively narrow and pointed, and give rise to blossom clusters with large primary leaves and long-stalked flowers which tend to set few fruits (Abbott, 1970, 1977, 1984; Luckwill, 1974).

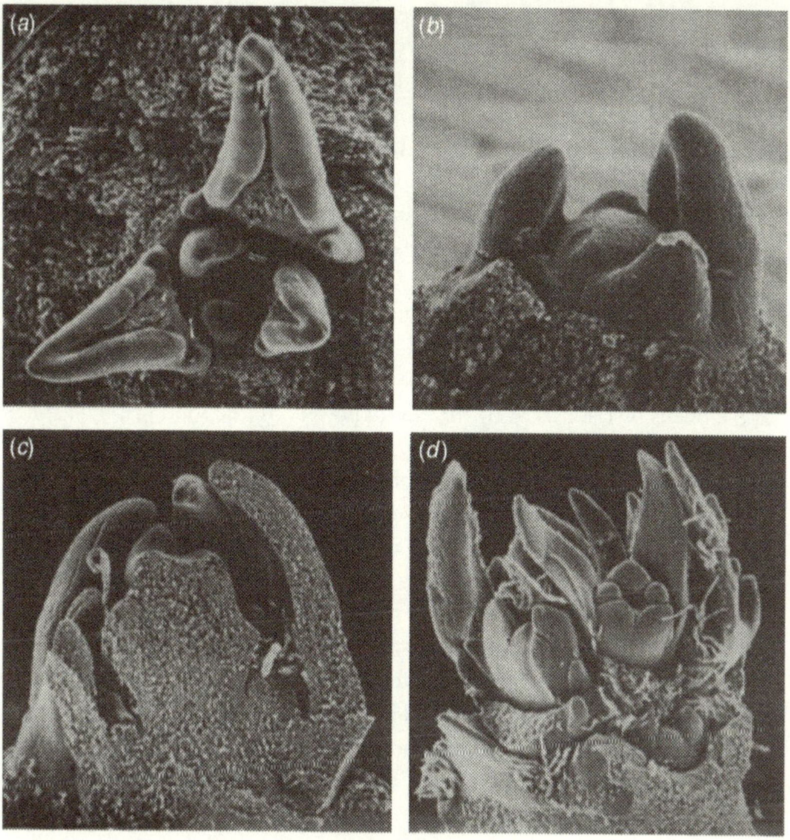

Figure 9.2 Scanning electron micrographs of stages of bud development. (*a*) Vegetative apex showing five leaf primordia – remainder removed (× 62), 1.8.74. (*b*) Apex becoming domed and ridges arising in axils of four leaf/bract primordia remainder removed (× 118), 15.8.74. (*c*) Longitudinal section at an early stage of flower development showing the 'king' flower at the apex (× 80), 21.8.74. (*d*) Floral apex showing bourse bud and three lateral flower primordia. The five sepals forming on the uppermost flower initial can be easily seen. Subtending leaves and 'king' flower removed (× 47), 19.9.74. (Photos by Jane Reeves). From Abbott (1977). Reproduced with permission.

Differentiation of the growing point

The first detectable change in the growing point of the bud after induction of flowering is increased synthesis of DNA and RNA (Buban and Faust, 1982; Faust, 1989). The first visible sign of differentiation is when the flat apical meristem becomes domed, then the central meristem is partitioned and the pith meristem develops. The reproductive meristem becomes a block-like structure and its subsequent development is relatively rapid. The order of appearance of the tissues is: sepals, petals, anthers and ovaries; and by leaf fall all parts are present in a large proportion of the flowers (Abbott, 1977; Faust, 1989). The stage of flower bud development at any given time during summer and autumn varies with type of bud (basal shoot bud, middle shoot bud, upper shoot bud and terminal bud) and with cultivar (Crabbé, 1984; Banno *et al.*, 1985b). Buds continue to develop through the winter: apple buds increase in size by 20–25% during December and January and by an additional 120–150% between mid-February and mid-March.

A considerable degree of differentiation of tissues takes place over this period (Faust, 1989). At budbreak, which follows satisfaction of the winter-chilling and heat-unit requirements discussed in Chapter 6, the apical or king flower of apple opens first, followed by the lateral flowers. In the southern hemisphere Bergh (1985a) found that flattening of the apex, marking the change from the vegetative to the reproductive phase, occurred during the first week of January in 'Starking' apple. Sepal primordia of the terminal and lateral flowers were formed simultaneously during the second week of January. The full complement of five sepals of the terminal flower had formed by the last week of January and development of the petal primordia and the first whorl of ten stamens of the terminal flower was completed during the second week of February. The sepals and bracts of lateral flowers were also evident at that time and carpel primordia of the terminal flower could be detected. The first whorl of ten stamens, first and second whorl of five stamens and developing carpels were found in all samples collected during the second week of March, well before leaf fall in May. Floral organs developed slowly during the mid-winter months of June, July and August. Carpels and pollen sacs developed rapidly in September and ovule primordia were distinguished towards the end of this period, approximately 21 days before anthesis. From the third week of September until anthesis in mid-October rapid swelling of the buds coincided with acceleration of elongation of the carpels, formation of pollen sacs and elongation of the filaments of stamens. Development of the bud from the green tip stage to the opening of the terminal flower coincided with further elongation of the floral organs and rapid development of the ovule. Stages of bud development are shown in Figure 9.2 and the structure of the mature flower in Figure 9.3.

Flower bud differentiation and development in Japanese pear (*Pyrus serotina* Rehder) follows a similar pattern to apple and to *P. communis* but Banno *et al.* (1986) reported relatively early initiation and differentiation. The first visible signs of flower bud differentiation on spurs of cv. 'Nijisseiki' in Japan (northern hemisphere) were on 25 June, petal and stamen primordia were initiated by late July and carpel primordia by mid-August.

The whole process of flower bud and flower development, taking place over two consecutive years, is influenced and modified by a large number of

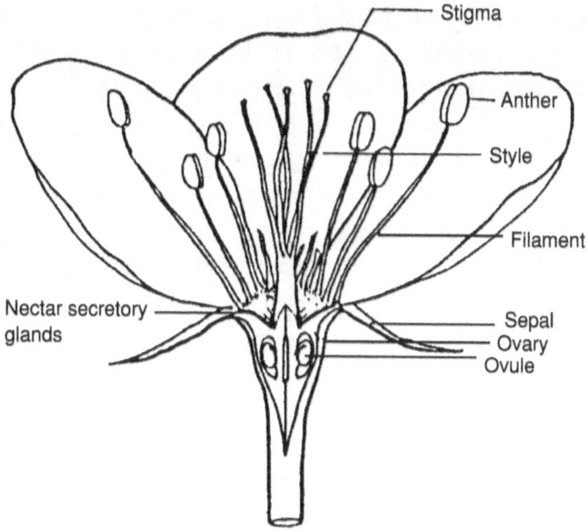

Figure 9.3 Apple flower split longitudinally and diagrammed to show its parts (MacDaniels and Heinicke, 1929).

factors. Variations in the number and 'quality' of flowers at blossom time are the main cause of year-to-year variations in yield in many apple and pear growing regions so much effort is devoted to understanding and controlling the causal factors.

Effects of fruits on flowering

Heavy cropping in one year can inhibit flower bud initiation and so reduce flowering in the following year. Some cultivars consequently tend to become biennial in cropping, with a heavily cropping 'on' year followed by an 'off' year in which crop is light or even absent.

In the orchard biennial bearing may develop over time but for many cultivars it tends to be triggered by frost damage with a consequent low crop and excessive flower bud initiation. In its most severe form it therefore is a whole-tree, indeed a whole-orchard, phenomenon. Heavy fruiting on one branch may reduce flower bud formation on adjacent ones, but the inhibitory effect is usually more localized and Parry (1974) showed that halves of an apple tree can be maintained so that one half is 'on' and one 'off' each year. Moreover, within a single branch fruits may inhibit fruit bud formation only on the part of the branch on which the fruits are borne or parts nearer to the trunk but not those nearer to the end of the branch (Fulford, 1962). Huet (1972, 1973) concluded that flower initiation in spurs is controlled mainly by factors, including presence of fruits, operating within each individual spur system of 'Williams' Bon Chrétien' pear. There is ample evidence of this in apple also; for example, Jonkers (1979) reported that only 2–5% of terminal buds initiate flowers in successive years. This results in a strong negative relationship between fruits per blossom cluster, on whole trees, in one season and number of blossom clusters in the following year (Landsberg, 1979).

The presence of seeded fruits for only 20–30 days after their pollination is enough to prevent about 50% of apple spurs from flowering in the following year, although flowering is further reduced if the fruits are left on for 150 days from pollination (Chan and Cain, 1967). An effect of fruits late in the season was also shown by Williams *et al.* (1980), who harvested 'Bramley's Seedling' apple at four different dates, from 23 August to 26 October. Fruit weight almost doubled between the first and last pickings. The number of primary blossom clusters in the following year was more than 34% higher on the trees harvested on 23 August and 15 September than on those harvested on 4 and 26 October, and the number of secondary blossoms several times as high. In 'Williams' Bon Chrétien' pear Huet (1973) found the greatest adverse effects of seeded fruits on flower bud development to be between 30 and 40 days after full bloom when the fruitlets were of about 15 mm diameter.

Seedless fruits do not have the same adverse effect as seeded fruits. Chan and Cain (1967) hand-pollinated half of a tree of the parthenocarpic apple cultivar 'Spencer Seedless' and adjusted the fruit load so that it did not differ between the seeded and seedless side of the tree. In the following year there was no blossoming on the previously seeded side of the tree, but profuse blossoming on the side where the fruits had been seedless. Similarly in 'Williams'' pear a heavy crop of seedless fruits was followed by heavy flowering, whereas this was totally inhibited by an equivalent crop of seeded fruits (Table 9.1). In Chan and Cain's study there was no difference in inhibition of flowering by different numbers of seeds and the effect of seed number, as contrasted with the effects of presence or absence, was small even in the third year.

The inhibitory effect of fruits on flower bud initiation is associated with a slowing of bud growth from a plastochron of 5 days between successive nodes to one of 18 days (Fulford, 1966c). It is accompanied by poorer vascular development. Izadyar (1997) found that in spurs with fruits there was greater differentiation of vascular bundles towards the fruit pedicels than towards the vegetative buds. These buds, which would normally be 'off', i.e. non-flowering, in the following year were not connected to the spur vascular bundles. In contrast the potentially 'on', i.e. potentially flowering, buds on spurs without fruits were connected with the vascular bundles of the spurs at an early stage. The 'off' buds may therefore suffer relative nutrient deficiency.

The lack of inhibiting effect of seedless fruits and the effects of the position of seeded fruits on their inhibitory influence led to the view that seed-produced

Table 9.1 *The effects of seeds and spur leaf area on floral initiation in 'Williams'' pear*

	Tree with seeded fruits	Tree with seedless fruits
Yield (kg)	26.1	26.9
Number of fruits	215	168
Seeds per fruit	7.6	0
% floral initiation on spurs with the following leaf area (cm^2)		
Less than 30	0	0
(4 leaves) 31–50	0	8.6
(5 leaves) 51–70	0	27.3
(6 leaves) 71–90	0	44.5
Greater than 90	0	87.0

From Huet (1973). Reproduced with permission.

hormones, especially gibberellins, play a major role (Fulford, 1962; Luckwill, 1970, 1974). Hoad (1978) found that more GA and other plant hormones move out of the fruitlets of biennial flowering cultivars than from those which flower regularly. Within any one cultivar, but not between cultivars, the amount of gibberellin in diffusates from excised fruitlets via their cut pedicels is dependent on the number of seeds in the fruitlet. Diffusion out of the fruits continues at a variable rate but without any clear trend with time for at least the first 11 weeks after full bloom (Hoad, 1980). Fruits of 'Spencer Seedless' export much less GA_3 than those of the seeded cvs. 'Elstar' and 'Golden Delicious', with it being undetectable at some dates in June. They can, however, export considerable amounts of GA_4 which does not inhibit flowering (Prang *et al.*, 1997). Exogenously applied GA can inhibit flowering, as discussed later. Hoad (1978) also found more auxin activity in fruit diffusates of the strongly biennial cv. 'Laxton's Superb' than in the less biennial 'Cox's Orange Pippin'. Grochowska *et al.* (1984) found no difference in auxin content between fruiting and non-fruiting (i.e. potentially flowering and non-flowering) spurs of 'Jonathan' and 'McIntosh' apples, but a high content of gibberellins in the former while cytokinins predominated in the latter. Grochowska (1973) also found less starch in fruit-bearing than non-bearing apple spurs and attributed this to an effect of hormones from the seeds.

Effects of leaves on flowering

Defoliation experiments have shown that leaves promote flower bud formation. Ryugo (1986) reported that no flowers formed on apple spurs that had been

Table 9.2 *Effects of imposed shade on fruit bud production of apple cv. 'Cox's Orange Pippin'/'M.26'*

Number of fruit bud clusters per tree in spring 1972	Percentage full daylight in 1971			
	100%	37%	25%	11%
Spur clusters	373	248	190	184
Axillary clusters	221	19	15	12
Terminal clusters	43	12	16	5

The number of each type of bud cluster at each level of shade was significantly less than the number on unshaded trees.
Data from Jackson and Palmer (1977b). Reproduced with permission.

defoliated within 6–10 weeks after full bloom but the number of flower buds formed on spurs defoliated after this period gradually increased. Defoliation after harvest does not reduce flowering of apple in the following year (Tustin *et al.*, 1997). Huet (1972, 1973) showed a very strong correlation between the total leaf area of the spur rosette and floral initiation in the terminal bud of the spurs (Table 9.1).

Fulford (1966c) concluded that the pattern of development of buds on non-fruiting trees following defoliation was very similar to that in which flower initiation was inhibited by fruiting. The effect of the leaves has been attributed to their effects on carbohydrate supply and on the flow of cytokinins (Jonkers, 1979; Ryugo, 1986). Introduction of zeatin and benzyladenine into cut petioles of de-bladed apple leaves enhances flower formation (Hoad, 1980; Ramirez and Hoad, 1981).

Effects of shade and light on flowering

The shaded inner parts of apple and pear trees tend to have very few flowers. This may be due in part to cumulative effects leading to weak growth and poor bud formation, but artificial shading of previously exposed trees reduces fruit bud formation and flowering in the following year (Table 9.2). Tromp (1984), using controlled environments with artificial lighting found that reduced light intensity for the 7 weeks following bloom halved flowering in the following year, while reducing light levels from weeks 7 to 16 had only a smaller and non-significant effect. This early-season response again suggests the effect is mediated by changes in the plastochron and attainment of an adequate number of nodes for flower formation.

Effects of shoot vigour and orientation on flowering

Within an apple or pear tree the branches which are well exposed to light and capable of average to vigorous growth usually produce the most, and the 'best quality', fruit buds. Shaded parts of trees and trees whose growth is generally below average for the particular stock/scion combination are usually poor in terms of flower production and crop (Feucht, 1976). Paradoxically, induction of flowering is often associated with a cessation or slowing of growth (Forshey and Elfving, 1989).

Huet (1973) found that the major factor controlling floral initiation on long shoots of 'Williams' Bon Chrétien' pear appeared to be the pattern of growth. Within any overall shoot length class the number of fruit buds which developed per shoot was inversely correlated with the relative growth rate of the shoot before harvest. He attributed this to the relative strength of apical dominance by the different shoots.

Training, bending or placing apple shoots so that they become horizontal both checks their growth and increases the proportion of the buds that become floral (Tromp, 1973; Robbie *et al.*, 1993). Banno *et al.* (1985a) found that within 10 days of shoot bending of Japanese pear in mid-June there was already an increase in the number of nodes within the axillary buds compared with controls. There was no effect on shoot growth at this time although it was evident after 10 more days. The first visible signs of flower initiation were a month earlier than on control shoots. The final proportion of buds becoming floral was about 60% following shoot bending but only 15% on control branches. Shoot bending led to an increase in the sorbitol, amino-acid, IAA and cytokinin content of lateral buds and a decrease in these in shoot tips. Bending was accompanied by a decrease in the gibberellin content of both shoot tips and axillary buds. Tromp (1973) found that shoots placed horizontally after cessation of their growth produced more flowers than when left vertical.

Effects of shoot type and pruning on flowering

Flower initiation in the axillary or terminal buds of long shoots generally does not begin until after extension growth ceases (Forshey and Elfving, 1989). It therefore usually occurs later than on spurs. Dencker and Hansen (1994) found that flower initiation in axillary buds of 'Elshof' (a colour mutant of 'Elstar') was delayed by about 40 days compared with spur buds. Treatments that stimulate growth and delay its termination may reduce flowering on long shoots (Forshey and Elfving, 1989). Dencker and Hansen, however, found that 'fertigated' trees, with nutrients supplied daily in the drip irrigation system and greater but earlier shoot growth, developed a greater proportion of, as well as more in total, floral buds on the shoots. They also found that the critical node number for axillary buds to become floral was 14 or 15 compared with

19 nodes in spur buds. Zhu *et al.* (1997) found a critical node number of 15 or 16 for terminal or lateral buds of 'Summerred' apple compared with about 18 for spur buds. As was discussed earlier, shoot orientation plays an important rôle in the induction of flowering on long shoots: fruiting is less dominant in its effects than on spurs where the presence of fruit on the individual spur is highly inhibitory. Very heavy cropping on the tree as a whole can, however, reduce fruit bud initiation on shoots. This is not just because of the reduction in shoot length, which is not necessarily great in the season of a heavy crop. In the year following a heavy crop all the axillary buds may prove to be vegetative (Blasco, 1976). Flowering on long shoots is much more important on some cultivars, e.g. 'Elstar' and 'Golden Delicious', than others and has become more important with the adoption of minimal pruning in the early years, branch tying-down and, probably, the shift of production to areas with longer growing seasons.

In general, old apple spurs are less likely to initiate flowers than young ones and pruning systems are usually designed to give a preponderance of spurs on young fruiting wood. In general pruning that stimulates growth depresses flowering. This is especially important in young trees. Increasing severity of pruning usually decreases flowering and cropping (Forshey and Elfving, 1989), and 'heading back' methods of pruning may convert potential fruiting spurs into shoots. Inadequate pruning may, however, result in excessive within-tree shade and inhibition of flowering (Feucht, 1976).

Crabbé (1984) considered bud development on extension shoots in relation to apical dominance and concluded that pinching out of the shoot tip could determine the development of sub-terminal buds. If done too early, too close to growth cessation, the buds below the pinching point break out and sprout as leafy shoots. If carried out too late, some weeks after growth cessation, the buds remain inactive and vegetative to the next season. If at the right time, they develop as flower buds. In practice, attempts to achieve this have met with variable success. Pruning somewhat later in the summer by partial removal of the current season's shoots gives variable effects on flowering depending on the type of dormant season pruning on which it is superimposed (Wagenmakers, 1988).

In Japanese pear (*P. serotina*) most flower buds are produced on young wood, especially in cultivars which commonly produce flowers from lateral buds, and pruning systems are designed to maximize the proportion of this (Klinac *et al.*, 1995).

Cultivar differences in flowering

Apple and pear cultivars differ greatly in the proportion of their buds which become floral. This has been studied in most detail in relation to biennial bearing caused by the inhibition of flower initiation by fruits. This is generally measured as a biennial bearing index, *I* (Hoblyn *et al.*, 1936), calculated as

$$I = 100 \times \frac{\text{difference between two successive crops}}{\sum \text{of two successive crops}}$$

which varies between 0 (completely regular) and 100 (biennial).

A number of cultivars are so strongly biennial that they have been much used in studies on biennial bearing, e.g. 'Laxton's Superb' and 'Miller's Seedling' apples. Among commercially important apple cultivars Jonkers (1979) concluded that 'Elstar', 'Idared' and 'Jonagold' showed no biennial tendency whereas 'Granny Smith', 'Starking' and 'Golden Delicious Spur' showed medium susceptibility. Of the numerous cultivars classed as strongly susceptible only 'Cox's Orange Pippin' is now of any importance and there are differences of opinion about the strength of its biennial tendency. Jonkers placed 'Golden Delicious' in the first, regular bearing, category but with a note that opinions differed on this. Lauri *et al.* (1995), from an experimental study of the fate of individual buds, categorized 'Jonagold', 'Royal Gala' and 'Granny Smith' as having a low 'alternation to fruit' index, 'Golden Delicious' as being intermediate and 'Braeburn', 'Fuji' and, especially, 'Oregon Spur Delicious' as having a high index. Bernhard (1961) and Looney and Lane (1984) showed that many spur-type cultivars have a strong alternate-bearing tendency.

Rootstock effects on flowering

Apple rootstocks have a major effect on the proportion of buds that become floral. This is very pronounced when the trees are young, influencing precocity of cropping as discussed in Chapter 2. It is also so for mature trees, Blasco (1976) reporting 48, 36, 28 and 32% of all 'Cox' buds to be floral on 'M.9', 'M.26', 'M.7' and 'MM.106', respectively (SED = 2.3 at $p = 0.001$). 'Quince C' rootstock induces flower bud formation in pear. Although those rootstocks that check scion growth most tend to enhance fruit bud production, this is not an invariable relationship.

Plant hormone and growth regulator effects on flowering

Guttridge (1962) first showed that applied gibberellins inhibit apple flowering, and many subsequent studies have confirmed this. Tromp (1982) found that GA_3 and GA_7 were particularly active whereas GA_4 had little effect. Both reduced flowering of spur buds of 'Cox's Orange Pippin' when applied at full bloom and GA_7 had negative effects when applied 2 or 4 weeks later. GA_7 reduced fruit bud formation on shoots when applied in July. Ramirez and Hoad (1981) found GA_3 or GA_{4+7} applied via cut petioles to spurs at 4 or 8 weeks after full bloom inhibited flower induction, but application at 12 weeks after full bloom did not. These results suggest that gibberellins inhibit the early stages of flower initiation, but Luckwill and Silva (1979) were not able to find any effect on the plastochron.

Injection of the cytokinins zeatin and benzyladenine 4, 8 or 12 weeks after full bloom into cut petioles on spurs increased flowering in the following year, zeatin being particularly effective (Ramirez and Hoad, 1981).

The growth retardant SADH (succinic acid 2, 2-dimethyl hydrazide) is very effective in both checking shoot extension growth and increasing the number of flower buds formed. These effects appear to be independent (Tromp, 1972). SADH application does not reduce the amount of extractable GA in apple shoots and fruits (Hoad, 1980; Ramirez and Hoad, 1981). Application of SADH after cessation of shoot growth can increase flowering in the following year (Tromp, 1972, 1973) and it seems likely that the effect is on late phases of the flower bud formation process. SADH was found by Luckwill and Silva (1979) to increase the number of floral buds without influencing the rate of node production in the buds.

Paclobutrazol inhibits three steps in the oxidation of the gibberellin precursor *ent*-kaurene to *ent*-kaurenoic acid (Hedden and Graebe, 1985). It retards extension growth and also stimulates flowering. Dheim and Browning (1988) found that its application to shoot tips was as effective in initiating flowering as were whole tree sprays. They concluded that active shoot tip meristems slowed node production in the axillary and spur buds to below the threshold rate for floral initiation and that paclobutrazol, by slowing shoot apical activity, reduced this inhibitory effect. Their results were thus consistent with paclobutrazol increasing flower initiation by reducing apical dominance. Application of paclobutrazol can increase flowering in the subsequent year both on spurs and on one-year-old extension shoots (Buban, 1986; Dheim and Browning, 1988).

Nitrogen effects on flowering

Flower production can be reduced when nitrogen fertilizer application prolongs the period of extension shoot growth and delays terminal bud formation. Excessive stimulation of shoot growth and leaf production can also have adverse effects on flower production as a result of shading.

Conversely, N deficiency results in poor leaf development on spurs early in the season and in reduced flower initiation. Good leaf development throughout the season as a result of adequate nitrogen supply can be expected to enhance root activity and cytokinin production and the upward flow of cytokinins and nutrients in the transpiration stream. Faust (1989) quotes data from Buban *et al.* (1978) to show very large effects of N nutrition on the cytokinin (zeatin) concentration in the xylem sap of apple. This can be expected to have positive effects on flowering.

Williams (1963) found that under conditions of sub-optimal N supply, application of nitrogenous fertilizer in summer after the formation of terminal buds led to increased flowering in the following year. When fertilizer was applied at the same time to a cultivar that had not ceased extension growth this was further stimulated and there was no effect on flowering. Nitrogen application after terminal bud formation also led to the development of flowers with much enhanced embryo sac longevity.

Water stress effects on flowering

Excessive rainfall or irrigation which maintains extension growth late into summer will often reduce flower bud formation, but this is inhibited under moderate drought and nearly completely prevented under severe stress (Salter and Goode, 1967).

Exposure to low (45–55%) as contrasted with high (80–100%) relative humidity for 9 weeks from full bloom increased the proportion of apple buds becoming floral from 12% to 54% (Tromp, 1984), as well as greatly reducing shoot growth. With 'Bartlett' pear (*P. communis*), irrigating to replace 23%, 47% and 92% of calculated evaporation was followed by, respectively, 2.39, 2.01 and 1.36 flower clusters per cm^2 of branch cross-section in the following spring (Mitchell *et al.*, 1984). Again, this followed effects in the opposite direction on shoot growth. However, Caspari *et al.* (1994) found that low irrigation rates (regulated deficit irrigation) early in the season had very adverse effects on flower production by Asian pear (*P. serotina*). Goode and Ingram (1971), in a long-term apple irrigation experiment, found that the vigorous growth induced by irrigation did not affect fruiting adversely and Wample (1982) concluded that apple trees which have not been stressed enough to stop growth consistently produce more flowers.

Technologies that promote flowering, e.g. dwarfing rootstocks, appropriate pruning and tying down of branches, may in many cases overcome the potentially negative effects of vigour-enhancing treatments. The effects of water and nutrient supply on flower initiation and development may therefore vary with circumstances.

Plant virus effects on flowering

Virus-free apple rootstocks generally induce more vigorous growth in scions grafted on them than does the virus-infected 'parent' material (Parry, 1980). Initially trees on these virus-free rootstocks were less floriferous in relation to their size but adoption of minimal pruning and branch bending in the first years after planting overcame this problem (Van Oosten, 1979).

Temperature effect on flowering and flowers

The reported effects of temperature on flower initiation are contradictory. From controlled environment studies, Abbott (1984) concluded that warm temperatures advance flower initiation and cool temperatures retard it but otherwise do not have any direct effect. Indirectly, however, temperature influences the intensity of initiation by its effects on the rates of shoot and fruit growth, high temperature stimulating shoot growth and so influencing flowering negatively. (Abbott, 1984; Tromp, 1984; Zhu *et al.*, 1997). The overall effect of temperature seems to be a balance between positive (direct) and negative (indirect) influences.

Apple fruit buds developed under warm conditions throughout the growing season are, however, later to break in spring (Abbott, 1984). The development of buds of the *Pyrus* cvs. 'Comice', 'Concorde' and 'Conference' and their dates of full bloom in spring are also delayed if the trees are subjected to October and November temperatures typical of a warm autumn (Atkinson and Taylor, 1994; Atkinson and Lucas, 1996; Atkinson *et al.*, 1997). It seems that warm (mild) conditions in autumn and early winter may delay entry into dormancy and the processes of accumulation of chilling units and bud development.

Bud development continues through the winter (Abbott, 1970) but the rate at which it does so is very dependent on cultivar. Cole *et al.* (1982) found that buds of *P. calleryana* have a much higher rate of development during autumn, when temperatures are declining, than buds of *P. communis*. This is attributed to the relatively higher rate of alternate pathway (cyanide-resistant) respiration under low temperature conditions in *P. calleryana*, hence more energy for growth and development. As a consequence *P. calleryana* enters winter with much larger, better developed buds than *P. communis* and flowers much earlier in spring.

In many fruit-growing areas temperature effects on emergence from dormancy, i.e. the meeting of winter-chilling and thermal time requirements to achieve budbreak, are of dominant importance. These have been discussed in Chapter 6.

In cool-temperate climates temperatures in the early spring, prior to budbreak and independent of frost damage effects, largely predetermine yield (Jackson and Hamer, 1980; Jackson *et al.*, 1983 (Figure 9.4); Lakso, 1994). This is at least in part the consequence of effects on ovule development and longevity and, as a result, on fruit set (pp. 294, 303–4).

Frost damage to buds and blossoms is a major constraint on apple and pear production in many areas. During late autumn dormant flower buds develop a relatively high degree of resistance to cold. As they develop in spring they become more vulnerable to damage by frost (Table 9.3). In the earliest stages of bud development, a temperature of about −15 °C is required to kill 50% of the buds but during the stages around full bloom temperatures of −3 °C

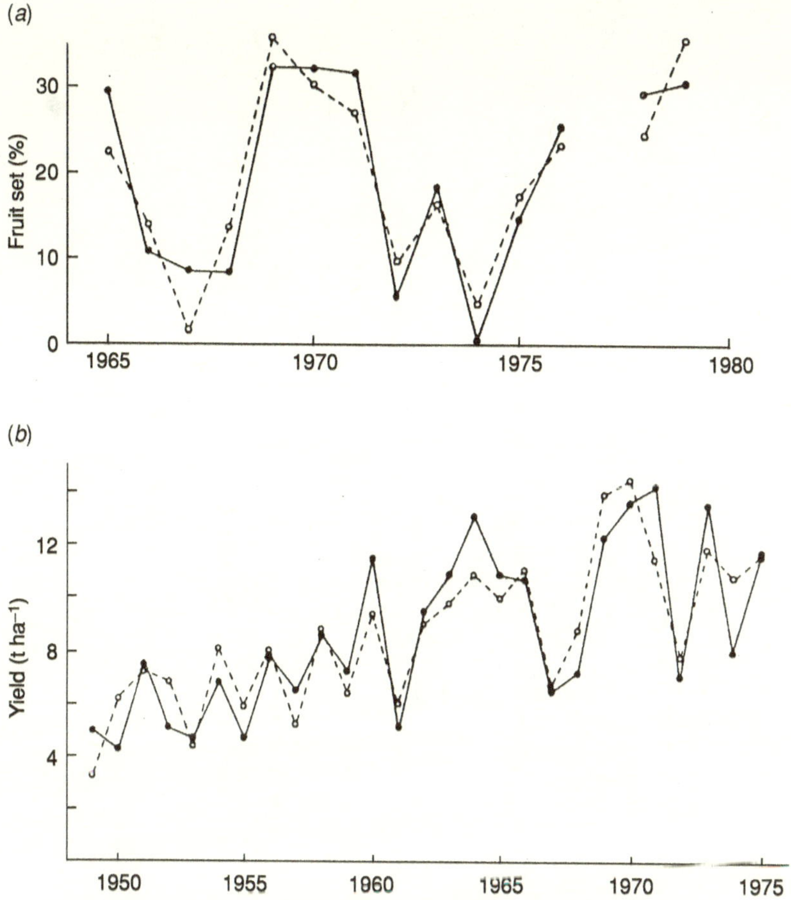

Figure 9.4 Effects of temperatures on apple fruit set and yield. (*a*) Percentage fruit set (*S*) of hand-pollinated 'Cox' flowers at East Malling, UK and that estimated from the equation $S = 114.0 - 9.0T_{maxFMA} - 2.0PT$ whoro T_{maxFMA} = mean of daily temperature maxima in February, March and April; PT = calculated days to complete pollen-tube growth. ●—●, Actual percentage fruit set; ○- - -○, estimated percentage fruit set. (*b*) UK average annual yields (t ha^{-1}), of cv. 'Cox' and those estimated from the equation: $Y = 4.91 + 0.254T - 1.127T_{maxFMA} - 0.622PT + 0.732T_{maxJ}$ where T = years from 1948; T_{maxFMA} = mean of daily temperature maxima in February, March and April; PT = calculated days to complete pollen-tube growth; T_{maxJ} = mean of daily temperature maxima in June. ●—●, actual yield; ○- - -○, estimated yield. Adapted from Jackson *et al.* (1983). Reproduced with permission.

Table 9.3 *Mean temperature values (°C) at which 10% (T_{10}) 50% (T_{50}) and 90% (T_{90}) of buds or flowers were killed in controlled freezing tests, Prosser, Washington, USA*

(a) 'Delicious' apple

Date Bud stage	18/3 Silver tip	24/3 Green tip	30/3 ½″ green	5/4 Tight cluster	12/4 First pink	18/4 Full pink	24/4 First bloom	30/4 Full bloom	5/5 Post bloom
T_{10}	−11.9	−7.5	−5.6	−3.9	−2.8	−2.7	−2.3	−2.9	−1.9
T_{50}	−15.1	−12.4	−8.6	−6.1	−4.4	−3.4	−3.3	−3.7	−2.7
T_{90}	−17.6	−15.7	−11.7	−7.9	−5.9	−4.6	−3.9	−4.7	−3.0

(b) 'Bartlett' pear

Date Bud stage	16/3 Scales separating	24/3 Bl. buds exposed	30/3 Tight cluster	7/4 First white	12/4 Full white	14/4 First bloom	19/4 Full bloom	27/4 Post bloom
T_{10}	−8.6	−7.3	−5.1	−4.3	−3.1	−3.2	−2.7	−2.7
T_{50}	−14.5	−12.2	−9.3	−7.3	−4.6	−4.8	−3.6	−3.2
T_{90}	−17.7	−15.4	−12.6	−9.4	−6.4	−6.9	−4.9	−4.0

Data from Proebsting and Mills (1978). Reproduced with permission.

to −4 °C can have a similar effect. For apple, buds at first bloom and pink bud appear to be rather more sensitive than those at full bloom (Proebsting and Mills, 1978; Hamer, 1983). In general the decline in frost resistance as buds develop is related to their increasing water content and the influence of this in decreasing their ability to supercool (Modlibowska 1962, 1975a). Low temperature in itself is not lethal and supercooling, i.e. when the liquids in plant tissues fall below the normal freezing point without crystallization, protects from frost killing. The death of cells results from intracellular ice formation causing mechanical damage and excessive dehydration leading to irreversible physico-chemical changes in the protoplasm.

Exposure to low temperatures and dry conditions prior to the incidence of frost induces a degree of hardening and resistance to frost. Hardened flowers are characteristically small, with small cells, low water content and high content of sugars, starch, hemicellulose and glucosides (Modlibowska, 1975a).

The styles are usually highly sensitive and after fertilization damage to the placenta and ovules is of most importance. Damage is characteristically shown by browning and blackening made obvious when the flowers are cut longitudinally. Death of only a proportion of the ovules results in misshapen fruits due to the uneven supply of seed-produced hormones. Pre-blossom frosts may affect the structure and elasticity of the skin, giving lines or bands of russet, 'frost eyes' and green blotches. Ice formation under the skin of the receptacle may protect the latter from freezing but result in blemishes on the fruit later.

Frosts are generally more frequent and more severe the earlier it is within the spring season (Hamer and Jackson, 1975), so one method of reducing the risk of frost damage is to delay budbreak and blossoming. This is most effectively done by selection of late-flowering cultivars, including cultivars that can crop heavily from late-blossoming axillary flowers on one-year-old shoots. These characteristics are highly heritable and can be bred for (Spiegel-Roy and Alston, 1979). Late-flowering 'sports' of a number of cultivars have also arisen by mutation and been adopted commercially. Some degree of blossom delay can be achieved with plant growth regulators. Paclobutrazol delays budbreak and full bloom by 4–6 days on apple trees treated in a previous season (Miller and Swietlik, 1986) and this chemical increases fruit bud formation on the later flowering one-year-old wood of both apple (Buban, 1986) and pear (Dheim and Browning, 1988). Day-time sprinkling with water in late winter and spring can delay blossoming as a result of evaporative cooling (Anderson and Seeley, 1993), but may fail to reduce frost damage because it increases the water content of the buds at each developmental stage (Hamer, 1981, 1983).

Direct physical protection from frost damage depends on maintaining the bud temperatures above that at which the buds will freeze. On cloudy, windy nights bud temperatures are near to air temperatures but under clear conditions typical of radiation frosts they may be about 1.5 °C lower because of their heat loss by radiation. On a small scale, use of smoke or above-canopy material may impede heat loss and provide effective frost protection. Drawing down warmer air with 'wind machines' from above a 'frost inversion' where the coldest air is nearest to the ground, is widely used in appropriate circumstances. Other frost control measures depend on heat inputs to the buds either by use of orchard heaters or by water sprinkling so that water freezing on the buds releases latent heat of fusion. Bud heat loss, hence the heat requirement, is a function of orchard net radiation and windspeed (Landsberg *et al.*, 1974). It also depends on the heat transfer properties of the buds, which are influenced by bud size *per se* and the effective changes in this as ice accumulates on the buds during sprinkling. Hamer (1985, 1986) has published data on bud heat transfer coefficients and their use to estimate the water requirements for frost protection by water sprinkling.

An alternative approach to frost problems is to spray the buds, blossoms or fruitlets with gibberellic acid or mixtures of hormones. These treatments substitute for the hormones normally produced by seeds and lead to the development of fleshy tissues to form seedless fruits. This is discussed in detail in the section on fruit set and fruit growth.

Pollination

At the time of blossoming the anthers open and pollen grains escape, this event being referred to as anthesis. The embryo sacs in the ovules usually mature at the same time or a few days later. In order to accomplish fertilization the pollen grain must be transferred to the surface of a stigma and germinate. The pollen tube must then grow down the style, the generation nucleus dividing mitotically into two sperm nuclei, and the ovule must be fertilized before it degenerates. Most apple and pear cultivars are not self-fertile or have only a limited degree of self-fertility. In the typical orchard cross-pollination between different cultivars is the norm.

Pollen production

If pollination is to occur naturally it is very important that the flowering periods of the cross-pollinating cultivars are matched. These are established by long-term observation in each fruit-growing area. It cannot be assumed that the relationships will hold good over areas of widely differing climate: the different cultivars may well have different winter-chilling and thermal time requirements for budbreak. Even within a single area the relative dates of flowering may vary from year to year to such an extent that two or more pollenizer cultivars may be needed for a main commercial cultivar (Church and

Table 9.4 *Pollen production by different cultivars*

Cultivar	1973 + 1974			1979	
	Flowers per 305 mm branch	Pollen (g) per 1000 flowers	Percentage germination *in vitro*	Flowers, per 305 mm branch	Flowers per m^3 canopy
'Egremont Russet'	29.5	1.45	71	12	1133
'Golden Delicious'	28.5	1.50	67	13	1438
'Golden Hornet'[a]	89.0	1.20	69	79	6604
'Hillieri'[a]	46.5	1.10	59.5	49	5161
'WinterGold'[a]	84.0	0.75	57	68	4980

[a] Ornamental *Malus.*
Representative data taken from Church and Williams (1983a).

Williams, 1983b). Cultivars with a long flowering season, e.g. those that flower profusely both on one-year-old long shoots and on spurs, may be especially useful as pollenizers. This trait can be accentuated by the type of pruning.

The quantity of pollen produced by a cultivar depends on its flower production and the yield of pollen per flower. Church and Williams (1983a) found that some ornamental *Malus* cultivars produced twice as much, or more than twice as much, viable pollen per metre of branch as cultivars in general use as pollenizers for 'Cox', while the dessert cultivars also differed greatly among themselves (Table 9.4). Triploid cultivars may produce large amounts of pollen but much of this may be defective as a result of chromosomal imbalance.

Low winter temperatures may reduce both the number of pollen grains produced and their viability. High temperatures in spring often result in sterile pollen but the severity of this problem varies with cultivar (Faust, 1989). Anther dehiscence can occur as low as 5 °C (Percival, 1955).

Pollen transfer

Apples and pears are insect-pollinated, predominantly by hive bees (*Apis mellifera* L.), bumble bees (*Bombus terrestris* L.), other wild bees and hover flies (*Syrphidae*) (Smith, 1970; Free, 1993). The efficiency of pollen transfer by insects depends on their abundance, the relative attractiveness of the apple and pear flowers to them, their mode of operation and climatic conditions.

Hive (honey) bees are usually the most important, especially on pear (Eijnde, 1996), although they are more adversely affected by low temperatures than bumble bees or wild bees. In general there is little insect activity below 10 °C (50 °F). Flower visiting by insects, except bumble bees, usually ceases during heavy rains or showers. Windspeeds above 15–20 m.p.h. inhibit bee flight.

Windbreaks result in greatly increased insect populations in the sheltered zones (Lewis and Smith, 1969). Other flowers may compete for the attention of bees (Benedek and Nagy, 1996) and may need to be controlled in the orchard or the neighbourhood.

The particular flower structure of a cultivar may reduce the effectiveness of pollination. 'Delicious' flowers have basal gaps between stamen filaments through which bees can extract nectar without touching any stigma (Dennis, 1979; Schneider *et al.*, 2002).

The effectiveness of pollination by bees in general can be increased by inserting pollen packs in hives in such a way as to dust the bees with pollen on leaving (Williams, 1970a). Alternatively, pollen may be applied to flowers by dusting them with a mixture of pollen and inert material.

Pollen germination and pollen tube growth

The pollen grain is a dormant, resistant structure that contains lipid reserves for its germination and early growth but is dehydrated and must absorb water to germinate when it reaches the stigma. In apples and pears the stigma has a wet surface composed of extracelluar secretions of its papilla cells and of the contents of the cells themselves which collapse after anthesis (Sedgley, 1990). This surface, while wet, provides the moist environment for pollen germination (Heslop-Harrison, 1976). After the hydrated pollen grain has germinated in the secretion pool on the stigma surface, the emergent pollen tube begins to grow through the interstitial material of the transmitting tract. 'Recognition' of incompatibility or otherwise takes place here.

Pollen germination rate depends on temperature and varies with the source of pollen. Petropoulou and Alston (1998) found 'Spartan' pollen to give a higher germination percentage than that of 'Cox' or 'Idared' at all temperatures tested, but the difference was likely to be most important at 8–10 °C and 14–16 °C. At these latter temperatures pollen of 'Redsleeves' was outstanding in its germination percentage and initial pollen tube growth (Table 9.5). Among Japanese pears Rohitha and Klinac (1994) found 'Shinsui' pollen germination to have a temperature optimum at 12 °C whereas for the cultivars 'Hosui', 'Kosui', 'Nijisseiki' and 'Shinseiki' the optimum temperature was between 16 °C and 20 °C.

The major factors influencing the speed of pollen tube growth are genetic compatibility (see below) and temperature, factors which interact. Modlibowska (1945) found that high temperature, i.e. 25–30 °C, speeds up the incompatibility reaction and thus results in an early inhibition of the growth of incompatible tubes. The growth of compatible tubes, in contrast, increased over the range from 10 °C to 30 °C for both diploids and triploids although the latter had a lower growth rate at any given temperature.

Table 9.5 *Pollen germination percentage of different cultivars after 3 and 6 hours at various temperatures*

Cultivar	8–10 °C	14–16 °C	19–21 °C	24–26 °C	28–30 °C
'Cox'	3	19	40	58	67
'Idared'	5	24	39	47	80
'Spartan'	14	34	50	67	83
'Redsleeves'	45	90	95	95	95

Data from Petropoulou and Alston (1998). Reproduced with permission.

Table 9.6 *Pollen tube growth rates for typical pollinators of 'Cox' apple at different temperatures. An accumulated total of 100 indicates completion*

Mean temp. (°C) over 24 h period	5	6	7	8	9	10	11	12	13	14	15
Pollen tube growth index per 24 h	8	9	10	11	12	14	17	20	25	35	50

From Williams (1970b). Reproduced with permission.

Williams (1970b) used data from several cultivars to calculate the time for pollen tubes to penetrate to the ovary for a diploid apple such as 'Cox' over the relatively cool range of temperatures typical of England (Table 9.6).

Incompatibility

In apples and pears, as in most flowering plants, the male and female organs are in close proximity within the same flower (Figure 9.3). Self-incompatibility has developed as one of the mechanisms that prevents successive self-fertilizations and deleterious inbreeding. It has long been recognized that most apple and pear cultivars are effectively self-incompatible, or very largely so, and that fruit set usually depends on cross-pollination between genetically different cultivars.

The site of incompatibility is in the style or ovary, because of physiological reactions occurring between the pollen tube and the stylar and ovarian tissue (Modlibowska, 1945). In general incompatible pollen tubes grow slowly down the style. They show heavy deposits of callose tissue along and at the end of the tube. In some cases the reaction occurs at an early stage, pollen tube growth being inhibited in the style. In others it takes place later, even when the tubes have reached ovarian tissue. Compatible pollen tubes, on the other hand, grow rapidly down the style, are characterized by small, widely-spaced, intermittent callose plugs and the absence of a terminal plug, and can effect fertilization (Modlibowska, 1945; Stott, 1972). On the basis of pollen penetration in the apple style, Kobel (1954) classified pollination responses as indicating compatibility (0), incompatibility (1) and half-compatibility (1/2). In the last case about half of the pollen tubes penetrate to the base of the style (Spiegel-Roy and Alston, 1982).

Self-incompatibility in apple is under the control of an S gene system. Kobel *et al.* (1939) proposed that this is based on S gene alleles. The essential feature of the incompatibility system is that pollen is inhibited in a style or ovary containing the same incompatibility alleles. The single, multi-allelic gene encodes ribonucleases (S-RNases) which are present in the pistil of the mature flower where they recognize and inhibit developing pollen. Several alleles of the S gene of apple have been cloned and characterized (Broothaerts *et al.*, 1996) and stylar ribonuclease isoenzymes corresponding to a large number of S-alleles detected (Boskovic and Tobutt, 1999).

Table 9.7 shows the incompatibility genotypes of many of the currently important cultivars of apple and others representing types likely to be important as parental material for the future. It is to be expected that where the cultivars have identical stylar ribonuclease patterns, corresponding to identical S-alleles, they will be mutually incompatible. Where the cultivars have one S-allele in common only 50% of the pollen grains will be compatible, i.e. the cultivars will be semi-compatible. Where both alleles differ between diploid pollen donor and acceptor cultivars then they will be fully compatible.

In general the pistils of triploids are more likely to have one or more alleles in common with the donor pollen because they have more alleles. Cultivars that are related are likely to have at least one incompatibility allele in common. 'Jonagold', a triploid arising from a cross between 'Jonathan' and 'Golden Delicious' has two S-alleles in common with 'Golden Delicious' and one with 'Jonathan'. 'Elstar' ('Ingrid Marie' × 'Golden Delicious') has only one incompatibility allele in common with 'Golden Delicious'.

Self-compatibility varies from very low (highly self-incompatible) to appreciable levels. 'Red Delicious' and its sports are generally self-unfruitful (Dennis, 1979; Kemp, 1996). De Witte *et al.* (1996) found 'Fuji' and 'Golden Delicious' to give only 1% and 1.8% set, respectively, following self-pollination under conditions where pollination with the ornamental apple *Malus Baskatong* gave 24% and 25% set. This low level of self-fertility is important in view of the predominance of these cultivars both in production and as parental material. De Witte *et al.* (1996) found much higher levels of self-fertility in 'Idared' (12.3%) and 'Elstar' (7%) although, importantly, very few of the fruits so produced were seeded. 'Cox's Orange Pippin' has given very variable set when self-pollinated, ranging from 0.7% (of a sample of over 12 000 flowers) found by Modlibowska (1945) to 17% (under conditions where *M. Baskatong* pollination gave 34%) found by De Witte *et al.* (1996). In the latter case the 'Cox' fruits from self-pollination averaged 4.9 seeds per fruit.

Table 9.7 *Incompatibility genotypes of apple cultivars*

Cultivar	Parentage	Genotype according to Broothaerts *et al.* (1995, 1996) or Janssens *et al.* (1995)	Genotype deduced from stylar ribonuclease bands
'Alkmene'	'Gehemreith Dr Oldenburg' × 'Cox'	S_5-	S_5S_{22}
'Arlet'	'Golden Delicious' × 'Idared'	S_2S_7	S_2S_7
'Braeburn'	Open-pollinated	S_9-	S_9S_{24}
'Cox's Orange Pippin'	Open-pollinated	S_5S_9	S_5S_9
'Delbard Jubilé'	'Golden Delicious' × 'Lundbytorp'	S_2-	S_2S_{23}
'Elstar'	'Golden Delicious' × 'Ingrid Marie'	S_3S_5	S_3S_5
'Fiesta'	'Cox' × 'Idared'	S_3S_5	S_3S_5
'Fuji'	'Ralls Janet' × 'Delicious'		S_1S_9
'Gala'	'Kidd's Orange' × 'Golden Delicious'	S_2S_5	S_2S_5
'Golden Delicious'	Open-pollinated	S_2S_3	S_2S_3
'Granny Smith'	Open-pollinated	S_3-	S_3S_{10}
'Idared'	'Jonathan' × 'Wagener'	S_3S_7	S_3S_7
'Jonathan'	Open-pollinated	S_7S_9	S_7S_9
'Kent'	'Cox' × 'Jonathan'	S_3S_9	S_3S_9
'Merlijn'	'Jored' × 'Liberty'	S_3-	S_3S_{25}
'Prima'	Complex including 'Golden Delicious'	S_2-	S_2S_{10}
'Summered'	'McIntosh' × 'Golden Delicious'	S_2S_7	S_2S_9
'Telamon'	'McIntosh Wijcik' × 'Golden Delicious'	S_3-	S_3S_{25}
'Jonagold'	'Golden Delicious' × 'Jonathan'	$S_2S_3S_9$	$S_2S_3S_9$
'Mutsu' (syn. 'Crispin')	'Golden Delicious' × 'Indo'		$S_2S_3S_{20}$

'Jored' is a sport of 'Jonagold'.
Based on genotype data given by Boskovic and Tobutt (1999). Reproduced with permission.

Most 'sports' show a level of self-fertility that is consistent across clones and with the 'parent' cultivar. This is not the case with 'Cox', where Stott and Campbell (1971) found appreciable variation in self-compatibility in clones of this cultivar. Clones with a high level of self-fertility and that could also pollinate the standard 'Cox EMLA' were subsequently selected from trees that had been subjected to gamma-irradiation to increase the incidence of mutations (Campbell and Lacey, 1982; Lacey *et al.*, 1982). Alston (1996) found evidence that when the 'self-fertile' 'Cox' is self-pollinated the pollen tubes grow the full length of the style but fertilization only takes place between gametes with different S-alleles.

Cross-compatibility between cultivars had been considered to be the norm (Modlibowska, 1945), but cross-incompatibility has become more prevalent as more inter-related cultivars are grown. Alston (1996) showed that although 'Cox' and 'Idared' are fully compatible, much lower levels of fruit set are achieved following crossings between their progeny with fully incompatible pairs of alleles in common. These crosses between siblings also gave much lower levels of fruit set than when they were back-crossed with the 'Cox' parent with which they had only one incompatibility allele in common. Even cultivars of uncertain ancestry may have incompatibility alleles in common, to the detriment of cross-pollination. 'Jonathan' is less effective than 'Golden Delicious' as a pollinator of 'Topred' (a 'Red Delicious' sport) as a result of a common incompatibility allele (S9) in both 'Jonathan' and 'Topred' (Goldway *et al.*, 1999). Under many circumstances, however, semi-compatible cultivars give adequate pollination for each other.

Triploid apple cultivars are less frequently self-incompatible and the self-incompatibility is usually less strongly expressed than in diploids. For this reason selfed triploids usually set more fruits than selfed diploids even though they often show gametic sterility (Modlibowska, 1945). They produce both seeded and seedless fruits.

Diploid European pear cultivars are almost invariably self-sterile but set parthenocarpic fruits more readily than do apples, either naturally or following hormone sprays. They are generally cross-fertile.

Japanese pears are largely self-incompatible and require cross-pollination, the self-incompatability being controlled by S-alleles in the pollen and pistil. Different cultivars have different S-alleles (Hiratsuka *et al.*, 1995). A mutant of 'Nijisseiki', 'Osa-Nijisseiki', is self-fertile as a result of stylar-part-mutation of the S_4-allele (Sassa *et al.*, 1992).

Female flower fertility

Successful fertilization depends on the pollen grains reaching the ovules before they degenerate. Williams (1966) defined the concept of an effective pollination period (EPP) determined by the longevity of the egg apparatus minus the time necessary for pollen tube growth. This is measured by carrying out controlled pollination at daily intervals from anthesis for as long as the stigmas remain receptive.

Cultivars differ in EPP duration. Although the EPP varies with a number of other factors differences between EPPs of cultivars under broadly comparable conditions have been found in a number of studies. Williams (1966) reported EPPs of about 2.5, 5.5 and 6.5 days for 'Cox', 'Jonathan' and 'Laxton's Superb' apples, respectively, and of 1, 7, 7 and 7–10 days for 'Doyenné du Comice', 'Packham's Triumph', 'Williams' Bon Chrétien' and 'Conference' pear, respectively. Tromp and Borsboom (1996) noted a much shorter EPP for 'Comice' than for 'Golden Delicious' which showed high levels of fruit set when pollinated to 6 days after bloom. 'Red Delicious' has a short EPP, Hartman and Howlett (1954) noting a linear decline in its fruit set when pollination was delayed from 0 to 48 hours after anthesis. Some cultivars also suffer from consistently asynchronous flower part development. Herrero (1983) found that 'Agua de Aranjuez' pear flowers have delayed maturation of the megagametophytes in relation to anthesis. Stigmas were not receptive at anthesis and no mature embryo sacs were present then, most of them being fully mature 5 days later.

Although largely controlled by genotype, female fertility also varies greatly from year to year (Williams, 1966) and has been shown to be influenced by a number of separate factors operating at different times, from flower initiation and development in the previous year to anthesis and beyond.

The application of nitrogenous fertilizer after the cessation of apple shoot growth leads to the production of 'strong' blossoms that develop earlier than those on control trees (Williams, 1965). The ovules of these flowers remain capable of fertilization for almost twice as long as those of 'normal' flowers.

Heavy cropping in the previous season results in smaller flowers, less well expanded papillae on stigmas and thinner styles of 'Cox' apple (Buszard and Schwabe, 1995). Such flowers failed to set fruit if pollinated at anthesis and no fruits set after day 7, i.e. the EPP was 6 days, whereas flowers on previously defruited trees set fruit when pollinated on any day from anthesis to day 10, i.e. had a 10-day EPP.

Flowers on young wood and on young trees of apple have shorter EPPs and higher proportions of immature and of degenerate ovules than those on older wood and trees (Robbie and Atkinson, 1994). These effects seem to be related to the timing of flower initiation and development, physiologically 'young' flowers being of lower quality than older ones. Flowers on horizontal branches have a higher proportion of healthy ovules at anthesis and later and have a longer EPP than those on vertical branches (Robbie *et al.*, 1993). This quite large effect could be achieved by bending branches that had been vertical into a horizontal position in April during flowering. If branches that had been horizontal were moved to the vertical in April, this resulted in a short EPP. These responses to changes in branch position at flowering time suggest that effects on ovule fertility and EPP are controlled by current hormonal relationships rather than, in this case, effects via influences on flower development in the previous year.

Temperatures simulating a warm English February, March and April resulted in EPPs for 'Cox' apple of 4 and 6 days in consecutive years, whereas under conditions simulating a cool English spring the EPP exceeded 8 days in each year (Miller, 1988). In general, low temperatures in the post-anthesis period increase ovule longevity (Sedgley, 1990).

Pollination increases embryo sac viability in such a way that even those pollen grains that do not fertilize ovules play an important part in assisting other pollen grains to do so. It stimulates a wave of cytoplasmic and biochemical activity in the pistil and also stimulates development within the ovary, readying it for fertilization. Herrero and Gascon (1987) showed that in unpollinated pear flowers the ovules degenerate between 12 and 21 days after anthesis whereas in cross-pollinated flowers the degeneration is postponed by about 10 days. This increase in EPP can also be induced by GA_3, which may explain why GA_3 treatment can increase the set of seeded fruits and suggests that GA_3 may be involved in the pollination effect on ovule fertility.

Flowers of 'Comice' pear treated with paclobutrazol at full bloom contain twice as many degenerating embryo sacs as untreated flowers (Dheim and Browning, 1987). This leads to complete inhibition of fruit set, which can be reversed if the flowers are treated with gibberellic acid within 3 days.

Putrescine applied at full bloom delays ovule senescence and can extend the EPP of 'Comice' pear by up to 4 days (Chrisosto *et al.*, 1992). Putrescine is a polyamine, a class of chemicals apparently present in cells of all tissues capable of normal growth and development and with a regulatory role as distinct from a purely nutritional requirement (Galston and Kaur-Sawhney, 1995).

Fruit set

Fruit set is the process that is necessary for pre-anthesis growth of the flower parts to be followed by sustained post-anthesis growth of the fruit. At anthesis all the chemical factors for growth – auxin, kinins and gibberellins as well as ABA – are already present in the flower (Martin *et al.*, 1982; Miki *et al.*, 1982). Despite this, further growth and continued development after anthesis are usually limited to only a small proportion of the flowers. Typically only 5–10% of the flowers give harvestable fruits, although the proportion may vary from more than 30% to less than 5%. The rest fail to set and are shed.

The abscission process

Typically flower and fruit shed occur in four waves (Murneek, 1954). The first is of flowers with no fertilized ovules and the second, often overlapping with it, is of fruitlets with 2–4 endosperm nuclei. Most shed occurs at these stages. The third wave is of fruits with endosperm and embryos with up to 16 cells, and the fourth is of fruits with completely developed endosperm and embryos of various sizes, but usually fewer or smaller than those of retained fruits. The fruitlets shed in the third and fourth waves are relatively large and conspicuous and these waves, usually occurring in late May and early June in the northern hemisphere, are collectively known as the June drop. Subsequent shed is numerically slight up to the time of fruit maturity and abscission. Stages of fruit and seed development are shown in Figure 9.5.

The flowers and fruitlets, and later the mature fruits, show a constriction at the base of the pedicel (the individual flower stalk) and apex of the peduncle (inflorescence stalk) to which it is joined. In this constricted zone the cells of all tissues, except those of the pith, are smaller than adjacent ones (McCown, 1943). The zone is usually 20–30 cells in width in the cortical tissue, less wide in the xylem and phloem. Flowers and immature fruits abscise following the differentiation of an abscission layer in the basal portion of the pedicel as a result of cell division. This abscission layer is usually differentiated within the limits of the constriction zone but it may also appear in the pedicel distal to the constriction zone. Cells in all tissues undergo division and the resulting abscission layer is six to eight cells deep. A slight swelling of the middle lamellae and primary walls, followed by disintegration of their pectic compounds, precedes cell separation. The vessels are ruptured, apparently as a result of mechanical force. If all flowers or immature fruits shed, then abscission layers form so that the entire peduncle is shed. Following the abscising of the pedicel or the peduncle, cork is formed by a cambium which is initiated a few cells below the surface of the scar.

Two contrasting assumptions underlie attempts to explain why flowers, fruitlets and fruits shed. One is that it is the natural tendency for the fruit to remain attached in the absence of nutritional or other stresses leading to abscission. The other is that fruitlets have a tendency to abscise and are only prevented from doing so by specific hormonal stimuli originating in the fruit itself (Luckwill, 1953).

It is now generally accepted that flowers are pre-programmed to shed after anthesis unless they receive a new stimulus to trigger continued growth. This stimulus is commonly given by pollination and fertilization, with the result that resources are subsequently allocated only to those fruitlets likely to give fruits with viable seeds. Sometimes pollination alone, or the application of plant growth substances during or shortly after flowering, can provide the trigger

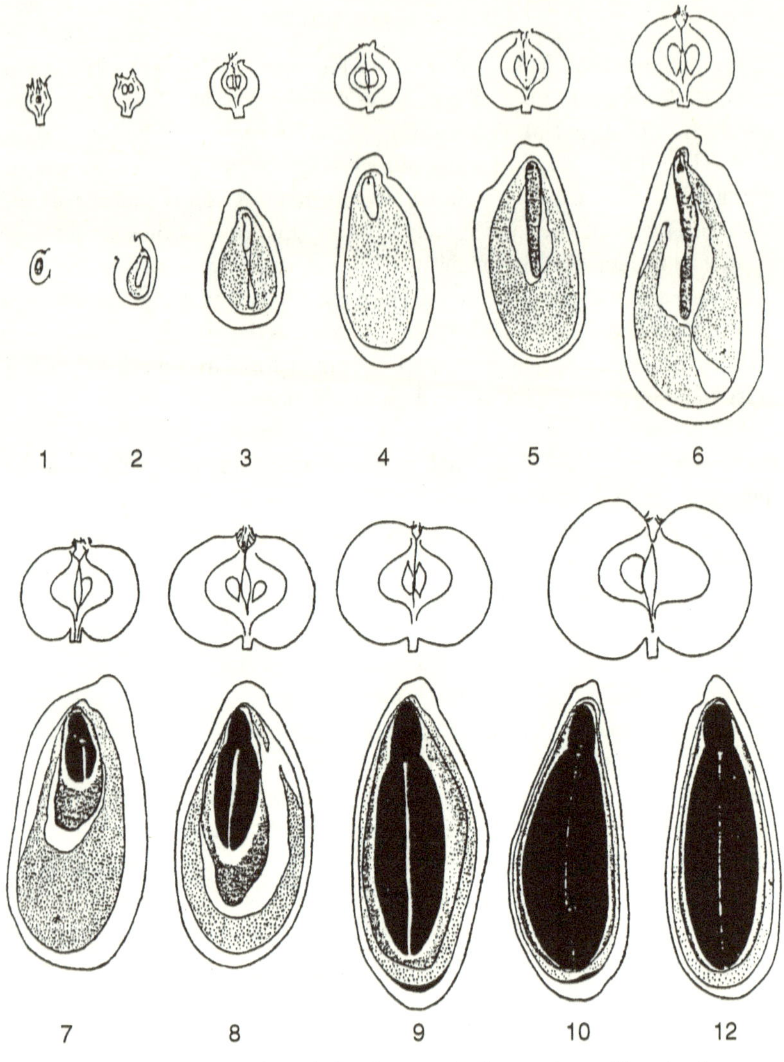

Figure 9.5 Fruit and seed development in the apple cv. 'Beauty of Bath'. The numbers are days from petal-fall. Lightly stippled tissue in the seed is nucellus; heavily stippled tissue, endosperm; solid black, embryo. Scale for seeds is 14 × scale for fruits. From Luckwill (1948). Reproduced with permission.

(Hedden and Hoad, 1985). In apples and pears the 'quality' of the flower, which may be influenced by factors throughout its development including those operating in the previous year, has a major influence on its ability to respond to the trigger.

The continued growth of the fruitlet or fruit then depends on its ability to compete with other fruits and shoots for metabolites. The concentration of such metabolites influences the consequences of this competition and if the fruit growth is checked, even for as little as two days (Fukui *et al.*, 1984a), it will shed.

The process of abscission is usually considered within the scheme proposed by Reid (1995). In this it is considered that shed is prevented by keeping the abscission zone in a non-sensitive state as a result of a gradient of auxin from the potentially shedding organ to the plant axis: this gradient being maintained by factors which inhibit senescence such as auxins, cytokinins, light and good nutrition. Reduction or reversal of the auxin gradient by application of auxin proximal to the abscission zone, by shading or by poor nutrition hastening senescence, allows the abscission zone to become sensitive to ethylene. Once sensitized the cells of the abscission zone respond to low concentrations of ethylene by secretion of hydrolytic enzymes and the organ is shed. Recent work suggests that carbohydrate supply is especially important in preventing the abscission process being triggered in the abscission zone.

Either de-fruiting but leaving the pedicel with its abscission zone intact attached to the tree, or steam-girdling the pedicel between the fruit and the abscission zone, results in abscission whenever carried out (Berüter and Droz, 1991). Steam-girdling the pedicel between the tree and the abscission zone induces fruit drop if carried out at 15 or 28 days after full bloom, but leads to a cessation of growth without abscission if done at 49 days after full bloom. This shows that up to the end of the June drop abscission can be induced by treatments which reduce or block nutrient supply from the leaves to the abscission zone. After that period the growing fruit has become a storage organ which is resistant to abscission. The fact that seed removal after June drop does not stimulate activation of the abscission zone in the pedicel, but fruit removal or phloem ringing between the fruit and the pedicel does, leads to the conclusion that substances produced or stored in the fruit flesh control post-June-drop separation at the abscission layer. Berüter and Droz (1991) concluded that the glucose concentration in the pedicel is a key factor controlling abscission. In citrus, fruitlet abscission induced by carbon shortage appears to be regulated by ABA (abscisic acid) and ACC, an ethylene precursor (Gómez-Cadenas *et al.*, 2000). ABA may act as a sensor of the intensity of the carbohydrate shortage that modulates the levels of ACC and ethylene. There is, as yet, no validation of this hypothesis for pome fruits. Effects of shade on fruit abscission are discussed on pp. 301–2 and of chemical thinning agents on p. 313.

Effects of seeds on fruit shed

Following fertilization the presence of immature, developing seeds prevents fruitlet shed and enables continued growth. Abbott (1959) found that if seeds were removed within 4–5 weeks of petal fall from apple fruits of 'Cox's Orange Pippin' and 'Crawley Beauty', the fruits would shed. Similar results were obtained with 'Golden Delicious' by Berüter and Droz (1991). Removal of seeds after the June drop period, e.g. at 42 days after full bloom, did not induce any fruit shed and the fruits continued to grow.

This effect of developing seeds is attributed to the hormones that they produce. Dennis (1967) found that whereas untreated, unpollinated flowers of 'Wealthy' apple almost all shed, those treated with an extract of immature 'Wealthy' seeds gave normal-sized but seedless fruits. The seed extract was found to contain gibberellin activity and a similar effect was obtained by application of the potassium salt of gibberellic acid at 5×10^{-3} KGA_3. Apple fruitlets showing the reduced rate of growth characteristics of those about to shed have lower gibberellin activity in their seeds than those showing the growth patterns of persistent fruits (Fukui *et al.*, 1985). High concentrations of biologically active GAs (GA_1, GA_3, GA_4 and GA_7) are present in apple seeds at about 20 days after anthesis, when the growth rate of the fruit is at a maximum (Hedden and Hoad, 1985). More than 30 gibberellins have been identified in immature apple seeds using GCMS (gas chromatography/mass spectometry) (Garcia-Martinez and Hedden, 1997).

Immature pear seeds of cultivars of *P. communis*, *P. ussuriensis* and *P. serotina* also contain gibberellins. Removal of the seeds 3 weeks after full bloom causes the shed of all fruitlets, but this is largely or completely prevented when gibberellins are applied to the cut surfaces at the time of seed extraction (Yuda *et al.*,1984). The most effective treatment is with GA_{4+7} even though the principal GA in the immature seeds is GA_3.

Apple seeds contain auxins and cytokinins as well as gibberellins. Concentrations of IAA in the seeds, measured by physico-chemical methods, are, however, low and relatively constant at the endosperm stage, not increasing until mid-July with the cv. 'Sunset' in England (Hedden *et al.*, 1984; Hedden and Hoad, 1985). Kondo and Mizuno (1989) found the auxin-like activity in seeds of 'Starking Delicious' and 'Fuji' to be more or less constant from 10 to 60 days after full bloom. Fukui *et al.* (1984a), using histochemical techniques, found the content of indole derivatives to be very low until 50 days after full bloom, increasing between 50 and 90 days after bloom and staying high until harvest. They also found that fruitlets showing the check in growth characteristic of those about to shed did not differ in indole derivative concentration from those with a growth pattern characteristic of persistent fruitlets. They concluded that indole derivatives in seeds have no correlation with early fruit drop. Auxin can, however, prevent the abscission of apple fruitlets from which the seeds have been removed (Abbott, 1959) but does not stimulate their further growth. In this apples differ from a number of other fruits such as tomatoes and some *Rosa* species.

Table 9.8 *The effect of the previous season's crop on fruit set per 100 blossom clusters of 'Cox's Orange Pippin' apple*

	Treatments in 1973				
	Hand pollination	Control	Light thinning	Heavy thinning	SED
Fruits per tree 1973	333	221	177	114	24.0
Fruit set 1974	7	12	12	20	2.7

From data of Blasco (1976). Reproduced with permission.

Apple fruitlets with the growth characteristics of those about to shed have only 2.5–7% as much cytokinin activity as those likely to persist, as well as less gibberellin activity (Fukui *et al.*, 1985).

Artificial pollination by hand, followed by fertilization and seed development, usually increases fruit set to well above the natural level (e.g. Table 9.8). Knowledge that this is so has prompted the development of widely used and successful practices to facilitate pollination in the orchard as a means of increasing yields (Williams and Wilson, 1970).

Parthenocarpy

Apples and pears, as noted above, can develop fruits without fertilization and seeds. Parthenocarpic fruit set can be vegetative, without pollination or any other externally applied stimulus. Gorter and Visser (1958) tested 21 apple and 21 pear cultivars for their ability to set fruits from flowers whose petals, stamens and styles had been removed and which were then enclosed in paper bags. More than 10% of the flowers set and matured parthenocarpic fruits in one apple and four pear cultivars. All of the pear cultivars and 12 of the apple cultivars set some parthenocarpic fruits. Much more complete vegetative parthenocarpy is shown by 'Spencer Seedless' apple, the flowers of which lack petals and stamens and do not attract bees. Although they will set seeded fruits if hand-pollinated all of their fruits are normally seedless. The apetalous characteristic and associated parthenocarpy is controlled by a recessive gene (Tobutt, 1994). Immature seedless apple fruits contain GA_3 (Hayashi *et al.*, 1968). Seedless 'Bartlett' pear fruits have a higher content of GA-like substances than seeded ones (Gil *et al.*, 1972).

Stimulative parthenocarpy can be induced by pollination without fertilization, e.g. by irradiated pollen (Marcutti *et al.*, 1984) or by plant hormones, primarily gibberellins. Sprays of the latter are widely used to induce pear crops, either following frost damage or with cultivars which set poorly under local environmental conditions. 'Conference' and 'Spadona' can be induced to give substantial crops of parthenocarpic fruits with GA_3 (Modlibowska, 1975a; Herrero, 1984). Set of 'Abbé Fetel', 'Doyenné du Comice' and 'Williams'' is achieved by use of a commercial product containing GA_4, GA_7 and benzyladenine (Calzoni and Speranza, 1996). GA_{4+7} is also effective on the triploid 'Bramley's Seedling' apple (Modlibowska, 1975b) and, to a lesser extent, on 'Cox's Orange Pippin' (Goldwin, 1978). For the latter cultivar a three-hormone mixture of a gibberellin, an auxin and a cytokinin appears most effective (Kotob and Schwabe 1971; Goldwin, 1978). Applications at blossom time, within 2 weeks of this, or multiple applications, are most effective. Those cultivars with a strong tendency to natural parthenocarpy are the most responsive to hormone sprays.

GA effects on seeded fruit set

Application of GA_{4+7} at any time between one and 40 days after full bloom does not influence the initial fruit set of 'Cox's Orange Pippin' apples but greatly reduces the June drop of fruitlets from pollinated flowers (Wertheim, 1973). Application in July after June drop has started does not reduce this.

In a number of other trials on apple gibberellins applied alone or in combination with other hormones have resulted in more, rather then less, fruit shed.

Auxin and cytokinin effects on fruit set

When applied as an overall spray at blossom time or later, NAA usually increases fruit drop, so is widely used as a chemical thinning agent (Wertheim, 1997). This may result from its side effects on assimilate availability and is also in keeping with the concept that it is the auxin gradient across the abscission zone that is important in controlling shed. Application of 2,4,5-TP either in the previous autumn or the spring may, however, increase fruit set on some pears (Westwood *et al.*, 1968; van Zyl and Strydom, 1982).

Cytokinin sprays may also increase rather than reduce fruit shed (Wertheim, 1997). This may be a consequence of stimulating other growing points.

Ethylene and fruit set

There is a general association between ethylene and fruit shedding but less certainty as to whether increases in ethylene production pre-date the first stages of shedding and are a direct cause of this. Kondo and Takahashi (1989) found that 'Starking Delicious' apple fruitlets, which had a high rate of ethylene output, shed heavily with up to 60% June drop. Ethylene output from 'Fuji' was low and there was little or no fruitlet drop. 'Tsugaru' was intermediate in both respects.

Table 9.9 *Direct, residual and cumulative effects of shade on fruit set (fruits harvested per 100 blossom clusters) 'Cox's Orange Pippin'/'M.26' apple*

	Light (% daylight) in 1970 and 1971										SED
Light 1970	100	100	100	100	37	25	11	37	25	11	
Light 1971	100	37	25	11	100	100	100	37	25	11	
Fruit set 1971	98	61	60	39	96	99	51	96	91	32	9.09

SED is the SE of the difference from the never-shaded control.
Data from Jackson and Palmer (1977b). Reproduced with permission.

They found that exogenous ethephon, an ethylene-generating compound, increased the rate of seed abortion and fruit abscission while application of AVG (aminoethoxyvinyl glycine), which inhibits ethylene biosynthesis, reduced the rate of seed abortion, cellulase activity in the abscission zone, and fruit abscission as well as preventing ethylene evolution. Other studies showed AVG to reduce endogenous ethylene production and increase fruit set of 'Delicious' (Greene, 1980) and fruit set of 'Cox' apples (Child and Williams, 1983) and 'Comice' pear (Lombard and Richardson, 1982). However, Fukui *et al.* (1984b) found that for 'McIntosh Red' apples subjected to high night temperatures, beginning on the 10th day after full bloom, the check to growth that preceded fruit drop also preceded ethylene evolution and the formation of an abscission layer. They concluded that ethylene was not the primary factor in inducing the drop. Also, Kondo and Takahashi (1987) did not find much effect of AVG on shade-induced apple fruitlet shed. Fruit size and shape are adversely affected by early AVG application, so it has not become a commercial fruit-setting agent although used as a pre-harvest 'stop-drop'.

Ethephon application frequently induces fruitlet shed and although effects are inconsistent it is widely used as a fruit thinning agent, applied at blossom time or later (Wertheim, 1997).

Effects of shade and photosynthesis on fruit set

Artificial shading reduces fruit set of apples both in the year of shading and in the following year (Table 9.9; Jackson and Palmer, 1977b). Fruit set is reduced linearly as light intensity within a mature apple tree canopy declines over the range from 54% to 6% of full sunlight (Rom, 1991).

Heavy shading for only a few days within the period from about 14 to 38 days after full bloom induces fruit shed (Kondo and Takahashi, 1987; Berüter and Droz, 1991; Byers *et al.*, 1991; Stopar, 1998). Shading earlier or later is ineffective. Fruitlets induced to abscise by shading stop growth within 1–6 days of shading and shed 7–12 days later (Byers *et al.*, 1991). Application of AVG does not reduce the effect of short-term shading on shed. There is a pronounced interaction between unshaded and shaded parts of trees. There is much less abscission on a single shaded branch on an otherwise unshaded tree than on the branches of a completely shaded tree and more shed on an unshaded branch on an otherwise shaded tree than on a corresponding branch on a totally unshaded tree (Berüter and Droz, 1991; Byers *et al.*, 1991). Lowered natural light intensity as a result of cloudy conditions prior to June drop is associated with increased fruit shed both in Akita, Japan and Virginia, USA (Kondo *et al.*, 1987; Byers *et al.*, 1991).

Shading reduces photosynthesis, and both photosynthesis and fruit set can also be reduced by application of terbacil (3-tert-butyl-*S*-chloro-6-methyluracil) or other inhibitors of photosynthesis. These have dramatic effects on fruit set when applied to leaves between 5 and 15 days from full bloom, but little effect if applied later and no effect if applied to fruits alone (Byers *et al.*, 1990a, b).

Effects of leaf removal on fruit set

Removal of either spur or bourse shoot leaves within 2 weeks of full bloom reduces fruit set of 'Cox' and 'Golden Delicious' apples (Proctor and Palmer, 1991). The effect of reducing both spur and bourse leaves is particularly severe.

Defoliation 10 or 31 days after harvest reduces both initial and final set of apples in the following year (Tustin *et al.*, 1997), presumably by reducing carbohydrate reserves. The effects of early defoliation may be partly attributable to its negative effect on subsequent spur leaf size.

Effects of fruit and shoot competition on set

Removal of a large proportion of the flowers results in an increased percentage set of those left on the tree (Blasco, 1976; Knight *et al.*, 1987; Lauri and Térouanne, 1999). In some experiments (Knight *et al.*, 1987) removal of a third of the flower clusters or a third of the flowers within clusters did not reduce the number of fruits finally harvested, therefore must have led to the retention of fruits that would otherwise have shed. Reduction of competition leads to the retention of fruits on inflorescences with a low number of leaves, e.g. on one-year-old wood (Lauri and Térouanne, 1999).

Removal of shoot tips starting 15 days after full bloom also reduces June drop of fruitlets (Quinlan and Preston, 1971; Makino *et al.*, 1986). Removal of shoot tips leads to diversion of photosynthates into the fruitlets.

Signalling through auxin production and movement may be involved in these competitive effects. Terminal (king) fruitlets are usually less likely to shed in either the first drop or the June drop than lateral fruitlets within the cluster. Their removal leads to an increase in set of the lateral fruitlets. The king fruitlets have more diffusible IAA than the lateral ones but removal of the dominant king flowers stimulates diffusible IAA in lateral fruits. Similarly, early excision of bourse shoot tips results in higher IAA export from nearby lateral fruits (Gruber and Bangerth, 1990).

Seedless fruits of many cultivars all shed when seeded fruits are present even though substantial numbers of seedless fruits can be retained if all pollination is prevented (Goldwin and Schwabe, 1975). This suggests that the number of fruits retained to maturity is not simply a function of available resources but is also influenced by hormonal signals. This is consistent with the view that hormones amplify differences between sinks to intensify competition and intervene to abort the weakest sinks before resources are wasted on them (Browning, 1989).

There are also residual effects on fruit set of crop load in the previous year. Blasco (1976) found that flowers on trees that had been induced to crop heavily in the previous year were smaller and set less well than those on trees that had borne lighter crops (Table 9.8).

Early harvesting, either by its effect on reserves or effects on fruit bud development and 'flower quality', also increases fruit set in the following year (Williams *et al.*, 1980).

Flower quality and 'tree factor' effects on set

It has long been observed that 'strong' buds and 'bold' blossoms are the ones most likely to set fruit. Blasberg (1943) found a direct linear relationship between the size (diameter) of spur buds and fruit set even when all the buds were on vigorous branches.

Factors shown earlier to enhance 'flower quality', in terms of demonstrated effects on ovule longevity or blossom size, generally increase fruit set. These include treatment with nitrogenous fertilizer in the previous autumn (Williams, 1965) and horizontal training of branches (Robbie *et al.*, 1993). Flower clusters on one-year-old wood have smaller flowers than those on 2- and 3-year-old wood and set less well (Robbie and Atkinson, 1994). Very young trees also show much poorer fruit set than mature trees (Robbie and Atkinson, 1994). Flowers on old, weak spurs set poorly.

Temperature effects on set

National average yields of 'Cox's Orange Pippin' apple in England are low in seasons with high February, March and April maximum temperatures and fruit set of hand-pollinated flowers in such years is also low (Jackson *et al.*, 1983; Figure 9.4). Fruit set of 'Cox' under controlled conditions is lower when

Table 9.10 *Effect of simulated warm and cold spring conditions on fruit set percentage after hand pollination in 'Cox's Orange Pippin'*

Pre-blossom temperature	'Warm Spring'	Ambient	'Cold Spring'	LSD 0.05
Fruit set percentage	11.6	12.6	17.2	1.9

Data from Jackson *et al.* (1983). Reproduced with permission.

these are typical of a warm rather than a cool English spring (Table 9.10) and in Canada fruit set of 'Macspur McIntosh' and 'Summerland' apples was lower the higher the daytime temperatures in controlled environments over the range 15°, 20° and 25 °C (Crowe, 1984). These effects are not associated with winter-chilling requirements: in the Canadian study all the trees had been in cold storage for at least 18 months and night temperatures (10 °C) were the same for all treatments. Effects on ovule longevity and asynchronous development of flower parts seem to be involved. Evaporative cooling of buds by water sprinkling also increases set (Jackson *et al.*, 1983).

High temperatures at blossom time reduce ovule longevity and can have adverse effects on initial set. Tromp and Borsboom (1994) found that EPP was extended and fruit set increased when 'Golden Delicious' apples were grown at a post-pollination temperature of 13 °C instead of 19 °C. With 'Doyenné du Comice' pear EPP and fruit set were higher (from February onwards) at 13 °C than at 17 °C.

In Akito, Japan heavy early fruit drop, from the 36th day after full bloom, of cvs. 'Starking Delicious', 'Mutsu' and 'Redgold' was found to be statistically correlated with high minimum temperatures and low sunshine levels around the 30th day after full bloom (Kondo *et al.*, 1987). Controlled environment studies on 'Starking Delicious' (Kondo and Takahashi, 1987) showed that high night temperatures (20 °C) over the four days starting 27 days from full bloom led to fruit abscission starting 8 days after treatment, with a final shed of 34% under conditions where there was no shed under outdoor night temperatures (mean 15 °C). High night temperatures earlier were ineffective and from 34 to 38 days after full bloom they induced only 12% shed. In subsequent studies high night temperatures for 4 days from 23 or 30 days after full bloom were consistently followed by severe shedding of 'Starking Delicious' fruits whereas there was less shedding of 'Tsugaru' and in three out of four years none at all of 'Fuji' (Kondo and Takahashi, 1989).

Studies on fruit set of apples in North America and Europe in relation to temperatures over different post-bloom periods have given rather variable results (Dennis, 1979). They include clear evidence from controlled-temperature studies that warm temperatures (>20 °C) during post-bloom to June drop period reduce set of a number of cultivars, including 'Delicious' and 'Golden Delicious' (Lu and Roberts, 1952; Grauslund and Hansen, 1975). Olien *et al.* (1995) found that 78% of the site-to-site variation in 'Starkspur Supreme Delicious' apple yields in rootstock trials over ten American states was accounted for by a highly significant negative linear regression on mean maximum temperatures. Mean site (state) maxima over the May–June period ranged from 20 °C to 29 °C. Results from California were excluded from the regression but were in conformity with the general association between lower post-bloom temperatures over this range and high yield.

It should be noted that experimental data can prove the reality of, for example, an effect of temperature on fruit set. Regression studies based on field and climatic data establish whether the relationship is a limiting one, and how important it is, in specific circumstances. In a climate in which blossom-time temperatures are frequently sub-optimal, variation in this factor may be very important: in climates frequently exceeding the optimal temperature over the later fruit set period these effects may predominate.

Water stress effects on set

Fruit set is very sensitive to water stress. Mature 'Cox' apple trees subjected to artificial drought by use of soil covers from late March to mid-June in England set only about half as many fruits as rainfed control trees. They showed the effect on fruit set before there were any obvious effects on shoot extension growth or trunk diameter expansion. Trees irrigated for 3 weeks (18 May to 10 June) shortly after flowering had more than 50% higher set than control trees, although during the period of irrigation soil water potentials in irrigated and control plots differed only in the upper soil levels (75 cm from the surface), and even there by only 0.2 bar, and the average plant water potentials differed by only about 0.5 bar (Powell, 1974).

Cultivar and strain differences in fruit set

Even within the same environment there are large differences between cultivars in their percentage fruit set. For some, e.g. 'Cox's Orange Pippin' and 'Red Delicious' apples and 'Comice' pear, inadequate fruit set is commonly a problem. For others, e.g. 'Gala' apple, excessive fruit set may lead to a requirement for regular flower or fruitlet thinning. There may even be large differences between clones of the same cultivar: four different 'sports' of 'Red Delicious' gave fruit set per 100 flower clusters ranging from 34 for 'Starking' to 106 for 'Idaho Spur' (Westwood *et al.*, 1967).

Several factors appear to be involved. Differences in fruitlet shed between 'Fuji', 'Tsugaru' and 'Starking Delicious' were positively related to their ethylene production (Kondo and Takahashi, 1989). However, Rahemi *et al.* (1997) did not find differences in set between some other cultivars to be related to their ethylene production. The 'Red Delicious' group as a whole shows a number of other features associated with their fruit setting problems (Dennis, 1979).

Poor fruit set in 'Packham's Triumph' pear is primarily related to excessive competition between shoots and fruits in young, vigorously growing trees and fruit–fruit competition in older 'spur-bound' trees (van Zyl and Strydom, 1982). In this cultivar in the absence of appropriate pruning to reduce the number of growing points and spurs, very heavy flowering can be associated with unsatisfactory fruit set. Similarly, severe pruning of 'Comice' improves its fruit set (Parry, 1976) but this cultivar, unlike 'Packham's Triumph', also has a short EPP. For a number of pear cultivars with a natural tendency towards parthenocarpy, set can usually be improved by the use of gibberellin sprays to increase the set of seedless fruits.

Fruit growth

A fruit can be defined as 'the edible product of a plant or tree, consisting of the seed and its envelope, especially the latter when juicy and pulpy' (*Oxford English Dictionary*). Understanding fruit growth, however, requires consideration of the growth of the flower prior to seed formation and must also encompass the case of parthenocarpic, seedless fruits.

Two separate processes are involved in fruit growth: cell division and cell expansion.

Cell division

There are two million cells in the flesh of an apple at anthesis and 40 million at harvest (Pearson and Robertson, 1953). To achieve this number, 21 doublings are required before anthesis and only 4.5 doublings after this (Coombe, 1976). All these figures must be regarded as approximations but they illustrate a general point.

Most of the post-anthesis cell divisions occur within the first few weeks after blossoming. Schechter *et al.* (1993a, b) found that the percentage of dividing cells in 'Idared' apples peaked at about 15 days from full bloom, when about 85% of the cells were dividing. It then decreased for about 25 days to 8–10%, which continued up to 100 days from full bloom. Bain and Robertson (1951) reported that cell division in 'Granny Smith' apple is completed within 3 weeks of full bloom in Australia, and Denne (1960) found it to last for 6–7 weeks in 'Cox' and unthinned 'Miller's Seedling' but for about 12 weeks in 'Miller's Seedling' thinned at the pink-bud stage in England.

In pear, Toyoma and Hayashi (1957) reported cell division in early cultivars to last for 25–30 days post-bloom and for 45 days for late cultivars. Sterling (1954) found cell division in 'Bartlett' pear to last for 6–8 weeks in the bulk flesh and 12 weeks at the periphery.

Cell expansion

Cell enlargement begins soon after pollination, continues through the cell division period and, at a diminishing rate, until harvest (Denne, 1960). The increase in cell size involves deposition of new cell wall material, influx of solutes, influx of water which depends on solute concentration, cell wall plasticity, water availability, transpiration and plant water potential, and the constraints imposed on the flesh by the extensibility of the skin and other surrounding layers.

During and shortly after anthesis the ovary is dependent on the overall level of reserves for its nutrient supply and has limited capacity to attract these (Coombe, 1976). Once the fruit has commenced growth, hormones produced in the seeds and flesh are important in mobilizing resources. The main supply of organic compounds imported into the fruit is then from current photosynthesis. Apple leaves produce sorbitol and to a lesser extent sucrose, sorbitol being the main transport carbohydrate. The photosynthates imported into the fruit are used in the synthesis of structural polysaccharides for growth, respiration and storage of carbohydrate (Berüter and Studer Feusi, 1997). They are also metabolized into fructose, sucrose, malic acid and starch (Berüter *et al.*, 1997). Fructose makes the greatest contribution to the fruit solute potential of both 'Cox's Orange Pippin' and 'Golden Delicious' apples but sucrose is the major soluble carbohydrate on a dry weight basis in 'Cox's Orange Pippin' (Pavel and De Jong, 1995). Soluble carbohydrate contributes about 46% of the total fruit solute potential in 'Golden Delicious' early in the season and about 86% at harvest.

Overall fruit growth

Early studies, e.g. by Denne (1960, 1963) on 'Cox's Orange Pippin', showed a typical single-sigmoid growth curve. For convenience of comparison this could be divided into three phases: a slow increase in weight for the first 6–12 days after pollination, a phase of rapid exponential increase lasting for 3–4 weeks, and a phase of progressive slowing of the rate of increase leading up to harvest. Orlandini *et al.* (1999) found that increases in diameter always followed a sigmoid curve in 16 sets of data on 'Golden Delicious'.

Schechter *et al.* (1993a, b) found little evidence for slowing down of 'McIntosh', 'Delicious', 'Empire' and 'Idared' growth towards harvest and defined the pattern of dry matter accumulation as consisting of two linear phases. The first of these is of very low growth, lasting about 40 days, and the second of steady linear growth at a rate of 0.21–0.28 g dry weight per day up to harvest. Lakso *et al.* (1995) found that the growth of 'Empire' and 'Golden Delicious' apples showed a positive curvilinear growth curve in the early part of the season (from 32 to 74 days, depending on cultivar and environment) followed by a much more rapid linear growth phase up to harvest. This growth pattern can be described by an expolinear equation, two key elements in which are the time from anthesis to the intercept of the linear growth phase and the slope of this phase. They found the former to be temperature-sensitive and the latter to be dependent on the number of cells in the cortex, in the limited number of cases considered.

Tukey and Young (1942) found that in New York State the growth curve of the entire fruit is nearly linear up to ripening and harvest of the early summer apple 'Early Harvest'. For successively later cultivars, including 'McIntosh' and 'Rome', the curve flattens as the season progresses. The shape of the growth curve is thus a cultivar characteristic although it is modified by climate and the degree of between-fruit competition.

In general terms the major determinants of variability in fruit growth are cell division before anthesis and cell expansion after this, with cell division post-anthesis making a smaller contribution (Coombe, 1976).

Effects of pre-bloom factors

The importance of cell division before blossoming as a determinant of the total number of fruit cells suggests that pre-bloom factors should have a considerable effect on fruit size. This effect appears to provide a rationale for much of the between-season and within-tree variability in fruit growth and size.

EFFECTS OF THE PREVIOUS CROP

Bergh (1985b) compared flower and fruit development on 'Starking' apple trees which had been thinned to give a normal crop load with that on trees which had been unthinned and consequently bore very heavy crops. The number of cells in the cortical tissue at the base of the developing flowers was reduced by heavy cropping, the reduction being already evident 8 weeks after the onset of differentiation of the sepal primordia (Table 9.11). The effect on cell number at full bloom was followed by a somewhat smaller effect on fruit size at harvest. Although the decrease in cell numbers of fruits following heavy cropping was accompanied by larger cells, these did not compensate for the effect of reduction in cell numbers. In most cases this is not so and fruits in an 'off' year following a heavy crop are large, but not as large as would be expected based on the very light crop.

Table 9.11 *The effect of cropping levels in 1980/81 on cell number × 10³ in the floral tube of terminal flowers of 'Starking' apple through the southern hemisphere winter from autumn (April) to full bloom (October)*

	Crop in 1980/81	
Collection dates 1981	Heavy	Light
16 April	80	95
2 July	100	130
20 August	113	167
13 October	1080	1484

From Bergh (1985b). Reproduced with permission.

EFFECTS OF POST-HARVEST DEFOLIATION

Premature post-harvest defoliation reduces the size of apples of 'Royal Gala' produced in the following season. This effect is shown even when crop load is standardized by hand thinning and it cannot be fully explained through effects on spur leaf area (Tustin *et al.*, 1997). A role of carbohydrate reserves status, influencing flower and subsequently fruitlet development, seems probable.

EFFECTS OF SPUR SIZE AND POSITION OF FLOWER ON THE SPUR

Blossom size is positively related to the diameter of the bearing spur and differences in blossom size are maintained through fruitlet growth to harvest (Denne, 1963). Within a spur the 'king' blossom is the largest, with the size of the others increasing towards the base. Fruit diameter at harvest follows a similar pattern.

EFFECTS OF AGE OF BEARING WOOD

The receptacle diameters and lengths of lateral flowers on one-year-old shoots of 'Royal Gala' and 'Braeburn' apple are less than those of flowers on 2-year-old spurs (Volz *et al.*, 1994) and the fruits are much smaller at maturity. Fruits developed from lateral flowers on one-year-old shoots also give smaller fruits than those on 2-year-old spurs of 'Granny Smith' and 'Fuji' (Volz *et al.*, 1994) and 'Laxton's Superb' (Jackson, 1970a). The small size of lateral flowers on one-year-old wood is related to their late initiation and differentiation.

EFFECTS OF PRE-BLOSSOM TEMPERATURES

Low pre-blossom temperatures result in reduced fruit size whether achieved by conventional temperature modification (Miller, 1988) or evaporative cooling (Collins, *et al.*, 1978). This may be an effect of later flowering, resulting in a shorter growing season, of the improvement in fruit set and consequently greater fruit–fruit competition or of an effect on pre-bloom cell division. Hamer (1981) found that apple buds subjected to evaporative cooling have lower dry weights than those on control trees at the later stages of bud development.

Post-bloom effects

EFFECTS OF SEEDS AND PLANT HORMONES

Fruit growth appears to be dependent on seeds only for the first 7 weeks or so after petal fall (Abbott, 1959).

The effect of seeds on growth is particularly obvious in fruits with a very uneven shape associated with the presence of seeds only in locules on one side of the fruit. Similar asymmetry can be induced by localized GA application (Bukovac and Nakagawa, 1968).

Fruit growth and final fruit size may be greater the higher the number of seeds (Denne, 1963). The effect of seed number on the percentage increase in fruitlet diameter is confined to the first 3 weeks to 40 days after blossoming although proportional differences established by that time are carried through to harvest. The total number of cells per fruit increases with seed number but there is no consistent relationship between seed number and cell size. The ratio of fruit diameter to length increases with increasing seed number in both apple and pear (Denne, 1963).

In those cultivars which can be induced to set parthenocarpic fruits by exogenous hormone application the complete cycle of fruit growth, extending for several months, is obtained by a single application at around the time of anthesis. This suggests that the hormones simply act as inducing agents (Goodwin, 1978). Subsequent growth is associated with a pattern of hormone activity in the developing fruit which is similar in both parthenocarpic and seeded fruits (Gil *et al.*, 1972, 1973). Auxins, cytokinins and gibberellins have all been identified in apple and pear fruit flesh (Gil *et al.*, 1972, 1973; Goodwin, 1978; Letham and Williams, 1969). The cytokinins found in fruits are not necessarily synthesized there: they are found in xylem sap and it is possible that root-produced cytokinins are transported to fruits.

Application of gibberellins and of benzyladenine (6-BA) increases the ratio of apple length to diameter and the development of enlarged calyx lobes (Letham, 1969; Williams and Stahly, 1969; Greenhalgh *et al.*, 1977). Elongated apples with pronounced calyx lobes are considered to be typical, therefore desirable, for 'Red Delicious' and in warm areas fruits that are not hormone-sprayed tend to be too round.

SHADE AND CARBOHYDRATE RELATIONSHIPS

During the growing season the main factors influencing individual fruit growth are light intensity and leaf area development, controlling potential supply of photosynthate, and competition for this between shoots and fruits and between fruits.

The changes in light intensity with position in the canopy are the most important source of within-tree variation in fruit growth in most types of apple and pear trees. Jackson (1967) found that 40–44% of within-tree variation in weight of 'Cox' apples was accounted for by fruit location effects that were clearly related to shade. Where canopy structure was such as to make vertically-summed leaf area index relevant to shade, apples in the upper third of the size distribution were not found beneath an LAI greater than 1.5 (Jackson, 1970b). Part of this effect is a consequence of poor spur development under shaded conditions, such spurs having lower photosynthetic potential in terms of leaf area and specific leaf weight (Tustin *et al.*, 1992). Major responses to shade are, however, demonstrated by artificial shading of previously well-exposed trees or branches immediately post-anthesis.

Fruit size is reduced by imposed shade even though the number of fruits is also reduced. The reduction in fruit size is primarily due to reduction in cell division and cell number but also involves reduction in cell size. With light intensities of 100%, 34% and 13% of full daylight, respective final fruit weights were 95, 82 and 63 g, cell numbers 27, 23 and 19×10^6 per fruit and mean cell volumes 3.43, 3.30 and 3.08×10^{-3} mm^3 (Jackson *et al.*, 1977). Part of the effect of shade may be associated with reduced fruit temperatures (Thorpe, 1974), hence reduced fruit metabolism and sink-strength (Minchin *et al.*, 1997) as well as with effects on leaf and fruit photosynthesis. Shading fruits without shading the adjacent leaves can considerably reduce their growth and size at harvest (Schrader and Marth, 1931).

Sources of photosynthate for fruit growth change over the season. The early development of flower clusters depends on stored carbohydrate and bud and flower photosynthesis (Chapter 8). Spur leaf emergence starts before blossoming and at two weeks after bloom spur leaf assimilate is almost equally distributed between the spur leaves themselves and the bourse shoots and fruitlets which develop more or less simultaneously. Bourse shoots initially compete with the fruitlets for spur leaf assimilates and also retain and utilize their own assimilates (Tustin *et al.*, 1992). Subsequently more and more of the spur leaf assimilate feeds the developing fruitlets and the bourse shoot leaves become exporters to the fruitlets. The fruits then become dominant importers of assimilates from extension shoot, bourse shoot and non-fruiting as well as fruiting spur leaves. Proximity between sources and sinks, modified by vascular connection patterns, determines which demands are satisfied first and the surplus of supply over proximate demand influences export to more distant sinks. Teng *et al.* (1998) found that spurs of all ages exported assimilated ^{13}C at 6 weeks after full bloom, the amount exported increasing with spur age and the associated increase in leaf area. By 18 and 24 weeks after full bloom, when fruit demand was higher, no assimilated ^{13}C left spurs of any age. Corelli-Grappadelli *et al.* (1994) found that shading eliminated export of assimilated ^{14}C from extension shoot leaves to fruits at 3 weeks after full bloom and reduced export to fruit at 5 weeks from full bloom. Using ringed branch sections Hansen (1977) found a linear relationship between fruit growth from early July to mid-September and leaf area per fruit. The distance between leaves and fruits had no effect on this relationship. Palmer *et al.* (1991) found that when apple flower clusters were removed at full bloom, either uniformly throughout the tree canopy, on alternate branches or on whole sides of trees, mean fruit weight at harvest was linearly dependent on leaf area per fruit and on tree light interception per fruit (Figure 9.6). This is compatible with earlier work by Parry (1974), who found blossom removal on a half-tree basis to be as effective as individual cluster thinning throughout the tree volume in increasing fruit size. The clear implication is that, at least with dwarf trees, assimilate is mobile within the tree and individual fruit growth is greatly influenced by potential tree photosynthesis and the number of fruits competing for the assimilates.

EFFECT OF FRUIT THINNING

The fruits on a tree compete for available resources and competition between fruits is a major factor controlling their growth and size. Denne (1960) showed that, at harvest, fruits from trees of 'Miller's Seedling' apple thinned at the pink-bud stage of blossom were four times as heavy as those of control trees. The rate of cell division on thinned trees was higher from 3 or 4 weeks after full bloom. The number of cells per fruit in thinned trees continued to increase up to 12 weeks from full bloom but remained constant from 6 weeks in control trees. At harvest there were about three times as many cortical cells in the fruits of thinned as of unthinned trees and the individual cells were about 25% wider, implying that cell volume was about doubled.

In extreme cases, when the thinning is carried out early, e.g. pre-bloom, the yield of thinned trees may be very similar to that of unthinned trees, i.e. the increase in individual fruit weight fully compensates for the lower numbers (data of Parry, 1974). Any delay in fruit thinning reduces the effect on the growth of the remaining fruits because the opportunity for much increase in cell division is lost. Commercial thinning is often delayed until the level of fruit set is known, so the increase in fruit size does not fully compensate for the loss in numbers.

Although the effect of fruit thinning by reducing competition for assimilates operates essentially at a whole-tree level, selective thinning so as to retain the

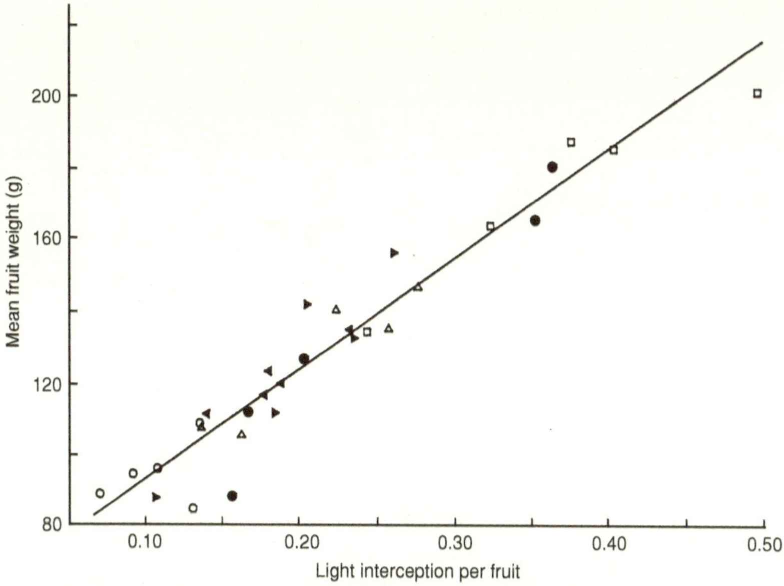

Figure 9.6 The dependence of mean fruit weight of 'Cox's Orange Pippin' apple at harvest on light interception per fruit for six flower thinning treatments. ○, control; △, alternate clusters removed; □, 3 out of 4 clusters removed; ◄, all clusters on E half removed; ►, all clusters on W half removed; •, all clusters on alternate branches removed. From Palmer *et al.* (1991). Reproduced with permission.

largest fruitlets or those likely to have a greater number of cells as a result of their previous history can give additional benefits.

Thinning to achieve the desired number of fruits is achieved by dormant-season pruning to reduce the number of spurs and fruit buds, by use of chemical thinners and by hand thinning. Details of practice depend on cultivar characteristics, e.g. the probability of over-set, environmental factors and the economic value of larger fruits even if acquired at the cost of total yield. Thinning practice also depends on the need to prevent overcropping so as to avoid biennial bearing.

Chemical thinning is used because of the excessive labour involved in hand thinning. It can be carried out at blossom time, using scorching chemicals that will damage open flowers and prevent successful pollination of these, or at the early stages of fruitlet development. Most thinning chemicals have been found by chance, although ethephon is sometimes used. Auxins are used as fruitlet thinners, possibly acting by temporarily reducing assimilate availability. Untiedt and Blanke (2001) found that NAA, naphthalene acetamine (Amidthin) and ethephon all reduce canopy photosynthesis. Some other photosynthesis inhibitors have shown promise as

Table 9.12 *Mean apple fruit expansion rates 10–40 days after full bloom and mean fruit weight at harvest on trees subjected to controlled temperature regimes over 10–40 days*

Cultivar	Max/Min temperature (°C)			
	13/3	16/6	19/9	22/12
Fruit expansion rate (mm day^{-1})				
'Delicious'	0.25 d	0.44 c	0.71 b	0.91 a
'Golden Delicious'	0.21 d	0.42 c	0.63 b	0.82 a
'Fuji'	0.25 d	0.52 c	0.77 b	0.94 a
Harvest fruit weight (g)				
'Delicious'	153*b*	157*b*	224*a*	219*a*
'Golden Delicious'	112*b*	132*b*	171*a*	186*a*
'Fuji'	178	192	194	199

Figures with different letters within rows significantly different at $P \leq 0.05$.
All fruits harvested when those at 22°/12 °C were commercially mature.
From Warrington *et al.* (1999). Reproduced with permission.

thinning agents (see p. 302), and carbaryl and cytokinins are used (Wertheim, 1997).

EFFECTS OF TEMPERATURE

In many traditional fruit-growing areas apple and pear fruits tend to be smaller in cooler than in average seasons, or when grown under cool conditions by reason of latitude or altitude. Controlled environment studies have shown that temperatures early in the season, from 10 to 40 days after full bloom, have a major controlling effect on fruit growth rates and on fruit weight at maturity (Table 9.12). The temperature regimes in Table 9.12 simulated natural diurnal patterns with temperature maxima and minima ranging from 13°/3 °C to 22°/12 °C, in other trials cooler (9°/3 °C) and warmer (25°/15 °C) regimes were more adverse and more favourable for fruit growth, respectively (Warrington *et al.*, 1999). Similar temperature regimes over the period from 40 to 80 days after full bloom showed much smaller and less consistent effects. This is consistent with the major effect of temperatures in this range being on the cell division process which is usually largely complete by 40 days after full bloom.

High night temperatures can be deleterious. Tukey (1956, 1960) found that although fruit growth was greater at 24 °C than at 9 °C at night, it was much poorer at 31 °C than at either of the lower temperatures. Tamura *et al.* (1981) found that keeping night temperatures at 23 °C and day temperatures 4 °C above ambient was stimulatory for the first 3 weeks after bloom and then was detrimental so that by 30–35 days the fruits were of the same size. The fruit growth stimulation was due to increased early cell division, which ended 7–10 days earlier than in the controls to give the same final cell number.

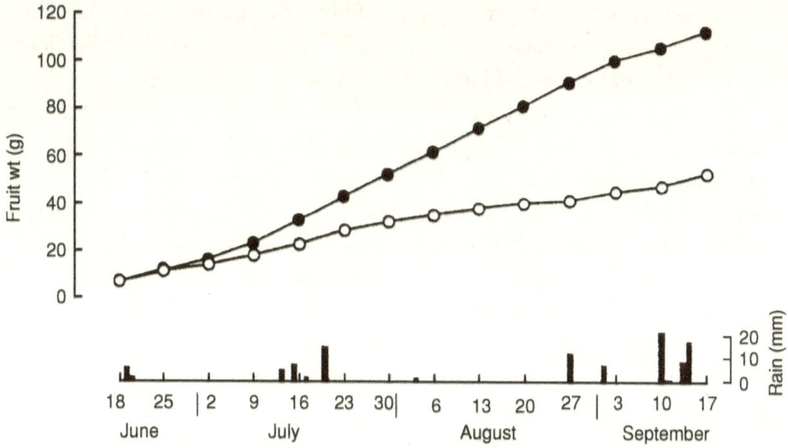

Figure 9.7 The effect of irrigation on apple fruit growth in a drought year (1976) in England. ●, trickle irrigated to provide full daily potential transpiration requirement; ○, unirrigated. Derived from Goode *et al.* (1978a) with permission.

Apples grown at high altitude and under cool conditions tend to be more elongated in shape (Greenhalgh and Godley, 1976; Eccher, 1986). This is commercially important for 'Red Delicious'.

EFFECTS OF WATER STRESS AND IRRIGATION

Irrigation in a dry season can have dramatic effects on fruit growth and final fruit size at harvest (Figure 9.7). Averaged over a number of years, Goode *et al.* (1978b) found that early irrigation, in June only, had no effect on mean fruit size whereas late irrigation (July to mid-September) increased this significantly even though it also (slightly) increased the number of fruits harvested.

This effect of early versus late irrigation, together with the contrasting fruit growth curves of irrigated and non-irrigated fruits shown in Figure 9.7, suggests that the effect of irrigation is on the cell expansion rather than the cell division stage of fruit growth. In field trials like the above, a complicating factor is that severe water stress seldom develops early in the season. However, early-season water stress imposed by regulated deficit irrigation under controlled conditions has only a temporary effect on fruit growth, reversible when irrigation is resumed (Behboudian and Mills, 1997). Fruits appear to be able to osmoregulate to maintain growth under early-season deficit irrigation conditions severe enough to inhibit vegetative growth (Chalmers *et al.*, 1986). Whereas induced water deficits for the period from 55 days after full bloom to harvest led to reduction in the weight of 'Braeburn' apples from 135 days after full bloom and to smaller fruits at harvest, deficits starting at 105 days after full bloom had no such effect (Mills *et al.*, 1996). The imposed deficits both reduced photosynthesis per unit leaf area, though the earlier the deficit was imposed the longer was the negative effect in this respect. Only the deficit imposed at 55 days after full bloom led to reduction in leaf area.

Effects of irrigation on fruit growth and final size are very dependent on crop load. Goode *et al.* (1978b) found that irrigation from July to mid-September only led to increased fruit size when crops were in excess of 16 t ha^{-1}. The response peaked at about 22 t ha^{-1} then declined at higher cropping levels. Naor *et al.* (1997) found the effect of increasing the number of fruits on fruit size was more severe at lower irrigation rates or, expressed differently, the effect of increasing irrigation on fruit size was greater the higher the number of fruits per tree. It seems probable that irrigation influences fruit size through its effect on leaf area, and hence assimilate production and carbon resources, as well as possibly in more direct ways.

EFFECTS OF GENOTYPE

Scion cultivars differ greatly in intrinsic fruit size. This is also modified by the rootstock used and, to a lesser extent, by the pollenizer cultivar.

Apple cultivars give fruits ranging from those the size of a cherry or plum, produced by different *Malus* species including crab apples, to fruits weighing over 500 g when grown in such a way as to meet local market demand (Fukuda, 1994).

Fruit size is under polygenic control (Brown, 1975). There has been rigorous selection for size over many years and, since there is a tendency for progeny to be smaller than the parents, it has been important to include large-fruited cultivars in breeding programmes (Brown, 1975). Triploids, e.g. 'Jonagold', usually have larger fruits than their diploid parents.

Pear fruits may be only 1 cm in diameter in some clones of *Pyrus betulifolia* and *P. calleryana*, while in some clones of *P. communis* and *P. pyrifolia* they may exceed 12 cm in diameter (Layne and Quamme, 1975). As with apples, fruit size in *P. communis* cultivars appears to be under polygenic control. There are some large-fruited triploid cultivars.

Smith (1940, 1950) found that cultivar differences in apple size were due to both differences in the amount of cell division (i.e. cell number) and cell size. The characteristic cultivar size was determined primarily by the amount of cell multiplication occurring after pollination. 'Bramley's Seedling', a triploid, was anomalous in that its large fruit size was related to both more numerous and larger cells, and a greater volume of tissue, at the pre-pollination flower stage. The average size of individual cells was related to the length of the growing season of the cultivar but there is no general relationship between the length of the growing season of a cultivar and its average fruit size.

Genotype also controls the pattern of growth within the fruit. The ratio of fruit length to diameter varies between cultivars and also between strains of 'Red Delicious' (Greenhalgh and Godley, 1976).

Rootstocks affect fruit size indirectly by influencing the proportion of fruits growing in high-light environments on branches and spurs well exposed to sunlight in both the current and the previous season. They also have intrinsic effects on fruit size which are not consequent on effects on tree vigour and size. Apple trees grown on 'MM.106' rootstock (semi-dwarfing) give smaller fruits than those on 'M.9' and 'M.26' even after taking into account effects of crop load per unit of tree size or of light interception (Blasco, 1976). Some recently selected rootstocks, e.g. 'AR 680–2' (Webster *et al.*, 1997) and 'Jork 9' (Callesen, 1997) induce fruits as large as or larger than those from trees on 'M.9'. Others, e.g. 'Mark' and 'PI' have adverse effects (Barritt *et al.*, 1997a).

The pollenizer cultivar can influence fruit size independently of effects on seed number. Use of 'Fuji' as a pollenizer has particularly beneficial effects (Keulemans *et al.*, 1996).

Fruit skin colour, russet and cracking

Apple and pear cultivars are, for purposes of marketing, largely defined by the appearance of their fruits, and colour and degree of russeting of the skin surface play a large part in their description. The major distinctions are between cultivars with red, partially red, brown or green skin and between those which are smooth-skinned, can have a degree of russeting or are mainly russeted.

Within cultivars the degree of red colour and of russet may also be used to define quality grade. This is based on visual attractiveness but also, within some cultivars, the fact that only those fruits most exposed to sunlight develop deep red colour has led to an association between colour, sweetness and other aspects of flavour and texture influenced by light climate. Russeted fruits may also be prone to water loss, both in storage and subsequently, with some consequent loss of crispness.

The apple skin

The term skin is usually used to refer to the outer protective tissues of the fruit consisting of cuticle, epidermis and hypodermis. At blossom time the cuticle is about one micrometre (1 μm) thick: at harvest its thickness varies between cultivars, being about 10 μm in 'Golden Delicious' (Meyer, 1944) and 25 μm in 'McIntosh Red' (Tukey and Young, 1942).

The epidermal cells have much larger radial than tangential dimensions at blossom time but the cells subsequently become rounder and after June become tangentially elongated (Figure 9.9). In some cultivars, e.g. 'Golden Delicious', the identity of the epidermis is lost as the fruit grows.

At blossom time the hypodermal cells are very similar to those of the cortex although slightly smaller (Tukey and Young, 1942). Later their cell walls thicken

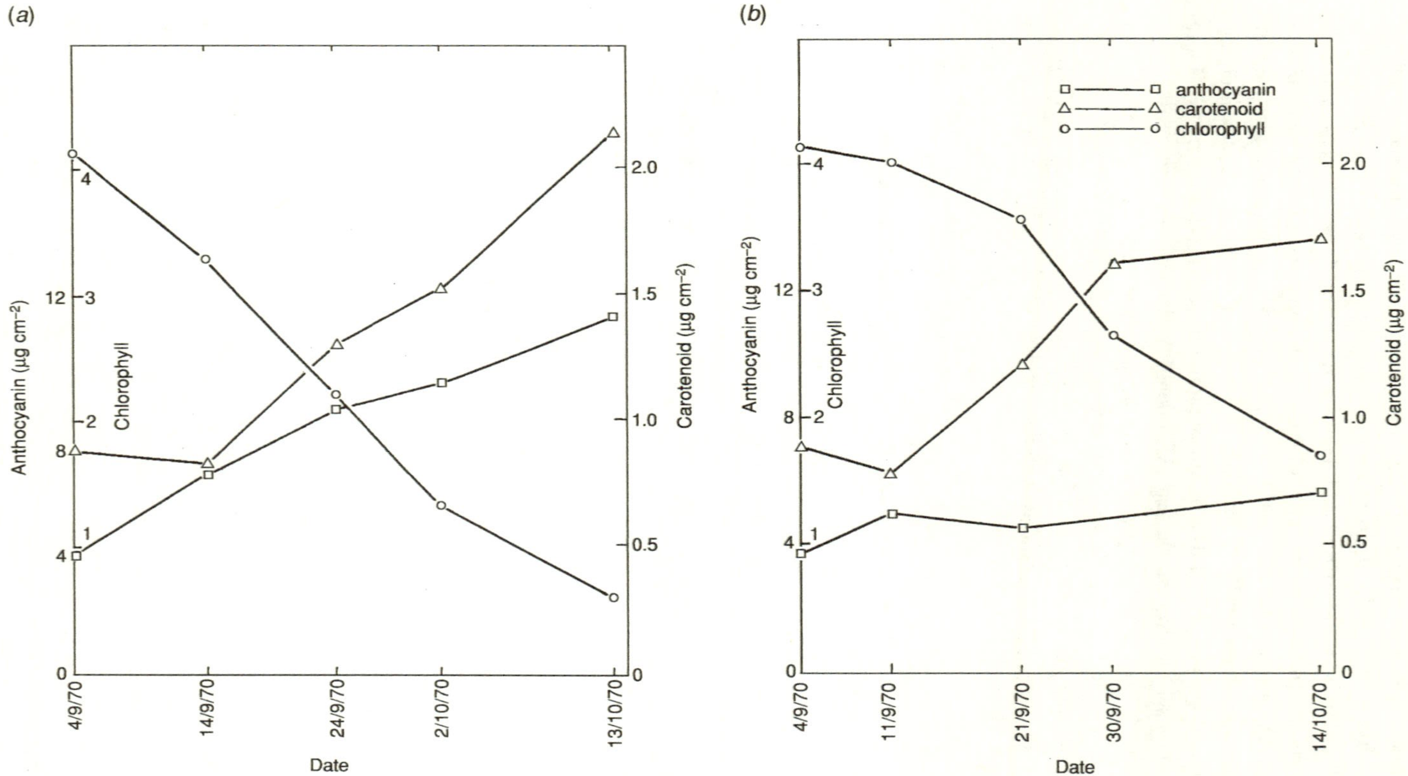

Figure 9.8 Anthocyanin, carotenoid and chlorophyll concentrations (µg cm^{-2}) in the peel of 'Cox's Orange Pippin', (*a*) ripening on the tree, (*b*) ripening in store at 12 °C. From Knee (1972). Reproduced with permission.

and the cells become elongated tangentially but the inner boundary between hypodermis and cortex may be hard to fix (Skene, 1962).

Fruit colour

The colour of the fruit skin depends on its background, or ground, colour caused by plastid pigments, chlorophylls and carotenoids, and to red coloration due to anthocyanins in the vacuoles.

The ground colour of immature fruits is usually dark green. As the fruit matures one of three things may happen, dependent primarily on cultivar. The green may fade until it has completely disappeared and the ground colour becomes cream to pale yellow; the green may fade less completely giving a greenish-yellow to yellowish-green ground colour; or the green may not fade. These colour changes reflect the disappearance of chlorophyll (green) and the unmasking or increased production of yellow carotenoids (Knee, 1972).

The potential for anthocyanin production is genetically determined, some cultivars producing very little of this even under the most favourable conditions, others developing a deep red colour over the whole surface except under the most unfavourable conditions, with others being intermediate. There are two peaks of anthocyanin formation. The first, which does not result in persistent red colour, is in fruitlets to the cell division phase. The second coincides with the ripening of red cultivars and may continue after harvest (Figure 9.8). The anthocyanin pigment may be located in any or all of the outer four to six, or even 11, layers depending on the cultivar (Dayton, 1959; Pratt *et al.*, 1975). The predominant anthocyanin is idaein (cyanidin-3-galactoside); others identified are cyanidin-3-arabinoside and cyanidin-7-arabinoside.

The shade of red that develops depends largely on the ground colour. The most brilliant red is produced when the ground colour is almost white and the dullest brown when the ground colour is green (Brown, 1975).

The colour development in fruitlet skin is not persistent and so has not been the subject of much research. Pre-harvest and immediately post-harvest red colour development are controlled by genotype, carbohydrate supply, direct effects of light on anthocyanin formation, temperature and nutritional factors.

EFFECTS OF GENOTYPE ON RED COLOUR DEVELOPMENT

Although green (e.g. 'Granny Smith') and yellow (e.g. 'Golden Delicious') apples have a very important place in the world markets there has been continued pressure of demand for increased extent or intensity of red coloration in those apples classed as red or part-coloured. One of the most powerful tools in meeting this demand has been the selection of colour-sports and their deliberate creation. Red sports are characterized by increased anthocyanin in the mature fruits, the increase being in the epidermis, the sub-epidermal cells, or both (Pratt *et al.*, 1975). Some red sports when used as parents pass on the red colour to their progeny. This is shown by comparing the progeny of crosses between the colour sport and a yellow-skinned cultivar with those of the original cultivar and the same yellow-skinned cultivar. Other colour mutants do not pass on the colour to their progeny. In the former case the mutation must have taken place in layer 2 of the meristem which produces the sub-epidermal cells and the gametes. The non-heritable colour mutation must have been in the superficial layer of the meristem which develops into the epidermis. 'Richared', a red sport of 'Delicious', is a good example of one in which increased anthocyanin production is heritable. It has the same number of pigmented cells in the epidermis but twice as many in the sub-epidermis as 'Delicious' (cf. review by Brown, 1975). The frequency of occurrence of colour sports and the ease with which they can be identified in the orchard, together with the heritability of the characteristic, has been particularly important in the 'Delicious' group of apples but is now also important in many other cultivars. Red-skinned mutants of several important pear cultivars, with anthocyanin production in either epidermal ('Starkrimson' mutation of 'Clapp's Favorite') or sub-epidermal ('Red Bartlett') layers, are also available.

EFFECTS OF CARBOHYDRATE SUPPLY ON COLOUR

The ratio of leaves to fruits has a marked effect on red colour development. Magness (1928) found that, for 'Delicious' and 'Winesap', fruits with only ten leaves per fruit had 23% of their surface red-coloured even with almost perfect light exposure, whereas fruits with 75 leaves had 58% red colour. The sugar contents of these fruits were 10% and 15%, respectively. Other fruit-thinning trials (reviewed by Walter, 1967a) showed similar effects on red colour. Faragher (1983) found that ripening, as measured by ethylene production, as well as skin anthocyanin production, was delayed in heavily cropping relative to lightly cropping trees. This may implicate internal carbohydrate conversion within the fruits, as well as gross carbohydrate supply, in the effect of leaf-to-fruit ratio.

DIRECT EFFECTS OF LIGHT ON COLOUR

If a small portion of the skin of a potentially red fruit is protected from light by an overlapping leaf or opaque adhesive tape, that protected area will not show anthocyanin development and will remain green. A number of enzymes involved in anthocyanin synthesis, including dihydroflavonal reductase (DFR) and phenylalanine ammonia-lyase (PAL), are light-dependent. However, there is no clear evidence that light effects on these control anthocyanin levels (Lister *et al.*, 1996; Ju *et al.*, 1997) although some correlations have been noted.

SHADE EFFECTS ON COLOUR

Artificial shading of small trees has a very marked effect both on the proportion of the skin developing a red colour and on the intensity of this colour

Table 9.13 *Effects of shade on fruit colour. 'Cox's Orange Pippin'/'M.26'*

	Shade level % full daylight			
	100	37	25	11
Average % surface coloured red	47.3	33.3	30.3	16.8
Mean colour intensity (0 = green, 8 = dark red)	5.04	3.13	2.55	1.56

Data from Jackson *et al.* (1977). Reproduced with permission.

(Table 9.13). The effect is proportional to the degree of shading. From associated measurements on the fruits (Jackson *et al.*, 1977) it is likely that the effect on red colour is due not only to direct effects of light intensity on the skin but also to effects on fruit carbohydrate content.

Within individual trees and in hedgerow orchards fruits from tree zones well exposed to light have a higher proportion of their fruit surface coloured red than fruits from shaded zones (Jackson 1970b, 1980; Jackson *et al.*, 1971). The concentration of anthocyanin in the most highly coloured skin areas is also a function of the light intensity under which the fruit is grown (Jackson *et al.*, 1977). Even under the high light intensities of Washington State, USA, 61% of within-tree variation in percentage red fruit colour was accounted for by a curvilinear regression on light intensity, with colour increasing up to approximately 50% full sunlight, i.e. 860 μmol m^2 s^{-1} (Barritt *et al.*, 1997b).

Reduction in fruit colour as a result of shade is the major factor determining depth of productive canopy, hence potential economic yield, in many fruit-growing areas. Jackson (1970b) found that under English conditions 'Cox' apples with more than 25% of their surface coloured red were not generally found below a leaf area index of more than 0.75, giving an effective canopy volume much smaller than that which can produce fruits of adequate size. In such cases colour sports requiring less light for adequate colour development are very valuable. Some cultivars are particularly sensitive to shade. Arakawa *et al.* (1986) found that 'Fuji' apples produce much less anthocyanin at any given light level than 'Starking Delicious' and at low light levels within the canopy 'Fuji' fails to develop adequate colour. If, however, colour sports produce good colour at very low light intensities the accepted link between amount and intensity of colour and fruit composition is broken and red colour no longer guarantees good eating quality (Seeley *et al.*, 1980).

TEMPERATURE EFFECTS ON COLOUR

The main effects of temperature on fruit colour occur late in the season when, in many fruit-growing areas, temperatures are falling. Curry (1997) summarized relevant interactions in terms of three factors in addition to light controlling red skin colour development. First, temperatures must be high enough, but not too high, and of sufficient duration in daylight hours to catalyse anthocyanin biosynthesis. Skin temperatures below 20 °C or above 35 °C are disadvantageous: under sunny conditions they may be much higher than air temperatures. Secondly, colour development varies with the stage of maturity. Once the climacteric phase is entered chlorophyll is degraded and anthocyanin is produced at a much lower rate. Thirdly, a period at cold temperatures stimulates anthocyanin production. A series of a few nights with temperatures in the range 2°–5 °C followed by warm sunny days promotes red colour development.

Creasy (1966) found that high daytime temperatures (31.6 °C) inhibited anthocyanin formation in 'McIntosh' apples even when the nights were cool. Faragher (1983) found that the optimum temperature for anthocyanin formation in unripe 'Jonathan' apples was 12 °C, and in ripe apples was 16–24 °C. At higher temperatures anthocyanin formation was much lower. Tan (1980) found that whole apples of 'Red Spy' receiving alternating periods at 6 °C and 25 °C in constant light or with darkness in the low-temperature period accumulated much more anthocyanin than apples receiving a constant 25 °C.

Some cultivars, e.g. 'McIntosh', 'Cortland' and 'Northern Spy', appear to be especially dependent on low temperatures for anthocyanin synthesis and 'Richared Delicious', 'Idared' and 'Spartan' tolerant of high temperature (Proctor, 1974). 'Delicious' and 'Fuji' have higher temperature optima than 'Gala' and 'Braeburn' (Curry, 1997). The highly coloured 'Scarlet Spur' sport of 'Delicious' has a similar rate of anthocyanin production to 'Oregon Spur' up to 24 °C but much higher rates at higher temperatures, even up to 33 °C.

Cooling apples by over-tree water sprinkling can substantially improve fruit colour. Unrath (1972) found that the amount of solid-red surface of 'Red Delicious' apples was doubled by this treatment.

EFFECTS OF COVERING FRUITS DURING DEVELOPMENT

In Japan fruits have traditionally been enclosed in paper bags from about one month after full bloom until shortly before harvest (Fukuda, 1994; Proctor and Lougheed, 1976). This practice, which protects from pest injury and disease, now has the main objective of producing smooth-skinned, vividly pink or red fruits.

For most cultivars bagging greatly reduces skin chlorophyll contents, which do not increase subsequently unless the fruits are exposed for 30–40 days prior to harvest. This lack of chlorophyll results in a unique pink or red colour being induced by anthocyanin (Proctor and Lougheed, 1976). Anthocyanin production occurs within 20–30 days of re-exposure to light. In most cultivars the previously bagged fruits produce more anthocyanin than unbagged ones.

The effect is particularly striking in cultivars such as 'Mutsu' which produce very little anthocyanin if not bagged.

EFFECTS OF MINERAL NUTRITION ON FRUIT COLOUR

High levels of nitrogen usually reduce the percentage of well-coloured fruits at harvest time, the effect being partly due to the shading effect of the extra foliage but primarily through a direct effect of N (Saure, 1990). Marsh *et al.* (1996) found evidence of a detrimental effect of high N on well-exposed fruits of 'Fuji', that they attributed to effects on fruit maturity and pigment development. High potassium levels are often associated with the development of red colour.

PLANT GROWTH REGULATOR AND HORMONAL EFFECTS ON FRUIT COLOUR

Ethephon, an ethylene-producing compound, has been widely used to promote anthocyanin formation, mainly in early and mid-season cultivars and their red sports. It appears to act mainly by accelerating the ripening process. Two drawbacks to its use are that it can induce red colour with insufficient light for fruit quality development (Saure, 1990), and that ethephon-treated fruits ripen and soften too quickly. There is some evidence that endogenous ethylene production during the ripening process increases the level of phenylalanine ammonia-lyase (PAL), a rate-limiting enzyme for anthocyanin formation (Faragher and Brohier, 1984). Some other factors influencing anthocyanin production do not appear to operate through effects on ethylene (Saure, 1990).

Fruit russeting and cracking

For a limited number of cultivars, e.g. the apple 'Egremont Russet', the skin has a rough but finely textured, predominantly light brown surface, that is regarded as characteristic of the cultivar. For most others russet is regarded as a blemish acceptable either not at all or to a limited extent, and is a major cause of downgrading.

Russet can be defined as a periderm that replaces the epidermis and forms a continuous layer of protective tissue. Its development in a russet cultivar is shown in Figure 9.9. As russet develops the cuticle ruptures, anthocyanin in the epidermis is lost and the underlying green tissues are obscured by brown quinones in the dead, ruptured, cells (Skene, 1982). The periderm may precede the rupture of the cuticle or be formed in a wound reaction to physical or chemical damage to the skin, e.g. by frost or spray damage. Russet is sometimes covered with shallow cracks up to 1 mm deep, when it is referred to as rough russet. Russeting of 'Cox' starts about mid-June and the percentage of fruit affected remains constant from mid-July to harvest (Skene, 1982).

Cultivars differ greatly in their propensity to russet: the characteristic is heritable but more than one factor seems to be involved (Brown, 1975). Some

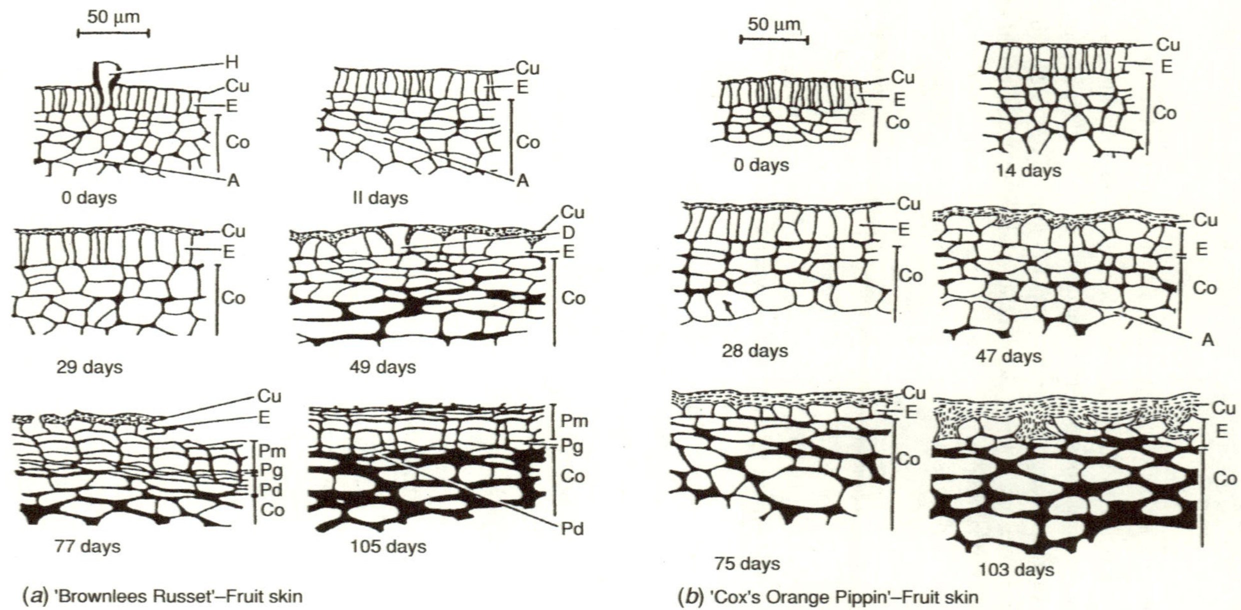

Figure 9.9 Radial long sections from the equator of the fruit cheek showing the skin structure at different days from blossoming of (*a*) a russet ('Brownlees Russet') and (*b*) a normal ('Cox's Orange Pippin') cultivar. H, hair base; Cu, cuticle; E, epidermis; Co, collenchyma or hypodermis; A, airspace; Pm, phellem or cork; Pg, phellogen; Pd, phelloderm. From Skene (1962). Reproduced with permission

sports of 'Golden Delicious', which is often downgraded because of russeting, are less sensitive than others.

Gibberellins, which increase the plasticity of fruitlet skin tissues (Taylor and Knight, 1986), play a major rôle in reduction of russeting. Eccher (1978) found that fruitlets from a russet sport of 'Golden Delicious' have less GA than those of standard Golden Delicious. He also found (Eccher, 1986) that 'Golden Delicious' fruits grown in Italy at high altitude (720 or 800 m) have less russet, a more elongated shape and a higher GA content than those grown at low altitude (300, 430 or 500 m). Application of gibberellins is a practical method of reducing russet, although with some risk of reducing flower bud formation (Taylor, 1975, 1978; Wertheim, 1982).

Shade, which increases cell elasticity, reduces russet (Walter, 1967b; Faust and Shear, 1972; Jackson *et al.*, 1977; Noè and Eccher, 1996). High humidity increases it (Walter, 1967b; Faust and Shear, 1972); benzyladenine also increases russet (Taylor, 1975).

Deep cracks occur simultaneously with the onset of russet, coinciding with high levels of stress in the developing fruit. This, for 'Cox', is first shown in early June when the fruits are about 15 mm in diameter and reaches a maximum in mid- to late June when the fruits are about 25 mm in diameter (Skene, 1982). It is possible that the periderm formed in localized russet reduces the elasticity of that part of the skin relative to un-russeted parts: this would reduce the capacity of the cuticle and epidermis to expand with fruit growth and cause the initially tiny cracks of rough russet to develop into splits.

Deep cracks and splits, becoming more severe as the season progresses, are a particularly serious problem on 'Cox', 'Gala' and 'Fuji' among major apple cultivars. Maintaining an adequate moisture supply and good leaf cover tends to reduce cracking, as do gibberellin sprays, but the effects of other factors are inconsistent (Opara *et al.*, 1997).

References

Abbott, D.L. (1959). The effects of seed removal on the growth of apple fruitlets. *Report of Long Ashton Research Station for 1958*, 52–6.

Abbott, D.L. (1970). The rôle of budscales in the morphogenesis and dormancy of the apple fruit bud. In *Physiology of Tree Crops*, ed. L.C. Luckwill and C.V. Cutting, pp. 65–82. London: Academic Press.

Abbott, D.L. (1977). Fruit-bud formation in Cox's Orange Pippin. *Report of Long Ashton Research Station for 1976*, 167–76.

Abbott, D.L. (1984). *The Apple Tree: Physiology and Management.* London: Grower Books.

Alston, F.H. (1996). Incompatibility alleles and apple pollination. *Acta Horticulturae* **423**, 119–24.

Anderson, J.L. and Seeley, S.D. (1993). Bloom delay in deciduous fruits. *Horticultural Reviews* **15**, 97–144.

Arakawa, O., Hori, Y. and Ogata, R. (1986). Characteristics of colour development and relationship between anthocyanin synthesis and phenylalanine ammonia-lyase activity in 'Starking Delicious', 'Fuji' and 'Mutsu' apple fruits. *Journal of the Japanese Society for Horticultural Science* **54**, 424–30.

Atkinson, C.J. and Lucas, A.S. (1996). The response of flowering date and cropping of *Pyrus communis* cv Concorde to autumn warming. *Journal of Horticultural Science* **71**, 427–34.

Atkinson, C.J. and Taylor, L. (1994). The influence of autumn temperatures on flowering time and cropping of *Pyrus communis* cv Conference. *Journal of Horticultural Science* **69**, 1067–75.

Atkinson, C.J., Lucas, A.S. and Taylor, L. (1997). Effects of warm autumns on the cropping of pear. *Acta Horticulturae* **451**, 743–7.

Bain, J.M. and Robertson, R.N. (1951). The physiology of growth in apple fruits. 1. Cell size, cell number and fruit development. *Australian Journal of Scientific Research B* **4**, 75–91.

Banno, K., Hayashi, S. and Tanabe, K. (1985a). Effects of SADH and shoot-bending on flower bud formation, nutrient components and endogenous growth regulators in Japanese Pear (*Pyrus serotina* Rehd). *Journal of the Japanese Society for Horticultural Science* **53**, 365–76.

Banno, K. Hayashi, S. and Tanabe, K. (1985b). Relationships between flower bud formation and endogenous growth regulators in Japanese Pear cultivars (*Pyrus serotina* Rehd). *Journal of the Japanese Society for Horticultural Science* **54**, 15–25.

Banno, K., Hayashi, S. and Tanabe, K. (1986). Morphological and histological studies on flower bud differentiation and development in Japanese Pear (*Pyrus serotina* Rehd). *Journal of the Japanese Society for Horticultural Science* **55**, 258–65.

Barritt, B.H., Barden, J.A., Cline, J., Granger, R.L., Kushad, M.M., Marini, R.P., Parker, M., Perry, R.L., Robinson, T., Unrath, C.R. and Dilley, M.A. (1997a). Performance of 'Gala' at year 5 with eight apple rootstocks in an 8-location North American NC-140 trial. *Acta Horticulture* **451**, 129–35.

Barritt, B.H., Drake, S.R., Konishi, B.S. and Rom, C.R. (1997b). Influence of sunlight level and rootstock on apple fruit quality. *Acta Horticulturae* **451**, 569–77.

Behboudian, M.H. and Mills, T.M. (1997). Deficit irrigation in deciduous orchards. *Horticultural Reviews* **21**, 105–31.

Benedek, P. and Nagy, C.S. (1996). Flower constancy of pollinating honeybees on some fruit tree species. *Acta Horticulturae* **423**, 65–71.

Bergh, O. (1985a). Morphogenesis of *Malus domestica* cv Starking flower buds. *South African Journal of Plant and Soil* **2**, 187–90.

Bergh, O. (1985b). Effect of the previous crop on cortical cell number of *Malus domestica* cv Starking Delicious apple flower primordia, flowers and fruit. *South African Journal of Plant and Soil* **2**, 191–6.

Bernhard, R. (1961). Mise à fleur et alternance chez les arbres fruitiers. *Congrès pomologiques. I.N.R.A.* Paris, October 1960 pp. 91–116.

Berüter, J. and Droz, Ph, (1991). Studies on locating the signal for fruit abscission in the apple tree. *Scientia Horticulturae* **46**, 201–14.

Berüter, J. and Studer Feusi, M.E., (1997). The effect of girdling on carbohydrate partitioning in the growing apple fruit. *Journal of Plant Physiology* **151**, 277–85.

Berüter, J., Studer Feusi, M.E. and Rüedi, P. (1997). Sorbitol and sucrose partitioning in the growing apple fruit. *Journal of Plant Physiology* **151**, 269–76.
Blasberg, C.H. (1943). The relation of size of McIntosh flower buds to the production of fruit. *Proceedings of the American Society for Horticultural Science* **42**, 220–2.
Blasco, A.B. (1976). Rootstock effects on growth and cropping of apples, with special reference to fruit quality. PhD thesis, University of London.
Boskovic, R. and Tobutt, K.R. (1999). Correlation of stylar ribonuclease isoenzymes with incompatibility alleles in apple. *Euphytica* **107**, 29–43.
Broothaerts, W. Janssens, G.A., Proost, P. and Broekaert, W.F. (1995). cDNA cloning and molecular analysis of two self-incompatibility genes in apple. *Plant Molecular Biology* **27**, 499–511.
Broothaerts, W., Verdoodt, L., Keulemans, J., Janssens, G.A. and Broekaert, W.F. (1996). The self-incompatibility gene in apple and determination of the S-genotype of apple cultivars by PCR. *Acta Horticulturae* **423**, 103–9.
Brown, A.G. (1975). Apples. In *Advances in Fruit Breeding*, ed. J. Janick and J.N. Moore, pp. 3–37. West Lafayette: Purdue University Press.
Browning, G. (1989). The physiology of fruit set. In *Manipulation of Fruiting*, ed. C.J. Wright, pp. 195–217. London: Butterworths.
Buban, T. (1986). Changes in growth properties of young apple trees treated with paclobutrazol, PP. 333. *Acta Horticulturae* **179**, 549–50.
Buban, T. and Faust, M. (1982). Flower bud induction in apple trees: internal control and differentiation. *Horticultural Reviews* **4**, 174–203.
Buban, T., Varga, A., Tromp, J., Knegt, E. and Bruisma, J. (1978). Effects of ammonium and nitrate nutrition on the levels of zeatin and amino nitrogen in xylem sap of apple rootstocks. *Zeitschrift für Pflanzenphysiologie* **89**, 289–95.
Bukovac, M.J. and Nakagawa, S. (1968). Gibberellin-induced asymmetric growth of apple fruits. *HortScience* **3**, 172–4.
Buszard, D. and Schwabe, W.W. (1995). Effect of previous crop load on stigmatic morphology of apple flowers. *Journal of the American Society for Horticultural Science* **120**, 566–70.
Byers, R.E., Barden, J.A. and Carbaugh, D.H. (1990a). Thinning of spur 'Delicious' apples by shade, terbacil, carbaryl and ethephon. *Journal of the American Society for Horticultural Science* **115**, 9–13.
Byers, R.E., Barden, J.A., Polomski, R.F., Young, R.W. and Carbaugh, D.H. (1990b). Apple thinning by photosynthetic inhibition. *Journal of the American Society for Horticultural Science* **115**, 14–19.
Byers, R.E., Carbaugh, D.H., Presley, C.N. and Wolf, T.K. (1991). The influence of low light on apple fruit abscission. *Journal of Horticultural Science* **66**, 7–17.
Callesen, O. (1997). Testing 20 apple rootstocks. *Acta Horticulturae* **451**, 137–45.
Calzoni, G.L. and Speranza, A. (1996). Pear and plum pollination: honey bees, bumble bees or both. *Acta Horticulturae* **423**, 83–90.
Campbell, A.I. and Lacey, C.N.D. (1982). Induced mutants of Cox's Orange Pippin apple with increased self-compatibility. 1. Production and selection. *Euphytica* **31**, 469–75.
Caspari, H.W., Behboudian, M.H., and Chalmers, D.J. (1994). Water use, growth and fruit yield of 'Hosui' Asian pears under deficit irrigation. *Journal of the American Society for Horticultural Science* **119**, 383–8.
Chalmers, D.J., Burge, G., Jerie, P.H. and Mitchell, P.D. (1986). The mechanism of regulation of 'Bartlett' pear fruit and vegetative growth by irrigation withholding and regulated deficit irrigation. *Journal of the American Society for Horticultural Science* **111**, 904–7.

Chan, B.G. and Cain, J.C. (1967). The effect of seed formation on subsequent flowering in apple. *Proceedings of the American Society for Horticultural Science* **91**, 63–8.

Child, R.D. and Williams, R.R. (1983). The effects of the interaction of aminoethoxyvinylglycine (AVG) with mixtures of growth regulators on fruit set and shape in Cox's Orange Pippin apple. *Journal of Horticultural Science* **58**, 365–9.

Chrisosto, C.H., Lombard, P.B., Richardson, D.G. and Tetley, R. (1992). Putrescine extends effective pollination period in 'Comice' pear (*Pyrus communis* L.) irrespective of post-anthesis ethylene levels. *Scientia Horticulturae* **49**, 211–21.

Church, R.M. and Williams, R.R. (1983a). Comparison of flower numbers and pollen production of several dessert apple and ornamental *Malus* cultivars. *Journal of Horticultural Science* **58**, 327–36.

Church, R.M. and Williams, R.R. (1983b). Comparison of flowering dates and pollen release characteristics of several *Malus* cultivars used as pollinators for Cox's Orange Pippin apple. *Journal of Horticultural Science* **58**, 349–53.

Cole, M.E., Solomos, T. and Faust, M. (1982). Growth and respiration of dormant flower buds of *Pyrus communis* and *Pyrus calleryana*. *Journal of the American Society for Horticultural Science* **107**, 226–31.

Collins, M.D., Lombard, P.B. and Wolfe, J.W. (1978). Effects of evaporative cooling for bloom delay on 'Bartlett' and 'Bosc' pear fruit maturity and quality. *Journal of the American Society for Horticultural Science* **103**, 187–9.

Coombe, B.G. (1976). The development of fleshy fruits. *Annual Review of Plant Physiology* **27**, 507–28.

Corelli-Grappadelli, L., Lakso, A.N. and Flore, J.A. (1994). Early season patterns of carbohydrate partitioning in exposed and shaded apple branches. *Journal of the American Society for Horticultural Science* **119**, 596–603.

Crabbé, J.J. (1984). Vegetative vigour control over location and fate of flower buds in fruit trees. *Acta Horticulturae* **149**, 55–63.

Creasy, L.L. (1966). The effect of temperature on anthocyanin synthesis in McIntosh apple skin. *Proceedings of New York State Horticultural Society* **111**, 93–6.

Crowe, A.D. (1984). Effect of prebloom temperature on fruit set in apple. *Annual Report 1984 Research Station Kentville, Nova Scotia, Canada*, 47–8.

Curry, E.A. (1997). Temperatures for optimum anthocyanin accumulation in apple tissue. *Journal of Horticultural Science* **72**, 723–9.

Dayton, D.F. (1959). Red colour distribution in apple skin. *Proceedings of the American Society for Horticultural Science* **74**, 72–81.

Dencker, I. and Hansen, P. (1994). Flower initiation and node development of axillary buds, growth partitioning, and bleeding sap production of young apple trees as affected by fertigation. *Journal of Horticultural Science* **69**, 869–76.

Denne, M.P. (1960). The growth of apple fruitlets and the effect of early thinning on fruit development. *Annals of Botany* **24**, 397–406.

Denne, M.P. (1963). Fruit development and some tree factors affecting it. *New Zealand Journal of Botany* **1**, 265–94.

Dennis, F.G. (1967). Apple fruit-set: evidence for a specific role of seeds. *Science* **156**, 71–3.

Dennis, F.G. (1979). Factors affecting yield in apple with emphasis on 'Delicious'. *Horticultural Reviews* **1**, 395–422.

De Witte, K., Vercammen, J., van Daele, G. and Keulemans, J. (1996). Fruit set, seed set and fruit weight in apple as influenced by emasculation, self-pollination and cross-pollination. *Acta Horticulturae* **423**, 177–83.

Dheim, M.A. and Browning, G. (1987). The mode of action of (2RS, 3RS)-paclobutrazol on the fruit set of Doyenne du Comice pear. *Journal of Horticultural Science* **62**, 313–27.
Dheim, M.A. and Browning, G. (1988). The mechanism of the effect of (2RS, 3RS)-paclobutrazol on flower initiation of pear cvs Doyenne du Comice and Conference. *Journal of Horticultural Science* **63**, 393–405.
Eccher, T. (1978). Russeting of Golden Delicious apples as related to endogenous and exogenous gibberellins. *Acta Horticulturae* **80**, 381–5.
Eccher, T. (1986). Russeting and shape of Golden Delicious apples as related to endogenous GA content of fruitlets. *Acta Horticulturae* **179**, 767–70.
Eijnde, J. van den (1996). Pollination of pear by bumblebees (*Bombus terrestris* L.) and honey bees (*Apis mellifera* L.). *Acta Horticulturae* **423**, 73–8.
Faragher, J.D. (1983). Temperature regulation of anthocyanin accumulation in apple skin. *Journal of Experimental Botany* **34**, 1291–8.
Faragher, J.D. and Brohier, R.L. (1984). Anthocyanin accumulation in apple skin during ripening: regulation by ethylene and phenylalanine ammonia-lyase. *Scientia Horticulturae* **22**, 89–96.
Faust, M. (1989). *Physiology of Temperate Zone Fruit Trees.* New York: John Wiley & Sons.
Faust, M. and Shear, C.B. (1972). Russeting of apples, an interpretive review. *HortScience* **7**, 233–5.
Feucht, W. (1976). *Fruitfulness in Pome and Stone Fruit.* Washington State University. *Extension Service Extension Bulletin* **665**, 1–32.
Forshey, C.G. and Elfving, D.C. (1989). The relationship between vegetative growth and fruiting in apple trees. *Horticultural Reviews* **11**, 229–87.
Free, J.B. (1993). *Insect Pollination of Crops.* London: Academic Press.
Fukuda, H. (1994). Apple. In *Horticulture in Japan*, ed. K. Konishi, S. Iwahori, H. Kitagawa and T. Yakuwa, pp. 23–7. Tokyo: Asakura Publishing Co. Ltd.
Fukui, H. Imakawa, S. and Tamura, T. (1984a). Histochemical observation and quantitative analysis of indole derivatives in apple seeds in relation to early drop of fruit. *Journal of the Japanese Society for Horticultural Science* **53**, 135–40.
Fukui, H. Imakawa, S. and Tamura, T. (1984b). Relation between early drop of apple fruit, ethylene evolution and formation of abscission layer. *Journal of the Japanese Society for Horticultural Science* **53**, 303–7.
Fukui, H., Imakawa, S. and Tamura, T. (1985). Relationship between early drop of apple fruit, cytokinin and gibberellin. *Journal of the Japanese Society for Horticultural Science* **54**, 287–92.
Fulford, R.M. (1962). The development of apple spur buds in relation to leaves and fruits. *XVIth International Horticultural Congress* **3**, 343–6.
Fulford, R.M. (1966b). The morphogenesis of apple buds III. The inception of flowers. *Annals of Botany* **30**, 207–19.
Fulford, R.M. (1966c). The morphogenesis of apple buds IV. The effect of fruit. *Annals of Botany* **30**, 597–606.
Galston, A.W. and Kaur-Sawhney, R. (1995). Polyamines as endogenous growth regulators. In *Plant Hormones. Physiology, Biochemistry and Molecular Biology*, ed. P.J. Davies, pp. 158–78. Dordrecht: Kluwer.
Garcia-Martinez, J.L. and Hedden, P. (1997). Gibberellins and fruit development. In *Phytochemistry of Fruit and Vegetables*, ed. F.A. Tomás-Barberan and R.J. Robins, pp. 263–85. Oxford: Clarendon Press.
Gil, G.F., Martin, G.C. and Griggs, W.H. (1972). Fruit set and development in the pear: extractable endogenous hormones in parthenocarpic and seeded fruits. *Journal of the American Society for Horticultural Science* **97**, 731–5.

Gil, G.F., Martin, G.C. and Griggs, W.H. (1973). Fruit set and development in the pear: diffusible growth substances from seeded and seedless fruits. *Journal of the American Society for Horticultural Science* **98**, 51–4.

Goldway, M., Shai, O., Yehuda, H., Matityahu, A. and Stern, R.A. (1999). 'Jonathan' apple is a lower-potency pollenizer of 'Topred' than 'Golden Delicious' due to partial S-allele incompatibility. *Journal of Horticultural Science and Biotechnology* **74**, 381–5.

Goldwin, G.K. (1978). Improved fruit setting with plant growth hormones. *Acta Horticulturae* **80**, 115–21.

Goldwin, G.K. and Schwabe, W.W. (1975). Parthenocarpic fruit in Cox's Orange Pippin apples, obtained without hormones. *Journal of Horticultural Science* **50**, 175–8.

Gómez-Cadenas, A., Mehouachi, J., Tadeo, F.R., Primo-Millo, E. and Talon, M. (2000). Hormonal regulation of fruitlet abscission induced by carbohydrate shortage in citrus. *Planta* **210**, 636–43.

Goode, J.E. and Ingram, J. (1971). The effect of irrigation on the growth, cropping and nutrition of Cox's Orange Pippin apple trees. *Journal of Horticultural Science* **46**, 195–208.

Goode, J.E., Higgs, K.H. and Hyrycz, K.J. (1978a). Abscisic acid applied to orchard trees of Golden Delicious apple to control water stress. *Journal of Horticultural Science* **53**, 99–103.

Goode, J.E., Higgs, K.H. and Hyrycz, K.J. (1978b). Nitrogen and water effects on the nutrition, growth, crop yield and fruit quality of orchard-grown Cox's Orange Pippin apple trees. *Journal of Horticultural Science* **53**, 295–306.

Goodwin, P.B. (1978). Phytohormones and fruit growth. In *Phytohormones and Related Compounds – a Comprehensive Treatise*, ed. D.S. Letham, P.B. Goodwin and T.J.V. Higgins, pp. 175–214. Amsterdam: Elsevier.

Gorter, C.J. and Visser, T. (1958). Parthenocarpy of apples and pears. *Journal of Horticultural Science* **33**, 217–27.

Grauslund, J. and Hansen, P. (1975). Fruit trees and climate. III. The effect of temperature on fruit set in apple trees. *Tidsskrift for Planteavl* **79**, 481–8.

Greene, D.W. (1980). Effect of silver nitrate, aminoethoxyvinylglycine and gibberellins A_{4+7} plus 6-benzylamino purine on fruit set and development of 'Delicious' apples. *Journal of the American Society for Horticultural Science* **105**, 717–20.

Greenhalgh, W.J. and Godley, G.L. (1976). Studies of fruit shape in apples: a survey of factors influencing the development of 'typiness' of the cultivar Red Delicious. *Australian Journal of Experimental Agriculture and Animal Husbandry* **16**, 592–5.

Greenhalgh, W.J., Godley, G.L. and Menzies, R. (1977). Studies of fruit shape in apples: response to gibberellin and cytokinin sprays. *Australian Journal of Experimental Agriculture and Animal Husbandry* **17**, 505–9.

Grochowska, M.J. (1973). Comparative studies on physiological and morphological features of bearing and non-bearing spurs of the apple tree. I. Changes in starch content during growth. *Journal of Horticultural Science* **48**, 347–56.

Grochowska, M.J., Karaszewska, A., Jankowska, B. and Mika, A. (1984). The pattern of hormones of intact apple shoots and its changes after spraying with growth regulators. *Acta Horticulturae* **149**, 25–38.

Gruber, J. and Bangerth, F. (1990). Diffusible IAA and dominance phenomena in fruits of apple and tomato. *Physiologia Plantarum* **79**, 354–8.

Guttridge, C.G. (1962). Inhibition of fruit-bud formation in apple with gibberellic acid. *Nature* **196**, 1008.
Hamer, P.J.C. (1981). The effects of evaporative cooling on apple bud development and frost resistance. *Journal of Horticultural Science* **56**, 107–12.
Hamer, P.J.C. (1983). Evaporative cooling of apple buds: the effect of timing of water application on bud development and frost resistance of the cv Cox's Orange Pippin. *Journal of the Horticultural Science* **58**, 153–9.
Hamer, P.J.C. (1985). The heat balance of apple buds and blossoms. I. Heat transfer in the outdoor environment. *Agricultural and Forest Meteorology* **35**, 339–52.
Hamer, P.J.C. (1986). The heat balance of apple buds and blossoms. II. The water requirements for frost protection by overhead sprinkler irrigation. *Agricultural and Forest Meteorology* **37**, 159–74.
Hamer, P.J.C. and Jackson, J.E. (1975). The incidence of late spring frosts. In *Climate and the Orchard*, ed. H.C. Pereira, pp. 63–5, Farnham Royal, UK: Commonwealth Agricultural Bureaux.
Hansen, P. (1977). Carbohydrate allocation. In *Environmental Effects on Crop Physiology*, ed. J.J. Landsberg and C.V. Cutting, pp. 247–58. London. Academic Press.
Hartman, F.O. and Howlett, F.S. (1954). Fruit setting of the Delicious apple. *Ohio Agriculture Experiment Station Research Bulletin* **745**.
Hayashi, F., Naito, R., Bukovac, M.J. and Sell, H.M. (1968). Occurrence of gibberellin A_3 in parthenocarpic apple fruit. *Plant Physiology* **43**, 448–50.
Hedden, P. and Graebe, J.E. (1985). Inhibition of gibberellin biosynthesis by paclobutrazol in cell-free homogenates of *Cucurbita maxima* endosperm and *Malus pumila* embryos. *Journal of Plant Growth Regulation* **4**, 111–22.
Hedden, P. and Hoad, G.V. (1985). Hormonal regulation of fruit growth and development. *B.P.G.R.G. Monograph* **12**, 211–21.
Hedden, P., Carter, K.J., Ward, D.A., Ward, G.C. and Bolton, M.A. (1984). Hormonal control of fruit development. *Report of the East Malling Research Station for 1983*, 57–8.
Herrero, M. (1983). Factors affecting fruit set in 'Agua de Aranjuez' pear. *Acta Horticulturae* **139**, 91–6.
Herrero, M. (1984). Effect of time of GA_3 treatment on 'Agua de Aranjuez' pear fruit set. *Acta Horticulturae* **149**, 211–18.
Herrero, M. and Gascon, M. (1987). Prolongation of embryo sac viability in pear (*Pyrus communis*) following pollination or treatment with gibberellic acid. *Annals of Botany* **60**, 287–93.
Heslop-Harrison, J. (1976). A new look at pollination. *Report of the East Malling Research Station for 1975*, 141–57.
Hiratsuka, S., Okada, Y., Kawai, Y., Tamura, F. and Tanabe, K. (1995). Stylar basic proteins corresponding to 5 self-incompatibility alleles of Japanese pears. *Journal of the Japanese Society for Horticultural Science* **64**, 471–8.
Hoad, G.V. (1978). The role of seed derived hormones in the control of flowering in apple. *Acta Horticulturae* **80**, 93–103.
Hoad, G.V. (1980). Growth regulators, endogenous hormones and flower initiation in apple. *Report of Long Ashton Research Station for 1979*, 199–206.
Hoad, G.V. (1984). Hormonal regulation of fruit-bud formation in fruit trees. *Acta Horticulturae* **149**, 13–23.
Hoblyn, T.N., Grubb, N.H., Painter, A.C. and Wates, B.L. (1936). Studies in biennial bearing. I. *Journal of Pomology* **14**, 39–76.

Huet, J. (1972). Etude des effets des feuilles et des fruits sur l'induction florale des brachyblastes du Poirier. *Physiologie Végétale* **10**, 529–45.
Huet, J. (1973). Floral initiation in pear trees. *Acta Horticulturae* **34**, 193–8.
Izadyar, A.B. (1997). Anatomical aspects of alternate bearing in apple trees. *Acta Horticulturae* **435**, 255–60.
Jackson, J.E. (1967). Variability in fruit size and colour within individual trees. *Report of the East Malling Research Station for 1966*, 110–15.
Jackson, J.E. (1970a). Individual fruit size in relation to age of bearing wood on 'Laxton's Superb' apple trees. *Report of the East Malling Research Station for 1969*, 83–5.
Jackson, J.E. (1970b). Aspects of light climate within apple orchards. *Journal of Applied Ecology* **7**, 207–16.
Jackson, J.E. (1980). Light interception and utilization by orchard systems. *Horticultural Reviews* **2**, 208–67.
Jackson, J.E. and Hamer, P.J.C. (1980). The causes of year-to- year variation in the average yield of Cox's Orange Pippin apple in England. *Journal of Horticultural Science* **55**, 149–56.
Jackson, J.E. and Palmer, J.W. (1977b). Effects of shade on the growth and cropping of apple trees. II. Effects on components of yield. *Journal of Horticultural Science* **52**, 253–66.
Jackson, J.E., Hamer, P.J.C. and Wickenden, M.F. (1983). Effect of early spring temperatures on the set of fruits of Cox's Orange Pippin apple and year-to-year variation in its yields. *Acta Horticulturae* **139**, 75–82.
Jackson, J.E., Palmer, J.W., Perring, M.A. and Sharples, R.O. (1977). Effects of shade on the growth and cropping of apple trees. III. Effects on fruit growth, chemical composition and quality at harvest and after storage. *Journal of Horticultural Science* **52**, 267–82.
Jackson, J.E., Sharples, R.O. and Palmer, J.W. (1971). The influence of shade and within-tree position on apple fruit size, colour and storage quality. *Journal of Horticultural Science* **46**, 277–87.
Janssens, G.A., Goderis, I.J., Broekaert, W.F. and Broothaerts, W. (1995). A molecular method for S-allele identification in apple based on allele-specific PCR. *Theoretical and Applied Genetics* **91**, 691–8.
Jonkers, H. (1979). Biennial bearing in apple and pear: a literature survey. *Scientia Horticulturae* **11**, 303–17.
Ju, Z., Yuan, Y., Liu, C., Wang, Y. and Tian, X. (1997). Dihydroflavonol reductase activity and anthocyanin accumulation in 'Delicious', 'Golden Delicious' and 'Indo' apples. *Scientia Horticulturae* **70**, 31–43.
Kemp, H. (1996). Pollination results of apple, *Malus*, pear, plum and cherry of the international working group on pollination. *Acta Horticulturae* **423**, 243–302.
Keulemans, J., Brusselle, A., Eyssen, R. Vercammen, J. and van Daele, G. (1996). Fruit weight in apple as influenced by seed number and pollinizer. *Acta Horticulturae* **423**, 201–10.
Klinac, D.J., Geddes, B. and Wright, S. (1995). Wood age and floral bud distribution on four nashi (*Pyrus serotina*) cultivars grown on pergola, Y-frame and centre-leader training systems in the Waikato region of New Zealand. *New Zealand Journal of Crop and Horticultural Science* **23**, 191–7.
Knee, M. (1972). Anthocyanin, carotenoid and chlorophyll changes in the peel of Cox's Orange Pippin apples during ripening on and off the tree. *Journal of Experimental Botany* **23**, 184–96.
Knight, J.N., Spencer, J.E., Looney, N.E. and Lovell, J.D. (1987). Chemical thinning of the apple cultivar Spartan. *Journal of Horticultural Science* **62**, 135–9.

Kobel, F. (1954). *Lehrbuch des Obstbaus auf physiologischer Grundlage*. Berlin: Springer-Verlag.

Kobel, F., Steinegger, P. and Anliker, J. (1939). Weitere Untersuchungen über die Befruchtungsverhaltnisse der Apfel und Birnsorten. *Landwirtschaftlichen Jarbuch der Schweitz* **53**, 160–91.

Kondo, S. and Mizuno, N. (1989). Relation between early drop of apple fruit and endogenous growth regulators, and effects of MCPB, GA_3 plus GA_4 and BA sprays on fruit abscission. *Journal of the Japanese Society for Horticultural Science* **58**, 9–16.

Kondo, S. and Takahashi, Y. (1987). Effects of high temperature in the nighttime and shading in the daytime on the early drop of apple fruit 'Starking Delicious'. *Journal of the Japanese Society for Horticultural Science* **56**, 142–50.

Kondo, S. and Takahashi, Y. (1989). Relation between early drop of apple fruit and ethylene evolution under high night-temperature conditions. *Journal of the Japanese Society for Horticultural Science* **58**, 1–8.

Kondo, S., Asari, M. and Kumagai, M. (1987). Effects of weather conditions, tree vigor and pruning on the early fruit drop of apple. *Journal of the Japanese Society for Horticultural Science* **55**, 415–21.

Kotob, M.A. and Schwabe, W.W. (1971). Induction of parthenocarpic fruit in Cox's Orange Pippin apples. *Journal of Horticultural* Science **46**, 89–93.

Lacey, C.N.D., Church, R.M. and Richardson, P. (1982). Induced mutants of Cox's Orange Pippin apple with apparent increased self-compatibility. 2. Laboratory and orchard assessment and testing. *Euphytica* **31**, 511–18.

Lakso, A.N. (1994). Apple. In *Handbook of Environmental Physiology of Fruit Crops*, Vol. I. *Temperate Crops*, ed. B. Schaffer and P.C. Anderson, pp. 3–42. Boco Raton, FL: CRC Press.

Lakso, A.N., Corelli-Grappadelli, L., Barnard, J. and Goffinet, M.C. (1995). An expolinear model of the growth pattern of the apple fruit. *Journal of Horticultural Science* **70**, 389–94.

Landsberg, J.J. (1979). From bud to bursting blossom: weather and the apple crop. *Weather*, **34**, 394–407.

Landsberg, J.J., Butler, D.R. and Thorpe, M.R. (1974). Apple bud and blossom temperatures. *Journal of Horticultural Science* **49**, 227–39.

Lauri, P.-E. and Térouanne, É. (1999). Influence of inflorescence removal on the fruit set of the remaining inflorescences and development of the laterals on one year old apple (*Malus domestica* Borkh) branches. *Journal of Horticultural Science & Biotechnology* **74**, 110–17.

Lauri, P.-E., Térouanne, É., Lespinasse, J.M., Regnard, J.L. and Kelner, J.J. (1995). Genotypic differences in the axillary bud growth and fruiting pattern of apple fruiting branches over several years – an approach to regulation of fruit bearing. *Scientia Horticulturae* **64**, 265–81.

Layne, R.E.C. and Quamme, H.A. (1975). Pears. In *Advances in Fruit Breeding*, ed. J. Janick and J.N. Moore, pp. 38–70. West Lafayette: Purdue University Press.

Letham, D.S. (1969). Regulators of cell division in plant tissues. VI. The effects of zeatin and other stimulators of cell division on apple fruit development. *New Zealand Journal of Agricultural Research* **12**, 1–20.

Letham, D.S. and Williams, M.W. (1969). Regulators of cell division in plant tissues. VIII. The cytokinins of the apple fruit. *Physiologia Plantarum* **22**, 925–36.

Lewis, T. and Smith, B.D. (1969). The insect faunas of pear and apple orchards and the effect of windbreaks on their distribution. *Annals of Applied Biology* **64**, 11–20.

Lister, C.E., Lancaster, J.E. and Walker, J.R.L. (1996). Phenylalanine ammonia-lyase (PAL) activity and its relationship to anthocyanin and flavonoid levels in New Zealand-grown apple cultivars. *Journal of the American Society for Horticultural Science* **121**, 281–5.

Lombard, P.B. and Richardson, D.G. (1982). Increase fruit set and cropping of 'Comice' pear trees with an ethylene inhibitor, amino-ethoxyvinylglycine. *Acta Horticulturae* **124**, 165–9.

Looney, N.E. and Lane, W.D. (1984). Spur-type growth mutants of McIntosh apple: a review of their genetics, physiology and field performance. *Acta Horticulturae* **146**, 31–46.

Lu, C.S. and Roberts, R.H. (1952). Effect of temperature upon the setting of 'Delicious' apples. *Proceedings of the American Society for Horticultural Science* **59**, 177–83.

Luckwill, L.C. (1948). The hormone content of the seed in relation to endosperm development and fruit drop in the apple. *Journal of Horticultural Science* **24**, 32–44.

Luckwill, L.C. (1953). Studies of fruit development in relation to plant hormones. I. Hormone production by the developing apple seed in relation to fruit drop. *Journal of Horticultural Science* **28**, 14–24.

Luckwill, L.C. (1970). The control of growth and fruitfulness in apple trees. In *Physiology of Tree Crops*, ed. L.C. Luckwill and C.V. Cutting, pp. 237–53. London: Academic Press.

Luckwill, L.C. (1974). A new look at the process of fruit bud formation in apple. *Proceedings of the XIXth International Horticultural Congress, Warsaw* **3**, 237–45.

Luckwill, L.C. and Silva, J.M. (1979). The effects of daminozide and gibberellic acid on flower initiation, growth and fruiting of apple *cv.* Golden Delicious. *Journal of Horticultural Science* **54**, 217–23.

MacDaniels, L.H. and Heinicke, A.J. (1929). Pollination and other factors affecting the set of apple. *New York (Cornell University) Agricultural Experiment Station Bulletin* **497**.

Magness, J.R. (1928). Observations on colour development in apples. *Proceedings of the American Society for Horticultural Science* **25**, 289–92.

Makino, T., Fukui, H., Imakawa, S. and Tamura, T. (1986). Relation between early drop of apple fruit and shoot growth. *Journal of the Japanese Society for Horticultural Science* **55**, 40–5.

Marcutti, M.C., Ragazzini, D. and Sansavini, S. (1984). The effects of gamma and laser rays on the functioning of apple pollen in pollination and mentor pollen experiments. *Journal of Horticultural Science* **59**, 57–61.

Marsh, K.B., Volz, R.K., Cashmore, W. and Reay, P. (1996). Fruit colour, leaf nitrogen level, and tree vigour in 'Fuji' apples. *New Zealand Journal of Crop and Horticultural Science* **24**, 393–9.

Martin, G.C., Horgan, R. and Nishijima, C. (1982). Changes in hormone content of pear receptacles from anthesis to shortly after fertilization as affected by pollination or GA_3 treatment. *Journal of the American Society for Horticultural Science* **107**, 479–82.

McCown, M. (1943). Anatomical and chemical aspects of abscission of fruits of the apple. *Botanical Gazette* **105**, 212–20.

Meyer, A. (1944). A study of the skin structure of Golden Delicious apple. *Proceedings of the American Society for Horticultural Science* **45**, 105–10.

Miki, N., Yukinaga, H., Hayashi, S. and Tanabe, K. (1982). Effect of Bendroquinone on development and endogenous growth regulators of flowers in 'Nijisseiki' pear. *Journal of the Japanese Society for Horticultural Science* **51**, 35–43.

Miller, J.M. (1988). Apple fruit set potential in relation to preblossom temperature. PhD thesis, Wye College, University of London.
Miller, S.S. and Swietlik, D. (1986). Growth and fruiting response of deciduous fruit trees treated with paclobutrazol. *Acta Horticulturae* **179**, 563–6.
Mills, T.M., Behboudian, M.H. and Clothier, B.E. (1996). Water relations, growth and composition of 'Braeburn' apple fruit under deficit irrigation. *Journal of the American Society for Horticultural Science* **121**, 286–91.
Minchin, P.E.H., Thorpe, M.R., Wünsche, J.N., Palmer, J.W. and Picton, R.F. (1997). Carbon partitioning between apple fruits: short- and long-term response to availability of photosynthate. *Journal of Experimental Botany* **48**, 1401–6.
Mitchell, P.D., Jerie, P.H. and Chalmers, D.J. (1984). The effects of regulated water deficits on pear tree growth, flowering, fruit growth and yield. *Journal of the American Society for Horticultural Science* **109**, 604–6.
Modlibowska, I. (1945). Pollen tube growth and embryo-sac development in apples and pears. *Journal of Pomology and Horticultural Science* **21**, 57–89.
Modlibowska, I. (1962). Some factors affecting supercooling of fruit blossoms. *Journal of Horticultural Science* **37**, 249–61.
Modlibowska, I. (1975a). Effects of frost on crop yield and quality. In *Climate and the Orchard*, ed. H.C. Pereira, pp. 73–81. Farnham Royal, UK: CAB.
Modlibowska, I. (1975b). Induction of parthenocarpic apples of Bramley's Seedling by low concentrations of gibberellins. *Journal of Horticultural Science* **50**, 21–2.
Murneek, A.E. (1954). The embryo and endosperm in relation to fruit development, with special reference to the apple, *Malus sylvestris*. *Proceedings of the American Society for Horticultural Science* **64**, 573–82.
Naor, A., Klein, I., Doran, I., Gal, Y., Ben-David, Z. and Bravdo, B. (1997). Irrigation and crop load interactions in relation to apple yield and fruit size distribution. *Journal of the American Society for Horticultural Science* **122**, 411–14.
Noè, N. and Eccher, T. (1996). 'Golden Delicious' apple fruit shape and russeting are affected by light conditions. *Scientia Horticulturae* **65**, 209–13.
Olien, W.C., Ferree, D.C., Bishop, B.L. and Bridges, W.C. Jr (1995). Prediction of site index and apple rootstock performance from environmental variables. *Fruit Varieties Journal* **49**, 179–89.
Opara, L.V., Studman, C.J. and Banks, N.H. (1997). Fruit skin splitting and cracking. *Horticultural Reviews* **19**, 217–62.
Orlandini, S., Moriondo, M., Cappellini, P. and Ferrari, P. (1999). Analysis and modelling of apple fruit growth. *Acta Horticulturae* **499**, 137–43.
Palmer, J.W., Cai, Y.L. and Edjamo, Y. (1991). Effect of part-tree flower thinning on fruiting, vegetative growth and leaf photosynthesis in 'Cox's Orange Pippin' apple. *Journal of Horticultural Science* **66**, 319–25.
Parry, M.S. (1974). The control of biennial bearing of Laxton's Superb apple trees. *Journal of Horticultural Science* **49**, 123–30.
Parry, M.S. (1976). Effect of pruning and partial disbudding on the cropping of Comice and Beurré Hardy pears. *Journal of Horticultural Science* **51**, 159–66.
Parry, M.S. (1980). Evidence of clonal variation and latent virus effects on the vigour of Cox's Orange Pippin apple trees on M9 rootstocks. *Journal of Horticultural Science* **55**, 439–40.
Pavel, E.W. and DeJong, T.M. (1995). Seasonal patterns of nonstructural carbohydrates of apple *Malus pumila* (Mill) fruits: relationship with relative growth rates and contribution to solute potential. *Journal of Horticultural Science* **70**, 127–34.

Pearson, J.A. and Robertson, R.N. (1953). Seasonal variation in cell size, nitrogen metabolism and respiration in developing Granny Smith apple fruits. *Australian Journal of Biological Science* **6**, 1–20.

Percival, M.S. (1955). The presentation of pollen in certain angiosperms and its collection by *Apis mellifera*. *New Phytologist* **54**, 353–68.

Petropoulou, S.P. and Alston, F.H. (1998). Selecting for improved pollination at low temperatures in apple. *Journal of Horticultural Science and Biotechnology* **75**, 507–12.

Powell, D.B.B. (1974). Some effects of water stress in late spring on apple trees. *Journal of Horticultural Science* **49**, 257–72.

Prang, L., Stephan, M., Schneider, G. and Bangerth, F. (1997). Gibberellin signals originating from apple fruit and their possible involvement in flower induction. *Acta Horticulturae* **463**, 235–41.

Pratt, C. (1988). Apple flower and fruit: morphology and anatomy. *Horticultural Reviews* **10**, 273–308.

Pratt, C., Way, R.D. and Einset, J. (1975). Chimeral structure of red sports of 'Northern Spy' apple. *Journal of the American Society for Horticultural Science* **100**, 419–22.

Proctor, J.T.A. (1974). Color stimulation in attached apples with supplementary light. *Canadian Journal for Plant Science* **54**, 499–503.

Proctor, J.T.A. and Lougheed, E.C. (1976). The effect of covering apples during development. *HortScience* **11**, 108–9.

Proctor, J.T.A. and Palmer, J.W. (1991). The role of spur and bourse leaves of three apple cultivars on fruit set and growth and calcium content. *Journal of Horticultural Science* **66**, 275–82.

Proebsting, E.L. and Mills, H.H. (1978). Low temperature resistance of developing flower buds of six deciduous fruit species. *Journal of the American Society for Horticultural Science* **103**, 192–8.

Quinlan, J.D. and Preston, A.P. (1971). The influence of shoot competition on fruit retention and cropping of apple trees. *Journal of Horticultural Science* **46**, 525–34.

Rahemi, M., Dennis, F.G., Andersen, R.L., Ozga, J.O. and Xia, R.X., (1997). The role of ethylene in apple fruit set. *Journal of Horticultural Science* **72**, 67–75.

Ramirez, H. and Hoad, G.V. (1981). Effects of growth substances on fruit-bud initiation in apple. *Acta Horticulturae* **120**, 131–6.

Reid, M.S. (1995). Ethylene in plant growth, development and senescence. In *Plant Hormones*, 2nd edn, ed. P.J. Davies, pp. 486–508. Dordrecht: Kluwer.

Robbie, F.A. and Atkinson, C.J. (1994). Wood and tree age as factors influencing the ability of apple flowers to set fruit. *Journal of Horticultural Science* **69**, 609–23.

Robbie, F.A., Atkinson, C.J., Knight, J.N. and Moore, K.G. (1993). Branch orientation as a factor determining fruit set in apple trees. *Journal of Horticultural Science* **68**, 317–25.

Rohitha, B.H. and Klinac, D.J. (1994). Some observations on the influence of temperature on the germination of pollen on excised nashi (*Pyrus serotina* Rehder var. *culta* Rehder) flowers. *New Zealand Journal of Crop and Horticultural Science* **22**, 339–42.

Rom, C.R. (1991). Light thresholds for apple tree canopy growth and development. *HortScience* **26**, 989–92.

Ryugo, K. (1986). Promotion and inhibition of flower initiation and fruit set by plant manipulation and hormones, a review. *Acta Horticulturae* **179**, 301–7.

Sachs, R.M. (1977). Nutrient diversion: an hypothesis to explain the chemical control of flowering. *HortScience* **12**, 220–2.

Salter, P.J. and Goode, J.E. (1967). Crop responses to water at different stages of growth. *Research Review No. 2. Commonwealth Bureau of Horticulture and Plantation Crops*, East Malling, Maidstone, Kent.

Sassa, H., Hirano, H. and Ikehashi, H. (1992). Self-incompatability related RNases in styles of Japanese pear (*P. serotina* Rehd.). *Plant Cell Physiology* **33**, 811–14.

Saure, M.C. (1990). External control of anthocyanin formation in apple. *Scientia Horticulturae* **42**, 181–218.

Schechter, I., Proctor, J.T.A. and Elfving, D.C. (1993a). Reappraisal of seasonal apple fruit growth. *Canadian Journal of Plant Science* **73**, 549–56.

Schechter, I., Proctor, J.T.A. and Elfving, D.C. (1993b). Characterization of seasonal fruit growth of 'Idared' apple. *Scientia Horticulturae* **54**, 203–10.

Schneider, D., Stern, R.A., Eisikowitch, D. and Goldway, M. (2002). The relationship between floral structure and honey bee pollination efficiency in 'Jonathan' and 'Topred' apple cultivars. *Journal of Horticultural Science and Biotechnology* **77**, 48–51.

Schrader, A.L. and Marth, P.C. (1931). Light intensity as a factor in the development of apple colour and size. *Proceedings of the American Society for Horticultural Science* **28**, 552–5.

Sedgley, M. (1990). Flowering of deciduous perennial fruit crops. *Horticultural Reviews* **12**, 223–64.

Seeley, E.J., Micke, W.C. and Kammereck, R. (1980). 'Delicious' apple fruit size and quality as influenced by radiant flux density in the immediate growing environment. *Journal of the American Society for Horticultural Science* **105**, 645–57.

Skene, D.S. (1962). Fruit skin structure in some tree fruits with special reference to russeting of apples. PhD thesis, University of London.

Skene, D.S. (1982). The development of russet, rough russet and cracks on the fruit of the apple of Cox's Orange Pippin during the course of the season. *Journal of Horticultural Science* **57**, 165–74.

Smith, B.D. (1970). Natural pollen transfer. In *Towards Regulated Cropping*, ed. R.R. Williams, and D. Wilson, pp. 29–40. London: Grower Books.

Smith, W.H. (1940). The histological structure of the flesh of the apple in relation to growth and senescence. *Journal of Pomology and Horticultural Science* **XVIII**, 249–60.

Smith, W.H. (1950). Cell multiplication and cell enlargement in the development of the flesh of the apple fruit. *Annals of Botany.* NS XIV. **53**, 23–38.

Spiegel-Roy, P. and Alston, F.H. (1979). Chilling and post-dormant heat requirements as selection criteria for late flowering pears. *Journal of Horticultural Science* **54**, 115–20.

Spiegel-Roy, P. and Alston, F.H. (1982). Pollination requirements of new apple cultivars. *Journal of Horticultural Science* **57**, 145–50.

Sterling, C. (1954). Sclereid development and the texture of Bartlett pears. *Food Research* **19**, 433–43.

Stopar, M. (1998). Apple fruitlet thinning and photosynthate supply. *Journal of Horticultural Science & Biotechnology* **73**, 461–6.

Stott, K.G. (1972). Pollen germination and pollen-tube characteristics in a range of apple cultivars. *Journal of Horticultural Science* **46**, 191–8.

Stott, K.G. and Campbell, A.I. (1971). Variation in self-compatibility in Cox clones. *Report of Long Ashton Research Station for 1970*, 23.

Tamura, T., Fukui, H., Imakawa, S. and Mino, Y. (1981). Effect of temperature at early stage of fruit development on development of fruit and seed in apple. *Journal of the Japanese Society for Horticultural Science* **50**, 287–96.

Tan, S.C. (1980). Phenylalanine ammonia-lyase and the phenylalanine ammonia-lyase inactivating system: effects of light, temperature and mineral deficiencies. *Australian Journal or Plant Physiology* **7**, 159–67.

Taylor, B.K. (1975). Reduction of apple skin russeting by gibberellin A_{4+7}. *Journal of Horticultural Science* **50**, 169–72.

Taylor, B.K. (1978). Effect of gibberellin sprays on fruit russet and tree performance of Golden Delicious apple. *Journal of Horticultural Science* **53**, 167–9.

Taylor, D.R. and Knight, J.N. (1986). Russeting and cracking of apple fruit and their control with plant growth regulators. *Acta Horticulturae* **179**, 819–20.

Teng, Y., Tanabe, K. Tamura, F. and Itai, A. (1998). Effect of spur age on the translocation of assimilates from spurs of 'Nijisseiki' pear (*Pyrus pyrifolia* Nakai). *Journal of the Japanese Society for Horticultural Science* **67**, 313–18.

Thorpe, M.R. (1974). Radiant heating of apples. *Journal of Applied Ecology* **11**, 755–60.

Tobutt, K.R. (1994). Combining apetalous parthenocarpy with columnar growth habit in apple. *Euphytica* 77, 51–4.

Toyoma, S. and Hayashi, S. (1957). Studies on the fruit development of Japanese Pears. *Journal of the Horticultural Association of Japan* **25**, 274–8.

Tromp, J. (1970). Shoot orientation effects on growth and flower-bud formation in apple. *Acta Botanica Neerlandica* **19**, 535–8.

Tromp, J. (1972). Effects of growth regulating substances and tree orientation on growth and flower-bud formation in apple. *Journal of Horticultural Science* **47**, 525–33.

Tromp, J. (1973). The interaction of growth regulators and tree orientation on fruit bud formation. *Acta Horticulturae* **34**, 185–8.

Tromp, J. (1982). Flower-bud formation in apple as affected by various gibberellins. *Journal of Horticultural Science* **57**, 277–82.

Tromp, J. (1984). Flower-bud formation in apple as affected by air and root temperature, air humidity, light intensity and day length. *Acta Horticulturae* **149**, 39–47.

Tromp, J. and Borsboom, O. (1994). The effect of autumn and spring temperature on fruit set and on the effective pollination period in apple and pear. *Scientia Horticulturae* **60**, 23–30.

Tromp, J. and Borsboom, O. (1996). Fruit set and the effective pollination period in apple and pear as affected by bloom and post-bloom temperature. *Acta Horticulturae* **423**, 193–9.

Tukey, H.B. and Young, J.O. (1942). Gross morphology and histology of developing fruit of apple. *Botanical Gazette* **104**, 3–25.

Tukey, L.D. (1956). Some effects of night temperatures on the growth of McIntosh apples. I. *Proceedings of the American Society for Horticultural Science* **68**, 32–43.

Tukey, L.D. (1960). Some effects of night temperature on the growth of McIntosh apples. II. *Proceedings of the American Society for Horticultural Science* **75**, 39–46.

Tustin, S., Corelli-Grappadelli, L. and Ravaglia, G. (1992). Effect of previous-season and current light environments on early-season spur development and assimilate translocation in 'Golden Delicious' apple. *Journal of Horticultural Science* **67**, 351–60.

Tustin, D.S., Stanley, C.J. and Adams, H.M. (1997). Physiological and phenological responses of apple trees to artificial reduction of the growth period from harvest to leaf fall. *Acta Horticulturae* **451**, 383–92.

Unrath, C.R. (1972). The quality of 'Red Delicious' apples as affected by overtree sprinkler irrigation. *Journal of the American Society for Horticultural Science* **91**, 58–61.

Untiedt, R. and Blanke, M. (2001). Effects of fruit thinning agents on apple tree canopy photosynthesis and dark respiration. *Plant Growth Regulation* **35**, 1–9.

van Oosten, H.J. (1979). Current developments in the Dutch fruit industry. *Compact Fruit Tree* **12**, 34–46.

van Zyl, H.J. and Strydom, D.K. (1982). Some factors affecting fruit set of Packham's Triumph pear. *Acta Horticulturae* **124**, 149–56.

Volz, R.K., Ferguson, I.B., Hewett, E.W. and Woolley, D.J. (1994). Wood age and leaf area influence fruit size and mineral composition of apple fruits. *Journal of Horticultural Science* **62**, 385–95.

Wagenmakers, P.S. (1988). Effects of planting system, tree shape and additional summer pruning on growth, production and flowering of apple trees at high plant densities. *Journal of Horticultural Science* **63**, 383–92.

Walter, T.E. (1967a). Factors affecting fruit colour in apples: a review of world literature. *Report of the East Malling Research Station for 1966*, 70–82.

Walter, T.E. (1967b). Russeting and cracking in apples: a review of world literature. *Report of the East Malling Research Station for 1966*, 83–95.

Wample, R.L. (1982). Water stress in fruit trees. In *Tree Fruit Water Management and Irrigation*. Proceedings of the short course 1982, ed. R.B. Tukey, pp. 95–108. Pullman: Washington State University.

Warrington, I.J., Fulton, T.A., Halligan, E.A. and de Silva, H.N. (1999). Apple fruit growth and maturity are affected by early season temperatures. *Journal of the American Society for Horticultural Science* **124**, 468–77.

Webster, A.D., Tobutt, K.R., James, D.J., Evans, K.M. and Alston, F.A. (1997). Rootstock breeding and orchard testing at Horticultural Research International – East Malling. *Acta Horticulturae* **451**, 83–8.

Wertheim, S.J. (1973). Fruit set and June drop in Cox's Orange Pippin apple as affected by pollination and treatment with a mixture of gibberellins A_4 and A_7. *Scientia Horticulturae* **1**, 85–105.

Wertheim, S.J. (1982). Fruit russeting in apple as affected by various gibberellins. *Journal of Horticultural Science* **57**, 283–8.

Wertheim, S.J. (1997). Chemical thinning of deciduous fruit trees. *Acta Horticulturae* **463**, 445–62.

Westwood, M.N., Kadivar, N. and Bjornstad, H.O. (1967). Differences in growth, chemical content and fruit set among four sports of Delicious apple. *Fruit Varieties and Horticultural Digest* **21**, 72–4.

Westwood, M.N., Lombard, P.B. and Brannock, L.D. (1968). Effect of seeded fruits and foliar applied auxin on seedless fruit set of pear the following year. *HortScience* **3**, 168–9.

Williams, M.W. and Stahly, E.A. (1969). Effect of cytokinins and giberellins on shape of "Delicious" apple fruits. *Journal of the American Society for Horticultural Science* **94**, 17–19.

Williams, R.R. (1963). The effect of nitrogen on the self-fruitfulness of certain varieties of cider apple. *Journal of Horticultural Science* **38**, 52–60.

Williams, R.R. (1965). The effect of summer nitrogen applications on the quality of apple blossom. *Journal of Horticultural Science* **40**, 31–41.

Williams, R.R. (1966). Pollination studies in fruit trees. III. The effective pollination period for some apple and pear varieties. *Report of Long Ashton Research Station for 1965*, 136–8.

Williams, R.R. (1970a). Artificial pollination of fruit trees in the United States. In *Towards Regulated Cropping*, ed. R.R. Williams and D. Wilson, pp. 23–8. London: Grower Books.

Williams, R.R. (1970b). Appendix. In *Towards Regulated Cropping*, ed. R.R. Williams and D. Wilson, pp. 57–61. London: Grower Books.

Williams, R.R. and Wilson, D. (eds.) (1970). *Towards Regulated Cropping*. London: Grower Books.

Williams, R.R., Arnold, G.M., Flook, V.A. and Jefferies, C.J. (1980). The effect of picking date on blossoming and fruit set in following year for the apple cv Bramley's Seedling. *Journal of Horticultural Science* **55**, 359–62.

Yuda, E., Matsui, H., Yukimoto, M. and Nakagawa, S. (1984). Effect of 15 β OH gibberellins on the fruit set and development of three pear species. *Journal of the Japanese Society for Horticultural Science* **53**, 235–41.

Zhu, L.H., Borsboom, O. and Tromp, J. (1997). The effect of temperature on flower-bud formation in apple including some morphological aspects. *Scientia Horticulturae* **70**, 1–8.

Zimmerman, R.H. (1972). Juvenility and flowering in woody plants: a review. *HortScience* **7**, 447–55.

10

Eating quality and its retention

Introduction

The market price of fruits depends on their attractiveness and eating quality, which need to be retained for as long as possible after harvest to facilitate the matching of supply to demand.

This market price is of predominant importance in apple and pear production economics because many of the costs are fixed per kilogram and fruits of low perceived value do not meet their costs of production. Fruit quality is therefore a key determinant controlling the cultivars that are grown. Attainment of a specific size and, where appropriate, degree of red colour does not guarantee that the fruits are ready to eat, either immediately or after a period of storage. Good appearance after removal from store does not guarantee that eating quality has been maintained. Eating quality is based on developmental processes of maturation, ripening and senescence (Watada *et al.*, 1984).

Maturation is the process leading to physiological or horticultural maturity. Physiological maturity is the stage of development when the fruit will continue ontogeny even if detached. Horticultural maturity is when the fruit meets the criteria laid down by consumers.

Ripening is the sum of the processes from the later stages of growth through to the early stages of senescence, that result in the attainment of the characteristic eating quality.

Senescence involves those processes following maturity that lead to death.

Fruit sensory quality

The biological basis of fruit sensory quality has been investigated to provide both understanding and quicker and more reproducible assessment of quality factors than can be achieved using taste panels. Key measurements are those of texture, juiciness, sugar and acid content, and aroma.

Texture

Consumer surveys in the United Kingdom suggest that the overall acceptability of apples is principally related to their texture (Smith, 1984). Immature fruits are too hard as well as having other undesirable characteristics. Fruits soften during ripening in a progressive way that leads to unacceptability. Initial firmness at harvest controls the effect of softening.

Softening can result from loss of turgor (Hatfield and Knee, 1988), degradation of starch and, most importantly, cell wall degradation (Tucker, 1993) and a weakening of the cohesive forces between cells.

Apple fruit cell walls consist mainly of cellulose and pectin, with some hemicellulose and very small amounts of extensin. The cellulose is resistant to degradation but the cell walls have a clearly distinguishable section known as the middle lamella which separates adjacent cells and serves as a bonding agent between cells. This section is rich in pectic polysaccharides and its cohesion is thought to depend on ionic bonds involving calcium ions and uronic acids (Bartley and Knee, 1982). The breakdown of the middle lamella may follow transport of exo-polygalacturonase and possibly pectinesterase to it. Cell separation proceeds in the apples until they develop a dry or mealy texture when the consumer's teeth pass between cells without breaking them, so that they fail to release juice (Knee, 1993).

The preferred texture of pears differs between species and markets. European-type (*Pyrus communis*) cultivars are expected to have a tender skin, soft, buttery, flesh and an absence of gritty stone-cells especially if the pears are to be used to produce purée for baby food. Chinese and Japanese pears (*P. ussuriensis* and *P. pyrifolia*) are gritty and a crisp, breaking, texture is preferred.

Fruit firmness can be measured using the resistance of peeled fruit flesh to the insertion of a plunger of known diameter, usually 8 mm for pear and 11 mm for apple, with a uniform crosshead speed set between 50 and 250 mm min^{-1}. The maximum force applied at each penetration at a depth of 8 mm is recorded in newtons or kilograms, the conversion factor from kg to N being $\times 9.81$ (Smith, 1985).

More complex analysis of results obtained with a commercial texture testing machine can discriminate between apple cultivars with respect to their fracturability (which correlates with crispness), hardness, cohesiveness, springiness and chewiness (Corrigan *et al.*, 1997). Tensile strength can also be measured as the force required to pull a cylinder of apple tissue into two halves (Poovaiah *et al.*, 1988). Acoustic (non-destructive) measurements of firmness as a 'stiffness factor' correlate well with compression and tensile rupture forces (Tu *et al.*, 1997).

Juiciness

Juiciness as perceived by the consumer is greater the more juice is released on chewing, the greater the force with which it is released, the higher its water content and the lower its viscosity and content of suspended solids. Juiciness requires a cohesive network of large, turgid, thin-walled cells. If the middle lamella is stronger than the cell walls, the cells fracture on biting and juice is released; if it is weaker, then the fracture is between cells, the juice is not released and the fruit is perceived as non-juicy (Szczesniak and Ilker, 1988; Boulton *et al.*, 1997; Tu *et al.*, 1997). Although juiciness is often associated with fruit firmness (Kingston, 1992; Tu *et al.*, 1997) apple cultivars with similar textural properties may differ in juiciness (Corrigan *et al.*, 1997). Juice content can be determined by expressing the juice from a known volume or weight of apple cortical tissue under known force and expressing the results as mg or ml per gram of tissue (Smith, 1985). The area of spread of juice from the freshly cut surface of apple slices on $CuSO_4$-treated filter paper is highly correlated with sensory panel ratings of juiciness (Boulton *et al.*, 1997).

Sugar and acid content

Apple taste is primarily related to the amount of sugar and acid in the fruit tissues and to the balance between these. In pears acidity is considerably lower than in apples and may be undetectable as a component of taste.

Fructose, glucose, sucrose and sorbitol are the main fruit sugars. Fructose is the main sugar in mature apples of most cultivars, including 'Golden Delicious' (Pavel and DeJong, 1995) and 'Fuji' (Drake *et al.*, 1997). Sucrose predominates in 'Cox's Orange Pippin' (Pavel and De Jong, 1995). European and Asian pears also accumulate high amounts of sucrose during fruit ripening (Moriguchi *et al.*, 1992).

Malic acid is the main organic acid in apples and pears but some apple cultivars also have appreciable amounts of citric acid (10–20% in 'Fuji' according to Hong *et al.*, 1997). Citric acid may exceed malic in some pear cultivars (Ulrich, 1970) and both apples and pears have smaller quantities of quinic, galactouronic, chlorogenic and other acids.

Total and reducing sugars are measured by standard chemical and enzymatic procedures (Smith, 1985). Alternatively, a few drops of juice are placed on the prism of a refractometer and results read as percentage soluble solids at 20 °C. Measurement of specific sugars is by HPLC. Acidity is best measured as titratable acidity but can also be recorded as pH using indicator paper strips.

Measured levels of percentage sugar and pH are closely correlated with taste assessments as long as the range is wide, e.g. between seedling cultivars, but taste does not discriminate well within extremes of sugar content or acidity.

Perceived sweetness is higher the lower the acidity (Visser *et al.*, 1968): a sport of 'Jonagored' characterized by its sweetness of taste was found to be identical in sugar content but much less acid than the parent type (Lott, 1965). For pears, which are less acid, perceived sweetness is higher the juicier the fruit. Sugar and acid content are inherited independently and change independently over time and in response to controlling factors. Therefore, although perceived taste depends on the balance of sugar and acids (Yahia, 1994) its variability and management are generally approached at the individual component, sugar and acid, level. There is no single desirable level of sugar, acid, or sugar/acid ratio that applies to all cultivars. Corrigan *et al.* (1997) showed soluble solids content to range from 11.0% to 15.7%, malic acid from 0.35% to 0.95% and the soluble solids to acid ratio from 12 to 36 between five cultivars of current commercial importance. This presumably shows that different consumers have different tastes, but consistency in matching the expected taste of a cultivar is important.

Aroma

Much of the character of apple and pear fruits depends on their aroma, resulting from trace amounts of volatile organic substances. Between 20 and 40 volatiles, out of more than 350 identified, are responsible for apple aroma (Yahia, 1994). Typical aroma volatiles include esters (e.g. hexyl hexanoate), lipid oxidation products (e.g. (E)-2–hexenal) and the terpenoid β-damascenone.

Different cultivars have different characteristic aroma volatiles. When amino-acid precursors are fed to different cultivars they are converted into different volatiles, for example 'Granny Smith' converts isoleucine to ethyl 2-methylbutanoate whereas 'Red Delicious' apples also show conversion to 2-methyl-2-butenyl esters (Rowan *et al.*, 1997).

Changes during maturation and ripening

Physical properties

Cell separation occurs during the growth of apples so that at maturity about 25% of the fruit volume is air space between the cells (Khan and Vincent, 1990). This increase in air space may account for the decline in firmness during growth (Knee, 1993) and may be difficult to separate from ripening-related changes in fruit texture. Hesse and Hitz (1938) and Smock (1948) found a slow decline in firmness, as measured by pressure tests, from late August to early October whereas the best eating quality was achieved by harvest in late

September. The firmness at maturity, i.e. at the optimum harvest date for immediate consumption or that for long term storage, varies from cultivar to cultivar, e.g. from 68 newtons for 'McIntosh' to 82 N for 'Delicious' in the same season (Lau, 1988), and also varies from season to season so is not a reliable sole indicator of the best date for harvesting. The decline in firmness continues after harvest in store, slowing this decline being a key objective of storage technology. Other things being equal, the firmness of fruits after storage is a linear function of their firmness at harvest (Johnson and Ridout, 2000).

Respiration rate

The rate of respiration per unit fresh weight is high early in the season, during the cell division phase of fruit growth and then declines to a very low level (Bepete and Lakso, 1997). Then, when the fruits reach physiological maturity and ripening processes are initiated, apples and European pears typically show a marked increase in respiratory activity resulting in increased evolution of CO_2 (Kidd and West, 1924; Rhodes, 1970). This increase, referred to as the respiration climacteric, precedes visible symptoms of ripening but once it has occurred the ripening process is irreversible. Subsequent to this climacteric rise, which can occur in apples left on the tree and in harvested fruits, respiration rates decline once more (Figure 10.1 *a*).

Japanese pears can show a respiration climacteric but some cultivars do not do so (Downs *et al.*, 1991).

Ethylene evolution

Soon after pollination ethylene production is high and gradually declines during cell division and expansion. In apples and pears showing a respiration climacteric ethylene production increases rapidly just prior to obvious signs of ripening. In 'climacteric' apples and pears the change in ethylene production may coincide with that in respiration but it may be significantly later (Reid, 1995). The time of the peak rate of ethylene production in pears coincides with that of respiration but in apples it comes later (Rhodes, 1980). The change in ethylene production is much greater than that in respiration. Preclimacteric apples produce 0.1 $\mu l\ kg^{-1}\ h^{-1}$ of ethylene, climacteric apples 100 $\mu l\ kg^{-1}\ h^{-1}$ (Knee, 1985). The rate of CO_2 evolution increases by only about 50–100% (Rhodes, 1980). Typical curves for apple ethylene evolution are shown in Figure 10.1 *b*. The rapid rise in ethylene production involves an autocatalytic effect. Ethylene is always present in the fruit tissues at a very low concentration but young fruitlets are not capable of responding to this endogenous, or to exogenous, ethylene by ripening. Once a critical point of

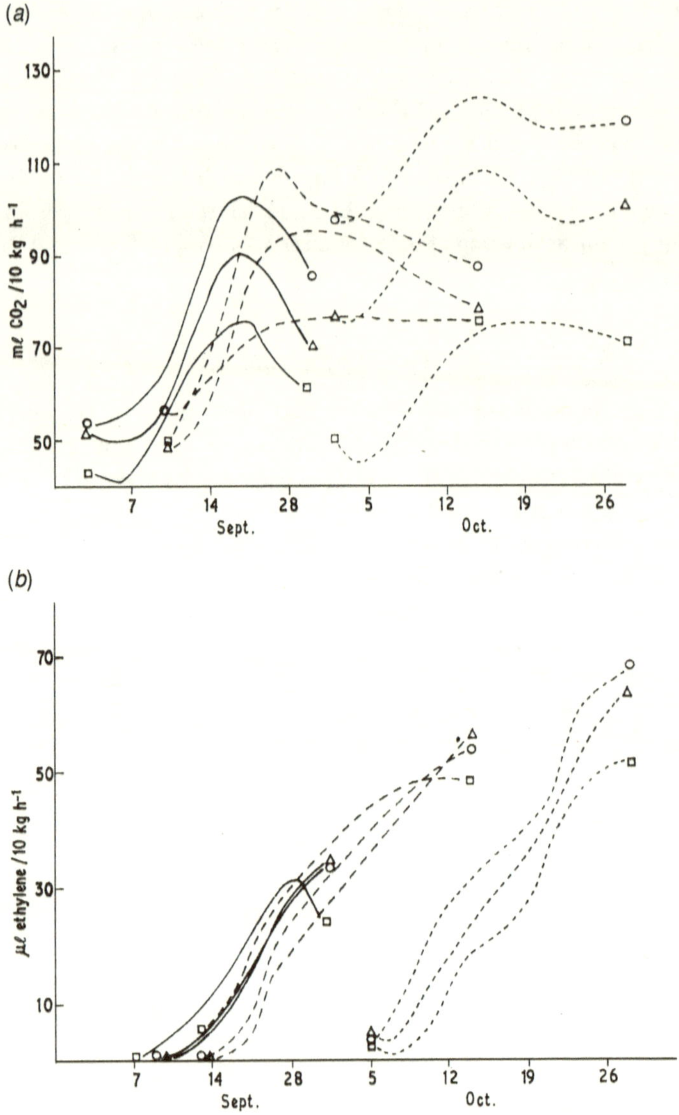

Figure 10.1 (*a*) Respiration and (*b*) ethylene production by apple fruits cv. 'Cox's Orange Pippin' grown under different levels of shade and harvested at different dates. ○, 100%; △, 37%; □, 11% daylight. ——, harvested 9 September; – – –, harvested 16 September; - - - harvested 1 October. Reproduced from Jackson *et al.* (1977), with permission.

development has passed ethylene promotes developmental processes that lead to obvious ripening, including further ethylene production.

In the Japanese pear cv. 'Nijisseiki', which does not show a respiration climacteric, there is little ethylene production and no evidence of an increase in this associated with the ripening which in other respects appears normal, whereas 'Chojura' shows a respiration and an ethylene climacteric (Downs *et al.*, 1991). 'Hosui' is also non-climacteric (Tian *et al.*, 1992).

'Fuji' apple also fails to show a classic climacteric ethylene emission (Jobling and McGlasson, 1995; Fan *et al.*, 1997; Fellman *et al.*, 1997). 'Gloster 69' apple, and a high proportion of progeny derived from this, have innately low ethylene production and the start of autocatalytic ethylene production can be delayed long into storage (Knee and Tsantili, 1988; Stow *et al.*, 1993).

It has been proposed that ethylene biosynthesis is regulated by two systems. System 1 is initiated or controlled by an unknown factor that is probably involved in the regulation of senescence. System 1 then triggers System 2 which, during the ripening of climacteric fruits, results in the production of large amounts of ethylene in an autocatalytic process with the production of ethylene triggering further production. Non-climacteric fruits do not have an active System 2.

It is thought that all ethylene production is via a common biosynthetic pathway (Adams and Yang, 1979; Tucker, 1993). This proceeds from methionine to *S*-adenosyl-methionine (SAM) and from SAM to 1-aminocyclopropane-1-carboxylic acid (ACC). The conversion of SAM to ACC by the enzyme ACC synthase is considered to be a rate-limiting step. ACC oxidase is the enzyme required to convert ACC to ethylene; it is sensitive to oxygen concentration. Apples also contain ACC malonyl transferase activity and malonylation may regulate the low rate of ethylene synthesis during growth of the fruit (Knee, 1993).

Exposure to low temperature stimulates ethylene synthesis in pears, both on the tree and when detached. A similar response is shown by 'Golden Delicious' apples with simultaneous increases in ACC, ethylene concentration in the gas spaces in the fruit and total ethylene production (Knee *et al.*, 1983). 'Cox's Orange Pippin' and 'Bramley's Seedling' do not show this effect. Production of ACC oxidase, in addition to ACC synthase, is induced by chilling pre-climacteric 'Granny Smith' apples (Leliévre *et al.*, 1995). A short period of cold stimulates ethylene biosynthesis in 'Royal Gala' and 'Starking Delicious' as well as 'Granny Smith' (Larrigaudiere *et al.*, 1997).

Starch and sugar content

Carbohydrate from photosynthesis is transported to developing pome fruits as sorbitol (see Chapter 8, p. 238). In the fruit it is converted mainly to fructose and starch with some sucrose and glucose. Starch concentration in young fruitlets declines from an initial level of less than 5 mg g^{-1} fresh weight to a minimum of less than 1 mg at 30 days after anthesis then increases to a maximum of above 25 mg g^{-1} at 110–130 days after anthesis. It declines to a very low level by 160–200 days from anthesis in 'Royal Gala' and 'Fuji', respectively (Brookfield *et al.*, 1997). In general starch hydrolysis begins in the later stages of fruit growth, usually 2 or 3 weeks before the increase of ethylene production (ethylene climacteric) in apples (Lau, 1985, 1988) and about 10 days before the respiration climacteric in European pears (North, 1971). Fruit abscission in some apple cultivars always occurs at a fixed starch concentration (Poapst *et al.*, 1959), suggesting a close linkage between starch content and natural ripening and senescence processes. The amount of starch present and its distribution can be assessed by exposing the cut surface of fruits cut in half equatorially for about 60 seconds to a 1% solution of iodine in 4% potassium iodide. The starch stains blue-black and the amount and distribution can be quantified by comparison with standard charts for the cultivar (North, 1971; Smith, 1985; Lau, 1988; Knee *et al.*, 1989; Kingston, 1992). The starch index values so obtained provide valuable guidance to the level of fruit maturity and the appropriate time of harvesting for immediate consumption or long-term storage.

The starch hydrolysis is accompanied by the appearance of sucrose but the amount of sucrose is much greater than can be accounted for solely by starch hydrolysis (Whiting, 1970). Sucrose is then slowly hydrolysed to form more glucose and fructose. Starch phosphorylase (EC 2.4.1.1), α-amylase (EC 3.2.1.1) and β-amylase (EC 3.2.1.2) are probably important. The concentration of sugars changes little during storage (Lott, 1965; Knee, 1975; Table 10.1).

Table 10.1 *Changes in some fruit characteristics during storage: 'Bramley's Seedling' apples stored in 8–10% CO_2 at 3.3 °C*

Variable	Days in store after harvest				SE
	0	78–109	164–195	253–284	(n = 8)
Peel chlorophyll ($\mu g\ cm^{-2}$)	7.6	6.9	5.9	6.2	0.20
Total sugar (mg g^{-1})	86.2	91.7	93.0	80.8	1.67
Titratable acid (mg g^{-1})	13.3	11.2	9.3	8.0	0.39
Firmness (kg)	4.4	4.3	4.0	3.8	0.26

Data from Knee (1975). Reproduced with permission.

Organic acid content

Organic acids per 100 g fresh weight in apples increase until about mid-July in the northern hemisphere and then decline progressively, following a smooth curve through the final stages of growth and through maturation, ripening and storage (Table 10.1) until the following April. In pear the decline is continuous from the early fruitlet stage to harvest (Ulrich, 1970). In apple both citric and malic acids show the mid-season rise in concentration but quinic acid falls from the beginning of the season.

The immediate precursors of organic acids are generally other organic acids or sugars. These could accumulate through metabolism of imported carbohydrate and amino acids and by fruit CO_2 fixation.

Organic acids provide the main substrate for respiration. The utilization of malate as a respiratory substrate is mediated by malic enzymes catalysing the reductive decarboxylation of malate to pyruvate, allowing carbon from malate to be fed into the Krebs tricarboxylic (TCA) cycle. Citrate can feed directly into this. The metabolism of malate by tissue slices increases as apples pass through the climacteric (Hulme and Rhodes, 1971; Knee, 1993). The utilization of organic acids during the post-harvest period is the main reason for the increase in sweetness in originally high-sugar, high-acid apples or insipidity and blandness when sugar and acid concentrations are initially low (Lott, 1965).

Pigments, lipids and flavour compounds

Anthocyanin synthesis occurs during fruit growth whereas colour changes during ripening depend mainly on the disappearance of chlorophylls *a* and *b* (see Chapter 9, Figure 9.8). Carotene declines during ripening but xanthophylls increase as mono- and diesters, mainly with palmitate and oleate. The appearance of these esters may precede the climacteric rise in ethylene synthesis.

Galactolipids and associated linolenyl moieties, which are typical chloroplast membrane constituents, are lost on ripening but phospholipids and fatty acyl groups remain constant or increase slightly. Lipid turnover increases. Traces of farnasene, thought to be implicated in the development of scald, are present on the surface of pre-climacteric fruits and the amounts increases rapidly on ripening.

The organic compounds on which apple and pear aroma and flavour depend are synthesized during the climacteric phase.

Readiness for harvest

Many of the factors that change as apples and pears mature are used as criteria to determine the optimal date of harvesting.

Calendar date is the simplest, and where growing conditions are relatively stable can be used to predict the likely harvest date for each cultivar in each locality based on past experience. However, the date of blossoming varies with pre-bloom temperatures from year to year (see Chapter 9) and the rate of fruit maturation is dependent on temperatures during the growing season especially in the early part of this (Eggert, 1960; Warrington *et al.*, 1999). Regression studies have shown that the optimum harvest date can be predicted in the majority of years from the date of 50% full bloom and summer temperatures (Luton and Hamer, 1983).

Attainment of appropriate levels of maturity is, however, affected by site factors, e.g. aspect and altitude, which can affect temperatures, and by tree factors including rootstock and pruning, which influence exposure to light and also the age of bearing wood. Site and tree factors influence the date of blossoming and, as Blanpied and Little (1991) have shown, the chronologically oldest fruits that develop from the earliest blooming flowers are the most advanced in maturity if all fruits are harvested at the same date. Assessment of the best time to harvest the fruits is therefore made on an orchard-by-orchard basis. South African procedures (van der Merwe, 1996a) illustrate methods used to determine harvest dates for apples and pears intended for long-term storage for export. Starting from 6 weeks before the historical optimum picking date for the cultivar, fruits are collected at first weekly, then twice-weekly, and seven maturity indices recorded:

Fruit firmness (kg)
Skin colour on a scale of 1 to 5 (background green colour)
Seed colour on a scale of 1 to 6
Titratable acids (g malic acid/100 g juice)
Total soluble solids (%)
Starch conversion (% white surface)
Days from full bloom

These are then evaluated against average maturity indices relating these characteristics to the optimum maturity for harvest (Table 10.2).

Control of ripening and senescence

Quality is optimized by harvesting at or near to ripeness for immediate consumption, or before this horticultural maturity in order to accommodate the developmental changes that will take place during storage, transport and marketing (Fidler, 1973).

Table 10.2 *Average maturity indices for some apple and pear cultivars at the optimum maturity stage*

	DFFB (days)	Firmness (kg)	Skin colour	TSS (%)	Titratable acid (g/100 g)	Starch %	Seed colour
Apples							
'Gala'/'Royal Gala'	121	7.90	3.3	12.1	0.41	35.1	4.7
'Golden Delicious'	131	7.63	2.3	12.9	0.54	18.0	5.0
'Starking'	136	8.04	2.5	11.7	0.31	12.4	4.8
'Granny Smith'	171	7.50	1.9	11.7	0.71	30.6	4.9
Pears							
'Williams' B.C.'	105	9.27	1.9	11.8	0.33		
'Packham's Triumph'	135	7.41	2.1	12.5	0.25		
'Forelle'	153	6.80	2.3	14.4	0.27		

DFFB = days from full bloom. Firmness with an 11.1 mm diameter plunger for apples and a 7.9 mm one for pears. Skin colour (background) from 1 = dark green to 5 = yellow.
TSS = Total Soluble Solids. Starch = % white surface. Seed colour on a scale from 1 to 6.
From van der Merwe (1996a). Reproduced with permission.

The general objective in storage is to slow down the ripening process. Fruit firmness throughout the storage period is largely predetermined by its level at harvest (Johnson and Ridout, 2000). Many of the most important approaches to improvement of texture after storage are those which improve texture at harvest. Sugar and acid concentrations after any given length of storage are also functions of those at harvest, while the development of aroma and those aspects of ripening consequent on rates of respiration and ethylene synthesis are very much influenced by factors already in place at the time of harvest. The control of pre-harvest as well as post-harvest factors is therefore essential to quality retention and development.

Three areas of fruit physiology provide the basis for most of the technologies for optimizing quality at any particular time. These are respiration, ethylene production and action, and nutrition, especially calcium nutrition but also including the supply of other mineral elements and carbohydrates to the fruit. Some of these are also involved in the post-harvest and post-storage ripening which has to be induced for some pears.

As well as being subject to progressive changes in eating quality, apple and pear fruits are also subject to disorders arising in a more localized way within the fruit pre-harvest or arising in response to specific, otherwise desirable, storage conditions. These also have to be addressed as aspects of fruit quality and its control.

Effects of respiration

Respiration is a key process in ripening. Most, probably all, climacteric apples would be totally inedible if they retained the high malic acid content characteristic of the unripe fruit rather than having it used as a substrate for respiration. The energy released in respiration is essential for the synthesis of ethylene and aroma volatiles and other physiological changes associated with ripening. Even after harvest the fruit is alive and its metabolic processes are maintained by respiration. The duration of its post-harvest life depends on the rate of this respiration and is primarily extended by slowing down metabolism and respiration.

Control of respiration

The first of the modern techniques of fruit storage, refrigerated storage in air, depends for its effectiveness on the fact that the rate of respiration varies directly with temperature (Table 10.3). A main-season apple held at 20 °C respires about three times as quickly as one at 10 °C which, in turn, respires about three times as quickly as one at 0 °C (Hardenburg *et al.*, 1986).

Different cultivars have very different rates of respiration. Smith (1940) concluded that the greater the number of cells per unit fruit weight the higher the respiration rate and, related to this, early maturing cultivars have much higher respiration rates than later maturing ones. In air at 12 °C the rates of CO_2 production, shortly beyond the peak of the climacteric rise in respiration, of 'Beauty of Bath' (very early), 'Worcester' (early) and 'Cox' were 2.9, 2.3 and 1.8 times as high, respectively, as those of the very late maturing 'Bramley's Seedling'. These differences also show up even in controlled atmosphere storage with a number of major international dessert cultivars having much lower respiration rates than 'Cox's Orange Pippin' (Table 10.4).

Table 10.3 *The effect of temperature on the respiration rates of apples and pears* (*mg* CO_2 kg^{-1} h^{-1})

	Temperature (°C)				
	0	4–5	10	15–16	20–21
Summer (early) apples	3–6	5–11	14–20	18–31	20–41
Autumn apples	2–4	5–7	7–10	9–20	15–25
'Bartlett' pear	3–7	5–10	8–21	15–60	30–70

From Hardenburg *et al.* (1986).

Table 10.4 *Rates of respiration (l CO_2/1000 kg/day) of some common cultivars of apple: autumn apples only*

Cultivar	Respiration rate
'Delicious'	18
'Golden Delicious'	20
'Jonathan'	33
'McIntosh'	35
'Bramley's Seedling'	40–45
'Cox's Orange Pippin'	42–80

Data on respiration under the recommended post-harvest storage conditions (low temperature, controlled CO_2 and O_2) for each cultivar (from Fidler, 1973).

Cultivars showing physiological disorders at very low temperatures (3–5 °C or less) show increased respiration at such temperatures.

The rate of deterioration in storage can be determined if the respiration rate is known (Findlay and Combrink, 1996). The storage life is related to temperature by the equation

$$\phi = ke^{-0.13t}$$

where ϕ is storage life in days, k is a constant for the specific cultivar (159 for 'Starking', 239 for 'Granny Smith') and t is temperature (°C). 'Starking', for example, will become mealy or senescent in 3 weeks at 20 °C but takes 18–20 weeks to do so at 0 °C.

For many cultivars storage life can be prolonged to the maximum achievable by temperature control, by cooling all the way down to just above the tissue freezing point. This is usually between −1 °C and −2 °C depending on the total soluble solids or sugar content. Cooling has rather less potential for some other cultivars because they suffer from physiological disorders induced at temperatures below, for example, 3–5 °C so cannot be stored at lower temperatures. Precise temperature regimes often reflect locality and cultural factors as well as cultivars.

It is essential that apples and pears are cooled as quickly as possible after harvest. A delay of one day at 21 °C after harvest reduces the potential storage life at 0 °C by 7–10 days.

Fruit firmness in store is a function of fruit storage temperature throughout the period of storage. Given similar initial firmness, an apple stored at 0 °C will be as firm after 120 days as one stored at 4 °C for 60 days or at 8 °C for 30 days (Fidler, 1973).

The rates of respiration, ripening and softening can be further slowed by storage at low temperatures in controlled atmospheres. The gas composition in a regular atmosphere (RA) store is 21% O_2, 0.04% CO_2 and about 78% N. If the oxygen content is lowered to 2% the respiration rate of apples will decrease by 64% and increasing the carbon dioxide concentration decreases the respiration rate of apples by 30%. In practice, the O_2 and CO_2 concentrations in controlled atmosphere (CA) stores vary with cultivar, O_2 ranging from 1.5% to 3% and CO_2 from 0.5% to 8% (Hardenburg *et al.*, 1986). Prolonged exposure to very low oxygen concentrations may result in anaerobic fermentation and production of alcohol. Excessive CO_2 concentrations cause internal injuries. The correct gas regime can, however, extend the storage life of the fruit by 40–60%, or even double it, compared with RA cold storage. Further benefits can be obtained by rapid CA storage in which optimum gas environments are obtained within 48 hours after sealing the store, and by low oxygen storage with 1–1.5% O_2, which inhibits the development of superficial scald, internal breakdown and fruit softening. The retention of firmness in store of 'Golden Delicious' apples and some pears is improved if they are exposed to 10–20% CO_2 for 4–7 days before adjustment to conventional CA conditions (Couey and Olsen, 1975; van der Merwe, 1996b). For some other apple cultivars this treatment is only effective if combined with low-ethylene storage (Stow, 1988). High CO_2 concentrations (5–10%) can halve respiration rates (Fidler, 1973) and lower ethylene production (Chavez-Franco and Kader, 1993), so their mode of action on firmness may be complex.

In general early harvesting, well in advance of the respiration climacteric, results in lower rates of respiration over the first few weeks of storage. Such fruits are firmer and maintain this advantage during storage but may have countervailing disadvantages in terms of size, colour, sweetness and aroma production (Song and Bangerth, 1996). Picking date also has a marked effect on physiological disorders and its optimization depends on a large number of factors other than effects on respiration.

Respiration rates can be reduced by high fruit calcium content and by control of ethylene production. These factors also have other major rôles regarding fruit quality.

Effects of ethylene

Ethylene has long been known to enhance fruit ripening and to increase the activity of a number of enzymes associated with this (Abeles, 1985). It regulates cellulase, chlorophyllase, invertase, laccase, malate dehydrogenase and polygalacturonase. The rise of ethylene production that accompanies the climacteric appears to be a key regulatory event in the development of most pome fruits, even though other regulatory events precede and follow this rise and there may be other aspects of regulation unrelated to ethylene (Knee, 1993). High ethylene concentrations can promote, but may not be required for, many ripening processes. They can ensure that all ripening processes occur synchronously under normal conditions. The low concentrations of ethylene characteristic of pre-climacteric or non-climacteric apples and pears can have major effects on ripening processes. A continuous treatment with ethylene at approximately 0.1 $\mu l\ l^{-1}$ can elicit almost 50% of the maximum effect on ethylene synthesis of 'Anjou' pear while 100% response is attained at around 1 $\mu l\ l^{-1}$. Softening of this pear is promoted by ethylene concentrations as low as 0.05 $\mu l\ l^{-1}$ (Knee, 1985). Theoretically there is no lower limit at which ethylene becomes inactive.

Knee (1985) concluded that ethylene acts at a high affinity site to initiate softening and chlorophyll degradation and that ethylene synthesis in mature climacteric fruits is triggered by a high affinity response. It may act at a low affinity site to stimulate respiration.

The higher the level of internal ethylene the lower the sensitivity to external ethylene, but as the fruit advances towards autonomous ripening there is an earlier response to any particular ethylene concentration.

The effects of ethylene on fruit ripening and fruit quality parameters have been inferred from the pattern of natural changes in these but more specifically demonstrated by the artificial supply of ethylene or ethylene-producing chemicals and by the use of chemicals that inhibit ethylene synthesis. The interpretation of any quantitative relationship between applied ethylene and responses in mature climacteric fruit may be complicated by the fact that endogenous ethylene production is stimulated by ethylene supply.

Smith *et al.* (1985) found that ethephon (2–chloroethyl phosphonic acid) applied as an ethylene source 2–3 weeks before harvest increased internal ethylene levels, fruit yellowing and starch loss and reduced fruit firmness and titratable acidity. These effects were accounted for as consequent on advancing the onset of ripening: there are no effects on the subsequent rates of change but initial differences are maintained.

AVG (aminoethoxyvinylglycine) regulates ethylene biosynthesis by inhibiting ACC synthase. Application in the orchard four weeks before the anticipated harvest date delays fruit maturity on the tree as shown by its effect in delaying the increases in internal ethylene concentration and soluble solids and the changes in starch, background colour and red colour (Johnson, 1998). The fruit can consequently be picked later. Throughout several months of storage under controlled atmosphere conditions, fruits from AVG-treated 'Cox' apple trees produced only about a tenth as much ethylene as fruits from control trees and were consistently firmer. Similar results of pre-harvest AVG sprays have been reported for 'Delicious', 'Golden Delicious', 'McIntosh' and 'Spartan' (Bangerth, 1978; Williams, 1980; Bramlage *et al.*, 1980a; Autio and Bramlage, 1982).

Diazocyclopentadiene (DACP) is a light-activated compound thought to inhibit ethylene binding. Applied to mature, pre-climacteric 'Red Delicious' apples, it results in lower internal ethylene concentration whether the fruits are kept at 21 °C or 0 °C. At 21 °C untreated apples had a flesh firmness of 46 newtons after 30 days, DACP-treated apples a flesh firmness of about 73 N. DACP-treated fruits also retained their firmness much better than controls when stored at 0 °C (Blankenship and Sisler, 1993).

Treatments during post-harvest storage so as to maintain internal ethylene levels below 0.1 ppm inhibit the onset of softening, delaying this by about 16–18 weeks (Stow *et al.*, 2000). Earlier studies (Smock, 1943; Forsyth *et al.*, 1969; Knee and Hatfield, 1981) had also shown ethylene removal during storage to aid in retention of firmness.

Ethylene is also involved in the development of the post-harvest disorder known as superficial scald. This is characterized by irregular shaped tan to dark brown blotches on the fruit skin. It is the most serious post-harvest disorder of many cultivars although some are highly resistant (Lau, 1993). Scald is thought to result from the oxidation of α-farnesene in fruit surface wax into conjugated trienes which cause injury to the epidermal cells. Removing ethylene during storage delays the production of α-farnesene and reduces the concentrations of its oxidation products and the development of scald (Knee and Hatfield, 1981; Knee, 1985; Dover, 1985).

There are some indications that inadequate ethylene concentrations may inhibit aroma production (Song and Bangerth, 1996).

Control of ethylene

Control of fruit quality can be effected through the control of synthesis, accumulation or action of ethylene or by combinations of these.

For climacteric apples and pears control is much more likely to succeed at the pre-climacteric stage before the large increase in ethylene concentration due to autocatalytic production occurs.

CONTROL OF SYNTHESIS

Synthesis rates increase as the fruits mature. Synthesis rates can be modified by chemical treatments in the orchard and by the use of high CO_2 treatment after harvest. They are also dependent on oxygen and temperature levels and vary greatly between cultivars.

AVG inhibits ACC synthase. Its application in the orchard a month before the expected date of harvest of 'Cox' apples reduces the ethylene concentration in the apples at harvest and over several months of storage at 1.2% O_2, <1% CO_2 and 3.5 °C. The fruits are appreciably firmer than controls after storage (Johnson, 1998).

Daminozide, which is no longer used, was also effective in delaying the onset of high ethylene production although it has no effect on ACC synthesis and the mechanism of its effect is uncertain (Knee, 1985). Its use pre-harvest resulted in firmer apples after storage under controlled atmosphere conditions with ethylene removal (Liu, 1985).

Treatment with 15% CO_2 for 10–15 days after harvest prior to storage at 3.5 °C in 2.5% O_2/<1% CO_2 delays the autocatalytic production of ethylene (Knee and Stow, 1985). High CO_2 treatment helps retain firmness in store if combined with removal of ethylene (Stow, 1988).

Oxygen is required for the conversion of ACC to ethylene and ethylene production in store is reduced as the O_2 content is reduced from 21% to 2% to 1% at temperatures from 0° to 4 °C (Johnson and Ertan, 1983).

Ethylene production by 'Cox' and 'Bramley' apples in both pre-climacteric and post-climacteric states declines with declining temperature with a Q_{10} value of about 2 (Knee, 1985).

Differences in ethylene production between cultivars may reflect differences in stage of maturity at harvest, for example 'Bramley's Seedling' is usually harvested much earlier relative to its climacteric than is 'Cox'. They may also be intrinsic. Blanpied *et al.* (1985) contrasted 'McIntosh' as a rapid ethylene producer with 'Empire' as a slow ethylene producer, and Jobling and McGlasson (1995) found 'Fuji' to have a maximum rate of ethylene production only one hundredth of that of 'Gala', which may be associated with its maintenance of flesh firmness. Knee (1985) noted that in 2% O_2 at 3.5 °C with continuous ethylene removal 'Crispin' ('Mutsu') and 'Gloster 69' took more than 150 days to reach 0.1 μl l^{-1} ethylene whereas 'Spartan' took only 20 days and other cultivars were intermediate. The slow production of ethylene by 'Gloster 69' is heritable (Stow *et al.*, 1993). It appears to be caused by enhancement of an inhibitory mechanism preventing initiation of autocatalytic ethylene production rather than lowered activity of a single enzyme in the ethylene biosynthesis pathway.

CONTROL OF ACCUMULATION

Control of accumulation is also a key element in control of synthesis since accumulation leads to enhanced production. Traditionally great care has been taken to avoid putting ripe apples with high ethylene production into a store with pre-climacteric fruits and to ensure good natural ventilation. In virtually gas-tight stores the internal ethylene concentration in the fruits can be minimized by removal of ethylene from the store atmosphere so that the gradient from the inside to the outside of the fruit, hence the rate of diffusion out of the fruit, is maximized. Removal of within-store ethylene also reduces the extent to which ripening in the entire population will be synchronized by the ethylene produced by the earliest ripening fruits.

Several methods of ethylene removal are available. They can involve oxidation with potassium permanganate (Smock, 1943; Liu, 1985; Blanpied *et al.*, 1985), ozone (Fidler and North, 1969), active oxygen (Scott *et al.*, 1971), catalysts such as platinum (Dover, 1985), or adsorption by charcoal or brominated charcoal (Fidler, 1973). Some of these methods, e.g. use of potassium permanganate, can be used in small sealed modified atmosphere packs (Smith *et al.*, 1987). Hypobaric storage has also been used.

The effectiveness of ethylene removal depends on the magnitude of the problem to be overcome. It is more likely to be effective with early than with late-picked fruits, with cultivars that are slow to produce much ethylene and with fruits that have been pre-treated either with chemicals that inhibit ethylene production or with high CO_2 concentrations with the same effect.

CONTROL OF SENSITIVITY AND OF ETHYLENE ACTION

Under several storage conditions high concentrations of ethylene have less severe effects than might be anticipated. Oxygen is required in the metabolic processes underlying some ripening changes including softening (Knee, 1982). Ethylene is unable to reverse this low oxygen effect so low oxygen storage of apples is commercially successful even though concentrations of several hundred $\mu l\ l^{-1}$ accumulate in the stores (Knee, 1985). Refrigerated air storage can also be very effective in prolonging post-harvest life even if the unripe apples are exposed to ethylene concentrations above 10 $\mu l\ l^{-1}$.

Calcium and fruit eating-quality

Low calcium concentrations in apple fruits first attracted attention because they were found to be associated with disorders such bitter pit and York spot. These may be obvious at harvest but develop in storage, and surface appearance, especially at harvest, may fail to reflect the severity of the disorders after storage. Low calcium concentration was subsequently found to predispose apples to other serious physiological disorders such as splitting, lenticel rupturing, watercore, breakdown, lenticel blotch pit and confluent pit or crinkle (Sharples, 1980; Bramlage *et al.*, 1980b; Perring, 1984; Perring *et al.*, 1985). Low calcium concentrations also increase the susceptibility of apples to storage rots caused by *Gloeosporium* spp. (Sharples, 1980) and to storage scald (Bramlage *et al.*, 1974). Some of these disorders are discussed in more detail later. Calcium appears to have fundamental effects not only on these disorders but also on eating quality and senescence through its influence on basic structural features and physiological mechanisms.

Calcium and cell wall and membrane characteristics

Calcium ions have been shown to bind pectin molecules and Knee and Bartley (1981) suggested that these ions form bridges between pectin molecules in the middle lamella and are responsible for cell cohesion.

Vacuum infiltration of 'Golden Delicious' apples with calcium chloride, either soon after harvest or after 3 months' cold storage, greatly reduces subsequent fruit softening. Electron micrographs of treated fruits show well-structured, darkly staining middle lamellae indicative of tightly packed polyuronides even after prolonged storage, whereas the corresponding untreated controls show the middle lamella to have degraded to the point of cell wall separation (Poovaiah *et al.*, 1988; Siddiqui and Bangerth, 1993).

Calcium acetate infiltration after a period of either cold storage or controlled atmosphere storage can even largely reverse softening (Table 10.5).

Although net degradation of cell walls takes place during ripening, incorporation of labelled methionine and inositol in apple cell wall polysaccharides indicates the synthesis of cell wall polymers during ripening (Knee, 1978; Poovaiah *et al.*, 1988).

Calcium also has pronounced effects on cell membranes and retards their breakdown during storage. Fuller (1976) studied membrane changes in 'Cox'

Table 10.5 *Effect of infiltration at harvest or after cold storage with glycerol/Tris buffer with and without calcium acetate on the firmness of fruit tissue from two apple cultivars*

Cultivar	Infiltration medium	Firmness (N) at harvest	Firmness after 10 weeks at 3 °C
'Cox'			
	None	32.3	13.4
	Glycerol	34.2	10.3
	Glycerol + Ca	35.2	30.9
'Gloster 69'			
	None	43.1	18.5
	Glycerol	41.7	18.3
	Glycerol + Ca	44.5	35.1

Glycerol = 400 mol m^{-3} glycerol in 100 mol m^{-3} Tris buffer, pH 8.0.
Glycerol + Ca = 200 mol m^{-3} glycerol + 200 mol m^{-3} Ca acetate in the same buffer.
From Stow (1993). Reproduced with permission.

apples from high Ca (6.4–7.7 mg/100 g fresh weight) and low Ca (3.1–3.7 mg/100 g fresh wt) populations in cold storage (0 °C and 3 °C) from late September to January. Electron micrographs showed that in November both high- and low- Ca fruits had intact plasma and vacuolar membranes, mitochondria, plastids and nuclei in the epidermal, hypodermal and many of the outer cortical cells. In mid-December the high-Ca fruits had intact membranes, mitochondria, plastids and nuclei in most of the epidermal and hypodermal cells and in many outer cortical cells. The low-Ca fruits had much disorganization of the cytoplasm and separation of the plasma membrane from the cell wall. By early January the differences were very pronounced, with relatively little membrane breakdown in the high-Ca fruits but much membrane breakdown and even complete cytoplasmic disorganization in the low-Ca fruits. During ripening, solute leakage from tissue discs increases but this leakage is retarded by elevated Ca concentrations (Sharples and Johnson, 1977; Ferguson and Watkins, 1989). Calcium affects physical properties of plant membranes by regulating their microviscosity or fluidity. Cellular senescence is accompanied by increases in microviscosity and the proportion of gel-phase lipid of membranes and Ca may diminish these trends (Ferguson, 1984).

Calcium, respiration and ethylene production

Calcium status influences respiration, respiration rates of untreated apples being inversely related to flesh calcium content (Faust and Shear 1972; Bramlage *et al.*, 1974), and calcium dips can reduce respiration (Watkins *et al.*, 1982). Bramlage *et al.* (1980b) considered that the accelerated respiration of Ca-deficient apples might be linked to effects of calcium on ADP/ATP ratios in the cell. Calcium ions are essential for the activity of a number of enzymes including the membrane-associated, calcium-dependent ATPases and calcium movement across the membrane can directly drive ATP synthesis.

Pyruvate kinase (pyruvate–ATP phosphotransferase, EC 2.7.1.40) exerts considerable control of respiration and the control of its activity is largely a function of the concentrations of Ca^{2+}, Mg^{2+} and K^{+} in its environment (Meli and Bygrave, 1972; Witney and Kushad, 1990). Pyruvate kinase activity in apple fruits declines throughout the period of fruit growth up to harvest, is reduced by $CaCl_2$ sprays and increased by $MgCl_2$ sprays, foreshadowing subsequent incidence of bitter pit (Witney and Kushad, 1990).

Calcium infiltration of 'Golden Delicious' apples significantly reduces ethylene production (Poovaiah *et al.*, 1988) and it has been suggested that microsomal membranes are the sites of interaction of calcium, and that ethylene biosynthesis is modulated through its binding with the membrane (Ben-Arie *et al.*, 1982).

Table 10.6 *The effects of calcium nitrate sprays during the growing season on fruit calcium, bitter pit and breakdown in storage in air at 3 °C of 'Cox's Orange Pippin' apples*

Trial		Fruit Ca (mg/100 g)	Bitter pit (%) External	Bitter pit (%) Internal	Breakdown (%)
1. Appleby, New Zealand	Control	1.7	55.8	53.5	60.5+
	6 sprays	2.8	21.5	51.5	11.3+
2. Teynham, Kent, UK	Control	3.3	42.8	36.6	13.6
	6 sprays	3.8	9.5	13.8	3.8
3. Hadlow, Kent, UK	Control	4.2	0	17.3	5.0
	5 sprays	4.8	0	3.5	0.3

Total $Ca(NO_3)_2$ applied in kg ha^{-1}: Trial 1, 130; Trial 2, 90; Trial 3, 110.
+ Included watercore and watercore-breakdown.
From Sharples (1980). Reproduced with permission.

Calcium and physiological disorders

There are a number of physiological disorders of apples, that can develop in the orchard or only become evident after a period of storage, which are associated with low fruit tissue calcium.

Bitter pit is perhaps the most important of these (Table 10.6). The primary symptom is discrete pitting of the cortical flesh. The pits are brown and become desiccated. They are mostly in the outer cortex, frequently just under the skin, and the collapse of the outermost cells causes small depressions. Pitting may also occur deep in the flesh and only become visible when the fruit is cut. Frequency of pitting is often greater towards the calyx end (Faust and Shear, 1968; Ferguson and Watkins, 1989). The first sign of bitter pit is collapse of the cell walls and by the time that pitting is visible to the naked eye plasmolysis of the cytoplasm has occurred. The pit cavities result from the collapse of several cells and are bound by the remains of the original cell walls (Smock and Van Doren, 1937). Within the fruits bitter pit is most common in those zones with the lowest Ca concentration although the lesions themselves do not show a low Ca content.

There is sometimes a closer correlation between the incidence of calcium-related storage disorders to the ratio of Ca to K or K + Mg than to Ca alone (Holland, 1980; Waller, 1980). Perring (1984) concluded that fruits with low Ca become more susceptible to bitter pit, lenticel blotch pit and cracking as potassium and magnesium concentrations increase. Low-oxygen stored fruit have a lower threshold level of Ca at which bitter pit is avoided than have air-stored fruit (Sharples and Johnson, 1987). Rowe (1980) stated that for 'Cox' in England avoidance of lenticel blotch pit, bitter pit, loss of firmness and breakdown required minima of 3.1, 4.5, 5.0 and 5.0 mg Ca/100 g fruit fresh weight respectively. In South Africa the minimum fruit Ca content to ensure freedom from bitter pit is 5.4 and 6.6 mg/100 g for unsprayed and calcium-sprayed fruit respectively (Terblanche *et al.*, 1980).

Calcium nutrition

Calcium is one of the major nutrient elements taken up by apple and pear trees, gross uptake being similar to, or greater than, that of N and K and many (*c*. 6–10) times as great as that of Mg and P (Greenham, 1980). The amount in the fruits, however, is very low in comparison with that of N and K and is only a few per cent of the total uptake by the trees (see Chapter 11, Table 11.1). Research on Ca nutrition has therefore emphasized not only ways of increasing uptake by the trees as such but also, very specifically, ways of increasing Ca supply to the fruits. Because of this, and the important interactions between Ca in fruits and fruit quality, Ca nutrition is discussed in this chapter instead of in Chapter 11.

In general the supply of mineral nutrients from the soil to leaf cells, and by analogy fruit cells, is considered to involve two stages involving mass flow, diffusion and ion-exchange mechanisms and two stages primarily dependent on active processes involving metabolic energy (Bowling, 1980; Shear, 1980).

Supply by the soil

The first stage in Ca uptake involves movement of Ca to the root surface. For this to proceed with maximum efficiency the ion concentration in the soil water must be high enough to enable the nutrients needed by the plant to reach the root by mass flow. Under good orchard soil management the exchange complex of the soil is dominated by Ca, and soil solution concentrations of Ca are usually the highest of all cations (Adams, 1974; Korcak, 1980). The Ca content in the soil solution in most temperate zone soils varies between 3.4 and 14 mM whereas at the root surface 0.1–1 mM appears adequate. Mass flow of soil solution, dependent on the transpiration rate, should provide ample Ca flux to the root surface under these conditions (Bangerth, 1979). Adequate soil moisture is, however, a pre-requisite for this. Liming may, however, be needed to counter excess soil acidity and increase the supply of Ca. Marks (1980) found apple leaf Ca concentrations are usually at or below optimum levels, in contrast to those of N, P and especially K which are commonly above the optima for satisfactory growth. In the Netherlands, on sandy soils, Van der Boon (1980) found that gypsum ($CaSO_4.2H_2O$) soil dressings increased leaf Ca concentration and slightly reduced bitter pit incidence in 'James Grieve' apples. A major problem even in temperate-zone soils is that incorporation of sufficient Ca to bring soil pH to a desirable level at orchard establishment may fail to maintain this, being followed by decreasing pH and a decline in exchangeable Ca to inadequate levels (Bolton, 1977; Korcak, 1980). Because of its divalent charge, surface applications of Ca salts often result in little downward movement in the soil.

In some important subtropical apple producing areas, e.g. Cape Province, South Africa (Kotzé and du Preez, 1988) and southern Brazil (Basso and Wilms, 1988), the predominant soils are strongly acidic. Exchangeable aluminium increases rapidly as pH (measured by the $CaCl_2$ method which records about 1 unit lower than the water-based method) declines below 5.0 (Terblanche *et al.*, 1980). This strongly suppresses Ca uptake leading to low leaf as well as fruit Ca. Pre-planting liming is essential to overcome this and liming to a level greater than that needed to eliminate Al effects is necessary to optimize Ca concentrations in the plant and control bitter pit. Annual liming is then needed to maintain soil pH and Ca supply. Without such treatments losses from calcium deficiency disorders, especially bitter pit, are very severe (Terblanche *et al.*, 1980; Kotzé and du Preez, 1988; Wilms and Basso, 1988; Kotzé, 1996). Calcitic or dolomitic lime, or gypsum, are used for soil rectification depending on acidity and Mg status. The balance of nutrients in the soil strongly influences Ca uptake. This is depressed by non-specific cation competition by K, Mg and NH_4 and enhanced by anions such as NO_3 and PO_4 (Bangerth, 1979). Post-planting applications of K to K-rich soils can reduce fruit Ca levels (Terblanche *et al.*, 1980). Supply of N as ammonium can greatly depress the Ca uptake by apple seedlings relative to that when N is supplied as NO_3 (Kotzé, 1979) and can give lower Ca content and a higher bitter pit incidence in 'Cox' apples (Ludders, 1979). It also depresses the Ca content of 'Fuji' apples (Motosugi *et al.*, 1995).

Uptake by the roots

The first step in Ca uptake from the soil solution is penetration of the apoplast or free space which occupies the cell wall network of the root cortex external to the cell membranes. This is very rapid in spruce (Kuhn *et al.*, 2000) with a half-time for ^{44}Ca entry in solution culture experiments of only 2–4 minutes and a plateau level in the cortex attained in about 20 minutes. This indicates very rapid equilibration in the entire apoplast of the cortex, and is compatible with Ca-binding to mobile and fixed binding sites being the major factor modulating the kinetics of Ca^{2+} entry.

In contrast, movement into the cell walls of the stele, in spruce, has a half-time of 100–120 minutes and the plateau level is not attained until about 1000 minutes from ^{44}Ca supply. This indicates the importance of the endodermis as a barrier. The endodermis typically has a suberized Casparian strip in the tangential walls of its cells. Calcium ions cannot move through it to the stele by mass flow, diffusion or ion exchange as they do through the free space, but must pass through at least two membranes. In general it seems that calcium movement through the symplast, i.e. the cytoplasm and connecting plasmodesmata, is much slower than that through the apoplast. Root tips and, temporarily, the sites of branch root formation where the development of the Casparian strip appears to lag behind endodermal cell division, may therefore be particularly effective in Ca uptake (Bangerth, 1979; Ferguson, 1979). As a consequence Ca uptake may depend largely on current root growth and branching.

There are, however, other plants in which the endodermis does not act as a barrier for Ca (Chino, 1979). For apple and pear the traditional view has been that the endodermis provides a major barrier and that most Ca uptake is by white, unsuberized, roots. Atkinson and Wilson (1980), however, found ^{45}Ca to be taken up to a similar extent by white and woody roots. They considered that the failure of the phellogen of woody roots to act as a barrier to Ca (and water) movement could be related to the deposition of the suberin on the inside of the cell walls rather than within these, so that the apoplastic pathway remains viable.

There is also evidence of involvement of processes dependent on metabolic energy in Ca uptake. Reduction in photosynthesis by use of photosynthetic inhibitors, or prevention of translocation of assimilates to the roots by ringing, decreases Ca uptake by apple root systems and is reversible by supplying sucrose (Faust, 1980).

Upward movement

Upward movement of Ca from the roots is at least primarily in the xylem. It is influenced by the transpirational flux but is not simply a matter of mass flow. Ferguson and Bollard (1976) found that movement of ^{45}Ca through excised pieces of apple stem was much slower than that of phosphate supplied at the same time. It appeared to be by exchange processes rather than mass flow. Some of the Ca entering the xylem becomes more firmly bound and some moves into phloem tissues. Bradfield (1976) found that about 50% of the Ca in the xylem sap of apple shoots was in the ionic form. The remainder was present as complexes with citric and malic acids. He concluded that the mobility of Ca in the xylem might be influenced by the supply of organic acids in the sap which could reduce the degree of adsorption at the negatively charged exchange sites in the xylem vessels. In the intact plant under conditions of steady water and Ca supply, Ca is likely to move with the mass flow of the transpiration stream in the larger conduits, particularly when exchange sites are saturated or in equilibrium with the Ca in the vessel volume. In smaller channels exchange will be much more important (Ferguson and Watkins, 1989).

One consequence of the stele acting partly as an ion exchange column is that it can act as a store for Ca. This is particularly important with respect to the supply of Ca during the initial stages of fruit growth. Führ and Wienecke (1974) found that ^{45}Ca supplied during one growing season is found in all new plant parts in the following spring. They estimated that about 18% of the total Ca in apple fruits in one season originated from Ca reserves deposited in the previous year, the precise sites of storage being uncertain.

Bell and Biddulph (1963) put forward the concept that calcium ascent is based not on the loss of water from the individual plant parts but on the metabolic removal of the ions from the exchange columns leading to the individual parts. This results in the various tissues acquiring nutrient ions in proportion to their metabolic utilization but also subject to the influence of mass flow in the major conduits and the supply of Ca, other cations and organic acids.

Some effects on fruit calcium content seem explicable in terms of factors controlling the general upward flux. The leaves and stems, as well as the fruits, in the upper and outer parts of the tree canopy tend to have less Ca per unit of dry matter than corresponding organs nearer to the roots, which is compatible with abstraction of Ca from the transport system *en route* (Preuschoff, 1968; Jackson *et al.*, 1971; Haynes and Goh, 1980; Barritt *et al.*, 1987). Many nutritional treatments increase both leaf and fruit Ca concentrations (Huguet, 1980; Terblanche *et al.*, 1980). Moreover, the accumulation of Ca by fruits is positively dependent on the area of primary and bourse shoot leaves subtending the fruit (Ferree and Palmer, 1982; Jones and Samuelson, 1983; Proctor and Palmer, 1991; Volz *et al.*, 1994, 1996a). Primary leaves are the more effective. Enclosing the leaves in polythene bags has a similar effect to their removal, confirming the effect of leaf transpiration on the supply of Ca to the fruits (Jones and Samuelson, 1983). It is obviously important to maximize early spur-leaf area.

Flux into the fruits

The final stages of movement into organs and tissues are at least partly under metabolic control.

Within the shoot system Ca is translocated preferentially towards the shoot apex even through the transpiration rate of young apple leaves is much lower than that of older leaves (Shear and Faust, 1970; Faust, 1989). Transport to the growing point is believed to be induced by IAA, synthesized in the shoot apex, stimulating a proton efflux pump in the elongation zones of the shoot apex. This increases the formation of cation exchange sites so that the growing tip becomes a centre for Ca accumulation (Faust, 1989). Movement into leaf cells is thought to be by active uptake at the plasmalemma depending on metabolic energy supplies (Bowling, 1980).

Reducing fruit transpiration of 'Gala' apples by bagging them reduces their Ca concentration (Tomala, 1997), but the effects of bagging 'Golden Delicious' apples are relatively small (Jones and Samuelson, 1983). Calculation of potential Ca flux into fruits of 'Golden Delicious' and, especially, 'Bramley' assuming mass flow of xylem sap to meet net water requirements for fruit growth and evaporation, underestimates Ca uptake into the apples early in the season and overestimates it in the month prior to harvest (Jones *et al.*, 1983). In fact net Ca uptake by the fruits is often restricted to the early part of the season, and there is little movement into fruits of ^{45}Ca supplied to the roots in early August (Ford and Quinlan, 1979) although it is detectable in the fruit stalk. In contrast, accumulation of dry matter and fresh weight usually continues up to harvest so the percentage of Ca in the fruits falls with time. Circumstances or treatments that enhance late-season fruit growth therefore tend to reduce fruit Ca concentration at harvest (Figure 10.2).

Late harvesting, and its associated larger fruits with lower Ca concentrations, results in more senescent breakdown, *Gloeosporium* rotting and watercore (Perring and Pearson, 1986), but effects on bitter pit incidence are inconsistent (Perring and Pearson, 1986; Ferguson and Watkins, 1989).

With fruits harvested at optimum maturity, in general the larger the fruit the lower the Ca concentration and the greater the incidence of all Ca deficiency disorders (Perring and Jackson, 1975; Terblanche *et al.*, 1980). Under South African conditions the maximum fruit diameter to ensure complete freedom from bitter pit is 61 mm (Terblanche *et al.*, 1980). Market demand for large fruits rules out control of fruit size as a technique for controlling Ca content. Instead, it is realized that procedures to increase fruit size, e.g. fruit thinning, will increase the risk of Ca deficiency disorders (Sharples, 1968; Johnson, 1992) and, with cultivars prone to Ca deficiency, must be accompanied by other practices to maintain fruit Ca at desirable concentrations. Exposure to high light intensities, which has beneficial effects on many aspects of quality, has negative effects on bitter pit. These are very largely accounted for by effects of exposure on fruit size (Figure 10.3) and leaf Ca concentrations, which result in lower fruit Ca concentrations (Jackson *et al.*, 1977).

The thinning-induced effect of fruit size on fruit calcium deficiency disorders is not specifically dependent on cell number or cell size but is an integrated effect (Sharples, 1968). Ca concentrations tend to be higher the greater the number of seeds in the fruit (Bramlage *et al.*, 1990; Tomala and Dilley, 1990). Consequent to this, supplementary pollination can increase seed number and the concentration of Ca even in fruits of the same size. Netting to give poor pollination reduces seed number and Ca concentration (Volz *et al.*, 1996b).

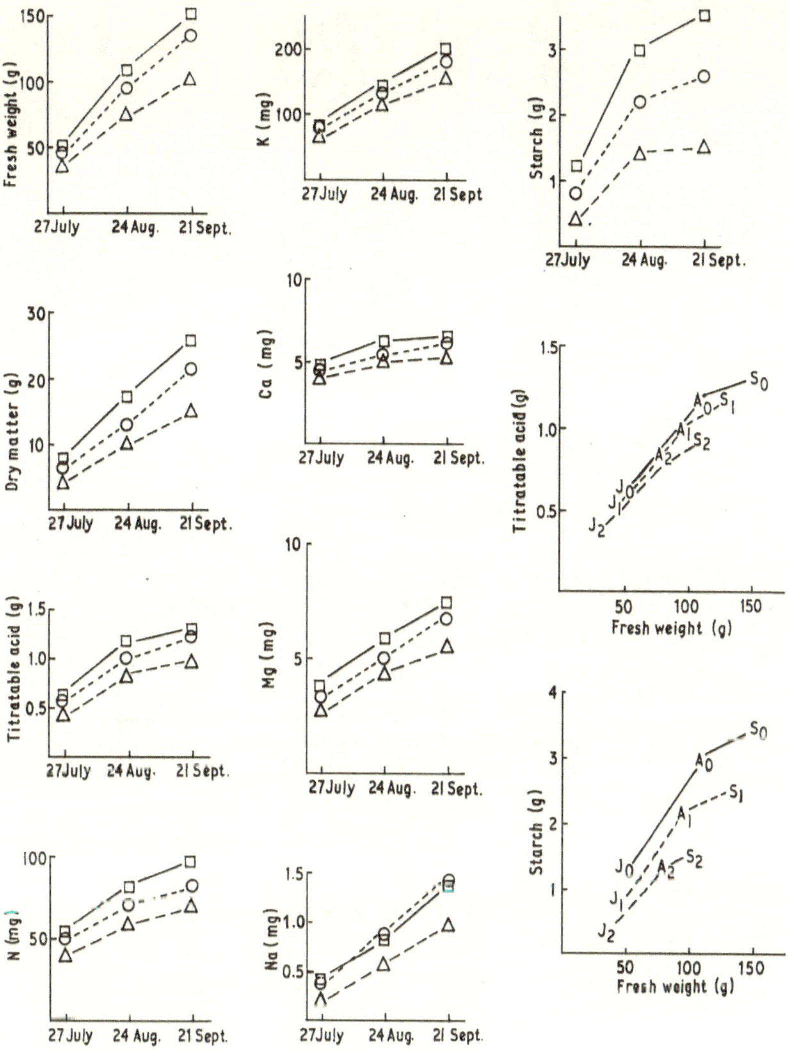

Figure 10.2 Effects of growing conditions, resulting from different levels of shading, on the chemical composition of fruits harvested at different times. □—□, 100% daylight; ○---○, 37% daylight; △---△, = 25% daylight; J, A, S harvested on 27 July, 24 Aug. and 21 Sept. 0, 1, 2, 100%, 37% and 25% daylight respectively. Reproduced from Jackson *et al.* (1977), with permission.

Apples induced to set by the use of gibberellins or gibberellins plus auxins have fewer seeds. The few-seeded fruits have particularly low Ca concentrations but the hormone treatment also depresses fruit Ca within all categories of fruit size and number of seeds (Jackson *et al.*, 1982).

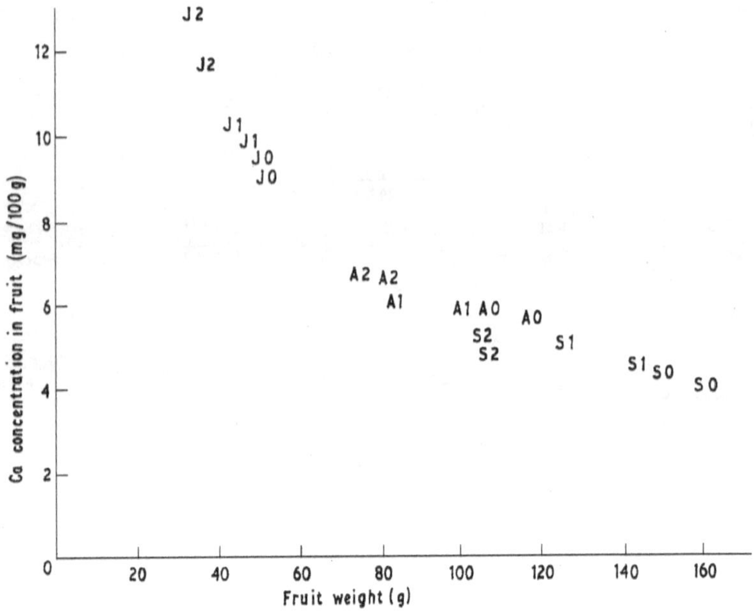

Figure 10.3 Relationship between fruit weight and calcium concentration in fruits harvested at different dates from trees subject to different levels of shading. Harvest dates J = 27 July, A = 24 August, S = 21 September. Shade treatments 0 = 100% daylight, 1 = 37% daylight, 2 = 25% daylight. Reproduced from Jackson *et al.* (1977), with permission.

Bangerth (1976) concluded that auxins produced by the seeds play a significant rôle in Ca translocation into fruits. The seeds themselves have a higher Ca concentration than other parts of the apple fruits and their Ca content is less affected by reduction in the Ca content of the medium around the roots than is that of the rest of the fruit (Huguet, 1980). When 2,3,5-triiodobenzoic acid (TIBA), which inhibits auxin transport, is sprayed on apple trees as early as 2 weeks after full bloom, Ca accumulation by the fruit is reduced (Stahly and Benson, 1976, 1982; Stahly, 1986). Lang (1990) and Lang and Ryan (1994) have shown that there is no increase in the number of xylem vessels in apple fruit pedicels from just after flowering onwards. There is a decline in xylem pedicel conductance as the season progresses and this is more severe in the cv. 'Cox', which is very subject to Ca deficiency disorders, than in the cv. 'Gala', which is less subject to these.

Terminal fruits of 'Granny Smith' and 'Royal Gala' have higher Ca concentrations than similar-sized or smaller fruits growing laterally on one-year-old shoots or 2-year-old spurs (Volz *et al.*, 1994).

Severe winter pruning, which stimulates vigorous shoot growth, results in lower fruit Ca concentrations and more bitter pit than summer pruning systems that result in numerous weak, fruiting twigs that cease growth early in the year (Schumacher *et al.*, 1980; Deckers and Missotten, 1993; Link, 1993). This is at least partly a consequence of the low cropping and large fruits that can be induced by heavy pruning but may also involve effects of competition.

Calcium can be withdrawn from fruits in the later stages of their growth, apparently in response to water stress. Fruits commonly shrink during the day as a result of water loss due to equilibration with transpiration-induced tensions in the xylem (Tukey, 1964, 1974; Tromp, 1979). Bitter pit, indicative of Ca deficiency, is frequently associated with vascular tips within the fruits and hot dry conditions leading to water stress (Smock, 1941; Perring, 1986). However, this water is replaced during the night and spur-wood xylem sap may have ten times as high a concentration of Ca as fruit xylem sap. Lang and Volz (1993, 1998), working with 'Royal Gala' which shows relatively little decline in xylem function late in the season, found evidence that this cycling of xylem sap out of and back into the fruit contributes to fruit Ca accumulation. This may also explain why leaf transpiration, leading to accumulation of Ca in conducting tissues, increases the Ca content of nearby fruits.

Movement within the fruits

Differences in Ca content in different parts of apple fruits and movement between these are well documented. The core, even after removing the seeds, has a much higher Ca concentration than the inner cortex which, in turn, has a higher concentration than the outer cortex (Perring and Pearson, 1986; Ferguson and Watkins, 1989). The peel concentration is intermediate between those in the core and the cortex (Perring and Pearson, 1986). It falls steeply, with increasing cell size, from the fruit surface to a depth of 2 mm. During the post-harvest period Ca may move from the middle and outer cortical tissues to the core zone. This is followed by development of bitter pit lesions.

Surface application to the fruits

Movement within the fruit is basic to the common use of orchard sprays and post-harvest dips with Ca salts to increase fruit Ca content and reduce the incidence of post-harvest disorders (Table 10.6; Ferguson and Watkins, 1989). Although orchard sprays increase the Ca content of both leaves and fruits (Van der Boon, 1980), bagging experiments have shown that the fruit must be directly exposed to the Ca sprays for its Ca content to rise (Ford, 1979). Ca intercepted by the leaves is not translocated to the fruit. Natural openings in the fruit surface such as stomata, lenticels (which are derived from stomata) and cracks in the cuticle provide the main avenues of entry. Some Ca transport can also occur across the cuticle itself (Ferguson and Watkins, 1989; Harker and Ferguson, 1991). Surfactants and 'sticking agents' may increase the uptake of applied Ca (Sharples *et al.*, 1979). Penetration into young fruits can be to a depth of 1 cm but only to a few millimetres under the skin of older fruits except via cracks. However, surface-applied Ca seems to move by diffusion in the free space and higher concentrations can be found in mid-cortical tissues after spraying. Movement within the fruit may, though, be relatively slow. Chittenden *et al.* (1969) painted one half of 'Cox' apples seven times in the month prior to harvest with 1 % calcium nitrate and found a 30% incidence of bitter pit on the untreated side compared with only 2 % on the treated side after 4 months' storage. Calcium chloride injected into the core cavity at harvest is as effective as that applied to the surface in reducing bitter pit in the outer and mid-cortex (Perring and Pearson, 1987).

The effectiveness of post-harvest application of Ca to the skin or to the core, in reducing Ca deficiency disorders in the fruit flesh and in reversing softening, suggests that the individual cells and their middle lamellae remain active as sinks for Ca well into their post-harvest life. This supports the concept that the decline in Ca importation into the fruit in the later stages of their growth in the orchard is a function of an impaired transport system through the stalk, not simply to a decline in demand with the cessation of cell division.

Effects of other nutrients on calcium flux into fruits

As well as influencing uptake from the soil, as noted earlier, other nutrients can influence the movement of Ca into fruits. Orchard sprays with zinc sulphate, especially early in the season, can result in increases in fruit Ca content, possibly by releasing bound Ca from various chelating and complexing agents such as lignin, organic acids and proteins for transport to the shoot (Shear, 1980). Copper can have similar effects. Addition of zinc chloride (0.2%) to post-harvest dips in 3% $CaCl_2$ can double the uptake of Ca from the latter, increasing fruit Ca in different cultivars by 0.62 to 1.35 mg/100 g compared with that following dipping in $CaCl_2$ alone (Testoni and Pizzocaro, 1980).

Cultivar effects on fruit Ca level

Clonal or seedling rootstocks provide the root systems of almost all commercially grown apple and pear trees and might be expected to affect Ca uptake and transport through the graft union. There is indeed some evidence for rootstock effects on the content of Ca in scion leaves (Kennedy *et. al.*, 1980).

The largest effects of rootstocks on fruit Ca are, however, mediated by their effects on crop load and fruit size, as discussed earlier. An exception to this is

'M.26'. 'Cox' apples of any given size on 'M.26' have lower Ca content than those on 'M.7', 'MM.106' and, especially, 'M.9'. The effect can be up to 1 mg/100 g. Fruits from trees on 'M.26' have correspondingly higher levels of bitter pit and senescent breakdown (Blasco and Jackson, 1974; Blasco, 1976). This effect is compatible with the observation that an 'M.26' interstock can lower Ca concentration in the xylem sap (Jones, 1976). The effect on xylem sap content was shown below as well as above the interstock, implying effects via root behaviour.

Scion cultivars differ greatly in fruit Ca concentration. Volz *et al.* (1994) showed apple Ca concentration in mg/100 g fresh wt to range from 6.8 to 10.5 on different ages of fruiting wood of 'Royal Gala' whereas that of 'Braeburn', 'Granny Smith' and 'Fuji' ranged from 3.6–4.4, 3.0–4.1 and 2.8–3.0, respectively. Johnson (2000), after adjusting for fruit weight, found 'Gala' fruits to have 7.5 mg/100 g compared with 5.2 for 'Jonagold'. Similarly, Perring (1979) showed that at any given fruit mass 'Mutsu' apples contained much more Ca than those of 'Cox' (*c.* 70% more in a 150 g apple). Testoni and Pizzocaro (1980) found 'Starking' to have 6.4 mg Ca/100 g fresh wt in comparison with 4.6 for 'Golden Delicious'. The relationship between the calcium concentration in the fruit of a cultivar and the incidence of Ca-related storage disorders is not absolute but holds in many cases, e.g. 'Gala' is generally considered resistant and 'Cox' fairly prone to bitter pit. Fruit Ca concentrations do not necessarily reflect those in leaves. Van der Boon (1980) found 'James Grieve' apple trees to have lower concentrations of Ca in the fruits though more in the leaves than 'Cox'. Johnson (2000) found 'Gala' to have much lower leaf Ca but higher fruit Ca than 'Jonagold'. Lang (1990) and Lang and Ryan (1994) attributed the lower concentrations of Ca in 'Cox' than in 'Gala' apples to a more severe malfunction of the xylem in the fruit pedicel of 'Cox' as the season progresses, with an increasing proportion of fruit in which the xylem is totally non-conducting.

There is evidence for two genes, Bp-1 and Bp-2, that control Ca accumulation and distribution, respectively, within apple fruits and control resistance to bitter pit (Korban and Swiader, 1984).

Other nutrients and fruit eating-quality

High nitrogen levels can result in decreasing fruit firmness at harvest and to an increase in breakdown during storage (Sharples, 1973). High N fertilizer rates can also reduce the concentration of alcohol-insoluble solids and malate but increase the concentration of sugar (Richardson, 1986). Fruits with high N content are more likely to be affected by cork spot and bitter pit and to develop more scald, bitter pit, internal browning and internal breakdown after storage

(Bramlage *et al.*, 1980b). They also show high respiration rates. However, if fruit nitrogen levels in 'Cox' fall below 50 mg/100 g susceptibility to breakdown is increased, especially if calcium levels are also low (Sharples, 1980). Johnson *et al.* (2001) found that a beneficial effect of late sprays of calcium nitrate on texture was associated with an increase in the ratio of N to C, rather than of Ca to C, in the primary cell wall and middle lamella. Many of the adverse effects of high N are associated with its effect in increasing fruit size and vigour of shoot growth, which adversely affect Ca concentrations, and also with the direct effects of the ammonium ion on Ca uptake.

High potassium concentrations have a strong positive effect on fruit acidity (Wilkinson, 1958; Johnson, 2000), which is an important aspect of the taste of some cultivars. Fruits high in K are, however, more susceptible to breakdown, bitter pit and other disorders associated with low calcium status. These effects relate to the interaction between K and Ca in cells and high K can induce Ca deficiency.

Magnesium concentrations in apple fruits tend to be proportional to those of K and have similar associations with fruit acidity, bitter pit, etc. Magnesium salt sprays or dips can greatly increase bitter pit incidence. Sharples *et al.* (1979) attributed this to direct competition with Ca at the cellular level. Hopfinger and Poovaiah (1979) found very high Mg concentrations in pitted tissue and concluded that bitter pit is due to a localized magnesium toxicity which can be prevented by calcium treatment, but Ferguson and Watkins (1989) considered this Mg-induced pitting to differ from true bitter pit.

Low phosphorus levels in fruits lead to increased senescent breakdown and low temperature breakdown. Phosphorus sprays can reduce the incidence of these disorders (Yogaratnam and Sharples, 1982).

Carbohydrate supply is important with respect to fruit carbohydrate concentration and associated quality factors. Shade reduces total fruit dry matter, alcohol-insoluble and alcohol-soluble matter and starch per unit fresh weight (Jackson *et al.*, 1977). Fruit thinning results in higher sugar (brix) concentration in 'Fuji' apples (Fukuda and Takishita, 1998). It also increases fruit firmness despite the fruits on thinned trees being larger and lower in Ca concentration (Johnson, 1992).

Recommended reading

Atkinson, D., Jackson, J.E., Sharples, R.O. and Waller, W.M. (eds.) (1980). *Mineral Nutrition of Fruit Trees.* London: Butterworths.

Combrink, J.C. (ed.) (1996). *Integrated Management of Post-harvest Quality.* Stellenbosch: Infruitec.

Ferguson, I.B. and Watkins, C.B. (1989). Bitter pit in apple fruit. *Horticultural Reviews* **11**, 289–355.

Fidler, J.C., Wilkinson, B.G., Edney, K.L. and Sharples, R.O. (1973). *The Biology of Apple and Pear Storage.* Research Review No. 3. Farnham Royal, UK: Commonwealth Agricultural Bureaux.

Hardenburg, R.E., Watada, A.E. and Wang, C.Y. (1986). *The Commercial Storage of Fruits, Vegetables and Florist and Nursery Stocks.* US Department of Agriculture, Agriculture Handbook No. 66 (revised). Washington, DC: Government Printing Office.

Shear, C.B. (1979). International symposium on calcium nutrition of economic crops. *Communications in Soil Science and Plant Analysis,* Vol. 10, Nos. 1 and 2.

References

Abeles, F.B. (1985). Ethylene and plant development: an introduction. In *Ethylene and Plant Development,* ed. J.A. Roberts and G.A. Tucker, pp. 1–8. London: Butterworths.

Adams, D.O. and Yang, S.F. (1979). Ethylene biosynthesis: identification of 1-aminocyclo-propane-1-carboxylic acid as an intermediate in the conversion of methionine to ethylene. *Proceedings of the National Academy of Sciences USA,* **76**, 170–4.

Adams, F. (1974). Soil solution. In *The Plant Root and its Environment,* ed. E.W. Carson, pp. 441–82. Charlottesville, VA: University Press of Virginia.

Atkinson, D. and Wilson, S.A. (1980). The growth and distribution of fruit tree roots: some consequences for nutrient uptake. In *Mineral Nutrition of Fruit Trees,* ed. D. Atkinson, J.E. Jackson, R.O. Sharples and W.M. Waller, pp. 137–50. London: Butterworths.

Autio, W.R. and Bramlage, W.J. (1982). Effects of AVG on maturation, ripening and storage of apples. *Journal of the American Society for Horticultural Science* **107**, 1074–7.

Bangerth, F. (1976). A role for auxin and auxin transport inhibitors on the Ca content of artificially induced parthenocarpic fruits. *Physiologia Plantarum* **37**, 191–4.

Bangerth, F. (1978). The effect of a substituted amino acid on ethylene biosynthesis, respiration, ripening and pre-harvest drop of apple fruits. *Journal of the American Society for Horticultural Science* **103**, 401–4.

Bangerth, F. (1979). Calcium-related physiological disorders of plants. *Annual Review of Phytopathology* **17**, 97–122.

Barritt, B.H., Rom, C.R., Guelich, K.R., Drake, S.R. and Dilley, M.A. (1987). Canopy position and light effects on spur, leaf, and fruit characteristics of 'Delicious' apple. *HortScience* **22**, 402–5.

Bartley, I.M. and Knee, M. (1982). The chemistry of textural changes in fruit during storage. *Food Chemistry* **9**, 47–58.

Basso, C. and Wilms, F.W.W. (1988). Nutritional status of apple orchards in southern Brazil. *Acta Horticulturae* **232**, 187–92.

Bell, C.W. and Biddulph, O. (1963). Translocation of calcium. Exchange versus mass flow. *Plant Physiology* **38**, 610–14.

Ben-Arie, R., Lurie, S. and Mattoo, A.K. (1982). Temperature-dependent inhibitory effects of calcium and spermine on ethylene biosynthesis in apple discs correlate with changes in microsomal membrane microviscosity. *Plant Science Letters* **24**, 239–47.

Bepete, M. and Lakso, A.N. (1997). Apple fruit respiration in the field: relationships to fruit growth rate, temperature, and light exposure. *Acta Horticulturae* **451**, 319–26.

Blankenship, S.M. and Sisler, E.C. (1993). Response of apples to diazocyclopentadiene inhibition of ethylene binding. *Postharvest Biology and Technology* **3**, 95–101.

Blanpied, G.D. and Little, C.R. (1991). Relationships among bloom dates, ethylene climacteric initiation dates, and maturity-related storage disorders of 'Jonathan' apples grown in Australia. *Postharvest Biology and Technology* **1**, 3–10.

Blanpied, G.D., Bartsch, J.A. and Turk, J.R. (1985). A commercial development programme for low ethylene controlled-atmosphere storage of apples. In *Ethylene and Plant Development*, ed. J.A. Roberts and G.A. Tucker, pp. 393–404. London: Butterworths.

Blasco, A.B. (1976). Rootstock effects on growth and cropping of apples, with special reference to fruit quality. PhD thesis, University of London.

Blasco, A.B. and Jackson, J.E. (1974). Rootstock influence on cropping and fruit quality of apples. In *Proceedings of the XIX International Horticultural Congress*, Vol. 1B, ed. R. Antoszewski, L. Harrison and C.C. Zych, p. 585. Warsaw: ISHS.

Bolton, J. (1977). Changes in soil pH and exchangeable Ca in two liming experiments on contrasting soils over 12 years. *Journal of Agricultural Science, Cambridge* **89**, 81–6.

Boulton, G., Corrigan, V. and Lill, R. (1997). Objective method for estimating sensory response to juiciness in apples. *New Zealand Journal of Crop and Horticultural Science* **25**, 283–9.

Bowling, D.J.F. (1980). Uptake and transport of nutrients. In *Mineral Nutrition of Fruit Trees*, ed. D. Atkinson, J.E. Jackson, R.O. Sharples and W.M. Waller, pp. 151–63. London: Butterworths.

Bradfield, E.G. (1976). Calcium complexes in the xylem sap of apple shoots. *Plant and Soil* **44**, 495–9.

Bramlage, W.J., Drake, M. and Baker, J.H. (1974). Relationships of calcium content to respiration and post harvest condition of apples. *Journal of the American Society for Horticultural Science* **99**, 376–8.

Bramlage, W.J., Drake, M. and Lord, W.J. (1980b). The influence of mineral nutrition on the quality and storage performance of pome fruits grown in North America. In *Mineral Nutrition of Fruit Trees*, ed. D. Atkinson, J.E. Jackson, R.O. Sharples and W.M. Waller, pp. 29–39. London: Butterworths.

Bramlage, W.J., Greene, D.W., Autio, W.R. and McLaughlin, J.M. (1980a). Effects of aminoethoxyvinylglycine on internal ethylene concentrations and storage of apple. *Journal of the American Society for the Horticultural Science* **105**, 847–51.

Bramlage, W.J., Weis, S.A. and Greene, D.W. (1990). Observations on the relationships among seeds number, fruit calcium and senescent breakdown in apples. *HortScience* **25**, 351–3.

Brookfield, P., Murphy, P., Harker, R. and MacRae, E. (1997). Starch degradation and starch pattern indices; interpretation and relationship to maturity. *Postharvest Biology and Technology* **11**, 23–30.

Chavez-Franco, S.H. and Kader, A.A. (1993). Effects of CO_2 on ethylene biosynthesis in 'Bartlett' pears. *Postharvest Biology and Technology* **3**, 183–90.

Chino, M. (1979). Calcium localization with plant roots by electron probe X-ray microanalysis. *Communications in Soil Science and Plant Analysis* **10**, 443–57.

Chittenden, E.T., Stanton, D.J. and Watson, J. (1969). Bitter pit in 'Cox's Orange' apples, *New Zealand Journal of Agriculture Research* **12**, 240–7.

Corrigan, V.K., Hurst, P.L. and Boulton, G. (1997). Sensory characteristics and consumer acceptability of 'Pink Lady' and other late-season apple cultivars. *New Zealand Journal of Crop and Horticultural Science* **25**, 375–83.

Couey, H.M. and Olsen, K.L. (1975). Storage response of 'Golden Delicious' apples after high-carbon dioxide treatment. *Journal of the American Society for Horticultural Science* **100**, 148–50.

Deckers, T. and Missotten, C. (1993). Influence of training method with or without chemical growth regulation on fruit quality of apples cv 'Jonagold', 'Schone van Boskoop' and 'Gloster'. *Acta Horticulturae* **326**, 39–48.

Dover, C.J. (1985). Commercial scale catalytic oxidation of ethylene as applied to fruit stores. In *Ethylene and Plant Development*, ed. J.A. Roberts and G.A. Tucker, pp. 373–83. London: Butterworths.

Downs, C.G., Brady, C.J., Campbell, J. and McGlasson, W.B. (1991). Normal ripening cultivars of *Pyrus serotina* are either climacteric or non-climacteric. *Scientia Horticulturae* **48**, 213–21.

Drake, S.R., Williams, M.W. and Raese, J.T. (1997). Interstem and its relationship to 'Fuji' apple quality. *Journal of Tree Fruit Production* **2**, 99–106.

Eggert, F.P. (1960). The relation between heat unit accumulation and length of time required to mature 'McIntosh' apples in Maine. *Proceedings of the American Society for Horticultural Science* **76**, 98–105.

Fan, X., Mattheis, J.P., Patterson, M.E. and Fellman, J.K. (1997). Optimum harvest date and controlled atmosphere storage potential of 'Fuji' apples. In *CA 1997 Proceedings*, Vol. 2. *Apples and Pears*, ed. E.J. Mitcham, pp. 42–9. Davis: University of California.

Faust, M. (1980). Interaction between nutrient uptake and photosynthesis. In *Mineral Nutrition of Fruit Trees*, ed. D. Atkinson, J.E. Jackson, R.O. Sharples and W.M. Waller, pp. 193–9. London: Butterworths.

Faust, M. (1989). *Physiology of Temperate Zone Fruit Trees*. New York: John Wiley & Sons.

Faust, M. and Shear, C.B. (1968). Corking disorders of apples: a physiological and biochemical review. *Botanical Reviews* **34**, 441–69.

Faust, M. and Shear, C.B. (1972). The effect of calcium on respiration of apples. *Journal of the American Society for Horticultural Science* **97**, 69–72.

Fellman, J.K., Mattinson, D.S., Fan, X. and Mattheis, J.P. (1997). 'Fuji' apple storage characteristics in relation to growing conditions and harvest maturity in Washington State. In *CA 1997 Proceedings*, Vol. 2. *Apples and Pears*, ed. E.J. Mitcham, p. 234. Davis: University of California.

Ferguson, I.B. (1979). The movement of calcium in non-vascular tissue of plants. *Communications in Soil Science and Plant Analysis* **10**, 956–60.

Ferguson, I.B. (1984). Calcium in plant senescence and fruit ripening. *Plant, Cell and Environment* **7**, 477–89.

Ferguson, I.B. and Bollard, E.G. (1976). The movement of calcium in woody stems. *Annals of Botany* **40**, 1057–65.

Ferguson, I.B. and Watkins, C.B. (1989). Bitter pit in apple fruit. *Horticultural Reviews* **11**, 289–355.

Ferree, D.C. and Palmer, J.W. (1982). Effect of spur defoliation and ringing during bloom on fruiting, fruit mineral level and net photosynthesis of 'Golden Delicious' apple. *Journal of the American Society for Horticultural Science* **107**, 1182–6.

Fidler, J.C. (1973). Conditions of storage. In *The Biology of Apple and Pear Storage*, ed. J.C. Fidler, B.G. Wilkinson, K.L. Edney and R.O. Sharples, pp. 1–61. Farnham Royal, UK: Commonwealth Agricultural Bureaux.

Fidler, J.C. and North, C.J. (1969). Production of volatile organic compounds by apples. *Journal of the Science of Food and Agriculture* **20**, 521–6.

Findlay, J.S. and Combrink, J.C. (1996). Cold chain management. In *Integrated Management of Post-harvest Quality*, ed. J.C. Combrink, pp. 79–90. Stellenbosch: Infruitec.

Ford, E.M. (1979). Effect of post-blossom environmental conditions on fruit composition and quality of apple. *Communications in Soil Science and Plant Analysis* **10**, 337–48.

Ford, E.M. and Quinlan, J.D. (1979). The distribution of ^{45}Ca in apple fruits when supplied to the roots at three times during the season. *Journal of Horticultural Science* **54**, 181–8.

Forsyth, F.R., Eaves, C.A. and Lightfoot, H.J. (1969). Storage quality of 'McIntosh' apples as affected by removal of ethylene from the storage atmosphere. *Canadian Journal of Plant Science* **49**, 567–72.

Führ, F. and Wienecke, J. (1974). Secondary translocation of ^{45}Ca to the fruit of apple trees the year after dormancy. In *Mechanisms of Regulation of Plant Growth*, ed. R.L. Bieleski, A.R. Ferguson and M.M. Cresswell. Bulletin 12, The Royal Society of New Zealand, pp. 171–5. Wellington, New Zealand.

Fukuda, H. and Takishita, F. (1998). Effect of fruit load on the average fruit size and quality of 'Fuji' apple trees on M9 dwarfing rootstock. *Bulletin of the National Institute of Fruit Tree Science (Japan)* 30–31, 63–73.

Fuller, M.M. (1976). The ultrastructure of the outer tissues of cold-stored apple fruits of high and low calcium content in relation to cell breakdown. *Annals of Applied Biology* **83**, 299–304.

Greenham, D.W.P. (1980). Nutrient cycling: the estimation of orchard nutrient uptake. In *Mineral Nutrition of Fruit Trees*, ed. D. Atkinson, J.E. Jackson, R.O. Sharples and W.M. Waller, pp. 345–52. London: Butterworths.

Hardenburg, R.E., Watada, A.E. and Wang, C.Y. (1986). *The Commercial Storage of Fruits, Vegetables, and Florist and Nursery Stocks.* US Department of Agriculture, Agriculture, Handbook No. 66 (revised), 130 pp. Washington, DC: Government Printing Office.

Harker, F.R. and Ferguson, I.B. (1991). Effects of surfactants on calcium penetration of cuticles isolated from apple fruit. *Scientia Horticulturae* **46**, 225–33.

Hatfield, S.G.S. and Knee, M. (1988). Effects of water loss on apples in storage. *International Journal of Food Science and Technology* **23**, 575–83.

Haynes, R.J. and Goh, K.M. (1980). Variation in the nutrient content of leaves and fruit with season and crown position for two apple varieties. *Australian Journal of Agricultural Research* **31**, 739–48.

Hesse, C.O. and Hitz, C.W. (1938). Maturity studies with Jonathan and Grimes Golden apples. *Proceedings of the American Society for Horticultural Science* **36**, 351–7.

Holland, D.A. (1980). The prediction of bitter pit. *In Mineral Nutrition of Fruit Trees*, ed. D. Atkinson, J.E. Jackson, R.O. Sharples and W.M. Waller, pp. 380–1. London: Butterworths.

Hong, Y.P., Choi, J.H. and Lee, S.K. (1997). Optimum CA condition for four apple cultivars grown in Korea. In *CA 97 Proceedings*, Vol. 2. *Apples and Pears*, ed. E.J. Mitcham, pp. 241–5. Davis: University of California.

Hopfinger, J.A. and Poovaiah, B.W. (1979). Calcium and magnesium gradients in apples with bitterpit. *Communications in Soil Science and Plant Analysis* **10**, 57–65.

Huguet, C. (1980). Effects of the supply of calcium and magnesium on the composition and susceptibility of 'Golden Delicious' apples to physiological and pathological disorders. In *Mineral Nutrition of Fruit Trees*, ed. D. Atkinson, J.E. Jackson, R.O. Sharples and W.M. Waller, pp. 93–8. London: Butterworths.

Hulme, A.C. and Rhodes, M.J.C. (1971). Pome fruits. In *The Biochemistry of Fruits and Their Products*, Vol. 2, ed. A.C. Hulme, pp. 333–73. London: Academic Press.

Jackson, J.E., Modlibowska, I., El-Senfaz, S.M., Wickenden, M.F. and Johnson, D.S. (1982). Effects of fruit setting hormones on calcium content, maturation and storage performance of 'Bramley's Seedling' apple. *Abstract 1093, Abstracts Vol 1, XXI*[st] *International Horticultural Congress Hamburg*. The Hague: ISHS.

Jackson, J.E., Palmer, J.W., Perring, M.A. and Sharples, R.O. (1977). Effects of shade on the growth and cropping of apple trees. III. Effects on fruit growth, chemical composition and quality at harvest and after storage. *Journal of Horticultural Science* **52**, 267–82.

Jackson, J.E., Sharples, R.O. and Palmer, J.W. (1971). The influence of shade and within-tree position on apple fruit size, colour and storage quality. *Journal of Horticultural Science* **46**, 277–87.

Jobling, J.J. and McGlasson, W.B. (1995). A comparison of ethylene production, maturity and controlled atmosphere storage life of 'Gala', 'Fuji' and 'Lady Williams' apples (*Malus domestica*, Borkh). *Postharvest Biology and Technology* **6**, 209–18.

Johnson, D.S. (1992). The effect of flower and fruit thinning on the firmness of 'Cox's Orange Pippin' apples at harvest and after storage. *Journal of Horticultural Science* **67**, 95–101.

Johnson, D.S. (1998). Effect of Retain™ sprays on storage quality of Cox apples. *Members Day Report March 1998, East Malling Research Association*, 13–15.

Johnson, D.S. (2000). Mineral composition, harvest maturity and storage quality of 'Red Pippin', 'Gala' and 'Jonagold' apples. *Journal of Horticultural Science and Biotechnology* **75**, 697–704.

Johnson, D.S. and Ertan, U. (1983). Interaction of temperature and oxygen level on the respiration rate and storage quality of 'Idared' apples. *Journal of Horticultural Science* **58**, 527–33.

Johnson, D.S. and Ridout, M.S. (2000). Effects on the quality of stored apple fruit. In *Fruit and Vegetable Quality: An Integrated View*, ed. R.L. Shewfelt and B. Brückner, pp. 67–84. Lancaster, PA: Basel Technomic Publishing Co. Inc.

Jones, H.G. and Samuelson, T.J (1983). Calcium uptake by developing apple fruits. II. The role of spur leaves. *Journal of Horticultural Science* **58**, 183–90.

Johnson, D.S., Dover, C.J., Samuelson, T.J., Huxham, I.M., Jarvis, M.C., Knox, J.P., Shakespeare, L. and Seymour, G.B. (2001). Nitrogen, cell walls and texture of stored 'Cox's Orange Pippin' apples. *Acta Horticulturae* **564**, 105–12.

Jones, H.G., Higgs, K.H. and Samuelson, T.J. (1983). Calcium uptake by developing fruits. I. Seasonal changes in calcium content of fruits. *Journal of Horticultural Science* **58**, 173–82.

Jones, O.P. (1976). Effect of dwarfing interstocks on xylem sap composition in apple trees: effect on nitrogen, potassium, phosphorus, calcium and magnesium content. *Annals of Botany* **40**, 1231–5.

Kennedy, A.J., Rowe, R.W. and Samuelson, T.J. (1980). The effect of apple rootstock genotypes on mineral content of scion leaves. *Euphytica* **29**, 477–82.

Khan, A.A. and Vincent, J.F.V. (1990). Anisotropy of apple parenchyma. *Journal of the Science of Food and Agriculture* **52**, 455–66.

Kidd, F. and West, C. (1924). The course of respiratory activity throughout the life of an apple. *Annual Report of the Food Investigations Board*, London, 27–32.

Kingston, C.M. (1992). Maturity indices for apple and pear. *Horticultural Reviews* **13**, 407–32.

Knee, M. (1975). Storage of 'Bramley's Seedling' apples. I. Effects of source of fruit, picking date and storage conditions on ripening and compositional changes. *Journal of Horticultural Science* **50**, 113–20.

Knee, M. (1978). Metabolism of polymethylgalacturonate in apple fruit cortical tissue during ripening. *Phytochemistry* **17**, 1261–4.

Knee, M. (1982). Fruit softening. III. Requirement for oxygen and pH effects. *Journal of Experimental Botany* **33**, 1263–9.

Knee, M. (1985). Evaluating the practical significance of ethylene in fruit storage. In *Ethylene and Plant Development*, ed. J.A. Roberts and G.A. Tucker, pp. 297–315. London: Butterworths.

Knee, M. (1993). Pome fruits. In *Biochemistry of Fruit Ripening*, ed. G. Seymour, J. Taylor and G. Tucker, pp. 325–46. London: Chapman and Hall.

Knee, M. and Bartley, I.M. (1981). Composition and metabolism of cell wall polysaccharides in ripening fruits. In *Recent Advances in the Biochemistry of Fruits and Vegetables*, ed. J. Friend and M.J.C. Rhodes, pp. 133–48. London: Academic Press.

Knee, M., and Hatfield, S.G.S. (1981). Benefits of ethylene removal during apple storage. *Annals of Applied Biology* **98**, 157–65.

Knee, M., and Stow, J.R. (1985). Effect of carbon dioxide pre-treatment on ethylene production by 'Cox' apples stored in low oxygen. *Report of the East Malling Research Station for 1984*, 191.

Knee, M. and Tsantili, E. (1988). Storage of 'Gloster 69' apples in the preclimacteric state for up to 200 days after harvest. *Physiologia Plantarum* **74**, 499–503.

Knee, M., Hatfield, S.G.S. and Smith, S.M. (1989). Evaluation of various indicators of maturity for harvest of apple fruit intended for long term storage. *Journal of Horticultural Science* **64**, 403–11.

Knee, M., Looney, N.E., Hatfield, S.G.S. and Smith, S.M. (1983). Initiation of rapid ethylene synthesis by apple and pear fruits in relation to storage temperature. *Journal of Experimental Botany* **34**, 1207–12.

Korban, S.S. and Swiader, J.M. (1984). Genetic and nutritional status in bitter pit-resistant and -susceptible apple seedlings. *Journal of the American Society for Horticultural Science* **109**, 428–32.

Korcak, R.F. (1980). The importance of calcium and nitrogen source in fruit tree nutrition. In *Mineral Nutrition of Fruit Trees*, ed. D. Atkinson, J.E. Jackson, R.O. Sharples and W.M. Waller, pp. 267–77. London: Butterworths.

Kotzé, W.A.G. (1979). Ionic interactions in the uptake and transport of calcium by apple seedlings. *Communications in Soil Science and Plant Analysis* **10**, 115–27.

Kotzé, W.A.G. (1996). Soil and nutritional factors affecting apple fruit quality. In *Integrated Management of Post-harvest Quality*, ed. J.C. Combrink, pp. 14–24. Stellenbosch, Infruitec.

Kotzé, W.A.G. and du Preez, M. (1988). Soil amelioration and fertilization for apple production in South Africa. *Acta Horticulturae* **232**, 167–76.

Kuhn, A.J., Schröder, W.H. and Bauch, J. (2000). The kinetics of calcium and magnesium entry into mycorrhizal spruce roots. *Planta* **210**, 488–96.

Lang, A. (1990). Xylem, phloem and transpiration flows in developing apple fruits. *Journal of Experimental Botany* **41**, 645–51.

Lang, A. and Ryan, K.G. (1994). Vascular development and sap flow in apple pedicels. *Annals of Botany* **74**, 381–8.

Lang, A. and Volz, R.K. (1993). Leaf area, xylem cycling and Ca status in apples. *Acta Horticulturae* **343**, 89–91.

Lang, A. and Volz, R.K. (1998). Spur leaves increase calcium in young apples by promoting xylem inflow and outflow. *Journal of the American Society for Horticultural Science* **123**, 956–60.

Larrigaudiere, C., Graell, J., Salas, J. and Vendrell, M. (1997). Cultivar differences in the influence of a short period of cold storage on ethylene biosynthesis in apples. *Postharvest Biology and Technology* **10**, 21–7.

Lau, O.L. (1985). Harvest indices for B.C. apples. *B.C. Orchardist* **7**(7), 1A–20A.

Lau, O.L. (1988). Harvest indices, dessert quality and storability of 'Jonagold' apples in air and controlled atmosphere storage. *Journal of the American Society for Horticultural Science* **113**, 564–9.

Lau, O.L. (1993). Scald and its control: the North American situation. *Acta Horticulturae* **326**, 225–30.

Lelièvre, J.M., Tichit, L., Fillion, L., Larrigaudière, C., Vendrell, M. and Pech, J-C. (1995). Cold-induced accumulation of 1-aminocyclopropane-carboxylate oxidase protein in 'Granny Smith' apples. *Postharvest Biology and Technology* **5**, 11–17.

Link, H. (1993). Influence of pruning on fruit quality and storage potentials of apples. *Acta Horticulturae* **326**, 29–38.

Liu, F.W. (1985). Low ethylene controlled-atmosphere storage of 'McIntosh' apples. In *Ethylene and Plant Development*, ed. J.A. Roberts and G.A. Tucker, pp. 385–92. London: Butterworths.

Lott, R.V. (1965). The quality, color and keepability characteristics of a low-acid 'Jonared' apple sport. *Proceedings of the American Society for Horticultural Science* **87**, 47–54.

Ludders, P. (1979). The effect of nitrogen nutrition on bitter pit in apples. *Communications in Soil Science and Plant Analysis* **10**, 401–15.

Luton, M.T. and Hamer, P.J.C. (1983). Predicting the optimum harvest dates for apples using temperature and full-bloom records. *Journal of Horticultural Science* **58**, 37–44.

Marks, M.J. (1980). The use of leaf analysis in advisory work in the U.K. *In Mineral Nutrition of Fruit Trees*, ed. D. Atkinson, J.E. Jackson, R.O. Sharples and W.M. Waller, p. 366. London: Butterworths.

Meli, J. and Bygrave, F. (1972). The role of mitochondria in modifying calcium-sensitive cytoplasmic metabolic activities: modification of pyruvate kinase activity. *Biochemical Journal* **128**, 415–20.

Moriguchi, T.M., Abe, K., Sanada, T., and Yamaki, S. (1992). Levels and role of sucrose synthase, sucrose-phosphate synthase, and acid invertase in sucrose accumulation in fruit of Asian pear. *Journal of the American Society for Horticultural Science* **117**, 274–8.

Motosugi, H., Gao, Y.P. and Sugiura, A. (1995). Rootstock effects on fruit quality of 'Fuji' apples grown with ammonium or nitrate nitrogen in sand culture. *Scientia Horticulturae* **61**, 205–14.

North, C.J. (1971). The use of the starch-iodine staining test for assessing the picking date for pears. *Report of the East Malling Research Station for 1970*, 149–51.

Pavel, E.W. and DeJong, T.M. (1995). Seasonal patterns of non- structural carbohydrates of apple (*Malus pumila* Mill) fruits: relationship with relative growth rates and contribution to solute potential. *Journal of Horticultural Science* **70**, 127–34.

Perring, M.A. (1979). The effects of environment and cultural practices on calcium concentration in the apple fruit. *Communications in Soil Science and Plant Analysis* **10**, 279–93.

Perring, M.A. (1984). Indirect influences on apple fruit mineral composition and storage quality. *Proceedings VIth International Colloquium for Optimization of Plant Nutrition 1984*. Montpellier, France **4**, 1199–206.

Perring, M.A. (1986). Incidence of bitter pit in relation to the calcium content of apples: problems and paradoxes, a review. *Journal of the Science of Food and Agriculture* **37**, 591–606.

Perring, M.A. and Jackson, C.H. (1975). The mineral composition of apples. Calcium concentration and bitter pit in relation to mean mass per apple. *Journal of the Science of Food and Agriculture* **26**, 1493–502.

Perring, M.A. and Pearson, K. (1986). Incidence of bitter pit in relation to the calcium content of apples: calcium distribution in the fruit. *Journal of the Science of Food and Agriculture* **37**, 709–18.

Perring, M.A. and Pearson, K. (1987). Suppression of bitter pit by calcium injected into the core cavities of harvested apples. *Journal of Horticultural Science* **62**, 303–4.

Perring M.A., Samuelson, T.J., Caverly, D.J. and Marks, M.J. (1985). Boron and calcium deficiency symptoms in apple. *Report of the East Malling Research Station for 1984*, 223–4.

Poapst, P.A., Ward, G.M. and Phillips, W.R. (1959). Maturation of 'McIntosh' apples in relation to starch loss and abscission. *Canadian Journal of Plant Science* **39**, 257–63.

Poovaiah, B.W., Glenn, G.M. and Reddy, A.S.N. (1988). Calcium and fruit softening: physiology and biochemistry. *Horticultural Reviews* **10**, 107–52.

Preuschoff, H.G. (1968). Mineralstoffgehalte in den Blattern eines Apfelbaumes. *Erwerbsobstbau* **10**, 193–6.

Proctor, J.T. and Palmer, J.W. (1991). The role of spur and bourse leaves of three apple cultivars on fruit set and growth and calcium content. *Journal of Horticultural Science* **66**, 275–82.

Reid, M.S. (1995). Ethylene in plant growth, development and senescence. In *Plant Hormones*, 2nd edn, ed. P.J. Davies, pp. 486–508. Dordrecht: Kluwer.

Rhodes, M.J.C. (1970). The climacteric and ripening of fruits. In *The Biochemistry of Fruits and their Products*, Vol. 1, ed. A.C. Hulme, pp. 521–33. London: Academic Press.

Rhodes, M.J.C. (1980). The maturation and ripening of fruits. In *Senescence in Plants*, ed. K.V. Thimann, pp. 157–205. Boca Raton, FL: CRC Press.

Richardson, A. (1986). The effect of herbicide soil management systems and nitrogen fertilizer on the eating quality of 'Cox's Orange Pippin' apples. *Journal of Horticultural Science* **61**, 447–56.

Rowan, D.D., Allen, J.M., Fielder, S. and Hunt, M.B. (1997). Deuterium labelling to study aroma biosynthesis in stored apples. In *CA 1997 Proceedings*, Vol. 2. *Apples and Pears*, ed. E.J. Mitcham, pp. 227–33. Davis: University of California.

Rowe, R.W. (1980). Future analytical requirements in the fruit industry. In *Mineral Nutrition of Fruit Trees*, ed. D. Atkinson, J.E. Jackson, R.O. Sharples and W.M. Waller, pp. 399–405. London: Butterworths.

Schumacher, R., Frankhauser, F., and Stadler, W. (1980). Influence of shoot growth, average fruit weight and daminozide on bitter pit. In *Mineral Nutrition of Fruit Trees*, ed. D. Atkinson, J.E. Jackson, R.O. Sharples and W.M. Waller, pp. 83–91. London: Butterworths.

Scott, K.J., Wills, R.B.H. and Patterson, B.D. (1971). Removal by ultra-violet lamp of ethylene and other hydrocarbons produced by bananas. *Journal of the Science of Food and Agriculture* **22**, 496–7.

Sharples, R.O. (1968). Fruit thinning effects on the development and storage quality of 'Cox's Orange Pippin' apple fruits. *Journal of Horticultural Science* **43**, 359–71.

Sharples, R.O. (1973). Orchard and climatic factors, Part IV. In *The Biology of Apple and Pear Storage*, ed. J.C. Fidler, B.G. Wilkinson, K.L. Edney and R.O. Sharples. *Research Review Commonwealth Bureau of Horticulture and Plantation Crops* No.3. pp. 175–225. Farnham Royal, UK: CAB.

Sharples, R.O. (1980). The influence of orchard nutrition on the storage quality of apples and pears grown in the United Kingdom. In *Mineral Nutrition of Fruit trees*, ed. D. Atkinson, J.E. Jackson, R.O. Sharples and W.M. Waller, pp. 17–28. London: Butterworths.

Sharples, R.O. and Johnson, D.S., (1977). The influence of calcium on senescence changes in apple. *Annals of Applied Biology* **85**, 450–3.

Sharples, R.O. and Johnson, D.S. (1987). Influence of agronomic and climatic factors on the response of apple fruit to controlled atmosphere storage. *HortScience* **22**, 763–6.

Sharples, R.O. Reid, M.S. and Turner, N.A. (1979). The effects of post harvest mineral element and lecithin treatments on the storage disorders of apples. *Journal of Horticultural Science* **54**, 299–304.

Shear, C.B. (1980). Interaction of nutrition and environment on mineral composition of fruits. In *Mineral Nutrition of Fruit Trees*, ed. D. Atkinson, J.E. Jackson, R.O. Sharples and W.M. Waller, pp. 41–50. London: Butterworths.

Shear, C.B. and Faust, M. (1970). Calcium transport in apple trees. *Plant Physiology* **45**, 670–4.

Siddiqui, S. and Bangerth, F. (1993). Studies on cell wall- mediated changes during storage of calcium-infiltrated apples. *Acta Horticulturae* **326**, 105–13.

Smith, S.M. (1984). Improvement of aroma of 'Cox's Orange Pippin' apples stored in low oxygen atmospheres. *Journal of Horticultural Science* **59**, 515–22.

Smith, S.M. (ed.) (1985). Measurement of the quality of apples. *Recommendations of an EEC Working Group.* Luxembourg. Commission of the European Communities.

Smith, S.M., Geeson, J. and Stow, J. (1987). Production of modified atmospheres in deciduous fruits by the use of films and coatings. *HortScience* **22**, 772–6.

Smith, S.M., Johnson, D.S. and Mori, J. (1985). The effect of the growth regulator ethephon on the quality of 'Discovery' apples during simulated marketing. *Journal of Horticultural Science* **60**, 305–10.

Smith, W.H. (1940). The histological structure of the flesh of the apple in relation to growth and senescence. *Journal of Pomology and Horticultural Science XVIII*, 249–60.

Smock, R.M. (1941). Studies on bitter pit of apples. *Cornell University Agricultural Experiment Station Memoir 234*.

Smock, R.M. (1943). *Bulletin 799 Cornell University Agricultural Experiment Station*, Ithaca, New York.

Smock, R.M. (1948). A study of maturity indices for 'McIntosh' apples. *Proceedings of the American Society for Horticultural Science* **52**, 176–82.

Smock, R.M. and Van Doren, A. (1937). The histology of bitter pit in apples. *Proceedings of the American Society for Horticultural Science* **35**, 176–9.

Song, J. and Bangerth, F. (1996). The effect of harvest date on aroma compound production from 'Golden Delicious' apple fruit and relationship to respiration and ethylene production. *Postharvest Biology and Technology* **8**, 259–69.

Stahly, E.A. (1986). Time of application of calcium sprays to increase fruit calcium and reduce fruit pitting of apples sprayed with TIBA. *HortScience* **21**, 95–6.

Stahly, E.A. and Benson, N.R. (1976). Calcium levels of 'Golden Delicious' apples as influenced by calcium sprays, 2,3,5-triiodobenzoic acid and other plant growth regulator sprays. *Journal of the American Society for Horticultural Science* **101**, 120–2.

Stahly, E.A. and Benson, N.R. (1982). Seasonal accumulation of calcium and potassium in the cortex of 'Golden Delicious' apple fruit sprayed with 2,3,5-triiodobenzoic acid. *HortScience* **17**, 781–3.

Stow, J. R. (1988). The effects of high carbon dioxide pre-treatments and ethylene removal on the storage performance of apples 'Cox's Orange Pippin'. *Scientia Horticulturae* **35**, 89–97.

Stow, J.R., (1993). Effect of calcium ions on apple fruit softening during storage and ripening. *Postharvest Biology and Technology* **3**, 1–9.

Stow, J.R., Alston, F., Hatfield, S. and Genge, P. (1993). New apple selections with inherently low ethene production. *Acta Horticulturae* **326**, 85–92.

Stow, J.R., Dover, C.J. and Genge, P.M. (2000). Control of ethylene biosynthesis and softening in 'Cox's Orange Pippin' apples during low-ethylene, low-oxygen storage. *Postharvest Biology and Technology* **18**, 215–25.

Szczesniak, A.S. and Ilker, R. (1988). The meaning of textural characteristics-juiciness in plant foodstuffs. *Journal of Texture Studies* **19**, 61–78.

Terblanche, J.H., Gürgen, K.H. and Hesebeck, I. (1980). An integrated approach to orchard nutrition and bitter pit control. In *Mineral Nutrition of Fruit Trees*, ed. D. Atkinson, J.E. Jackson, R.O. Sharples, and W.M. Waller, pp. 71–82. London: Butterworths.

Testoni, A. and Pizzocaro, F. (1980). Calcium absorption in apples and pears treated postharvest with calcium, zinc or urea. In *Mineral Nutrition of Fruit Trees*, ed. D. Atkinson, J.E. Jackson, R.O. Sharples and W.M. Waller, pp. 329–31. London: Butterworths.

Tian, M.S., Hewett, E.W. and Lill, R.E. (1992). 'Hosui' fruit in New Zealand is nonclimacteric. *New Zealand Journal of Crop and Horticultural Science* **20**, 371–5.

Tomala, K. (1997). Orchard factors affecting nutrient content and fruit quality. *Acta Horticulturae* **448**, 257–64.

Tomala, K. and Dilley, D.R. (1990). Some factors influencing the calcium level in apple fruits. *Acta Horticulturae* **274**, 481–6.

Tromp, J. (1979). The intake curve for calcium into apple fruits under various environmental conditions. *Communications in Soil Science and Plant Analysis* **10**, 325–35.

Tu, K., Waldron, K., Ingham, L., DeBarsy, T. and De Baerdemaeker, J. (1997). Effect of picking time and storage conditions on 'Cox's Orange Pippin' apple texture in relation to cell wall changes. *Journal of Horticultural Science* **72**, 971–80.

Tucker, G.A. (1993). Introduction. In *Biochemistry of Fruit Ripening*, ed. G.B. Seymour, J.E. Taylor and G.A. Tucker, pp. 1–51. London: Chapman and Hall.

Tukey, L.D. (1964). A linear electronic device for continuous measurement and recording of fruit enlargement and contraction. *Proceedings of the American Society for Horticultural Science* **84**, 653–60.

Tukey, L.D. (1974). Some relationships in the growth and development of apple fruits. *Proceedings of the 19th International Horticultural Congress* **3**, 35–45.

Ulrich, R. (1970). Organic acids. In *The Biochemistry of Fruits and Their Products*, Vol. 1, ed. A.C. Hulme, pp. 89–118. London: Academic Press.

Van der Boon, J. (1980). Dressing or spraying calcium for bitter pit control. In *Mineral Nutrition of Fruit Trees*, ed. D. Atkinson, J.E. Jackson, R.O. Sharples and W.M. Waller, pp. 309–15. London: Butterworths.

van der Merwe, J.A. (1996a). Maturity standards for deciduous fruit. In *Integrated Management of Post-harvest Quality*, ed. J.C. Combrink, pp. 25–32. Stellenbosch: Infruitec.

van der Merwe, J.A. (1996b). Controlled and modified atmosphere storage. In *Integrated Management of Post-harvest Quality*, ed. J.C. Combrink, pp. 104–12. Stellenbosch: Infruitec.

Visser, T., Schaap, A.A. and DeVries, D.P. (1968). Acidity and sweetness in apple and pear. *Euphytica* **17**, 153–67.

Volz, R.K., Ferguson, I.B., Hewett, E.W. and Woolley, D.J. (1994). Wood age and leaf area influence fruit size and mineral composition of apple fruits. *Journal of Horticultural Science* **62**, 385–95.

Volz, R.K., Tustin, D.S. and Ferguson, I.B., (1996a). Mineral accumulation in apple fruit as affected by spur leaves. *Scientia Horticulturae* **65**, 151–61.

Volz, R.K., Tustin, D.S. and Ferguson, I.B. (1996b). Pollination effects on fruit mineral composition, seeds and cropping characteristics of 'Braeburn' apple trees. *Scientia Horticulturae* **66**, 169–80.

Waller, W.M. (1980). Use of apple analysis. In *Mineral Nutrition of Fruit Trees*, ed. D. Atkinson, J.E. Jackson, R.O. Sharples and W.M. Waller, pp. 383–94. London: Butterworths.

Warrington, I.J., Fulton, T.A., Halligan, E.A. and de Silva, H.N. (1999). Apple fruit growth and maturity are affected by early season temperatures. *Journal of the American Society for Horticultural Science* **124**, 468–77.

Watada, A.E., Herner, R.C., Kader, A.A., Romani, R.J. and Staby, G.L. (1984). Terminology for the description of developmental stages of horticultural crops. *HortScience* **19**, 20–1.

Watkins, C.B, Harman, J.E., Ferguson, I.B. and Reid, M.S. (1982). The action of lecithin and calcium dips in the control of bitter pit in apple fruit. *Journal of the American Society for Horticultural Science* **107**, 262–5.

Whiting, G.C. (1970). Sugars. In *The Biochemistry of Fruits and their Products*, Vol. 1, ed. A.C. Hulme, pp. 1–32. London: Academic Press.

Wilkinson, B.G. (1958). The effect of orchard factors on the chemical composition of apples. II. The relationship between potassium and total titrateable acidity, and between potassium and magnesium, in the fruit. *Journal of Horticultural Science* **33**, 49–57.

Williams, M.W. (1980). Retention of fruit firmness and increase in vegetative growth and fruit set of apples with aminoethoxyvinylglycine. *HortScience* **15**, 76–7.

Wilms, F.W.W. and Basso, C. (1988). Soil amelioration for apple production in southern Brazil. *Acta Horticulturae* **232**, 193–203.

Witney, G.W. and Kushad, M.M. (1990). Correlation of pyruvate kinase activity with bitter pit development in apple fruit. *Scientia Horticulturae* **43**, 247–53.

Yahia, E.M. (1994). Apple flavor. *Horticultural Reviews* **16**, 197–234.

Yogaratnam, N. and Sharples, R.O. (1982). Supplementing the nutrition of Bramley's Seedling apple with phosphorus sprays. II. Effects on fruit composition and storage quality. *Journal of Horticultural Science* **57**, 53–9.

11

Mineral nutrition

Introduction

Aspects of nutrient uptake by roots and of mineral nutrition in relation to vegetative growth and fruiting have been discussed in earlier chapters. Fruit mineral content in relation to storage and eating quality was considered in Chapter 10 and calcium uptake, transport and effects on cell structure and metabolism were discussed in particular detail because of its dominant rôle with respect to fruit firmness and the incidence of some pre- and post-harvest physiological disorders. In the present chapter more general aspects of the uptake, transport and redistribution of nutrients are dealt with.

Nutrient requirements

A first approximation of the necessary supply of major elements for apple and pear tree growth is obtained by measurement of the mineral content of well-grown and productive trees. Relevant figures from Washington State, USA are given in Table 11.1. These are for an old-style orchard at maturity, with a similar cropping level but most probably more vegetative dry matter than many modern orchards on dwarfing rootstocks. Most of the nutrients removed from the soil and not returned to it are in the fruits so the need to replace nutrients is largely a function of crop yield. Where yields are much higher, e.g. in South Africa and, especially, New Zealand the replacement needs will be much greater. Losses by leaching can be appreciable under conditions of high rainfall or irrigation on appropriate soil types. Haynes and Goh (1980) estimated that leaching losses were about 40% of the amount of nitrogen supplied in the irrigation water or as fertilizer in New Zealand, and exceeded irrigation and fertilizer inputs of Ca and Mg. The amount of N lost by drainage depends on irrigation level and the

Table 11.1 *Estimated annual utilization (kg ha^{-1}) of major nutrients by a mature apple orchard (cv. 'Delicious') with yields of 44.8 t ha^{-1}*

	N	P	K	Ca	Mg
Removed in fruit including seeds	20.8	6.3	56.6	4.4	2.2
In framework of trees including roots	18.4	4.2	14.3	45.8	2.3
Total (A) = Net uptake	39.2	10.5	70.9	50.2	4.5
Leaf-fall	47.6	3.3	52.4	85.8	18.1
Dropping blossoms and fruitlets including thinning of fruits	11.9	1.7	14.8	3.7	1.1
Prunings	11.8	2.3	3.6	28.0	1.7
Total (B) = Return to soil	71.3	7.3	70.8	117.5	20.9
Total (A + B) = Gross uptake	110.5	17.8	141.7	167.7	25.4

From Greenham (1980), summarized from Batjer *et al.* (1952). Reproduced with permission.

presence or absence of grass on the orchard floor (Stevenson and Neilsen, 1990).

The net uptake of K is almost twice that of N and many times as high as that of P and Mg. A remarkably similar balance of uptake is shown by Japanese pear (*Pyrus serotina* Rehder). Buwalda and Meekings (1990) found the nutrient content of deciduous parts (fruits and leaves) to be equivalent to 52 kg N, 10 kg P, 90 kg K and 8 kg Mg per hectare, and considered that these figures would be exceeded in commercial orchards which have 15–50% higher yields than the experimental one. These figures for nutrient removal are very much lower than those, for example, of cereals and vegetable crops. Greenham (1980) quotes N, P and K removal, in kg ha^{-1}, of 180, 26 and 163 for wheat and 285, 45 and 450 for carrots, respectively.

The annual requirement for fertilizer application depends on total requirements and on the natural supply from the soil, both of which are variables. Estimation of the balance of these is made almost impossible by the complexity of the factors involved (Klein and Weinbaum, 2000). Soil analysis pre-planting and at regular intervals is very important in order to identify potential needs and problems but changes in soil nutrients do not, in general, provide a guide for fertilizer practice. This is especially so because of the difficulty in determining the effective tree rooting zones, the importance of carry-over of nutrient reserves in the tree from one year to the next and the importance of the supply of nutrients to specific organs rather than gross uptake. Leaf and fruit nutrient concentrations have, however, been found to reflect nutrient status and requirements. Desirable levels established empirically are summarized in Table 11.2. These figures are based on much original

Table 11.2 *Leaf nutrient concentrations in apple and pear*

	Apple		Pear	
Nutrient element	Deficient	Normal	Deficient	Normal
N%	<1.5	1.7–2.5	<1.8	1.8–2.6
P%	<0.13	0.15–0.3	<0.11	0.12–0.25
K%	<1.0	1.2–1.9	<0.7	1.0–2.0
Ca%	<0.7	1.5–2.0	<0.7	1.0–3.7
Mg%	<0.25	0.25–0.35	<0.25	0.25–0.90
Mn ppm	<25	25–150	<14	20–170
Fe ppm		40–500?		100–800
B ppm	<20	20–60	<15	20–60
Cu ppm	<4	5–12	<5	6–20
Zn ppm	<14	15–200	<16	20–60
Mo ppm	<0.05	0.10–0.20		

Apple leaves taken from the mid-portion of terminal shoots at about the time of cessation of terminal growth (July). Pear leaves are spur leaves in August except for Zn for which samples were taken in September. Mn concentrations of more than 150 ppm in 1-year-old bark of apple indicate toxicity but this critical level may vary depending on the level of Ca, B and other elements.
Data from Shear and Faust (1980).

data summarized by Shear and Faust (1980). Rowe (1980), citing data for nutrient levels in mid-extension shoots taken in late August, concluded that rather higher levels of N and K (2.4–2.8% and 1.4–1.7%, respectively) are needed to show sufficiency and that more than 100 ppm of Zn or Mn indicates toxic levels.

The rootstock can influence the susceptibility of the tree to nutrient deficiencies. Apple trees on 'MM.106' are highly susceptible to magnesium deficiency, trees on 'M.9', 'M.26' and 'MM.111' are susceptible and those on 'M.2', 'MM.104' and 'M.16' are resistant (MAFF, 1972). Pear trees on quince rootstocks on alkaline (high pH) soils generally suffer from lime-induced chlorosis resulting from iron deficiency (Lombard and Westwood, 1987), although some new quince selections are tolerant (Webster, 1997). Pear trees on *Pyrus* rootstocks are generally tolerant.

Scion cultivars may have different nutrient requirements depending on the quality characteristics required. High nitrogen concentrations promote dark green skin colour, which is desirable for cultivars like 'Bramley's Seedling' but not for cultivars like 'Worcester Pearmain'. Different cultivars also have large differences in fruit acidity, which relates to potassium content so indicates likely differences in K requirement.

Symptoms of deficiency or excess

Nitrogen (N)

Nitrogen deficiency is characterized by reduced top growth with short upright spindly shoots bearing pale yellowish-green leaves. The symptoms become more severe as the season progresses and fruits are usually smaller than on trees with adequate N concentrations. Excessive N content reduces the area and intensity of red colour on red cultivars of apple, delays maturity, shortens storage life and increases susceptibility to some storage disorders, especially those associated with low Ca concentrations. Trees with excessive N levels show very vigorous growth and dark green leaves.

Phosphorus (P)

Phosphorus deficiency is uncommon in most fruit-growing areas. It is evidenced by very dark green leaves with some purplish discoloration on the lower sides, especially on the margins and main veins. Active vegetative growth is inhibited. Excess P leads to interference with zinc, copper, iron or manganese uptake and may be one cause of symptoms of deficiency of these.

Potassium (K)

The most characteristic symptom of potassium deficiency is a reddish-brown scorching of leaf margins, often preceded by a greyish-green discoloration. Excess K may increase the fruit symptoms of Ca deficiency and the leaf symptoms of Mg deficiency.

Calcium (Ca)

Calcium deficiency is shown by the upward cupping of the margins of the youngest leaves. The expanding leaves develop uniform veinal and interveinal chlorosis, and the very youngest leaves may become entirely chlorotic. Margins of older leaves may shatter and terminal shoots die back. Bitter pit may develop as slight indentations in the skin in the orchard or in store. These areas turn brown, and desiccated tissue develops below the spots. Cork spot develops early as small blushed areas of skin above hard brown spots. Lenticel breakdown shows as pale areas, later white halos that turn brown or black, around the lenticels. Sunburn on fruits may be more pronounced when they are Ca-deficient. Calcium effects are discussed in much more detail in Chapter 10.

Magnesium (Mg)

Magnesium-deficient leaves of apple show a colour-fading around the margins followed by interveinal chlorosis from the margins inwards leading to a 'herring-bone' pattern. Tissues around the margins, especially towards the tips, may die and the older leaves, which show the symptoms first, may shed. In pear the interveinal tissues on both sides of the midrib may develop as dark purplish islands surrounded by chlorotic bands while the leaf margins remain green. Fruits may drop early. Excessive Mg can result in the appearance of Ca or K deficiency symptoms.

Manganese (Mn)

Manganese deficiencies tend to occur on soil with a high lime content, also where there is a high water table or very light soil. Chlorosis between the main veins and extending towards the midrib is typical. Manganese toxicity is commonly found on very acid soils. It is characterized by the development of 'measles' in which pimples erupt on the bark of 2-year-old shoots. These pimples enlarge and erupt over the years, producing sunken areas surrounded by callus. The sunken areas then coalesce, the bark becomes rough, cracked and scaly and the branch may die.

Iron (Fe)

Iron deficiency is most prevalent where the lime content of the soil is high or where a drainage problem exists. The youngest leaves on Fe-deficient trees are very pale between the veins, ranging from pale green to white, but the veins remain green or become green even if initially paler. Dieback of shoots and branches may occur. Excessive Fe, though rare, can result in Mn deficiency.

Boron (B)

The fruits of boron-deficient apple and pear trees are misshapen with dark green depressed areas underlaid by hard corky tissue which turns brown on exposure to air. Cork may also develop inside the fruit. Apple fruits may crack, especially if Ca content is also low. Sometimes a multiplicity of small cracks may callus over, giving a russeted appearance. Shoot tips may die and weak new shoots emerge from below the tips. Leaves on B-deficient trees are dark green, thick and brittle and shed early. In pears the wilting and death of blossoms (blossom blast) may indicate B deficiency. This may also contribute to the development of 'measles' on apple shoots. Excessive B can result in early maturity and short storage life in apples.

Copper (Cu)

The first symptom of copper deficiency is necrosis of the terminal leaves of actively growing shoots in mid-summer. In pear the leaf tips turn black; in apple the young leaves develop reddish necrotic spots before the scorched tips and margins turn brown. The shoots begin to die back from the tips, the scorched leaves shed and the drying tips curve downwards. Multiple shoot emergence from below the affected part gives a bushy appearance. Later a bark necrosis with rectangular cracks may develop. Cu deficiency tends to be associated with specific soils in limited areas.

Zinc (Zn)

Zinc deficiencies are fairly widespread and may be severe in some environments. The initial symptom is a chlorotic mottling of the leaves. This is followed by a rosetting or tufting of the leaves on the terminal growth. There is little or no development of leaves on 2-year-old wood. Those leaves that develop, either in the rosettes or laterally on the shoots, are small and strap-like as well as chlorotic, hence the common name 'little-leaf disease'. Crops on Zn-deficient trees are reduced and the fruits are smaller than normal. An excess of Zn usually appears as Fe chlorosis.

Aluminium (Al)

Aluminium is not thought to be essential for apples or pears. It is, however, very toxic especially below pH 4.7 leading to root malformation, malfunction and death. The leaf and fruit symptoms are those of Ca deficiency.

Nitrogen nutrition

The nitrogen available to fruit trees comes from the mineralization of soil organic matter and from atmospheric deposition in rain. The balance needed to meet demand is provided by fertilizers, applied to the soil or to the leaves. Fertilizer practice is often governed by the need to compensate for N removal by the orchard surface cover crop, usually grass, and also by the requirement for additional N in the tree at flowering time. Fertilizer N is most usually supplied in inorganic form as nitrates or ammonium compounds or in the organic form as urea.

Under English, humid temperate, conditions on clean cultivated unfertilized orchard soil the amount of mineral nitrogen may increase by 50–70 kg N ha^{-1} yr^{-1} and in a Danish orchard soil annual nitrate production was 56 kg N ha^{-1} (Greenham, 1980). The annual supply of N in rainfall in Southeast England may be in the range of 8–17 kg ha^{-1}. Under these circumstances there may be little or no need for, or response to, N fertilizers. Indeed the practice of 'grassing down' orchards was introduced to remove soil N and improve fruit firmness and red colour.

In soils low in organic nitrogen trees commonly respond to fertilizer N. Many such soils are found in areas of low rainfall and the N required is often supplied in the water delivered through drip or trickle irrigation systems.

Uptake and transformations in the roots

Movement of nitrogen to the root surface is thought to be mainly by mass flow together with water, with diffusion playing a smaller part (Neilsen and Neilsen, 1997). The uptake of both nitrate and ammonium is continuous throughout the year in mild climates, with a relatively high peak during the summer (Titus and Kang, 1982). This pattern, associated with leaf development, may indicate a dependence of N uptake on transpiration. Organic N compounds such as urea are also readily absorbed by apple roots.

Some nitrate, especially under conditions of high nitrate supply, may be transported as such to the aerial tissues of apple and nitrate reductase has been found in apple and pear leaves. Most nitrate is, however, normally reduced to ammonium in the roots, with the greatest concentration of nitrogen reductase in the fine roots. The processes of reduction of nitrate to nitrite and nitrite to ammonium require energy from the respiration of carbohydrates.

Ammonium ions, whether produced by nitrate reduction or absorbed directly from ammonium fertilizers, are then metabolized to synthesize amino acids. It is thought likely that, as in other higher plants, glutamine is the first product of NH_4^+ assimilation, catalysed by glutamine synthase (GS). Glutamate is then formed from glutamine in the presence of glutamate synthase (GOGAT) using 2-oxoglutarate. Aspartate synthesis from glutamate and oxaloacetate is catalysed by glutamate–oxaloacetate transaminase (EC 2.6.1.1). Asparagine is produced by the transfer of amide N of glutamine to aspartate. Arginine is synthesized from glutamate. These steps in synthesis are considered in detail by Titus and Kang (1982).

Upward movement and accumulation in leaves

Upward movement of nitrogenous compounds is apparently mainly in the xylem. The nitrogen in the xylem sap consists of a range of amino acids, with aspartate and glutamate and their amides making up to 90% of total N (Bollard, 1957, 1960; Hill-Cottingham and Bollard, 1965; Tromp and Ovaa, 1967). There is some evidence of radial movement of N compounds from the xylem to the phloem but this is slower than the upward movement.

The leaves act as an active sink for nitrogenous compounds. The greater the N supply the higher the proportion found in the leaves. Millard and Neilsen (1989) found only 33% of the N content of unfertilized apple rootstocks to be in the leaves, whereas 43% and 52% were found in plants given progressively more N. The proportion in roots showed the reverse pattern. Amino acids transported to the leaves are synthesized into proteins, primarily into the photosynthetic enzyme ribulose 1, 5-bisphosphate (RUBP) carboxylase. In mid-July this can amount to more than 50% of total protein in apple leaves (Kang and Titus, 1980), while more than 90% of total amino acid and protein N is in proteins.

Export from senescing leaves

Much leaf nitrogen is exported from the leaves before their abscission. When abscission is induced by low temperatures, 70% of the initial leaf N is lost before leaf shed (Shim *et al.*, 1972). Millard and Thomson (1989) found a similar percentage loss from unfertilized 'MM.106' rootstock leaves between 1 September and 1 November, with a smaller relative loss from leaves of plants well supplied with N. Earlier studies showed the amount of N lost during senescence to be from 23% to 50% (Kang and Titus, 1980). Leaf N begins to decline from the onset of senescence, the time varying with availability of nutrients, crop load and climatic conditions, especially temperature. The loss of leaf N results from loss of leaf protein. Up to 87% of the soluble protein lost was identified as being ribulose 1, 5-bisphosphate carboxylase–oxygenase (RuBPC/O), there being preferential loss of this compared with other soluble leaf proteins (Millard and Thomson, 1989). The measured decline in protein represents the difference between synthesis and degradation over the period of senescence. It is generally accepted, though without direct evidence, that amides are the major forms of nitrogenous compounds transported from senescing leaves. This leaf N migrates back into spurs and branches but is eventually translocated to the older wood and root system (Murneek and Logan, 1932, cited by Titus and Kang, 1982).

Storage over winter

After the cessation of shoot extension in summer there is a gradual increase of nitrogen in wood and bark of all parts of the apple tree including roots (Mason and Whitfield, 1960). The autumn application of N in late October and early November increases the N content mainly in the roots during the late autumn and winter (Tromp, 1970). The seasonal course of N reserves in apple roots (Tromp, 1983) is shown in Figure 11.1. The main storage in the roots is in a soluble form with arginine as the main component of the soluble N fraction. The amides, especially asparagine, also appear to be important sources of N reserves, especially in the roots (Hill-Cottingham and Cooper, 1970; Cooper *et al.*, 1976).

Both bark and wood of stems also store nitrogen over the winter, with bark storage probably being the most important with respect to remobilization (Mason and Whitfield, 1960). Much storage is in the form of proteins.

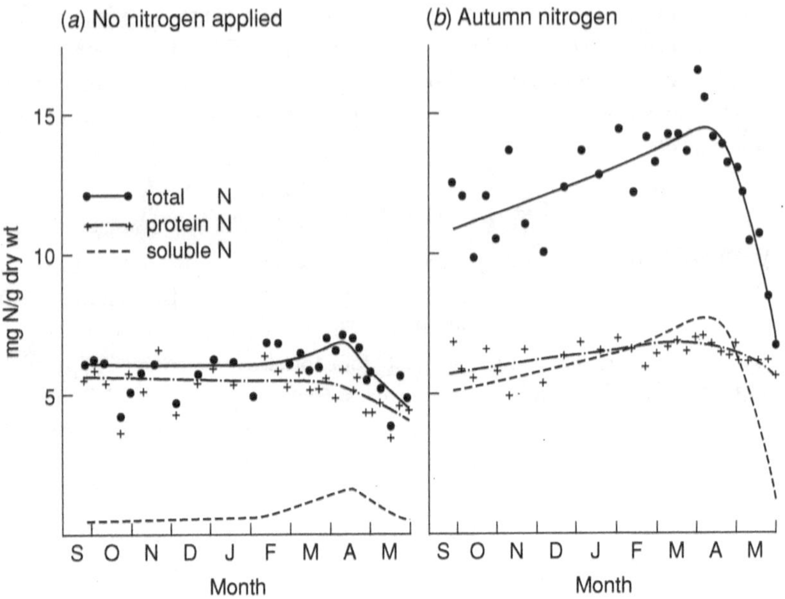

Figure 11.1 Total protein and soluble N in roots of 1-year-old 'M.7' apple rootstocks throughout autumn, winter and spring. (*a*) Unfertilized, (*b*) N applied at the end of September. The difference between total N and protein N is considered as soluble N. From Tromp (1983). Reproduced with permission.

Mobilization in spring

In orchard trees protein levels in shoot bark and wood decrease rapidly before the beginning of budbreak in spring (Hennerty *et al.*, 1980). The decrease in proteins is accompanied by increased levels of soluble N in bark and wood tissues, especially amino-nitrogen and arginine. The movement of soluble N from the roots in spring is indicated by Figure 11.1.

This N mobilized from storage is critically important for new shoot development, occurring as it does when soil conditions for N uptake are usually poor. There is a highly positive correlation between amounts of storage N and the extent of new shoot growth in both apples and pears in the following spring (Harley *et al.*, 1958; Taylor *et al.*, 1975). The amount of storage N remobilized depends on the amount of N in store and is unaffected by the concurrent supply of N from the soil (Millard, 1995). Ringing experiments have shown that upward translocation of previously stored N over the 6 weeks from budbreak, is in the phloem (Tromp and Ovaa, 1971). The major soluble amino compounds transported are asparagine and arginine.

Foliar application of nitrogen

Above-ground plant parts have a basic capability to absorb mineral nutrients and water although this may be reduced by barriers to water loss. The usefulness of absorption of nutrients by leaves may, however, be limited by lack of ability to direct them subsequently. For example, Ca absorbed by leaves from foliar sprays is not translocated to the fruits where it is needed. The value of foliar sprays of nitrogenous compounds was greatly enhanced when it was realized that N supplied to leaves just prior to leaf fall might move into storage tissues as part of the natural cycling of N within the trees (Oland, 1960, 1963; Shim *et al.*, 1972). Nitrogen applied to leaves post-harvest prolongs their period of photosynthetic activity but does not give unwanted further shoot growth or fruit enlargement and softening. It can also be applied at much higher concentrations, giving more uptake, than if applied to buds, flowers and developing fruits which might show phytotoxicity. Foliar applications of urea after harvest of apples and pears has been found to add to tree reserves of N (Shim *et al.*, 1972; Sanchez *et al.*, 1990) and to have generally appreciable effects on apple fruit set and shoot growth in the following spring. Effects on blossom fertility have been discussed in Chapter 9.

Apple leaf cuticles are traversed by polysaccharide microfibrils which may facilitate penetration, and the cutin itself is not totally impermeable (Swietlik and Faust, 1984). Nutrients may enter relatively easily through stomatal pores, in which the cuticle is hydrated and wax-free, and trichomes. Cuticles are 10–20 times more permeable to urea than to inorganic ions (Yamada *et al.*, 1965b). Urea is absorbed more rapidly by intact leaves than any mineral nutrient, and facilitates the penetration of other nutrients through isolated cuticles and into intact leaves (Yamada *et al.*, 1965a).

After passing through the cuticle there is some evidence for passage of urea through the cellulose cell walls by way of thread-like structures called ectodesmata (Swietlik and Faust, 1984).

Absorption by the lower surface of apple leaves is very rapid within the first 24 hours and then levels off, whereas the upper surface absorbs urea more slowly and steadily (Boynton *et al.*, 1953). Shim *et al.* (1972) found 75% of a 5% solution of urea applied to senescing apple leaves was absorbed in 24 hours.

Leaf-absorbed N must be metabolized before it can be utilized. This involves the hydrolysis of urea, reduction of nitrate and incorporation of ammonium into amino acids. Absorbed urea can be broken down in the leaves by urease (Oland, 1960). The N is then incorporated into amino acids and proteins (Shim *et al.*, 1973a, b). The proteins in turn are converted to amino acids, transported to the storage tissue and re-assembled into proteins. The urea may also be exported as such. This export can take place very quickly. Boynton *et al.* (1953) reported that about 50% of the urea N absorbed by apple leaves on shoots was translocated out of the leaves within 24 hours.

About 62% of leaf-absorbed N from autumn application of foliar urea to apple trees is recovered in permanent tissues during dormancy and is evenly distributed among root and stem tissues of the stock and scion (Hill-Cottingham and Lloyd-Jones, 1975). Sanchez *et al.* (1990) found that post-harvest foliar application of urea to 'Comice' pear increased the concentration of N in the following winter and spring in the bark and wood of one-year-old shoots and in flower buds and blossoms.

Foliar sprays of urea are widely used as a supplementary source of nitrogen. Sprays of 0.5–1.2 kg urea per 100 l water may be applied before or after blossom or two to three times during the growing season beginning after flowering. Post-harvest spray with 5 kg urea per 100 l may be applied prior to leaf fall.

Phosphorus nutrition

The net uptake of phosphorus by apple trees is much lower than that of nitrogen, potassium and calcium (Table 11.1), and the concentration of P necessary in apple and pear leaves to avoid deficiency symptoms is much lower than that of the other major nutrient elements (Table 11.2). Notwithstanding the key rôle of P in the DNA and RNA macromolecules, in energy transfer involving ATP and in many enzymic processes, phosphorus deficiency is seldom observed in the orchard.

Concentrations of P in the soil solution are usually very much lower than those of N, K, Ca and Mg (Robson and Pitman, 1983). These elements are present at median values of 1500, 1300, 1500 and 4000 μM respectively whereas the corresponding value for P is 1 μM. More than for any other element, the uptake of P has to be a metabolically driven process. Its concentration in xylem sap is around 400 times higher than that in the soil solution (Bieleski and Ferguson, 1983) and its concentration in plant cells is up to 10 000 times as high.

Bhat (1983) found that in both young and mature apple trees P influx increased approximately linearly with concentration in the external solution up to about 10 mmol m^{-3}. At external concentrations below 0.25–0.5 mmol m^{-3} uptake is negligible. Uptake rates are much higher in summer than in spring (Asamoah, 1984). Harley *et al.* (1958) found that ^{32}P supplied in early spring was not detectable in leaves for at least a month and ringing experiments showed it to be translocated in the phloem. The phosphorus used in early leaf and shoot growth appears to come from reserves, mainly those in bark. Bark P content falls steeply in April, simultaneously with a fall in N, whereas branch wood concentration shows a slower decline and the concentration in trunk wood is fairly constant (Mason and Whitfield, 1960). The concentration of P in the bark, roots, and the previous year's extension wood rises sharply in October when P is exported from the leaves.

During summer the xylem sap contains appreciable amounts of P (Hansen, 1980) and it seems that both xylem and phloem transport are involved in the supply of P to the fruits. Accumulation of P in leaves reaches a maximum in July. That in fruit continues up to harvest (Wilkinson and Perring, 1964), closely following the weight increase of the fruit. Low fruit content of P predisposes apple fruits to low temperature breakdown in storage. Spraying the fruits with P compounds between mid-June and mid-July increases their P content at harvest and reduces the incidence of low temperature breakdown, senescent breakdown and, in some years, superficial scald and coreflush (Johnson and Yogaratnam, 1978; Yogaratnam and Sharples, 1981). The beneficial effects are, however, fairly limited and sprays with KH_2PO_4 may increase fruit K content and have adverse effects on bitter pit.

As noted in Chapter 3 (root systems), apple tree roots are mycorrhizal. Factors that reduce mycorrhizal infection, e.g. the use of herbicides overall in the orchard, can reduce P uptake (Atkinson, 1983) and the concentration of P in fruits (Johnson *et al.*, 1983).

Potassium nutrition

Apple trees have high potassium requirements in terms of both gross and net uptake (Table 11.1). In England young trees frequently show K deficiency symptoms which must be remedied if growth is not to be stunted. Subsequent fertilizer requirement depends largely on cropping levels. Excess soil potassium leads to inhibition of calcium and magnesium uptake. It also leads to high fruit K and increased problems if Ca concentrations are low. In many fruit-growing regions little or no K fertilization is needed. Possibly as a result of previous fertilizer practice, up to around 40% of orchards in Santa Catarina, Brazil showed above-normal leaf K (Basso and Wilms, 1988). A survey in Germany found 90% of orchard soils to contain levels of K which were excessive with regard to susceptibility to physiological disorders of apple fruit (Quast, 1980).

Potassium movement through the soil to the root surface is primarily by diffusion. The soil water content is a dominating factor for this and K uptake can be severely limited by drought. Although passive as well as 'active' processes participate in the movement of potassium within plants, the overall rate of K uptake by apple trees appears to be under metabolic control. Tromp (1980) subjected apple trees to a range of environmental conditions and found that K uptake was linearly related to growth, i.e. to metabolic demand. The main transport vehicle for potassium appears to be the transpiration stream in the xylem, but it also moves freely in the phloem. Fruits are very strong sinks for potassium (Hansen, 1980) and there is a strong positive correlation between fruit K and soluble dry matter or acid content (Wilkinson, 1958; Perring and

Preston, 1974). Potassium plays an important role in cell expansion and other phenomena, e.g. stomatal behaviour, which depend on cell osmotic potential. It is also the most abundant cation in the cytoplasm with important roles in enzyme activation, pH stabilization, protein synthesis, etc.

Calcium nutrition

This is dealt with in Chapter 10.

Magnesium nutrition

Apple trees have a greater demand for magnesium than many other fruit trees and readily show Mg deficiency symptoms. These can arise because of Mg deficiency in the soil or because uptake is depressed by competing cations. Of these K^+ is usually the most important but H^+ (low pH), NH_4^+, Ca^{2+} and Mn^{2+} also compete. In aerobic soils of neutral pH the Mg content of the soil solution is usually quite high (Marschner, 1995) and mass flow to the root surface should give an adequate supply (Robson and Pitman, 1983). It is said not to be transported in the symplast (Luttge, 1983), but unlike Ca it reaches much higher concentrations in the phloem than in the xylem (Robson and Pitman, 1983) and may be re-transported within the plant. During April and May when leaf growth is rapid, the amount of Mg in leaves increases sharply with corresponding withdrawals of the element from the bark and wood of shoots, branches and rootstock (Mason and Whitfield, 1960). The biggest withdrawal is from wood. Developing fruits are then able to withdraw Mg from neighbouring leaves (Shear and Faust, 1980). According to Forshey (1963), 37% of the Mg absorbed after foliar application is exported to permanent woody tissues and roots. Oland (1963), however, found that there is no change in leaf Mg content during senescence and Wittwer and Teubner (1959) classed Mg as being an immobile element. The complexity of results reported may reflect different modes of transport in relation to different processes. A high proportion of Mg in plants, often over 70%, is diffusible and is associated with inorganic anions and organic acid anions such as malate and citrate. About 10–20% of Mg is in the chloroplast. Less than half is bound to chlorophyll; the rest serves as an activator of ribulose-bisphosphate carboxylase. It is also important for protein synthesis and in the transfer of high energy phosphate in ATP metabolism (Faust, 1989). Magnesium is supplied to fruit trees as kieserite or as magnesian limestone to correct initial soil inadequacy. Where there are leaf symptoms three or more foliar sprays at two-weekly intervals with 2% magnesium sulphate (Epsom salts) beginning at petal fall can alleviate these, and can increase fruit set and fruit size and reduce fruit drop (Greenham and White, 1959; Ford, 1968). The residual effect of such sprays is minimal and soil application of $MgSO_4$ or of dolomitic (magnesian) limestone is essential for a longer term solution, especially on soils high in potassium.

Manganese nutrition

Although manganese deficiencies can occur, Mn toxicity is a much more widespread problem.

Manganese is usually found only in very low concentrations in the soil solution (<0.02 μM) where, like Fe, Cu, Zn and Co, it is present mainly in a complexed form (Robson and Pitman, 1983). The concentration of available Mn (Mn^{2+}) in the soil solution is dependent on the soil pH and its oxidation–reduction potential. Mn availability decreases very sharply as the pH rises to a certain level, below the neutral point, and deficiency can be induced by the application of large amounts of agricultural limestone. It may also occur on naturally alkaline soils. The necessary pH to avoid deficiency varies with soil type (Beyers and Terblanche, 1971 a). Exudates from plant roots may solubilize amorphous Mn oxides, bacterial activity in the rhizosphere may affect the solubility of these and use of NH_4^+ instead of NO_3^- fertilizers can also increase Mn availability.

Transport of Mn within the plant is in both the xylem and the phloem, in the cationic form in the former and largely in the cationic form in the latter (Robson and Pitman, 1983).

Manganese deficiency often occurs simultaneously with zinc deficiency, the symptoms shown reflecting the element in greatest deficiency. It is often controlled adequately by routine sprays of Dithane M-45 fungicide, which contains Mn. Additional control can be achieved by $MnSO_4$ sprays, either in the dormant season or as foliar applications. Soil treatment with Mn compounds is not effective on alkaline or heavily limed soil unless applied in large quantities, e.g. 3 kg $MnSO_4$ plus compost in holes around each tree. The soil can, however, be acidified by use of acid fertilizers or sulphur to reach desirable pH levels.

Manganese toxicity is expressed as leaf chlorosis, early leaf abscission, reduced growth and flower bud formation and internal bark necrosis (IBN). If IBN develops early in the life of the tree, it will fail to become productive (Ferree and Thompson, 1970). Toxicity is associated with very high Mn concentrations, up to 500 ppm, in the affected part of the bark and seems to be associated with high Mn availability under low Ca supply conditions (Domoto and Thompson, 1976). Calcium influences both Mn absorption and translocation (Fucik and Titus, 1965). These effects may involve ion exchange processes independent of pH. However, under conditions where Mn toxicity is likely to occur pH should be adjusted to 6.5 before planting. Under conditions of low Ca and high Mn the degree and incidence of IBN increase with increasing K concentrations (Domoto and Thompson, 1976).

Copper nutrition

Copper, in minute quantities, is essential as a constituent of plastocyanin, which functions as an electron donor in photosystem-1. Enzymes containing Cu include a number of oxidases. It is therefore involved in lignification and the development of xylem vessels (Marschner, 1983). Copper deficiency, although rare, can lead to severe shoot dieback and withering of the terminal portions to the extent that a normal tree framework cannot develop. Affected trees bear few fruits and, commonly, no crop at all (Beyers and Terblanche, 1971b).

Copper is generally found at typical trace element concentrations (<0.01–0.6 μM) in the soil solution, where it is largely (89–100%) complexed with low molecular weight organic compounds (Robson and Pitman, 1983; Faust, 1989). It is also found in complexed forms in the xylem. In other plant species it can be retranslocated from old to young leaves in parallel with nitrogen, so this may also be the case in fruit trees.

Copper deficiency tends to occur in acid sandy soils and occasionally also in black alluvial soil with a high humus content. Pre-planting application of organic matter and pH adjustment reduces the incidence of copper deficiency. Young trees can be sprayed in spring with copper oxychloride or copper oxysulphate at a rate of 250 g per 500 l of water. A second application may be needed a month later. Bearing trees are sprayed at the green-tip stage at 2 kg of either of the same compounds per 500 l. Sprays during the growing season can cause fruit russet. Post-harvest leaf sprays or dormant season pre-blossom sprays can also be used. Potential Cu deficiency may be prevented by the use of fungicidal sprays containing copper. Soil dressings by broadcasting are not recommended but a mixture of compost with up to 500 g $CuSO_4$ applied in holes around the tree may be effective.

Boron nutrition

The most important effects of boron on apple and pear trees are on fruit set and, secondarily, on fruit quality with the development of external and internal cork formation and cracking in B-deficient fruits. The B content of flowers of apple and pear is high (Crassweller *et al.*, 1981; Johnson *et al.*, 1955). If the necessary high levels are not attained then, especially in pears, the flowers wilt and die but persist on the tree (Batjer *et al.*, 1953). The incidence of B deficiency tends to be a problem at a regional level associated with soil characteristics, for example on acid, sandy soils with low humus content in South Africa (Beyers and Terblanche, 1971c) or on readily-leached soils in high rainfall areas.

The concentration of B in soil solutions ranges very widely, from 3 μM, a true trace element level, to 1000 μM. It is present in the inorganic form as H_3BO_3 and is probably taken up and found in the xylem in this form (Robson and Pitman, 1983). It is passively absorbed by roots into the free space (Tanaka, 1967) and forms complexes with polysaccharides. There is some active transport into root cells but it is thought that the net uptake of boric acid is influenced by the transpiration rate and transport in the xylem is probably directly proportional to the rate of transpiration (Raven, 1980). Water stress due to drought and scarcity of irrigation water can be followed by boron deficiency symptoms even on soils with sufficient B.

The B content of leaves and fruits increases throughout the season in apples and there is a straight-line relationship between fruit weight and B per fruit, indicating continuous B transport (Van Goor and Van Lune, 1980).

Accumulation in the bark follows a similar pattern to that in leaves and fruits and is reduced in dry summers (Johnson *et al.*, 1955). The supply of B to the flowers comes from reserves in the wood of the branches bearing them (Callan *et al.*, 1978) and the B content of flowers (of 'Italian' prune) is increased much more by B sprays in the previous autumn than by spraying immediately pre-bloom.

There is very little safety margin for B application. The normal range for apple and pear leaf content of B is 20–60 ppm (Shear and Faust, 1980) but toxicity can be shown at 70 ppm in apple (Faust, 1989). A soil application of borax or of Solubor (disodium actoborate tetrahydrate, containing 20% B) is effective on acid, sandy soils but may incur the risk of toxicity (Beyers and Terblanche, 1971c). Solubor sprays at a concentration of 500 g per 500 l for apple and pear can be effective at any time and should be applied at the first symptoms of deficiency. Blossom-time sprays should be followed up four weeks later. Autumn sprays, applied when symptoms such as external corking of fruits appear late in the season, should be followed by post-harvest sprays before leaf drop. This has a good residual effect in the following spring. An adequate supply of compost or manure, which contains a small quantity of boron, good drainage and irrigation to maintain optimum soil moisture, including after harvest, can prevent deficiency arising.

Iron nutrition

Although iron constitutes about 5% by weight in the earth's crust and is only found in trace element quantities in apples and pears, iron deficiency symptoms can be found in both these crops. This is mainly because the activity of soluble iron in soils is very low, particularly in calcareous soils or where drainage problems exist. Iron deficiency leads to chlorosis and loss of photosynthetic efficiency (Sun *et al.*, 1987). It cannot reliably be diagnosed from the Fe content of the leaves (Korcak, 1987) but can be proved by response to foliar-supplied Fe and diagnosed and predicted using a test based on peroxidaze enzyme, an iron haemoprotein. Pears are much more subject to iron chlorosis than apples. *Pyrus communis* rootstocks are more tolerant than *P. calleryana*, *P. ussuriensis* or quince (Korcak, 1987), although some quince rootstocks are said to be tolerant of high pH soils (Webster, 1997). The pear cultivars 'Bartlett', 'Nelis', and 'Comice' are more subject to iron chlorosis than 'Hardy' and 'Clairgeau'.

Iron can exist at either the ferric (Fe^{3+}) or the ferrous (Fe^{2+}) level. The oxidized ferric form gives highly insoluble ferric hydrate precipitates especially above pH 7. The activity of Fe^{3+} decreases 1000-fold for each unit increase in pH. Well-aerated acid soils have much more soluble Fe than calcareous soils. In calcareous soils the bicarbonate ion, rather than $CaCO_3$, content is the most important factor associated with lime-induced chlorosis and the bicarbonate ion level is increased by high soil water content. Iron deficiency chlorosis problems can arise when roots initially growing in surface soil penetrate underlying calcareous zones.

Before plants can utilize Fe^{3+} this must be solubilized and then reduced to Fe^{2+}. Solubilization is thought to be by malate as an organic acid supplied by the roots into the rhizosphere (Korcak, 1987; Sun *et al.*, 1987). Reduction to Fe^{2+} is thought to be exocellular. The uptake requires energy and needs both O_2 and photosynthate. It also responds to plant Fe status: roots under iron stress take up 7–10 times more Fe than unstressed roots. In barley, uptake is by the root zone 1–4 cm from the tip. With blueberry, it has been shown that fertilizer supply of NH_4-N gives a more acid rhizosphere than NO_3-N and this is thought to be relevant to Fe uptake (Korcak, 1987).

Iron is transported upwards in the xylem in a complexed form as a ferric-citrate chelate. Once in the leaf iron moves rapidly across the plasmalemma into the symplast via nicotianamine, a divalent cation chelator. It is found in a complexed form in the phloem at higher concentration than in the xylem (Robson and Pitman, 1983).

Temporary control of iron deficiency can be achieved by trunk injection with ferrous sulphate or ferric citrate. $FeSO_4$ applied to the roots is also useful and can be provided as a mixture of manure and ferrous sulphate inserted in holes around the tree. Chelated (EDTA) iron is effective on acid soils, with FeEDDHA being more effective on calcareous soils because of its greater stability at high pH. Chelated Fe can be used in much smaller quantities and gives higher foliar leaf levels than unchelated iron. Although there are a number of problems in its use due to fixation and adsorption (Korcak, 1987), soil application of FeEDDHA chelate is considered to be the best way to correct lime-induced Fe chlorosis (Beyers and Terblanche, 1971 d; Swietlik and Faust, 1984). Iron chelate can also be used as a foliar spray at petal fall and three or four times subsequently at 2-weekly intervals. If the Fe deficiency is associated with poor drainage, attention should be given to this.

Zinc nutrition

Apple and pear demand for zinc is very low and unless deficiency symptoms are shown application of additional zinc usually has no demonstrable effects. In those circumstances where Zn is deficient there is a strong positive relationship between Zn content of leaves within the range of 5.9 to 14.4 ppm dry weight and apple, cv. 'McIntosh', yield (Stiles, 1966). Treatments to ameliorate Zn deficiency are essential in a number of important areas of apple production, e.g. southern British Columbia, Canada. Susceptibility varies between cultivars, the low-chilling-requirement apple 'Anna', which is very widely grown in the tropics and subtropics, can become totally unproductive under conditions where it is the only cultivar to show zinc deficiency symptoms. The most important functions of Zn concern enzyme activation, carbohydrate metabolism, cell membrane integrity and auxin synthesis (Swietlik, 1999). The precise mechanism of the dramatic effects of Zn deficiency on shoot extension and foliage development is not established.

Zinc in soils is present (*a*) in the soil solution, (*b*) in an adsorbed exchangeable form associated with alumino-silicates, hydrous oxides of Al, Fe and Mn and solid organic matter, (*c*) associated with organic matter, e.g. by incorporation into organic molecules and by chelation, (*d*) in association with hydrous oxides and carbonates and (*e*) in soil minerals, e.g. silicates (Swietlik, 1999). The causes of deficiency, and the responses to soil treatments directed towards its correction, are correspondingly complex. The following seem to be most important.

1 The amount of zinc available in soils depends very much on the Zn content of the parent material. It is low in soils derived from granite and basalt (Tagwira, 1995) and in soils containing little clay or organic matter. Consequently Zn can be very low in sandy soils, including those derived from quartz which is low in this element, even when these are acidic.
2 In most circumstances water-soluble Zn, and thus total available Zn, decreases dramatically with increasing pH: there can be a hundred-fold reduction in Zn activity for each unit of pH increase (Swietlik, 1999). Soil acidification can increase leaf Zn under these circumstances. Availability is generally low in high-pH calcareous soils and deficiency can follow repeated heavy liming.

3 Biochemical substances in the rhizosphere increase Zn availability. Organic amendments generally increase bioavailability although there are exceptions, e.g. HCO_3^- from decomposing organic matter immobilizes Zn. Although total Zn is usually distributed universally with soil depth, extractable Zn declines with depth.
4 High levels of P in the soil can induce Zn deficiency symptoms without reducing total tissue Zn.

Zinc is predominantly taken up as a divalent cation Zn^{2+} at lower pH and as a monovalent cation $ZnOH^+$ at higher pH. Its uptake may be inhibited by Ca^{2+} and other divalent cations. It is bound in the apoplast as well as being mobile there. Zn is transported in xylem both as a free cation and in complexes with citric or malic acid. It is also found in the phloem, where the concentration is much higher (Robson and Pitman, 1983). It is classed as having intermediate mobility within the plant, but ^{65}Zn isotope applied to the leaves does not appear to move from them although there is some evidence for movement of Zn from roots and stems and remobilization of Zn from the flag leaf of wheat (Swietlik, 1999).

There are three main techniques for applying Zn to fruit trees: soil application, foliar or dormant-season spray application and trunk injection.

Soil treatments with zinc sulphate ($ZnSO_4$) can be effective (Neilsen and Hoyt, 1990; Swietlik, 1999). They may, however, need to be accompanied by soil acidification or to be applied in concentrated bands, piles or peat plugs. Toxicity may be induced. ZnEDTA (chelate) may be more effective but in some soil types the compound is converted to FeEDTA.

Some fungicides such as Dithane M45 and Zineb contain enough Zn for a regular fungicidal spray programme to control incipient deficiency. Cessation of such a programme may lead to Zn deficiency. Post-harvest and dormant-season Zn sprays are recommended in many areas, e.g. a $ZnSO_4$ spray at the silver-tip stage of bud development or one or more foliar sprays of zinc chelate. Zinc chelate is not as effective as $ZnSO_4$ in correcting zinc deficiency but can be used to maintain an adequate amount of Zn in the tree once this is within the optimum range. Post-harvest and dormant-season sprays alone seem to have little long-term effect owing to the limited translocation of the absorbed Zn.

Trunk injection, tried for other fruit trees, is not generally used for apples or pears.

Effects of aluminium on nutrition

Although aluminium is a major component of the earth's crust there is no evidence that it essential to the growth of any plant species. Very minute traces

Table 11.3 *Effects of 4 ppm aluminium in the nutrient solution on nutrient uptake by apple seedlings*

Nutrient solution		Ca	Mg	K	P	Zn	Cu	Mn	Fe
$NO_3 -$ Al	Uptake per plant[a]	5.7	2.3	12.4	1.9	72	30	46	615
	Conc^n in shoots[b]	0.82	0.31	1.37	0.20	60	23	81	185
$NO_3 +$ Al	Uptake per plant[a]	2.0	0.8	7.3	0.6	25	14	27	180
	Conc^n in shoots[b]	0.61	0.22	1.56	0.10	48	27	54	143

N concentration in nutrient solution 11.3 ppm.
[a] uptake in mg for Ca, Mg, K and P, in µg for Zn, Cu, Mn and Fe.
[b] concentration in % for Ca, Mg, K and P, in ppm for Zn, Cu, Mn and Fe.
Data from Kotzé *et al.* (1977). Reproduced with permission.

of Al may increase the growth of some plants in experimental situations but its importance both in general and for apple and pear trees in particular is through its toxicity and its adverse effects on the uptake of nutrient elements.

Aluminium is present in high concentrations in igneous rocks and their weathering products. Sesquioxide clays containing large quantities of the oxides or hydrated oxides of Al are widespread, especially in tropical and subtropical climates. The solubility of Al increases sharply below pH 5.5 and below pH 4.0 all soluble Al is present as Al^{3+}. Toxic Al concentrations are only likely to occur at soil pH values of 5 or below (Robson and Pitman, 1983).

Typically the growth of roots is adversely affected by toxic levels of Al. At low concentrations the primary root may elongate normally but lateral growth is inhibited. At higher Al concentrations even primary roots are stunted and thickened. In general plants grown in external media with high concentrations of Al may show symptoms of P or Ca deficiency arising through effects of Al on both absorption and utilization.

Aluminium toxicity has been studied on apples both under orchard conditions where it is a problem, e.g. in the southeastern United States and in South Africa, and in culture solution. At pH 4.5 in solution culture, 4 ppm Al reduced the growth of apple seedlings by 59% when NO_3^- was the N source and by 32% when NH_4^+ was the N source (Kotzé *et al.*, 1977). The presence of 4 ppm Al in the culture medium also reduced the total amounts of different nutrients taken up by the seedlings. The adverse effects on nutrient uptake were greater than those on growth, as shown by their reduced concentrations per unit of dry matter (Table 11.3). Al toxicity is prevented by correcting soil pH before planting. This is achieved automatically if the pH is adjusted so as to optimize Ca uptake, since liming to a soil pH greater than that needed to eliminate Al effects is advantageous from the point of view of Ca-related fruit quality factors (Kotzé and du Preez, 1988).

Interactions between nutrients

There are many situations in which one nutrient affects the absorption, distribution or function of another.

Soil acidity is involved in a number of interactions important in apple and pear nutrition. Compounds which increase soil pH, e.g. $CaCO_3$, can eliminate Al and Mn toxicity but if used to excess can reduce the availability of Fe, Mn and Zn to the extent of inducing deficiencies. Increasing the pH in the presence of Ca reduces P solubility.

Ions also interact in the processes of absorption by roots. Ca depresses Mn absorption but increases that of H_2PO_4 (Robson and Pitman, 1983). NH_4 depresses K absorption (Tromp, 1962) and K reduces the absorption of Ca and Mg (Ludders, 1980). Zn^{2+} uptake is inhibited by Cu^{2+} and Zn^{2+} and Mn^{2+} inhibits Fe absorption.

High concentrations of P can interfere with Fe transport, causing its precipitation in veins.

Nitrate application in summer can increase the content of K, Ca and Mg in apple shoots and fruits, but ammonium-N supply reduces these (Ludders, 1980).

Although low Ca concentration within fruits is very strongly associated with poor fruit texture and with the incidence of a number of post-harvest disorders, these problems are frequently greater if the low Ca levels are accompanied by low B, high K or high N. Potassium can displace Ca on membrane surface binding sites but cannot cross-link components. Boron can enhance Ca accumulation in fruits. High-N apples respire at an accelerated rate: this effect, which shortens storage life, can be overcome by increased concentrations of Ca or P (Bramlage *et al.*, 1980).

The concentration of any particular nutrient in a tissue is affected not only by its uptake and transport, each of which may be influenced by the competitive effects of other ions, but also by the effects of other nutrients on growth of the tissue. This is particularly relevant to Ca concentration in fruits, which may be depressed if fruit growth is stimulated while Ca import is unaffected.

Orchard management and tree nutrition

Two aspects of orchard management have very large effects on the mineral nutrition of apple and pear trees. The first is herbicide use to eliminate part or all of the grass in orchards; the second is irrigation according to systems giving different patterns of water distribution. These affect mineral nutrition and fertilizer usage both through their effects on soil nutrients and through their effects on root distribution.

Effects of herbicide use

Apple and pear orchard surfaces have, in different historical periods, been cultivated, had a surface cover of mown grass with small areas kept clear around the tree trunks, or had different proportions of the surface kept free from grass by the use of herbicides. With the development of high-density dwarf fruit tree systems in which the fruit trees were very closely spaced within the row (e.g. with 1–2 m between trees in the row and rows 3–4 m apart) the herbicide-strip system, in which a central alleyway was grassed to provide a working surface for tractors and the soil under the trees was kept clear of grass using herbicides, came into use. With multirow or bed systems the trees were in double or triple rows, or multirow beds, with just 1–1.5 m or less between the trees within the beds, this soil being kept clear of grass between the trees, and narrow grass alleyways kept between beds. Alternatively, irrespective of tree spacing and arrangement the entire surface of the soil was treated with herbicides in an overall-herbicide system.

The effects on nutrient consumption of changes between these systems are very large. In the Netherlands, overall grassed orchards were given about 200 kg N, 30 kg P and 150 kg K per hectare. After the change to herbicide strip culture annual dressings were reduced to around 65 kg N, 7 kg P and 40 kg K per hectare without adverse effects on tree nutrient status (Delver, 1980a). Clearly, under the 'grassed-down' system the major nutrient use was by the grass. A grass sward is very efficient at withdrawing nutrients from the soil and grass competition for both water and nutrients needs to be compensated for. A further complication is that in the herbicide strip system grass cuttings may be deposited on the herbicide strip, so effecting a transfer of nutrients from the alleyway to the under-tree area. The soil in the herbicided strips has pH levels about 1 unit lower, much higher available P and somewhat higher available K than that in the grassed alleys in the upper 7 cm (Atkinson and White, 1980).

Trees growing under overall-herbicide conditions have larger root systems than those in the herbicide-strip system which, in turn, have larger root systems than those grown in grass (Atkinson and White, 1980). Under both systems of herbicide management the tree roots are much more frequent in the surface layer of soil than in grassed-down orchards (Figure 11.2). In overall-herbicide treated orchards apple tree roots are much more evenly distributed throughout the row and alley than in grassed or herbicide-strip orchards (Atkinson and White, 1980). Uptake of ^{15}N into apple leaves is much less in grassed orchards than where herbicides are used. In the herbicide-strip orchards ^{15}N uptake is much greater from the strip than from the alley. In overall-herbicide orchards there is less effect of proximity of placement to the trunk. Uptake of ^{32}P is much lower in overall-herbicide treated orchards. This effect, which was

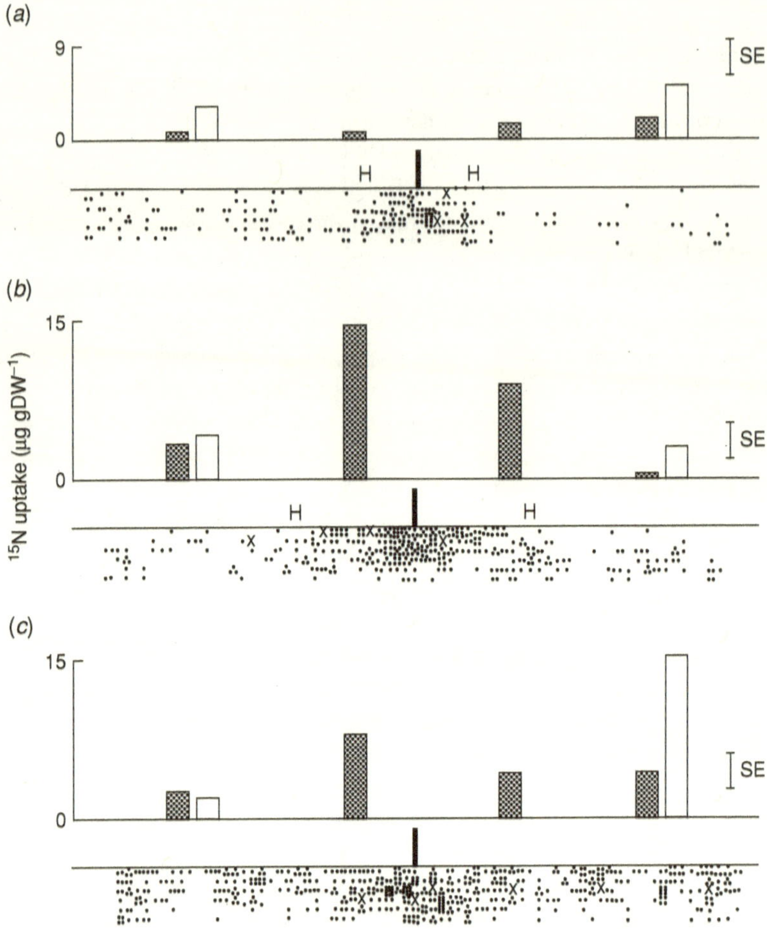

Figure 11.2 Root distribution between the orchard surface and 30 cm depth (*, roots <2 mm diameter; x, roots >2 mm diameter) and uptake into the leaves of ^{15}N from the row and inter-row positions: (*a*) grass, (*b*) herbicide strip, (*c*) overall herbicide. Shaded = at 10 cm depth, open = at 25 cm depth. From Atkinson and White (1980). Reproduced with permission.

unexpected in view of the increased soil P, may be linked to reduced populations of mycorrhizal fungi in overall-herbicide treated soil, which may be a result of elimination of grass roots there (Atkinson, 1986). The increase in soil N, P and K as a result of overall-herbicide management may lead to problems of excess, including Mg and Zn deficiency as a result of excess soil P and K (Hennerty, 1980). The problem of excess K uptake from herbicide strips can be reduced by cultivation to a depth of 6 centimetres (Delver, 1980b). This

Table 11.4 *Effects of water supply on the growth and nutrient content of fruiting trees of 'Golden Delicious'/'M.9'*

Water rate (%)	100	50	25
Root dry weight (kg/tree)	0.29	0.26	0.27
Total dry weight (kg/tree)	4.06	2.19	1.23
N content (g/tree)	30.2	15.1	10.0
P content (g/tree)	4.3	1.9	1.3
K content (g/tree)	30.6	14.5	8.7

Calculated from data of Buwalda and Lenz (1992).

suggests that enhanced rooting in the surface zone is a contributory cause of excess K uptake from herbicide strips.

Effects of water stress and irrigation: fertigation

Shortage of water may influence nutrient uptake in several different ways. Low soil water potentials in the rooting volume may adversely affect the supply of nutrients moving largely by diffusion, e.g. K and P (Marschner, 1995). The effective length of root can be considered as that in soil at a water potential greater than −0.05 MPa. The drying of unirrigated soil in England in spring and early summer under grass management can reduce 'effective' apple root density from 9.1 to 2.7 cm cm^{-2} by early June (Atkinson, 1986). Leaf expansion is, however, usually checked by less severe levels of water stress than are other plant processes including root growth. Plant growth is therefore checked by low water potentials through effects on leaf area and total photosynthesis. This is likely to reduce sink demand for nutrients by the shoots and therefore nutrient uptake. In lysimeter experiments reduction of water supply to 50% or 25% of that consumed by control trees of 'Golden Delicious'/'M.9' reduced N, P and K uptake in approximate proportion to the reduction in total growth (Buwalda and Lenz, 1992). This was despite the fact that root growth was not reduced at lower irrigation levels (Table 11.4).

On some soils excessive water supply by precipitation plus irrigation can cause very large amounts of N to be leached from the soil (Neilsen and Neilsen, 1997).

Apple root systems adjust very quickly to localization of water supply. Trees which had been irrigated for many years by surface irrigation, and had widespread root systems, adjusted their roots to a very small wetted volume of soil within one season of localized drip irrigation (Levin *et al.*, 1980). Vegetative growth and cropping were not reduced. The root distribution pattern of trees irrigated by tricklers depends mainly on the wetted soil volume, which may be

30–50% of that irrigated by surface irrigation. Goode *et al.* (1978) found fibrous root concentrations in the wet core under the trickle nozzle to be increased by four or five times compared with those in corresponding unirrigated (rainfed) zones. This concentration of roots makes it likely that local nutrient reserves will be depleted quickly and also offers the opportunity for very effective supply to the tree through the trickle irrigation system (fertigation).

In fertigation the fertilizer may be added to the irrigation water in small doses at frequent intervals (discontinuous fertigation) or by means of a fertilizer pump that injects a steady fertilizer concentration (continuous or proportional fertigation).

Because their chemical characteristics differ, mineral nutrients are not equally distributed in the soil when supplied by trickle irrigation.

Nitrates and urea do not react with soil exchange sites and are not held in soils, so, whether previously incorporated into the soil or added in the trickle irrigation stream, move with soluble salts to the wetted front unless intercepted by the roots (Elfving, 1982; Klein and Weinbaum, 2000). They are therefore best provided in small doses on a frequent or 'proportional' basis. Nitrate status in the soil at any time reflects addition in the irrigation water, removal by the plant and losses from leaching and denitrification. The ammonium form of N is not so subject to immediate leaching losses because it will fix temporarily on exchange sites in the soil. It is, however, practically immobile in high pH soils with a high cation exchange capacity (CEC) and calcium carbonate content. It is fixed very close to the soil surface under the drippers in such soils although it can penetrate to a depth of 70 cm in sandy soils. NH_3 may also lead to clogging of drippers as a result of forming precipitates with Ca or Mg salts. Nitrogen use efficiency is generally greater in fertigated trees than in those with furrow irrigation.

Potassium is less mobile than nitrate (Goode *et al.*, 1978), but moves both laterally and downwards and is subject to both binding and leaching. It is best supplied by continuous fertigation. Supplied in this way it can correct K deficiency more readily than by trenching large quantities of fertilizer into the soil every few years.

There can be asymmetric distribution of P, Ca and K and, to a lesser extent, Mg and N in the foliage of apple trees given fertigation in a humid zone climate where trickle irrigation is essentially supplementary (Goode *et al.*, 1978).

Iron chlorosis can be corrected within 7–10 days using small doses of FeEDDHA chelate in the fertigation water. Much less is needed than when applied with conventional soil application and sprinkler irrigation (Klein and Weinbaum, 2000).

Nutrients not intercepted by the roots may leach and reach aquifers, e.g. following winter rains, or may accumulate at the wetted front in very dry climates, leading to salinity problems. Other potential problems include soil acidification around the emitters when NH_4-N is applied to orchards on sandy soils (Edwards *et al.*, 1982; Parchomchuk *et al.*, 1993) and reduction in plant and soil K values as a result of N P fertigation (Neilsen *et al.*, 1995, 1997).

Recommended reading

Marscher, H. (1995). *Mineral Nutrition of Higher Plants*, 2nd edn. London: Academic Press. 889 pp.
Although this gives little information on apple and pear nutrition as such it provides a valuable reference work on nutrients in the soil and their uptake, transport and function in higher plants.

References

Asamoah, T.E.O. (1984). Fruit tree root systems: effects of nursery and orchard management and some consequences for growth, nutrient and water uptake. PhD thesis, University of London.

Atkinson, D. (1983). The growth, activity and distribution of the fruit tree root system. *Plant and Soil* **71**, 23–35.

Atkinson, D. (1986). The nutrient requirements of fruit trees: some current considerations. *Advances in Plant Nutrition* **2**, 93–128.

Atkinson, D. and White, G.C. (1980). Some effects of orchard soil management on the mineral nutrition of apple trees. In *Mineral Nutrition of Fruit Trees*, ed. D. Atkinson, J.E. Jackson, R.O. Sharples and W.M. Waller, pp. 241–54. London: Butterworths.

Basso, C. and Wilms, F.W.W. (1988). Nutritional status of apple orchards in southern Brazil. *Acta Horticulturae* **232**, 187–92.

Batjer, L.P., Rogers, B.L. and Thompson, A.H. (1952). Fertilizer application as related to nitrogen, phosphorus, potassium, calcium and magnesium utilization by apple trees. *Proceedings of the American Society for Horticultural Science* **60**, 1–6.

Batjer, L.P., Rogers, B.L. and Thompson, A.H. (1953). 'Blossom Blast' of pears: an incipient boron deficiency. *Proceedings of the American Society for Horticultural Science* **62**, 119–22.

Beyers, E. and Terblanche, J.H. (1971 a). Identification and control of trace element deficiencies. II. Manganese deficiency and toxicity. *The Deciduous Fruit Grower* **21**, 167–71.

Beyers, E. and Terblanche, J.H. (1971 b). Identification and control of trace element deficiencies. III. Copper deficiency. *The Deciduous Fruit Grower* **21**, 199–202.

Beyers, E. and Terblanche, J.H. (1971 c). Identification and control of trace element deficiencies. IV. Boron deficiency and toxicity. *The Deciduous Fruit Grower* **21**, 235–9.

Beyers, E. and Terblanche, J.H. (1971 d). Identification and control of trace element deficiencies. V. Iron deficiency. *The Deciduous Fruit Grower* **21**, 265–8, +282.

Bhat, K.K.S. (1983). Nutrient inflows into apple roots. *Plant and Soil* **71**, 371–80.

Bieleski, R.L. and Ferguson, I.B. (1983). Physiology and metabolism of phosphate and its compounds. In *Encyclopaedia of Plant Physiology*, New Series, Vol. 15A, ed. A. Läuchli and R.L. Bieleski, pp. 422–49. Berlin and New York: Springer-Verlag.

Bollard, E.G. (1957). Composition of the nitrogen fraction of apple tracheal sap. *Australian Journal of Biological Sciences* **10**, 279–287.

Bollard, E.G. (1960). Transport in the xylem. *Annual Review of Plant Physiology* **11**, 141–66.

Boynton, D., Margolis, D. and Gross, C.R. (1953). Exploratory studies on nitrogen metabolism by McIntosh apple leaves sprayed with urea. *Proceedings of the American Society for Horticultural Science* **62**, 135–46.

Bramlage, W., Drake, M. and Lord, W.J. (1980). The influence of mineral nutrition on the quality and storage performance of pome fruits grown in North America. In *Mineral Nutrition of Fruit Trees*, ed. D. Atkinson, J.E. Jackson, R.O. Sharples and W.M. Waller, pp. 29–39. London: Butterworths.

Buwalda, J.G. and Lenz, F. (1992). Effects of cropping, nutrition and water supply on accumulation and distribution of biomass and nutrients for apple trees on 'M9' root systems. *Physiologia Plantarum* **84**, 21–8.

Buwalda, J.G. and Meekings, J.S. (1990). Seasonal accumulation of mineral nutrients in leaves and fruit of Japanese Pear (*Pyrus serotina* Rehd). *Scientia Horticulturae* **41**, 209–22.

Callan, N.W., Thompson, M.M., Chaplin, M.H., Stebbins, R.L. and Westwood, M.N. (1978). Fruit set of 'Italian' prune following fall foliar and spring boron sprays. *Journal of the American Society for Horticultural Science* **103**, 253–7.

Cooper, D.R., Hill-Cottingham, D.G. and Lloyd-Jones, C.P. (1976). Distribution and identity of labelled products following autumn application of ^{15}N labelled urea or potassium nitrate to apple trees. *Journal of the Science of Food and Agriculture* **27**, 266–72.

Crassweller, R.M., Ferree, D.C. and Stang, E.J. (1981). Effects of overtree misting for bloom delay on pollination, fruit set and nutrient element concentration of 'Golden Delicious' apple tree. *Journal of the American Society for Horticultural Science* **106**, 53–6.

Delver, P. (1980a). Uptake of nutrients by trees grown in herbicide strips. In *Mineral Nutrition of Fruit Trees*, ed. D. Atkinson, J.E. Jackson, R.O. Sharples and W.M. Waller, pp. 229–40. London: Butterworths.

Delver, P. (1980b). In *Mineral Nutrition of Fruit Trees*, ed. D. Atkinson, J.E Jackson, R.O. Sharples and W.M. Waller, p. 254. London: Butterworths.

Domoto, P.A. and Thompson, A.H. (1976). Effect of interactions of calcium, potassium and manganese supply on 'Delicious' apple trees as related to internal bark necrosis. *Journal of the American Society for Horticultural Science* **101**, 44–7.

Edwards, J.H., Bruce, R.R., Horton, B.D., Chesness, J.L. and Wehunt, E.J. (1982). Soil cation and water distribution as affected by NH_4NO_3 applied through a drip irrigation system. *Journal of the American Society for Horticultural Science* **107**, 1142–8.

Elfving, D.C. (1982). Crop response to trickle irrigation. *Horticultural Reviews* **4**, 1–48.

Faust, M. (1989). *Physiology of Temperate Zone Fruit Trees*. New York: John Wiley and Sons.

Ferree, D.C. and Thompson, A.H. (1970). Internal bark necrosis of the apple as influenced by calcium placement and soil manganese. *Maryland Agriculture Experiment Station Bulletin* A-166.

Ford, E.M. (1968). The response to Epsom salt sprays of mature apple trees of three varieties on two contrasting rootstocks. *Journal of Horticultural Science* **45**, 505–17.

Forshey, C.G. (1963). The effect of nitrogen status of 'McIntosh' apple trees in sand culture on the absorption of magnesium from Epsom salt sprays. *Proceedings of the American Society for Horticultural Science* **83**, 21–31.

Fucik, J.E. and Titus, J.S. (1965). Split root studies on calcium and manganese absorption and translocation in seedling apple trees. *Proceedings of the American Society for Horticultural Science* **86**, 12–22.

Goode, J.E., Higgs, K.H. and Hyrycz, K.J. (1978). Trickle irrigation of apple trees and the effects of liquid feeding with NO_3^- and K^+ compared with normal manuring. *Journal of Horticultural Science* **53**, 307–16.

Greenham, D.W.P. (1980). Nutrient cycling: the estimation of orchard nutrient uptake. In *Mineral Nutrition of Fruit Trees*, ed. D. Atkinson, J.E. Jackson, R.O. Sharples and W.M. Waller, pp. 345–52. London: Butterworths.

Greenham, D.W.P. and White, G.C. (1959). The control of magnesium deficiency in dwarf pyramid apples. *Journal of Horticultural Science* **34**, 238–47.

Hansen, P. (1980). Crop load and nutrient translocation. In *Mineral Nutrition of Fruit Trees*, ed. D. Atkinson, J.E. Jackson, R.O. Sharples and W.M. Waller, pp. 201–12. London: Butterworths.

Harley, C.P., Regeimbal, L.O. and Moon, H.H. (1958). The role of nitrogen reserves in new growth of apple and the transport of P^{32} from roots to leaves during early-spring growth. *Proceedings of the American Society for Horticultural Science* **72**, 57–65.

Haynes, R.J. and Goh, K.M. (1980). Distribution and budget of nutrients in a commercial apple orchard. *Plant and Soil* **56**, 445–57.

Hennerty, M.J. (1980). Some effects of orchard soil management. In *Mineral Nutrition of Fruit Trees*, ed. D. Atkinson, J.E. Jackson, R.O. Sharples and W.M. Waller, p. 254. London: Butterworths.

Hennerty, M.J., O'Kennedy, B.T. and Titus, J.S. (1980). Conservation and reutilization of bark proteins in apple trees. In *Mineral Nutrition of Fruit Trees*, ed. D. Atkinson, J.E. Jackson, R.O. Sharples and W.M. Waller, pp. 369–77. London: Butterworths.

Hill-Cottingham, D.G. and Bollard, E.G. (1965). Chemical changes in apple tree tissues following application of fertilizer nitrogen. *New Zealand Journal of Agriculture Research* **8**, 778–87.

Hill-Cottingham, D.G. and Cooper, D.R. (1970). Effect of time of application of fertilizer nitrogen on the distribution and identity of the nitrogenous constituents of young apple trees. *Journal of the Science of Food and Agriculture* **21**, 172–7.

Hill-Cottingham, D.G. and Lloyd-Jones, C.P. (1975). Nitrogen-15 in apple nutrition investigations. *Journal of the Science of Food and Agriculture* **26**, 165–73.

Johnson, D.S. and Yogaratnam, N. (1978). The effects of phosphorus sprays on the mineral composition and storage quality of 'Cox's Orange Pippin' apples. *Journal of Horticultural Science* **53**, 171–8.

Johnson, D.S., Stinchcombe, G.R. and Stott, K.G. (1983). Effects of soil management on mineral composition and storage quality of Cox's Orange Pippin apples. *Journal of Horticultural Science* **58**, 317–26.

Johnson, F., Allmendinger, D.F., Miller, V.L. and Polley, D. (1955). Fall application of boron sprays as a control for blossom blast and twig dieback of pears. *Phytopathology* **45**, 110–14.

Kang, S.M. and Titus, J.S. (1980). Qualitative and quantitative changes in nitrogenous compounds in senescing leaf and bark tissues of the apple. *Physiologia Plantarum* **50**, 285–90.

Klein, I. and Weinbaum, S.A. (2000). Fertilization of temperate-zone fruit trees in warm and dry climates. In *Temperate Fruit Crops in Warm Climates*, ed. A. Erez, pp. 77–100. Dordrecht: Kluwer.

Korcak, R.F. (1987). Iron deficiency chlorosis. *Horticultural Reviews* **9**, 133–86.

Kotzé, W.A.G. and du Preez, M. (1988). Soil amelioration and fertilisation for apple production in South Africa. *Acta Horticulturae* **232**, 167–76.

Kotzé, W.A.G., Shear, C.B. and Faust, M. (1977). Effect of nitrogen source and aluminium in nutrient solution on the growth and mineral nutrition of apple and peach seedlings. *Journal of the American Society for Horticultural Science* **102**, 279–82.

Levin, I., Assaf, R. and Bravdo, B. (1980). Irrigation, water status and nutrient uptake in an apple orchard. In *Mineral Nutrition of Fruit Trees*, ed. D. Atkinson, J.E. Jackson, R.O. Sharples and W.M. Waller, pp. 255–64. London: Butterworths.

Lombard, P.B. and Westwood, M.N. (1987). Pear rootstocks. In *Rootstocks for Fruit Crops*, ed. R.C. Rom and R.F. Carlson, pp. 145–83. New York: John Wiley and Sons.

Ludders, P. (1980). Effect of time and amount of nutrient additives on nutrient status and distribution and on fruit quality. In *Mineral Nutrition of Fruit Trees*, ed. D. Atkinson, J.E. Jackson, R.O. Sharples and W.M. Waller, pp. 165–72. London: Butterworths.

Luttge, U. (1983). Import and export of mineral nutrients in plant roots. In *Inorganic Plant Nutrition*, Part A, ed. A. Läuchli and R.L. Bieleski, pp. 181–211. Berlin: Springer-Verlag.

MAFF, (1972). Apples. *Bulletin 207, Ministry of Agriculture, Fisheries and Food.* London: Her Majesty's Stationery Office.

Marschner, H. (1983). General introduction to the mineral nutrition of plants. In *Inorganic Plant Nutrition*, Part A, ed. A. Läuchli and R.L. Bieleski, pp. 5–60. Berlin: Springer-Verlag.

Marschner, H. (1995). *Mineral Nutrition of Higher Plants.* London: Academic Press.

Mason, A.C. and Whitfield, A.B. (1960). Seasonal changes in the uptake and distribution of mineral elements in apple trees. *Journal of Horticultural Science* **35**, 34–55.

Millard, P. (1995). Internal cycling of nitrogen in trees. *Acta Horticulturae* **383**, 3–14.

Millard, P. and Neilsen, G.H. (1989). The influence of nitrogen supply on the uptake and remobilization of stored N for the seasonal growth of apple trees. *Annals of Botany* **63**, 301–9.

Millard, P. and Thomson, C.M. (1989). The effect of the autumn senescence of leaves on the internal cycling of nitrogen for the spring growth of apple trees. *Journal of Experimental Botany* **40**, 1285–9.

Murneek, A.E. and Logan, J.C. (1932). Autumnal migration of nitrogen and carbohydrate in the apple tree with special reference to leaves. *Bulletin Missouri Agricultural Experiment Station.* 171.

Neilsen, G.H. and Hoyt, P.B. (1990). A comparison of methods to raise zinc concentration of apple leaves. *Canadian Journal of Plant Science* **90**, 599–603.

Neilsen, G.H. and Neilsen, D. (1997). Orchard nutrition to maximize crop quality and minimize environmental degradation. *Acta Horticulturae* **448**, 365–73.

Neilsen, G.H., Parchomchuk, P. and Berard, R. (1995). NP fertigation and irrigation affect potassium nutrition of newly planted apple trees. *Acta Horticulturae* **383**, 57–65.

Neilsen, G.H., Parchomchuk, P. and Neilsen, D. (1997). Distribution of soil P and K as affected by NP-fertigation in high density apple orchards. *Acta Horticulturae* **448**, 439–47.

Oland, K. (1960). Nitrogen feeding of apple trees by post-harvest urea sprays. *Nature* **185**, 857.

Oland, K. (1963). Changes in the dry matter and major nutrient elements of apple foliage during senescence and abscission. *Physiologia Plantarum* **16**, 682–94.

Parchomchuk, P., Neilsen, G.H. and Hogue, E.J. (1993). Effects of drip fertigation of NH_4-N and P on soil pH and cation leaching. *Canadian Journal of Soil Science* **73**, 157–64.

Perring, M.A. and Preston, A.P. (1974). The effect of orchard factors on the chemical composition of apples. III. Some effects of pruning and nitrogen application on 'Cox's Orange Pippin' fruit. *Journal of Horticultural Science* **49**, 85–93.

Quast, P. (1980). Mineral nutrition and physiological fruit disorders in the Niederelbe region of Germany. In *Mineral Nutrition of Fruit Trees*, ed. D. Atkinson, J.E. Jackson, R.O. Sharples and W.M. Waller, pp. 101–2. London: Butterworths.

Raven, J.A. (1980). Short and long distance transport of boric acid in plants. *New Phytologist* **84**, 231–9.

Robson, A.D. and Pitman, M.G. (1983). Interactions between nutrients in higher plants. In *Encyclopedia of Plant Physiology*, New Series, Vol. 15A, ed. A. Läuchli and R.L. Bieleski, pp. 147–80. Berlin and New York: Springer-Verlag.

Rowe, R.W. (1980). Future analytical requirements in the fruit industry. In *Mineral Nutrition of Fruit Trees*, ed. D. Atkinson, J.E. Jackson, R.O. Sharples and W.M. Waller, pp. 399–405. London: Butterworths.

Sanchez, E.E., Righetti, T.L., Sugar, D. and Lombard, P.B. (1990). Response of 'Comice pear' trees to a postharvest urea spray. *Journal of Horticultural Science* **65**, 541–6.

Shear, C.B. and Faust, M. (1980). Nutritional ranges in deciduous tree fruits and nuts. *Horticultural Reviews* **2**, 142–63.

Shim, K.K., Splittstoesser, W.E. and Titus, J.S. (1973a). Changes in urease activity in apple trees as related to urea application. *Physiologia Plantarum* **28**, 327–31.

Shim, K.K., Titus, J.S. and Splittstoesser, W.E. (1972). The utilization of post-harvest urea sprays by senescing apple leaves. *Journal of the American Society for Horticultural Science* **97**, 592–6.

Shim, K.K., Titus, J.S. and Splittstoesser, W.E. (1973b). The fate of carbon and nitrogen from urea applied to foliage of senescing apple trees. *Journal of the American Society for Horticultural Science* **98**, 360–6.

Stevenson, D.S. and Neilsen, G.H. (1990). Nitrogen additions and losses to drainage in orchard-type irrigated lysimeters. *Canadian Journal of Soil Science* **70**, 11–19.

Stiles, W.C. (1966). Micronutrient studies in Maine orchards. *Proceedings of New York State Horticultural Society* **111**, 105–12.

Sun, X.P., Wang, S.Y., Tong, Y.A., Korcak, R.F. and Faust, M. (1987). Metabolic changes in iron-deficient apple seedlings. *Journal of Plant Nutrition* **10**, 1021–30.

Swietlik, D. (1999). Zinc nutrition in horticultural crops. *Horticultural Reviews* **23**, 109–78.

Swietlik, D. and Faust, M. (1984). Foliar nutrition of fruit crops. *Horticultural Reviews* **6**, 287–356.

Tagwira, F. (1995). Chemical and physical factors associated with micronutrient status and availability in Zimbabwean soils. In *Dambo Farming in Zimbabwe*, ed. R. Owen, K. Verbeek, J. Jackson and T. Steenhuis, pp. 141–8. Harare: University of Zimbabwe Press.

Tanaka, H. (1967). Boron absorption by excised sunflower root. *The Journal of Soil Science and Plant Nutrition* **13**, 77–82.

Taylor, B.K., van den Ende, B. and Canterford, R.L. (1975). Effects of rate and timing of nitrogen applications on the performance and chemical composition of young pear trees cv Williams Bon Chretien. *Journal of Horticultural Science* **50**, 29–40.

Titus, J.S. and Kang, S.M. (1982). Nitrogen metabolism, translocation and recycling in apple trees. *Horticultural Reviews* **4**, 204–46.

Tromp, J. (1962). Interactions in the absorption of ammonium, potassium and sodium by wheat roots. *Acta Botanica Neerlandica* **11**, 147–92.

Tromp, J. (1970). Storage and mobilization of nitrogenous compounds in apple trees with special reference to arginine. In *Physiology of Tree Crops*, ed. L.C. Luckwill and C.V. Cutting, pp. 143–59. New York: Academic Press.

Tromp, J. (1980). Mineral absorption and distribution in young apple trees under various environmental conditions. In *Mineral Nutrition of Fruit Trees*, ed. D. Atkinson, J.E. Jackson, R.O. Sharples and W.M. Waller, pp. 173–82. London: Butterworths.

Tromp, J. (1983). Nutrient reserves in roots of fruit trees, in particular carbohydrates and nitrogen. *Plant and Soil* **71**, 401–13.

Tromp, J. and Ovaa, J.C. (1967). Seasonal variations in the amino acid composition of xylem sap of apple. *Zeitschrift für Pflanzenphysiologie* **57**, 11–21.

Tromp, J. and Ovaa, J.C. (1971). Phloem translocation of storage nitrogen in apple. *Physiologia Plantarum* **25**, 407–13.

Van Goor, B.J. and Van Lune, P. (1980). Redistribution of potassium, boron, iron, magnesium and calcium in apple trees determined by an indirect method. *Physiologia Plantarum* **48**, 21–6.

Webster, A.D. (1997). A review of fruit tree rootstock research and development. *Acta Horticulturae* **451**, 53–73.

Wilkinson, B.G. (1958). The effect of orchard factors on the chemical composition of apples. II. The relationship between potassium and titratable acidity, and between potassium and magnesium, in the fruit. *Journal of Horticultural Science* **33**, 49–57.

Wilkinson B.G. and Perring, M.A. (1964). Changes in the chemical composition of apples during development and near picking time. *Journal of the Science of Food and Agriculture* **15**, 146–52.

Wittwer, S.H. and Teubner, F.G. (1959). Foliar absorption of mineral nutrients. *Annual Review of Plant Physiology* **10**, 13–27.

Yamada, Y., Jyung, W.J., Wittwer, S.H. and Bukovac, M.J. (1965*a*). The effect of urea on ion penetration through isolated cuticular membranes and ion uptake by leaf cells. *Proceedings of the American Society for Horticultural Science* **87**, 429–32.

Yamada, Y., Wittwer, S.H. and Bukovac, M.J. (1965*b*). Penetration of organic compounds through isolated cuticular membranes with especial reference to ^{14}C urea. *Plant Physiology* **40**, 170–5.

Yogaratnam, N. and Sharples, R.O. (1981). Supplementing the nutrition of 'Bramley's Seedling' apple with phosphorus sprays. II. Effects on fruit composition and storage quality. *Journal of Horticultural Science* **57**, 53–9.

12

Water relations

Introduction

Most of the water taken up by fruit trees is in response to evaporative loss (transpiration) through the leaf pores or stomata. This is a consequence of the need for stomata to be open to admit CO_2 entry for photosynthesis.

Over any given period the fruit tree usually takes up more than 100 times as much water as it produces dry matter (Lenz, 1986). A cropping orchard in Washington State, USA requires up to 1 m of irrigation per year (Evans, 1982). This is 10 000 t ha^{-1} as compared with about 42.5 t of water in the fruits of a 50 t ha^{-1} crop. The close correlation between water use and crop yield that is generally observed is not a result of water use in the actual production of the crop. It is primarily a consequence of the close correlation between CO_2 assimilation and water loss, as a result of the dependence of both of these on leaf area and stomatal behaviour.

However, the ability to keep stomata open for CO_2 assimilation and to avoid consequent desiccation depends on the adequacy of the supply of water as well as mechanisms for controlling water stress. To this extent water supply is a truly limiting factor to crop production.

The water flux through the soil–tree–air system is under tension, i.e. negative pressure. Tree tissues and cells equilibrate with this tension and different aspects of growth, development and function respond to this in different ways.

Smaller water flows in the tree are the results of gradients in osmotic potential due to solutes moving into conducting elements. This is especially important in the transport of water and solutes in the xylem to developing tissues in spring, before there is any major transpirational flow. Osmotic forces also determine the mass flow of water and solutes in the phloem.

Many aspects of water relations, e.g. uptake by roots and effects of water availability on growth, fruiting and nutrition have been discussed earlier. This chapter emphasizes the factors controlling water availability and requirements on an orchard basis, the water status of trees and their constituent parts, and the overall effects of irrigation and water status on tree performance.

Soil water availability

Knowledge of the factors controlling soil water availability is essential to the understanding of fruit tree water use and irrigation requirements.

The amount of water available in the soil depends on the amount supplied (by rainfall or irrigation), the amount lost by run-off from the surface or by drainage to below the rooting zone, and the amount retained in the rooting zone until taken up by the trees.

Run-off from the surface depends on slope and on soil texture and structure. The latter factors influence the speed with which surface-applied water infiltrates into the soil. Data on infiltration rates into different soil types are published in irrigation reference books and reviews (Anon., 1982; Miller, 1982). Infiltration into sands can be at rates of more than 100 mm h^{-1}, into clays at less than 5 mm h^{-1}. The compactive effect of rain or sprinkler drops on a bare soil surface can result in a low porosity crust on the soil surface which reduces infiltration (Miller, 1982). Jackson *et al.* (1989) found that the infiltration of water into herbicided alleyways in apple orchards was only 14.5 mm h^{-1} while that into the corresponding positions in grassed alleyways was 96.5 mm h^{-1}. Hamer (1985) found that 100% of summer rainfall could be accountable for in the centre of grassed alleyways but only 50% in the corresponding, rooting zone, depth in herbicide-treated land. Mulching can be used to ameliorate poor water penetration and retention under bare soil conditions (Rom, 1972, quoted by Stiles, 1982). Where the herbicide treatment is initiated by killing a grass sward the soil is re-wetted by rainfall to a much greater extent than where it is initiated on cultivated soil (Atkinson and White, 1981).

Drainage to below the rooting zone also varies with soil type, the soil water content at which the downward flux of water becomes slow being called the field capacity. This can be influenced by layering in the soil profile. In comparison with a uniform soil, any profile layer with a different pore size distribution will increase the water content above the layer whether this is of finer or coarser soil. This is an important feature in some major apple production areas e.g. Washington State, USA (Miller, 1982).

The wilting point is the soil water content below which plants growing in that soil remain wilted even when transpiration is nearly eliminated. This varies with soil type: soils with more than 35% clay can be at the permanent wilting point when they have 25–40% of water on a volume/volume basis, coarse-textured sands reach the same point at less than 10% water content.

As an approximation the water retained against a pressure of 15 bar usually corresponds to the permanent wilting point.

Available water capacity is the difference between field capacity and the permanent wilting point. The soil moisture deficit is the difference between the amount of water held at field capacity and the amount held at the time considered.

The above concepts are simplest in their application to uniform soils in which the roots are evenly distributed to a standard depth and may give misleading results if this is assumed to be the case when it is not. It is essential to characterize both the soil and the root distribution.

Cover crops and weeds can use a large part of total water resources. In the early orchard years, when the soil area around the base of the trees receives very high levels of solar irradiation, grass or weeds growing there can compete very effectively for water. Holloway and White (1967) found that even a sparse cover of clipped annual weeds allowed to extend up to the trunk of newly planted apple trees reduced their shoot extension growth to only a third of that of clean-cultivated controls, irrespective of fertilizer treatment. Atkinson and Thomas (1985) showed that for 3-year-old trees the soil water deficit in the surface 75 cm, 50 cm from the trunk, was about 3 cm under overall herbicide management but almost 9 cm under grass by mid-September in England.

Evaporation and evapotranspiration

Evaporation (E_0)

Evaporation is the conversion of water to water vapour. This requires energy, known as the latent heat of vaporization. The rate of evaporation varies with incoming energy and some other factors, so a key step in determining potential crop water needs is to characterize evaporative water consumption at the relevant sites.

Direct observations of evaporation can be made using evaporation pans. The small capacities and shallow depths of these allow proportionately more advected heat to be absorbed than in the natural situation, so pan evaporation gives an overestimate, and a pan coefficient specific to the pan design has to be applied. Generally the standard British pan has a coefficient of 0.92 and that of the US Weather Bureau Class A pan is about 0.75, but there can be wide variation.

Alternatively potential evaporation from an open water surface (E_0) can be calculated from widely available meteorological data (Penman, 1948). The key factors determining evaporation rate are as follows. Evaporation increases as solar radiation increases. As water vaporizes, the boundary layer between the evaporating surface and the air becomes saturated and must be replaced by drier air if evaporation is to proceed. This replacement is a function of windspeed. The higher the relative humidity of the air at any given temperature, the lower the evaporation. Evaporation is also increased the higher the ambient temperatures of the surface and the air. This is through the supply of heat energy and also because as air temperature rises its capacity to absorb water vapour increases. Evaporation is driven by gradients in atmospheric vapour pressure deficit (VPD).

Evapotranspiration by a standard reference crop (ET_0)

Potential transpiration by vegetation completely shading the ground and adequately supplied with water is generally less than potential evaporation because stomata close at low light intensities, so the effect of nightfall is greater than that due to the reduced energy receipts alone and because of the greater reflectivity of vegetation than water. For a standard reference crop of grass 8–15 cm tall the evapotranspiration (ET_0, the combined transpiration and evaporation) is usually about 80% of E_0. ET_0 can be calculated from weather data and assumed values of bulk surface resistance (70 S m^{-1}) and albedo (0.23) using the FAO Penman–Monteith equation (Allen *et al.*, 1997).

$$ET_0 = \frac{0.408\Delta(R_n - G) + \gamma \frac{900}{T+273} u_2(e_a - e_d)}{\Delta + \gamma(1 + 0.34u_2)} \qquad (12.1)$$

where ET_0 measured is in mm d^{-1}; R_n is net radiation and G is soil heat flux density (MJ m^{-2} d^{-1}) for use with 24-h calculation time steps. T is mean daily temperature (°C), Δ is the slope of the saturation vapour pressure curve at the mean daily air temperature (kPa °C^{-1}), γ is the psychometric constant (kPa °C^{-1}), u_2 is mean 24-h windspeed at 2 m (m s^{-1}), e_a is mean daily saturation vapour pressure at air temperature (kPa) and e_d is saturation vapour pressure at dewpoint temperature (kPa) at 1.5–2 m height.

The modified Blaney–Criddle method can be used to estimate evapotranspiration from a standard reference crop when neither pan evaporation nor radiation data are available (Doorenbos and Pruitt, 1977; James *et al.*, 1982; Wilson, 1990). A monthly water use factor (f), expressed as mm per day, is calculated from the mean of the daily maximum and minimum temperatures of the month considered (t) and the mean daily percentage of the annual daytime hours (p).

$$f = p(0.46t + 8.13) \qquad (12.2)$$

ET_0 values are then estimated from these f values at different levels of relative humidity, windspeed and cloudiness using standard tables or graphs.

Table 12.1 *Crop coefficient (K_c) values for apple orchards in sub-humid conditions*

	Early season	Mid-season	End season[a]
Without ground cover	0.45	0.95	0.70
With ground cover	0.50	1.20	0.95

[a] Prior to leaf fall. After leaf fall K_c is 0.20 for bare dry soil or dead ground cover and 0.50 to 0.80 for actively growing ground cover (which will increase its evapotranspiration when the trees lose their leaves).
Data from Allen *et al.* (1997).

Evapotranspiration by orchards

To calculate the evapotranspiration for other crops (ET_c), including orchard crops, the reference ET_0 value is multiplied by a crop coefficient K_c.

$$ET_c = K_c\, ET_0 \tag{12.3}$$

K_c depends on (a) crop height which affects roughness and aerodynamic resistance; (b) crop–soil surface resistance, which is affected by leaf area, the fraction of the ground covered by vegetation, leaf age and condition, stomatal control and soil surface wetness; and (c) the albedo (reflectance) of the crop–soil surface, which is affected by the fraction of ground covered by vegetation and by the soil surface.

In general the ratio of ET_c to ET_0 increases as windspeed increases and relative humidity decreases. K_c for a tall crop, e.g. 2–3 m high, increases by as much as 40% when changing from a calm humid climate to a windy arid climate.

Table 12.1 gives estimates of the K_c value for apple orchards in early, mid-, and end of season conditions, assuming mature orchards with trees 4 m tall and taking into account soil evaporation and ground-cover crop evapotranspiration (Allen *et al.*, 1997). Similar results for Washington State conditions are shown in Figure 12.1. The increase in K_c in the early part of the season reflects the early leafing out of spur buds and rapid completion of shoot growth (Chapter 7). These values apply to mature orchards. At planting the fraction of the ground covered by tree canopy is very low and the number of years needed to attain final canopy dimensions depends on planting density and vigour of growth. Orchard ground cover has been variously estimated as the percentage of ground vertically beneath the tree canopy and by LAI. These are both likely to introduce inaccuracy, the most relevant measurement being that of the fraction of radiant energy intercepted, which is not a simple function of LAI in orchard crops.

The estimates of potential water use calculated in this way provide a basis for planning the relative irrigation requirements in different areas. Optimizing irrigation water application, especially where this is by localized (trickle)

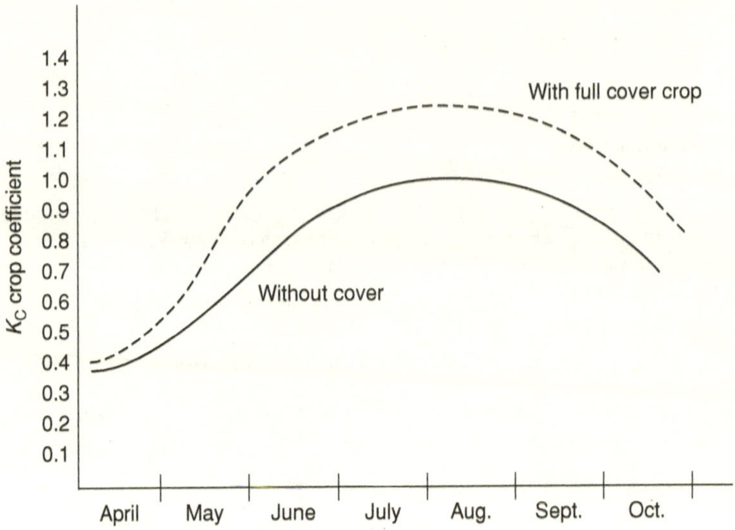

Figure 12.1 Approximate crop coefficients for apple orchards in Central Washington relative to alfalfa-based evapotranspiration. Reproduced from Evans (1982), credit Washington State University Co-operative Extension. Based on data given by James *et al.* (1982).

irrigation to only part of the root zone, or where water is an expensive or very limited resource, depends on a much more detailed knowledge of tree water relations.

Basic concepts in tree water relations

Water movement is governed by driving forces, originating from differences in solute concentration and pressure, and by the conductance of the flow pathway (Boyer, 1985). The driving forces can all be expressed in units of pressure. They are described in units of water potential (ψ_w) which is the chemical potential (free energy mol^{-1}) divided by the partial molal volume of liquid water (18.0 cm^3 mol^{-1}). This water potential is expressed in units of bars (1 bar $= 10^6$ dynes cm^{-2} $= 10^6$ ergs cm^{-3} $= 0.987$ atmospheres). It is also quoted in megapascals (MPa), 1 MPa equalling 10 bar. Measurements of ψ_w are always expressed relative to pure free water at atmospheric pressure and the same temperature as the system being measured. This reference potential of pure water is taken as 0. Other water potentials are governed by equation 12.4.

$$\psi_w = \psi_s + \psi_p \qquad (12.4)$$

ψ_w can be measured as the water potential in particular organs or tissues, e.g. in leaf (ψ_l), xylem (ψ_x) or fruit (ψ_{fr}). ψ_s is the solute or osmotic potential and its value is always negative. ψ_p represents forces resulting from external pressures, i.e. turgor (ψ_t) and other local pressures, and can be positive or negative depending on whether the pressure is above or below atmospheric.

Water moves by bulk flow or by diffusion. Within xylem cell walls and within protoplasts, where there is a continuous liquid phase, bulk flow predominates. It also seems to be the major mechanism of water movement through membranes, bulk flow having been shown to be the major mechanism of water movement through artificial membranes having conductivities similar to higher plant membranes (Boyer, 1985).

During daylight hours the main driving force for water uptake is the tension (negative pressure) developed in the xylem following evaporation of water in the leaves as a result of the vapour pressure gradient between the internal leaf surfaces, which are assumed to be fully wet, and the atmosphere. The water potentials, expressed in bars, are typically −142 to −1242 in the external atmosphere (at 20 °C 90% RH and 20 °C 40% RH respectively), −5 to −28 in leaves, −5 to −6 in root cells and −0.1 in soil water (Kennedy and Fujii, 1982). Water, together with dissolved solutes moves up through the tree in a cohesive stream. The flux of water through any part of a transpiring plant is a function of the drop in water potential across that part and the resistance to flow, i.e.

$$F = (\psi - \psi^1)/r \tag{12.5}$$

where F is transpirational flux, $\psi - \psi^1$ the drop in potential and r the resistance to flow in the pathway. For the soil to leaf water flux of the whole plant this can be written as

$$E_l = (\psi_{soil} - \psi_l)/R_{sp} \tag{12.6}$$

where E_l is the transpiration rate, ψ_{soil} is the soil water potential, ψ_l is the leaf water potential and R_{sp} is the total hydraulic resistance in the soil plant pathway. Resistance is the reciprocal of conductance.

An additional driving force for water uptake and flux comes from the energy-requiring accumulation of solutes, especially ions, in root tissues and their transfer to the xylem. The xylem sap has a higher concentration of solutes than the soil solution, with a consequently lower osmotic potential. This results in a gradient of osmotic potential between the soil solution and the xylem and a flow of water from the former to the latter. This water movement gives rise to the phenomenon of root pressure and the 'bleeding' of decapitated stumps.

The water flux into and out of cells is also determined by water potential gradients (equations 12.4, 12.5). The osmotic potential (ψ_s) is related to cell sap concentration measured in number of moles of solute. As water potential in the transpiration stream is lowered water should flow from adjacent tissues into it. This actually happens, with measurable trunk diameter shrinkage during the day followed by expansion at night. Such tissue water losses, with potentially unfavourable reductions in cell turgor, are minimized in, for example, mature apple leaves by active osmotic adjustment in which there is active accumulation of solutes in cells (Lakso, 1994). This can result from the production of sorbitol, glucose and fructose from starch (Wang and Stutte, 1992).

If there is no osmotic adjustment then turgor potential (ψ_t) (the cellular version of ψ_p in equation 12.4) changes directly with ψ_w; if there is osmotic adjustment it changes to a lesser extent. Although changes in ψ_w in themselves are unlikely to affect cell enzymes or metabolism, changes in ψ_t in the guard cells have a large effect on stomatal aperture, which controls photosynthesis as well as transpiration, and on cell expansion and growth. The relative water content (RWC, earlier known as Relative Turgidity), can be measured directly as the water content of the tissue as a proportion of its water content at full hydration.

The growth of tissues, e.g. leaves, depends on cell division followed by irreversible cell expansion. The expansion phase of this growth is governed by equation 12.7.

$$\text{Expansion Growth} = M(P - Y) \tag{12.7}$$

where M is cell wall extensibility and is the slope of the curve relating expansion to turgor, P is the turgor pressure and Y is the minimum turgor for growth, or yield threshold (Taylor and Davies, 1985). The tissue extensibility is often described as a process of cell wall loosening and is sometimes referred to as plasticity.

The leaf water potential (ψ_l) which is the usual starting point for plant water relations studies depends on three factors, the soil water potential (ψ_{soil}), the resistance of the plant–soil system to the flow of water (R_{sp}) and the rate of evaporation from the leaves (E_l), i.e. equation 12.6 can be re-written:

$$\psi_l = \psi_{soil} - E_l R_{sp} \tag{12.8}$$

The key plant factors controlling this are the resistances to flow in the pathway for liquid water movement and control of the stomatal aperture.

Roots and tree water relations

Water uptake and flow through roots

Apple root systems can explore all the space between the trees to a depth of at least 1.6 m (Hughes and Gandar, 1993) but commonly, especially in the case of young trees, they explore only a small part of the available soil volume (Atkinson, 1980). The rooting density within the exploited soil volume is generally low. Hughes and Gandar (1993) found root length densities (L_V values) in the top metre of soil to average only 0.1–0.2 cm cm^{-3} even in the excellent growing conditions of New Zealand. Expressed as m m^{-2} ground surface (L_A values), the rooting density of apples is generally less than 10×10^2 whereas that of *Gramineae* can be up to 40×10^4 (Landsberg and Jones, 1981). This low root density is likely to lead to local depletion of soil moisture and relatively high adverse effects of resistance to water flow in the soil. The mycorrhizal nature of the roots and the fact that they can proliferate in moisture-rich soil zones several metres below the surface may compensate.

Apple roots also become concentrated in other areas of high moisture status, e.g. near trickle irrigation drippers. Levin *et al.* (1979) found about three times as many roots per m^2 at 10 or 30 cm from a trickle line than at a distance of 110 cm.

There is evidence that some roots preferentially supply water and nutrients to particular parts of the shoot system. However, experiments with split root systems of apple showed that when only a quarter of the root system was supplied with water the transpiration rate was 63% of that when all four root quarters has access to water (West *et al.*, 1970). Use of tritium-enriched water showed lateral transport in the tree to be initiated only after appreciable soil moisture deficits around the unwatered roots had built up.

In the older literature it was often assumed that the absorption of water occurs entirely through the younger regions of roots, e.g. root tips and the root hair zone. However, evidence has accumulated that all of the tree roots, woody as well as white, take up water (Atkinson and Wilson, 1980). There do, however, appear to be differences between different types of root. Baxter and West (1977) found that the water flow through intact root systems along a fixed pressure gradient was two or three times as high per unit root surface area when the root system included new white roots as when it did not (cf. p. 364).

The intact root system has a high resistance to water flow. This resistance is clearly dependent on membrane or symplast resistance because killing a root system under water results in a large increase in water uptake at low transpiration rates (Stoker and Weatherley, 1971). Under 50 and 100 kPa imposed pressure gradients, Baxter and West (1977) found that longitudinal resistance to water flow was very high in apple fibre roots with poorly developed secondary xylem, very low in main roots and low in stems. The capacity of trunks to transport water was more than 100 times that of the intact root systems under a low (200 kPa) applied pressure gradient.

As transpiration rates increase, so do the conductances (reciprocal of resistances) of many living root systems (Stoker and Weatherley, 1971; Steudle, 2000). There is some evidence that this is so for apple (Powell, 1974) but the majority of reports (Baxter and West, 1977; Jones *et al.* 1985) show an apparently constant resistance, with leaf water potential declining linearly as transpiration

Table 12.2 *The calculated rootstock shank, rootstock shank + graft union + scion ('Queen Cox') and scion hydraulic resistances derived from measurements of Lp and the hydraulic resistances of the graft union estimated by difference*

Rootstock	Rootstock shank Rr	Rootstock shank + graft union + scion Rrus	Scion Rs	Graft union Ru
'M.27'	3.26	10.25	0.66	6.33
'M.9'	1.82	5.19	0.48	2.90
'MM.106'	0.39	0.64	0.31	−0.07

All differences between rootstocks highly significant.
Reproduced from Atkinson *et al.* (2001) with permission.

rate increases (Landsberg *et al.*, 1975; Lakso, 1994). Apparent changes in conductance with increases in flux at low flux rates may be a consequence of the importance of osmotic forces at these rates (Boyer, 1985).

Rootstocks differ in their resistance to flow. This was deduced by Olien and Lakso (1984), who found that 'Empire' apples on 'M.9' and 'M.26' were under greater water stress at mid-day, as measured by stem water potential, than those on 'M.7', 'MM.106' and 'MM.104' despite lack of effect on transpiration, stomatal conductance or the ability of the scion stem to conduct water. Part of this effect could be attributable to the smaller root systems of dwarfing rootstocks and lower root/shoot ratios of trees budded on them (McKenzie 1956; see also pp. 148–9). Part is due to differences in specific conductivity between rootstocks. Atkinson *et al.* (2001) showed the resistance of rootstock shanks to be greater the more dwarfing the rootstock (Table 12.2), and the graft unions between 'M.27' and 'M.9' and 'Queen Cox' to provide a large resistance to flow whereas that for the union between 'MM.106' and 'Queen Cox' did not do so.

The more dwarfing rootstocks also increase the resistance to flow of scions worked on them, though the effect is less pronounced than the changes in rootstock resistance *per se*.

Variations in the size of trees on the same rootstock are accompanied by variations in root resistance, larger trees having higher root conductivity. The slope of this relationship is, however, less on 'M.9' and 'M.26' than on more vigorous rootstocks (Olien and Lakso, 1986). The rootstock effect on conductivity cannot, therefore, be fully accounted for by a simple linear relationship between tree size (and consequently root size) and root system conductivity.

Xylem vessel number and radius markedly affect resistance; flow through capillaries of the dimensions of xylem vessels is proportional to the fourth power of the capillary radius (Hagen–Poiseuille flow equation). In general, the more dwarfing the rootstock the lower the proportion of xylem tissues in the roots (Beakbane and Thompson 1939; Rogers and Beakbane 1957; McKenzie 1961). Moreover, although there are some anomalies, dwarfing rootstocks have fewer vessels and these are smaller. Vigorous rootstocks may have 25% of the xylem tissue composed of vessels whereas the roots of the same age (2–3 years) from dwarf trees have only 5% (Rogers and Beakbane, 1957). The mean cross-sectional area of vessels of unworked 'M.9' rootstock roots was only 16–50% of that of more vigorous rootstocks in one study (McKenzie, 1961). Scions grafted on 'M.9' rootstocks have somewhat smaller vessels than when on more invigorating rootstocks (McKenzie, 1961).

Fruiting trees have smaller xylem vessels than deblossomed ones (McKenzie, 1961). The effects of some rootstocks on precocity and on relative fruitfulness are therefore likely to affect the resistance to flow of their root systems.

Anatomical differences between graft unions of different stock–scion combinations and the high proportion of non-functional xylem in graft unions involving some dwarfing rootstocks were discussed earlier (pp. 145–6).

Root signals controlling leaf growth and stomatal behaviour

Roots appear to control plant water use and plant water status by sensing soil water status and sending a chemical signal to the shoots when the soil is dry. In a split-root experiment on apple, Gowing *et al.* (1990) subjected half of the roots to soil drying until leaf initiation and expansion in the shoots was inhibited. Excision of the roots in the dry soil resulted in a recovery of leaf growth rate, the effect being almost as large as that of rewatering them (Figure 12.2). When roots are in contact with drying soil they produce ABA in increased quantities. The ABA enters the xylem and is transported to the leaves. This can inhibit stomatal opening even before the shortage of soil moisture causes any measurable change in the water status of the leaves (Davies and Zhang, 1991, Mansfield and McAinsh, 1995) and so reduce transpiration and water stress. ABA may also increase the permeability of roots to water, hence their hydraulic conductivity, and may limit the growth of shoots and enhance that of roots at low water potentials (Mansfield and McAinsh, 1995).

Drought stress generally reduces the production and transport of cytokinins from roots. This can be expected to reduce stomatal apertures and transpiration (Davies and Zhang, 1991).

The fruit tree root system is fairly sparse and is distributed through large soil volumes with varying water status. The roots are therefore subject to different soil water potentials, especially under drying conditions. Shoots are therefore likely to receive steadily increasing signals of moisture deficit and to adapt in

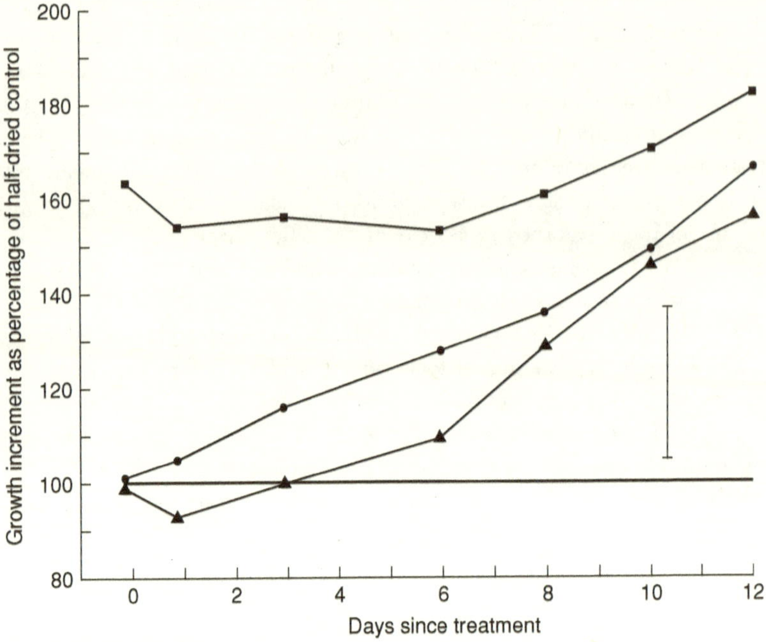

Figure 12.2 Daily growth increment of apple plants following treatment on the 24th day after water was withheld. Plants with half of their roots in dry soil [—]; rewatered plants [•]; plants with roots in dry soil excised [▲] and plants well watered throughout [■]. Reproduced from Gowing *et al.* (1990) with permission.

a number of ways while there is still appreciable available water in some soil zones.

Stem water relations

The longitudinal resistance to water flow in stems is small, being least in the main trunk and increasing towards the branch ends (Landsberg and Jones, 1981).

At high levels of tension, shown by very negative leaf water potentials, the water columns in xylem vessels can snap, leading to embolisms. Jones and Pena (1986) found that the cross-sectional area of wood that was conducting was markedly reduced by a single drought to a leaf water potential of −3.0 MPa. This effect may be partially offset by continued differentiation of new vessels throughout the season.

Water stored in the trunk is withdrawn during the day and replenished at night. This has a buffering effect, reducing the impact of transpirational losses. It can be used as an indicator of transpirational flux and thus water requirement.

Leaf water relations

Water flux through leaves

Total transpiration by a fruit tree is the sum of the transpiration of the individual leaves plus a much smaller amount of transpiration from fruits, stems, sepals, etc. It is therefore primarily controlled by leaf area and leaf conductance.

Apple leaves have thick, waxy, cuticles with very low vapour conductances, so most transpiration is via the stomata. It seems likely that most of the liquid water moving through the leaf goes to evaporation sites close to the underside of the stomata. The movement of the water vapour out of the leaf then follows well-known laws of diffusion (Boyer, 1985). The rate of transpiration per unit leaf area therefore depends on the physical factors controlling evaporation and the degree of opening of the stomatal pores.

The main physical factors controlling evaporation were discussed earlier, among them being wind, which increases evaporation by removing humid air from the leaf surface and keeping the vapour pressure gradient steep. In some circumstances, however, increasing windspeed reduces transpiration because cooling reduces the vapour pressure of water in the intercellular spaces (Mansfield and McAinsh, 1995). This effect can be important in apple. Beukes (1984) found that the mean transpiration rates of 'Granny Smith' apple trees in South Africa were negatively correlated with windspeed. This significant negative effect was a major determinant of the transpiration rate. The other determinants were leaf temperature and available soil water, which were positively associated with transpiration, and the stomatal resistance of the leaves: the higher the resistance the lower the transpiration. Stomatal resistance in apple, as discussed later, increases as the leaf-to-air vapour pressure gradient increases.

Control of stomatal aperture

BASIC MECHANISMS

Stomatal opening and closing results from alterations in the turgor of the two guard cells surrounding the pore. Empirical evidence shows it to be influenced by plant hormones, CO_2, crop load, light, humidity (atmospheric vapour pressure deficit) and windspeed, as well as obvious water stress. These factors may interact with, or be dependent on, one another for their effects.

Changes in guard cell turgor are generally driven by fluxes of cations and anions, notably K^+ balanced either by Cl^- or malate, across the plasma membrane and tonoplast. Stomatal opening reflects a net accumulation of K^+ into guard cells, stomatal closing a net loss of K^+. ABA both inhibits the inward movement of K^+ and activates its efflux (Mansfield and McAinsh, 1995). It appears that ABA need not enter the cytosol of the guard cells to induce these fluxes. The guard cells seem to have two sites of ABA perception involved in the regulation of stomatal aperture, one located at the plasma membrane and one intracellularly (Leung and Giraudat, 1998).

Goode *et al.* (1978a) found that spray application of ABA to apple trees resulted in a very large but short-lived reduction in stomatal conductance and a corresponding reduction in transpiration. The leaf water potential was increased by 2–5 bar in both unirrigated and trickle-irrigated trees but the effects were of too short duration to be of practical value.

The ABA concentration in seedling apple leaves can be increased severalfold by imposed water stress (Landsberg and Jones, 1981). Fernandez *et al.* (1997) found increased ABA in the leaves of droughted trees of 'Imperial Gala' apple, especially when on 'M.9 EMLA' rootstock.

It is noteworthy that ABA is produced in leaves as well as roots and the ABA content of excised leaves of a number of plant species increases at critical levels of water deficit (Davies *et al.*, 1981).

With other plants it has been found that the inhibitory action of ABA can be overcome by IAA and by the cytokinins kinetin and zeatin. Paradoxically synthetic auxins, e.g. NAA, applied as foliar sprays reduce stomatal CO_2 conductance and transpiration of apple (Stopar *et al.*, 1997) as well as other plants.

EFFECTS OF LEAF WATER POTENTIAL (ψ_l)

In general stomata are insensitive to reduction in water potential until a threshold is passed, then they close rapidly and almost completely. In apple the threshold is usually between −1.9 and −2.5 MPa (West and Gaff, 1976; Warrit *et al.*, 1980, Atkinson *et al.*, 2000). Landsberg *et al.* (1975), however, concluded that stomatal conductance, k_s, decreased at least 50% when ψ_l fell below −1.2 to −1.3 MPa, and West and Gaff (1976) found no effect of changes in leaf water potential above −3.0 MPa in CO_2-free air. The relationship between g_l and ψ_l varies between different genotypes (rootstock cultivars) of apple, and also with leaf age and leaf preconditioning to water deficits (Atkinson *et al.*, 2000). Lakso *et al.* (1984) found that the water potential at which stomata closed (to a conductance of 0.1 cm s^{-1}) was above −2 MPa in June but around −4 MPa in September in a dry year (Figure 12.3). This change was associated with increasing osmotic potential.

EFFECTS OF CARBON DIOXIDE

The intracellular concentration of CO_2 in the leaf is a major factor controlling stomatal aperture. In controlled environment studies, Warrit *et al.* (1980) found

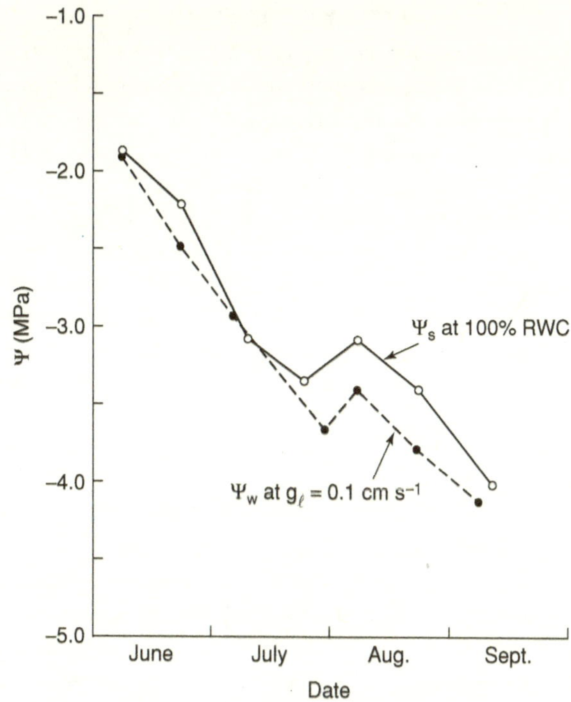

Figure 12.3 Seasonal changes in mature apple leaf osmotic potential (ψ_S) at 100% relative water content (RWC) and in the water potential (ψ_W) required to close stomata to a conductance of 0.1 cm s^{-1}. Reproduced from Lakso *et al.* (1984) with permission.

a fairly steady increase in stomatal conductance of 'Golden Delicious' apple leaves as ambient CO_2 concentration was reduced from about 750 μl l^{-1} to about 50 μl l^{-1} but for some other cultivars there was a step-change with very low conductances at more than 550 μl l^{-1} and high conductances at less than about 350 μl l^{-1}. West and Gaff (1976) found that whereas leaf resistance in normal air (around 300 μl l^{-1} CO_2) increased rapidly as water potential fell below −1.9 MPa in CO_2-free air it was independent of water potential to −3.0 MPa.

EFFECTS OF LIGHT

The long-established effect of light on stomatal opening may operate, at least in large part, through its effects on internal CO_2 concentration. West and Gaff (1976) found maximum conductance in 'Granny Smith' apple leaves to be attained at 15 W m^{-2} (about 3% of full sunlight) with no response to further increases in intensity. Others have found light-saturation at about 10% of full sunlight (Lakso, 1994).

EFFECTS OF CROP LOAD

Hansen (1971) observed that the uptake of water by fruiting trees of 'Golden Delicious' apple was about 80% more than that of non-fruiting trees in August and 3.8 times as great in October. In late August the average size of the stomatal apertures, measured microscopically, was 2.6 times as great in the fruiting as in the non-fruiting trees. Buwalda and Lenz (1992) found the water uptake per unit root weight was more than twice as high in fruiting as in non-fruiting trees of apple cvs. 'Golden Delicious', 'Cox's Orange' and 'Gloster'. These crop load effects may involve effects on internal CO_2 concentrations and the lower ABA concentrations found in leaves of fruiting than non-fruiting trees. Giuliani *et al.* (1997) found canopy conductances of fruiting trees of 'Smoothee Golden Delicious'/'Pajam 2' to be much higher at all VPD levels than those of de-fruited trees and fruiting trees to transpire at up to twice the rate of de-fruited trees.

EFFECTS OF TEMPERATURE

Warrit *et al.* (1980) found that the stomatal conductance of 'Golden Delicious' apple was not usually affected by leaf temperature over the range 13–29 °C, provided the humidity gradient was kept constant. West and Gaff (1976) found a maximum conductance at 23 °C and then a progressive pronounced decline up to 40 °C. Giuliani *et al.* (1997), however, found canopy conductance to increase with temperature over the range 20–35 °C with a positive linear relationship at atmospheric VPD of both 0.5 and 1.0 kPa. The negative relationship between atmospheric VPD and canopy conductance was pronounced at 20, 25, 30 and 35 °C for fruiting trees but was only shown at 35 °C by non-fruiting trees.

EFFECTS OF LEAF—AIR HUMIDITY GRADIENT

Stomatal conductance declines as the leaf-to-air vapour pressure difference increases (Warrit *et al.*, 1980; Fanjul and Jones, 1982). The effects are large, conductances at a 2.5 kPa vapour pressure difference being only about a fifth of those at a 0.5 kPa difference (Fanjul and Jones, 1982). This results in transpiration initially rising with increases in the vapour pressure gradient and then declining with increasingly severe vapour pressure deficits as the effects on stomatal closure outweigh those on evaporation *per se* (West and Gaff, 1976). The stomatal response to changes in atmospheric humidity is very rapid, most of it occurring within 1 minute and some within 15 seconds (Fanjul and Jones, 1982). The leaf therefore seems to have some mechanism for sensing the ambient humidity and triggering a rapid stomatal response. It is possible that stomata respond to direct cuticular loss of water from guard and subsidiary cells, not passing through the stomata, which will be directly related to ambient humidity (Farquhar, 1978).

This mechanism appears to be much more effective in apple than in Asian pear. Higgins *et al.* (1992) found that transpiration of Asian pear leaves increased linearly with increases in leaf-to-air vapour pressure difference but transpiration of apple (which showed much greater changes in leaf conductance) was almost constant over a wide range of vapour pressure differences.

EFFECTS OF WINDSPEED

In general stomata partially close as windspeed increases (Mansfield and McAinsh, 1995). This is attributed to the fact that although wind takes water vapour away from the leaf surface, it brings CO_2 towards it which could result in partial stomatal closure. It also seems that plant hormones (ABA, IAA and cytokinins) can induce changes in the sensitivity to CO_2 which act to prevent excessive transpiration under windy conditions.

EFFECTS OF IRRIGATION AND SOIL MOISTURE STRESS

Irrigation of apple trees under conditions where water supply is otherwise suboptimal results in an increase in transpiration per unit leaf area and in stomatal conductance (Gowing *et al.*, 1990; Alleyne *et al.*, 1989; Fernandez *et al.*, 1997). Leaf conductance can be up to about twice as high in well-watered trees as in those subject to severe, but not irreversibly damaging, water stress.

Control of leaf water potential (ψ_l)

During the day, apple and pear leaf water potentials are lower when transpiration rates are higher. Landsberg *et al.* (1975) showed a linear negative relationship ($r^2 = 0.94$) applying to both control and droughted trees. This is largely a result of the high resistances in the soil–leaf continuum. The daily patterns are very similar over widely different environments, with minimum ψ_l values of -2 to -2.5 MPa in Griffith, Australia; East Malling, England (Figure 12.4), and Michigan, USA (Fernandez *et al.*, 1997). This indicates effective physiological control, primarily by stomatal control of transpiration.

Irrigation under moderate levels of evaporation does not influence ψ_l as long as the soil moisture potential is above -0.08 MPa. When soil moisture stress in unirrigated plots becomes more negative than this, ψ_l becomes 0.53–0.63 MPa more negative than that of irrigated trees under English conditions (Goode and Higgs, 1973). Fernandez *et al.* (1997) found the ψ_l of water-stressed trees to be up to about 0.5 MPa more negative than that of irrigated trees in Michigan.

Stomatal closure in dry conditions can reduce the effect of drought on leaf water potential, especially in younger trees. This stomatal closure may, unusually, be so effective in reducing water loss that droughted trees have higher leaf water potentials than irrigated trees at the end of a season (Jones *et al.*, 1983).

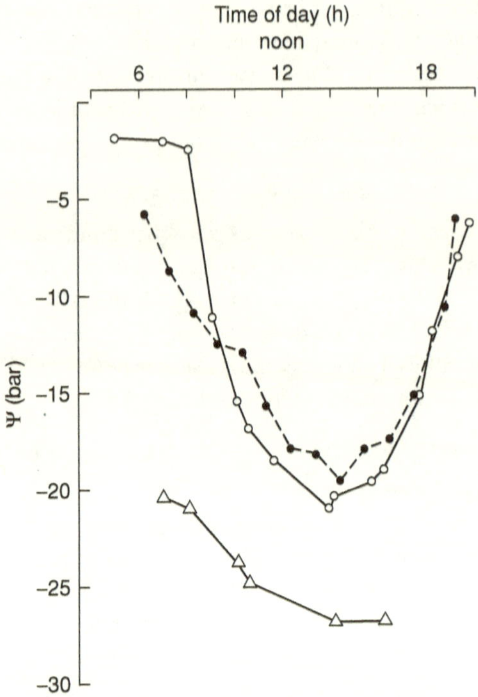

Figure 12.4 Diurnal changes in ψ_W (o–o) and ψ_S (Δ–Δ) of an irrigated apple tree on a sunny day in July in southeast England and diurnal change in ψ_W of an irrigated pear tree (•--•) growing at Griffith, NSW (from Klepper, 1968). Reproduced from Goode and Higgs (1973) with permission.

Control of leaf relative water content (RWC)

RWC is approximately linearly related to ψ_l. In orchard trees of 'Golden Delicious'/'M.9' in England it varied from nearly 100% at leaf (xylem) potentials of around −0.25 MPa or less, to 85–91% at −2.5 MPa (Jones and Higgs, 1979; Fanjul and Rosher, 1984). Previously water-stressed trees had higher RWC values at any given water potential. At Geneva, New York, RWC declined through the morning from around 80% to 62% as ψ_l declined from −1 MPa to −2.4 MPa. The changes were reversed in the afternoon (Davies and Lakso, 1979).

Control of leaf osmotic (ψ_s) and turgor (ψ_p) potential

As transpiration-stream water tensions increase, i.e. as its water potential ψ_w becomes more negative, cells lose water to it along gradients of potential. This may, in many plants, simply continue until the resultant increase in concentration of the cell sap raises its osmotic water potential enough for cell sap water potential to be in equilibrium with that of the transpiration stream. In this new equilibrium, attained by 'passive osmotic adjustment', the cells inevitably have lower turgor than before (equation 12.4).

In apple a second type of osmotic adjustment may take place in response to water stress: the cell osmotic potential may rise over and above that due to the concentration of sap by water loss. This 'active osmotic adjustment' can be calculated following measurement of the osmotic potential of expressed sap (Wang and Stutte, 1992). Carbohydrates appear to be the primary osmotic solutes that change with increasing water stress.

Sorbitol concentration increases linearly with increasing drought stress as measured by increasingly negative leaf water potentials. Sucrose and, especially, starch content declines. Sorbitol is a very effective osmoticum. It has a small molecule (it is a 6-carbon alcohol) so that it has a greater effect on osmotic potential per unit mass than compounds with larger molecules, and it has little effect on cell metabolism. The amount of active osmotic adjustment in mature leaves can be as much as 2.5 MPa (Lakso, 1994), although in many circumstances it is much lower.

This increase in osmotic potential results in an increase in leaf turgor potential at any given level of ψ_w. Fanjul and Rosher (1984), for example, found that the water potential at which there was zero turgor was 1.1 MPa more negative in field-grown trees grown under stress-inducing conditions than in those irrigated. At a typical 'stress' ψ_w of -2.4 MPa turgor potential of leaves on trees grown under drought conditions was about 1.1 MPa while that on trees which had not osmotically adjusted was about 0.6 MPa.

Control of leaf growth

Leaf expansion is a linear function of ψ_p (Figure 12.5), being negligible at $\psi_p =$ 0.25 MPa and increasing steadily up to $\psi_p = 1.5$ MPa (Davies and Lakso, 1979). Young leaves and shoot tips do not show active osmotic adjustment, as the season progresses, in the same way that mature leaves do (Figure 12.6). As a consequence shoot and leaf growth is checked under water stress conditions which have little or no effect on stomatal apertures of adapted mature leaves.

Fruit water relations

Fruits have a very high content of osmotica and show increasingly negative osmotic potentials as the season progresses. This enables maintenance of turgor potential in the face of declining water potential (Mills *et al.*, 1997). Fruit water potential itself is generally much higher than leaf water potential, maximum

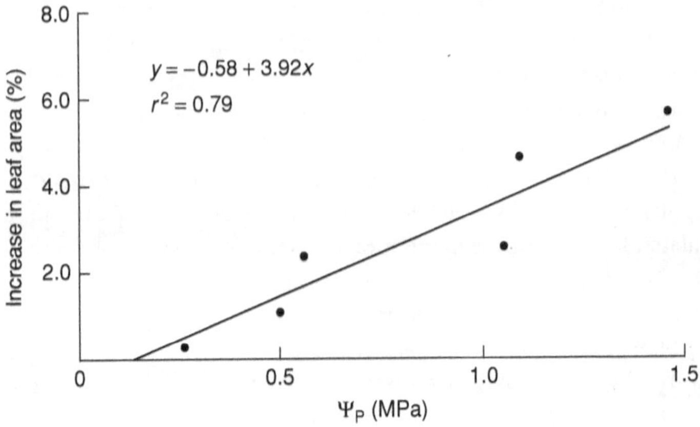

Figure 12.5 The relationship between the increase in area of apple leaves and turgor potential (ψ_P) over 24 h in the dark at 30 °C and 100% RH. Reproduced from Davies and Lakso (1979), with permission.

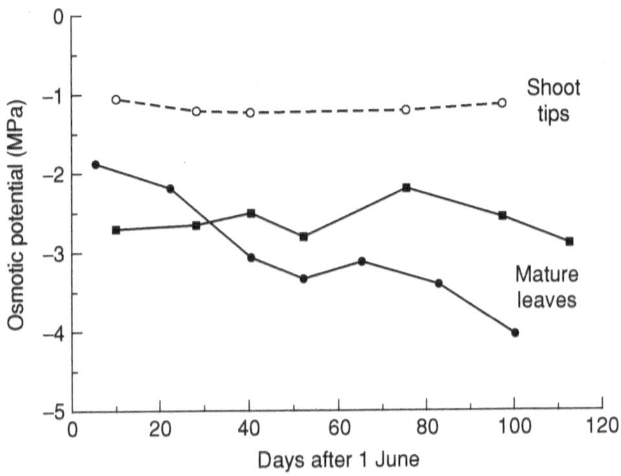

Figure 12.6 Seasonal patterns of osmotic potentials of expanding shoot tips (○) and mature leaves of apple trees in New York. The different patterns for mature leaves (●, ■) are from seasons with different rainfall patterns. Reproduced from Lakso (1994) with permission.

values found by Goode *et al.* (1979) being about −24 bar for leaves and −13 bar for fruits.

Apple fruits shrink during the day as leaf water potentials become more negative and then expand again as leaf water potential begins to increase.

Continuous rain prevents both the diurnal depression in ψ_w and contraction of fruits. Most studies have shown that the greater part of the shrinkage is a result of water moving out of the fruit back through the pedicel in equilibration with the water tensions in the conducting system (Lang and Volz, 1998). In some cases the major part of the water loss is by direct evaporation from the fruits themselves (Jones and Higgs, 1982) but the re-expansion at night must depend on the balance of water potentials.

Fruit pedicels have low hydraulic conductance. Especially for some cultivars and late in the season, the xylem becomes largely non-functional and water movement into the apple is largely through the phloem. This is important in relation to calcium flux and was discussed in Chapter 10.

Fruit surface (skin) conductance varies between cultivars, being higher in russeted cultivars than those with waxy skins.

Integrated effects of water stress

Fruit set can be very adversely affected by levels of water stress too mild to have any adverse influence on shoot extension growth (Powell, 1974). Under many environmental conditions this may, however, not cause a problem as soil moisture levels in the early part of the season are high because of winter rains or frost-protection irrigation. The severe effects of water deficits on fruit bud formation under some circumstances (Landsberg and Jones, 1981) may not be evident in many areas for the same reason.

The greatest effect of reduced water supply on the vegetative growth and dry matter production by fruit trees is that which arises through the reduction of leaf area per tree (Table 12.3). Root growth is much less affected so the root/shoot ratio is increased.

Table 12.3 *Effects of water supply on leaf area (m^2/tree) and dry weight of leaves, stems, fruits and roots (g) of fruiting trees of 'Golden Delicious'/'M.9'*

	Water rate (%)		
	100	50	25
Leaf area	6.34	2.44	1.86
Leaf dry weight	517	228	147
Stem dry weight	1188	608	375
Fruit dry weight	2070	1088	445
Root dry weight	286	264	270

Calculated from data of Buwalda and Lenz (1992).

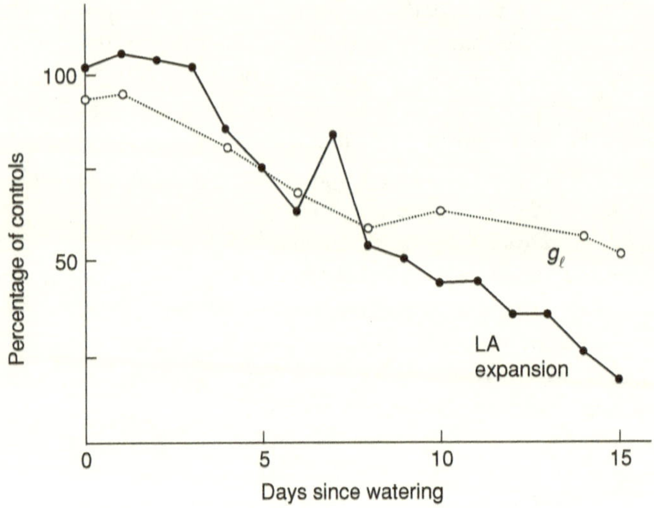

Figure 12.7 Responses of apple leaf stomatal conductance (g_l) and leaf area expansion to soil water stress on outside-grown potted trees. Reproduced from Lakso (1985), with permission.

The different responses of root and leaf growth to drought arise from their differing levels of osmotic adjustment and in the different consequences of their poor hydraulic linkages, due to immature xylem, in their different environments. For root tips, when water potential is suddenly reduced there is rapid osmotic adjustment. This allows for partial recovery of turgor. The cell walls also loosen and growth resumes. Just-unfolding leaves have very low petiole conductance and do not show rapid osmotic adjustment or cell wall loosening. Consequently, when subject to drought their growth is inhibited (Hsiao and Xu, 2000) and they also wilt readily.

The impact of these effects on shoot and leaf growth on orchard productivity depends on tree age, planting density and vigour of growth in the absence of serious water stress. The effects of checking the growth of young trees, before the orchard canopy has attained its final dimensions, will normally be very adverse. However, once the canopy has attained its optimal dimensions and leaf area on an orchard basis, some reduction in annual shoot growth is not necessarily disadvantageous.

Water stress effects on stomatal conductance are much less severe than those on leaf expansion (Figure 12.7). Stomatal conductance is a major factor controlling both transpiration and photosynthesis per unit leaf area but whereas water vapour conductances of approximately 0.3–0.6 cm s^{-1} permit maximum rates of photosynthesis (Lakso, 1985), transpiration continues to increase with stomatal conductance above these levels. Stomatal conductances in apple are usually in the range 0.3–0.8 cm s^{-1} but have fairly frequently been reported in the range 1.1–1.3 cm s^{-1} and occasionally up to 1.8 cm s^{-1} (Palmer, 1986). One-year-old apple trees subjected to water stress had a higher water use efficiency (CO_2 uptake/water loss) than controls at any given vapour pressure deficit (Flore *et al.*, 1985).

Under orchard conditions water stress effects on photosynthesis per unit leaf may only occur under severe stress conditions or be very limited. Fanjul (1982), for example, found little relationship between photosynthesis of 'Bramley' apple leaves and water potential over the range −0.75 to −2.0 MPa. Whole-tree photosynthesis, which incorporates effects of leaf area, shows much more consistent negative effects of imposed droughting than does individual leaf photosynthesis (Fernandez *et al.*, 1997).

The osmotic adjustment of mature leaves contributes to keeping their stomata open, so enabling photosynthesis to be maintained, while the concomitant check to shoot growth and new leaf production helps conserve water use. The maintenance of root growth enables the tree to tap ever greater soil volumes.

Fruiting simultaneously increases water use through its effects on stomatal conductance and through reducing root growth. This obviously accentuates the effect of drought, i.e. potential water stress, on vegetative growth. This may partly explain the adverse interaction between limitations to water supply and crop load with respect to fruit size (Goode *et al.*, 1978b). Effects of differences in fruiting may also contribute to differences between cultivars in some aspects of tree water relations (Higgs and Jones, 1991).

Irrigation

Irrigation is carried out to achieve economic benefit through the reduction of plant water stress. The amount and frequency of water application is based on assessment of crop needs, adjusted to take account of inefficiencies in the irrigation system (e.g. evaporation from open water channels and surface run-off) and effective natural precipitation, i.e. rainfall adjusted for losses from surface run-off and drainage.

Maintenance (depletion) irrigation

This is the traditional system based on full replenishment of soil moisture in the active rooting zone, to field capacity, at timings such that the soil moisture in the rooting zone is never depleted by more than a given percentage of available soil water. The objective is to supply sufficient water, while avoiding run-off or deep percolation or the creation of anaerobic conditions.

Frequency of irrigation under this system depends on (1) the rate of water depletion (2) the maximum allowable depletion (3) the volume of the active rooting zone, (4) the soil water holding capacity and (5) the soil infiltration rate.

Estimation of water depletion for scheduling purposes falls into three categories: (1) Calculation of ET_c from climatic data which is usually provided by local advisory services. (2) Measurement in changes in soil water content. Gravimetric and matric potential (e.g. tensiometer) measurements have been supplemented or replaced by, for example, neutron probe or electric capacitance measurements often carried out by specialist contractors. With experience the moisture content of the soil can be judged by the feel of a sample. For all soil measurements sampling representative of the rooting zone is essential. (3) Pan evaporation based scheduling. Evaporation data is provided by local advisory services. It is then adjusted by an empirically determined factor for the area, crop and soil type, e.g. 0.9 for apples with full ground cover in Washington State (Evans, 1982), and by the percentage of the land actually covered by the foliage of the trees being irrigated.

The maximum allowable depletion for apples has been assumed to be 50–60% (Evans, 1982). The volume of the effective rooting zone varies enormously with tree age, soil type and soil characteristics, and must be determined empirically. Soil water holding capacity and infiltration rates are functions of soil type and must be measured or estimated from the soil classification.

These concepts are also relevant to other forms of irrigation. Plant parameter based irrigation scheduling is currently being investigated.

It has not, so far, been possible to develop universally applicable systems to determine irrigation requirement from the measurement of plant water status. Osmotic adjustments by the leaves and fruits can influence the effects of any given leaf water potential not only on stomatal aperture and photosynthesis but also on fruit growth. A further complication is that stomatal closure, probably in response to root-sensed water stress, can actually increase leaf water potential especially towards the end of the season (Jones *et al.*, 1983).

Infra-red thermometry can be used to determine when plants are showing stomatal closure in response to water stress. Leaf temperature rises as stomata close and evaporative cooling decreases. Crop water stress indices (CWSIs) have therefore been calculated based on the difference between measured leaf temperatures and the typical canopy temperature of a well-watered crop. This technique is used as a basis for irrigation scheduling on a number of crops. Jones *et al.* (1997) have pointed out some problems in its application in variable climates. The fact that several key aspects of apple tree growth and fruiting are adversely affected by less severe water stress than that which causes stomatal closure also suggest limits to the applicability of this technique.

Trunk shrinkage, stem water potential, leaf water potential, stomatal resistance and the rate of apple fruit growth have been used with varying success as physiological parameters relevant to irrigation scheduling (Bravdo, 2000). Leaf shrinkage has been found to be the most sensitive indicator of water stress in other tree fruits. It can be measured as a change in leaf thickness which is linearly related to leaf turgor.

Traditional depletion or maintenance irrigation may be by surface systems or sprinklers. Surface irrigation is largely into basins around each individual tree. This is a very efficient way of supplying water to young, widely spaced trees. Furrow irrigation is also used but has been largely replaced by overhead sprinkler irrigation because of its ease of management and flexibility of use. Sprinklers apply water over most or all of the orchard surface and so can be wasteful. They are most valuable when multi-purpose, e.g. when also used for frost protection or tree cooling.

Trickle irrigation

Trickle irrigation (Elfving, 1982) differs from conventional bulk water application in that smaller volumes are applied to the plant at more frequent intervals and the application is to only a limited part of the soil surface area. Water is piped through plastic tubing and is released only at selected points through emitters (drippers). Typically the tubing will run on the soil surface down the centre line of a hedgerow, or to either side of it, with emitters at regular intervals. The more frequent the emitters, the less water is applied through each and the less the risk of water loss through drainage.

The shape of the irrigated soil volume under the emitter varies with the hydraulic characteristics of the soil. The wetted 'bulb' is always deeper than it is wide but heavy soils have a higher horizontal to vertical ratio of bulb dimensions than lighter (e.g. sandy) ones.

As discussed in Chapter 3, roots proliferate in the wetted volume to give a much higher rooting density, especially of fine roots, in that volume than is characteristic of rainfed or sprinkler-irrigated orchards. There is some lateral water movement from roots in the wetted bulb to roots outside this (Bravdo, 2000). This may contribute to the overall annual efficiency of root systems in areas with alternation of wet and dry seasons, or even just periods, and to tree anchorage, root hormone production, etc.

The irrigation water requirement is calculated much as for standard maintenance irrigation but application is based on high frequency, which maintains the wetted bulb at a high water potential, and relies much less on the soil as a water reservoir. The normal frequency in hot climates is once every 1–3 days, but in extreme hot weather and on sandy soil may be a few times each day (Bravdo, 2000). In humid areas where trickle irrigation is used to supplement rainfall, normal scheduling may aim to provide 50% of daily water needs leaving roots outside the irrigated zone to supply the rest (Anon., 1982). More may be applied on sandy soils and in drought years.

Trickle irrigation has a very high water use efficiency, typically 85–90%, because of the enclosed nature of the water delivery system and the restriction of the water supply to the soil zone with the highest density of fruit tree roots.

Deficit irrigation

Deficit irrigation (Behboudian and Mills, 1997) is a system of managing soil water supply to impose periods of plant water deficit in such a way as to be economically beneficial. It involves applying less than the calculated water need. Initially it was used in Australia to control tree vigour in high-density plantings by imposing a water deficit at a period of rapid shoot and slow fruit growth (Chalmers *et al.*, 1981, 1986). It has subsequently evolved into a technology for reducing water use while obtaining some benefits from control of canopy growth, e.g. less shade on fruits within the canopy, possibly less water demand because of the check to leaf growth, and reduced pruning costs.

Two types of deficit irrigation are under test. The term regulated deficit irrigation (RDI) is commonly used to describe deficit irrigation early in the season when shoot growth is rapid but before rapid fruit growth. Late-season RDI has also been used sometimes in an attempt to improve fruit quality.

The key to its successful use is a difference in the times during the growing season when fruit development and shoot growth are sensitive to water stress.

Fruit set and flower bud differentiation are very sensitive to water stress (Powell, 1974, 1976) and apple fruit growth is most sensitive to water stress late in the season. Regulated deficit irrigation is therefore typically applied to apple over the period from 55 to 100 days after full bloom. Even this may have an adverse effect on fruit growth during the deficit period and, conversely, late deficit irrigation may not have the expected adverse effects (Mills *et al.*, 1997).

On Asian pears (*Pyrus serotina*) deficit irrigation can reduce consumption of water without affecting either yield or fruit size, although when the RDI treatment was imposed from 42 days after full bloom, fruit bud formation (return bloom) was reduced (Caspari *et al.*, 1994). With 'Bartlett' pear (*P. communis*), RDI not only checked shoot growth but after normal irrigation was resumed fruit growth and size and consequently yield, were increased (Chalmers *et al.*, 1986). This was attributed to (1) osmotic adjustment of the fruits while under deficit irrigation and accelerated growth when this was replaced by normal irrigation, and (2) less negative leaf water potentials in the RDI trees once this was replaced by normal irrigation than in trees irrigated normally throughout.

This may have been a consequence of the relative reduction in leaf area and increase in roots as a result of water stress. It may also reflect a carried-forwards effect on stomatal conductance.

RDI is clearly more relevant to arid than humid climates where control of the period on deficit is difficult. The variability in responses suggests that specific regimes may have to be devised for cultivars differing in the period of flower bud differentiation and in the extent to which reduction in fruit size reduces market value.

Partial rootzone drying

Another approach to minimizing water use with little or no adverse effect on fruit growth and a useful checking of shoot growth is that of partial root drying (PRD). This seeks to exploit the signals sent by roots in dry soil to reduce leaf and shoot growth and induce partial stomatal closure (Davies *et al.*, 2000). ABA seems to be the major signalling chemical but is not necessarily the only one. From work on other fruit crops it seems that PRD may exert a particularly favourable differential effect on fruit and leaf growth where the fruits are relatively isolated from the hydraulics of the rest of the plant, ABA in the xylem sap being thought to be particularly effective. In apple, xylem flux into the fruits declines through the season and its relative magnitude varies with cultivar (Lang and Ryan, 1994) so some cultivars may respond better to PRD than others.

PRD has been furthest developed in grapes, where it improves water use efficiency (yield of crop per unit of water) by up to 50% without significant crop reduction (Stoll *et al.*, 2000). This is achieved by maintaining half of the root system in a dry or drying state while irrigating the other half. The wetted and dried sides of the root system are alternated on a 10–14-day cycle: this is easily done with trickle irrigation lines. PRD results in increased xylem sap ABA and a reduction in zeatin and zeatin riboside concentrations in roots, shoot tips and buds. There is a nocturnal net flux of water from wetter roots to the roots in dry soil. The technique is being evaluated for apples.

Over-tree irrigation for temperature and humidity control

This has four main objectives: to reduce heat-stress in summer; to reduce water-stress; to achieve evaporative cooling in winter or spring to prevent 'de-chilling' in warm winter areas and to delay blossoming in frost-prone areas; and to give frost protection in winter by utilizing the latent heat of fusion.

Unrath (1972a, b) found that apple fruits could be cooled by 8–12 °C, with consequent improvements in red coloration and sometimes improvements in size, shape and incidence of bitter pit and cork spot.

In a cooler climate Goode *et al.* (1979) found that over-tree misting increased leaf ψ_w and ψ_p and also fruit ψ_w. It led to an increase in fruit bud production, number of fruits and yield per tree but did not affect fruit size.

Uses of over-tree irrigation to prevent de-chilling, delay blossoming and prevent frost damage have been discussed earlier (pp. 185, 285–6).

Droughting

In the tropics droughting may be used to induce leaf fall and stimulate budbreak under conditions of inadequate winter chilling.

Effects of flooding

Flooding affects apple and pear trees by reducing availability of oxygen around their roots. It has little adverse effect when the trees are leafless and dormant but can have lethal effects after leafing out. These appear to be largely due to increases in root resistance, leading to water-stress and wilting of leaves (Lakso, 1994).

Rootstocks differ in their sensitivity to flooding. This was discussed in Chapter 2.

References

Allen, R.G., Smith, M., Pereira, L.S. and Pruitt, W.O. (1997). Proposed revision to the FAO procedure for estimating crop water requirements. *Acta Horticulturae* **449**, 17–33.

Alleyne, V., Larsen, F.E. and Higgins, S.S. (1989). Water relations of container-grown, virus-tested and common apple (*Malus domestica* Borkh). *Scientia Horticulturae* **38**, 117–29.

Anon, (1982). *Irrigation Reference Book 138*. London: Her Majesty's Stationery Office.

Atkinson, C.J., Else, M.A., Taylor, L. and Webster, A.D. (2001). The rootstock graft union: a contribution to the hydraulics of the worked fruit tree. *Acta Horticulturae* **557**, 117–22.

Atkinson, C.J., Policarpo, M., Webster, A.D. and Kingswell, G. (2000). Drought tolerance of clonal *Malus* determined from measurements of stomatal conductance and leaf water potential. *Tree Physiology* **20**, 557–63.

Atkinson, D. (1980). The distribution and effectiveness of the roots of tree crops. *Horticultural Reviews* **2**, 424–90.

Atkinson, D. and Thomas, C.M.S. (1985). The influence of cultural methods on the water relations of fruit trees. *Acta Horticulturae* **171**, 371–82.

Atkinson, D. and White, G.C. (1981). The effect of weeds and weed control on temperate fruit orchards and their environment. In *Pests, Pathogens and Vegetation*, ed. J.M. Thresh, pp. 415–28. London: Pitman.

Atkinson, D. and Wilson, S.A. (1980). The growth and distribution of fruit tree roots; some consequences for nutrient uptake. In *Mineral Nutrition of Fruit Trees*, ed. D. Atkinson, J.E. Jackson, R.O. Sharples and W.M. Waller, pp. 137–50. London: Butterworths.

Baxter, P. and West, D. (1977). The flow of water into fruit trees. I. Resistances to water flow through roots and stems. *Annals of Applied Biology* **87**, 95–101.

Beakbane, A.B. and Thompson, E.C. (1939). Anatomical studies of stems and roots of hardy fruit trees. II. The internal structure of the roots of some vigorous and some dwarfing apple rootstocks, and the correlation of structure with vigour. *Journal of Pomology and Horticultural Science* **17**, 141–9.

Behboudian, M.H. and Mills, T.M. (1997). Deficit irrigation in deciduous orchards. *Horticultural Reviews* **21**, 105–31.

Beukes, D.J. (1984). Transpiration of apple trees as related to different meteorological, plant and soil factors. *Journal of Horticultural Science* **59**, 151–9.

Boyer, J.S. (1985). Water transport. *Annual Review of Plant Physiology* **36**, 473–516.

Bravdo, B-A. (2000). Irrigation of temperate fruit trees in dry and warm conditions. In *Temperate Fruit Crops in Warm Climates*, ed. A. Erez, pp. 49–76. Dordrecht: Kluwer.

Buwalda, J.G. and Lenz, F. (1992). Effects of cropping, nutrition and water supply on accumulation and distribution of biomass and nutrients for apple trees on 'M.9' root systems. *Physiologia Plantarum* **84**, 21–8.

Caspari, H.W., Behboudian, M.H. and Chalmers, D.J. (1994). Water use, growth, and fruit yield of 'Hosui' Asian pears under deficit irrigation. *Journal of the American Society for Horticultural Science* **119**, 383–8.

Chalmers, D.J., Burge, G., Jerie, P.H. and Mitchell, P.D. (1986). The mechanism of regulation of 'Bartlett' pear fruit and vegetative growth by irrigation withholding and regulated deficit irrigation. *Journal of the American Society for Horticultural Science* **111**, 904–7.

Chalmers, D.J., Mitchell, P.D. and van Heek, L. (1981). Control of peach tree growth and productivity by regulated water supply, tree density, and summer pruning. *Journal of the American Society for Horticultural Science* **106**, 307–12.

Davies, F.S. and Lakso, A.N. (1979). Diurnal and seasonal changes in leaf water potential components and elastic properties in response to water stress in apple trees. *Physiologia Plantarum* **46**, 109–14.

Davies, W.J. and Zhang, J. (1991). Root signals and the regulation of growth and development of plants in drying soil. *Annual Review of Plant Physiology and Molecular Biology* **42**, 55–76.

Davies, W.J., Bacon, M.A., Thompson, D.S., Sobeih, W. and González Rodriguez, L. (2000). Regulation of leaf and fruit growth in plants growing in drying soil: exploitation of the plants' chemical signalling system and hydraulic architecture to increase the efficiency of water use in agriculture. *Journal of Experimental Botany* **51**, 1617–26.

Davies, W.J., Wilson, J.A., Sharp, R.E. and Osonubi, O. (1981). Control of stomatal behaviour in water-stressed plants. In *Stomatal Physiology*, ed. P.G. Jarvis and T.A. Mansfield. Society for Experimental Biology Seminar Series; 8, pp. 163–85. Cambridge: Cambridge University Press.

Doorenbos, J. and Pruitt, W.O. (1977). Guidelines for predicting crop water requirements. *FAO Irrigation and Drainage Paper 24*. Rome: UN Food & Agriculture Organization.

Elfving, D.C. (1982). Crop response to trickle irrigation. *Horticultural Reviews* **4**, 1–48.

Evans, R.G. (1982). Irrigation scheduling and monitoring water requirements. In *Tree Fruit Water Management and Irrigation. The Proceedings of the 1982 Pacific Northwest Tree Fruit Short Course*, ed. R.B. Tukey, pp. 201–30. Pullman: Washington State University.

Fanjul, L. (1982). Effects of water stress on photosynthesis in apple leaves. PhD thesis, London, Wye College/East Malling Research Station.

Fanjul, L. and Jones, H.G. (1982). Rapid stomatal responses to humidity. *Planta* **154**, 135–8.

Fanjul, L. and Rosher, P.H. (1984). Effects of water stress on internal water relations of apple leaves. *Physiologia Plantarum* **62**, 321–8.

Farquhar, G.D. (1978). Feedforward responses of stomata to humidity. *Australian Journal of Plant Physiology* **5**, 787–800.

Fernandez, R.T., Perry, R.L. and Flore, J.A. (1997). Drought response of apple trees on three rootstocks. II. Gas exchange, chlorophyll fluorescence, water relations and leaf abscisic acid. *Journal of the American Society for Horticultural Science* **122**, 841–8.

Flore, J.A., Moon, J.W. and Lakso, A.N. (1985). The effect of water stress and vapour pressure gradient on stomatal conductance, water use efficiency and photosynthesis of fruit crops. *Acta Horticulturae* **171**, 207–18.

Giuliani, R., Nerozzi, F., Magnanini, E. and Corelli-Grappadelli, L. (1997). Influence of environmental and plant factors on canopy photosynthesis and transpiration of apple trees. *Tree Physiology* **17**, 637–45.

Goode, J.E. (1975). Water storage, water stress and crop responses to irrigation. In *Climate and the Orchard*, ed. H.C. Pereira. Research Review No 5, pp. 51–62. Farnham Royal, UK: C.A.B.

Goode, J.E. and Higgs, K.H. (1973). Water, osmotic and pressure potential relationships in apple leaves. *Journal of Horticultural Science* **48**, 203–15.

Goode, J.E., Higgs, K.H. and Hyrycz, K.J. (1978a). Abscisic acid applied to orchard trees of 'Golden Delicious' apple to control water stress. *Journal of Horticultural Science* **53**, 99–103.

Goode, J.E., Higgs, K.H. and Hyrycz, K.J. (1978b). Nitrogen and water effects on the nutrition, growth, crop yield and fruit quality of orchard-grown 'Cox's Orange Pippin' apple trees. *Journal of Horticultural Science* **53**, 295–306.

Goode, J.E., Higgs, K.H. and Hyrycz, K.J. (1979). Effects of water stress control in apple trees by misting. *Journal of Horticultural Science* **54**, 1–11.

Gowing, D.J.G., Davies, W.J. and Jones, H.G. (1990). A positive root-sourced signal as an indicator of soil drying in apple, *Malus* × *domestica* Borkh. *Journal of Experimental Botany* **41**, 1535–40.

Hamer, P.J.C. (1985). Water balance in orchards. *Report of the East Malling Research Station for 1984*, 129–30.

Hansen, P. (1971). The effect of fruiting upon transpiration rate and stomatal opening in apple leaves. *Physiologia Plantarum* **25**, 181–3.

Higgins, S.S., Larsen, F.E., Bendel, R.B., Radamaker, G.K., Bassman, J.H., Bidlake, W.R. and Al Wir, A. (1992). Comparative gas exchange characteristics of potted, glass-house grown almond, apple, fig, grape, olive, peach and Asian pear. *Scientia Horticulturae* **52**, 313–29.

Higgs, K.H. and Jones, H.G. (1991). Water relations and cropping of apple cultivars on a dwarfing rootstock in response to imposed drought. *Journal of Horticultural Science* **66**, 367–79.

Holloway, R.I.C. and White, G.C. (1967). Can apples be grassed-down in the year of planting? *Report of the East Malling Research Station for 1966*, 209–10.

Hsiao, T.C. and Xu, L-K. (2000). Sensitivity of growth of roots versus leaves to water stress: biophysical analysis and relation to water transport. *Journal of Experimental Botany* **51**, 1595–1616.

Hughes, K.A. and Gandar, P.W. (1993). Length densities, occupancies and weights of apple root systems. *Plant and Soil* **148**, 211–21.

Jackson, J.E., Hamer, P.J.C. and Jackson, B. (1989). Water balance and soil water relations studied in a mixed tree/grass/bare-soil system. In *Meteorology and Agroforestry*. Proceedings of an international workshop on the application of meteorology to agroforestry systems and management, ed. W.S. Reifsnyder and T.O. Darnhofer, pp. 431–42. Nairobi: ICRAF.

James, L.G., Erpenbeck, J.M., Bassett, D.L. and Middleton, J.E. (1982). Irrigation requirements for Washington. Estimates and Methodology. Washington State University. Co-operative Extension Bulletin EB 1513. Pullman:

Jones, H.G. and Higgs, K.H. (1979). Water potential–water content relationships in apple leaves. *Journal of Experimental Botany* **30**, 965–70.

Jones, H.G. and Higgs, K.H. (1982). Surface conductance and water balance of developing apple (*Malus pumila* Mill) fruits. *Journal of Experimental Botany* **33**, 67–77.

Jones, H.G. and Pena, J. (1986). Relationships between water stress and ultrasound emission in apple. *Journal of Experimental Botany* **37**, 1245–54.

Jones, H.G., Aikman, D. and McBurney, T.A. (1997). Improvements to infra-red thermometry for irrigation scheduling in humid climates. *Acta Horticulturae* **449**, 259–65.

Jones, H.G., Lakso, A.N. and Syvertsen, J.P. (1985). Physiological control of water status in temperate and sub-tropical fruit trees. *Horticultural Reviews* **7**, 301–44.

Jones, H.G., Luton, M.T., Higgs, K.H. and Hamer, P.J.C. (1983). Experimental control of water status in an apple orchard. *Journal of Horticultural Science* **58**, 301–16.

Kennedy, R.A. and Fujii, J.A. (1982). Apple tree physiology, water relations, photosynthesis and carbohydrate mobilization. In *Water Management and Irrigation of Tree Fruits. The Proceedings of the 1982 Pacific Northwest Tree Fruit Short Course*, ed. R.B. Tukey, pp. 109–32. Pullman: Washington State University.

Lakso, A.N. (1985). The effect of water stress on physiological processes in fruit crops. *Acta Horticulturae* **171**, 275–90.

Lakso, A.N. (1994). Apple. In *Handbook of Environmental Physiology of Fruit Crops*, Vol. 1. *Temperate Crops*, ed. B. Schaffer and P.C. Andersen, pp. 3–42. Boco Raton, FL: CRC Press.

Lakso, A.N., Geyer, A.S. and Carpenter, S.G. (1984). Seasonal osmotic relations in apple leaves of different ages. *Journal of the American Society for Horticultural Science* **109**, 544–7.

Landsberg, J.J. and Jones, H.G. (1981). Apple orchards. In *Water Deficits and Plant Growth*, Vol. VI, ed. T.T. Kozlowski, pp. 419–69. New York: Academic Press.

Landsberg, J.J., Beadle, C.L., Biscoe, P.V., Butler, D.R., Davidson, B., Incoll, L.D., James, G.B., Jarvis, P.G., Martin, P.J., Neilson, R.E., Powell, D.B.B.,

Slack, E.M., Thorpe, M.R., Turner, N.C., Warrit, B. and Watts, W.R. (1975). Diurnal energy, water and CO_2 exchanges in an apple (*Malus pumila*) orchard. *Journal of Applied Ecology* **12**, 659–84.

Lang, A. and Ryan, K.G. (1994). Vascular development and sap flow in apple pedicels. *Annals of Botany* **74**, 381–8.

Lang, A. and Volz, R.K. (1998). Spur leaves increase calcium in young apples by promoting xylem inflow and outflow. *Journal of the American Society for Horticultural Science* **123**, 956–60.

Lenz, F. (1986). Fruit effects on transpiration and dry matter production in apples. In *The Regulation of Photosynthesis in Fruit Trees*, ed. A.N. Lakso and F. Lenz, pp. 101–4. Symposium Proceedings Publication. New York State Agricultural Experiment Station. Geneva, New York.

Leung, J. and Giraudat, J. (1998). Abscisic acid signal transduction. *Annual Reviews of Plant Physiology and Molecular Biology* **49**, 199–222.

Levin, I., Assaf, R. and Bravdo, B. (1979). Soil moisture and root distribution in an apple orchard irrigated by tricklers. *Plant and Soil* **52**, 31–40.

Mansfield, T.A. and McAinsh, M.R. (1995). Hormones as regulators of water balance. In *Plant Hormones*, 2nd edn, ed. P.J. Davies, pp. 598–616. Dordrecht: Kluwer.

McKenzie, D.W. (1956). Interactions between rootstocks and scions with special reference to tissue structure and development. PhD thesis, University of London.

McKenzie, D.W. (1961). Rootstock–scion interactions in apples with special reference to root anatomy. *Journal of Horticultural Science* **36**, 40–7.

Miller, D.E. (1982). Physical characteristics of soils as they relate to irrigation and water management. In *Water Management and Irrigation of Tree Fruits. The Proceedings of the 1982 Pacific Northwest Tree Fruit Short Course*, ed. R.B. Tukey, pp. 3–28. Pullman: Washington State University.

Mills, T.M., Clothier, B.E. and Behboudian, M.H. (1997). The water relations of 'Braeburn' apple fruit grown under deficit irrigation. *Acta Horticulturae* **449**, 385–92.

Olien, W.C. and Lakso, A.N. (1984). A comparison of the dwarfing character and water relations of five apple rootstocks. *Acta Horticulturae* **146**, 151–8.

Olien, W.C. and Lakso, A.N. (1986). Effect of rootstock on apple (*Malus domestica*) tree water relations. *Physiologia Plantarum* **67**, 421–30.

Palmer, J.W. (1986). Seasonal variation of light saturated photosynthetic rate of Golden Delicious apple leaves as influenced by leaf type and crop load. In *The Regulation of Photosynthesis in Fruit Trees*, ed. A.N. Lakso and F. Lenz, Symposium proceedings pp. 30–3. Geneva, New York: New York State Agricultural Experimental Station.

Penman, H.L. (1948). Natural evaporation from open water, bare soil and grass. *Proceedings of the Royal Society A* **193**, 120–45.

Powell, D.B.B. (1974). Some effects of water stress in late spring on apple trees. *Journal of Horticultural Science* **49**, 257–72.

Powell, D.B.B. (1976). Some effects of water stress on the growth and development of apple trees. *Journal of Horticultural Science* **51**, 75–90.

Rogers, W.S. and Beakbane, A.B. (1957). Stock and scion relations. *Annual Review of Plant Physiology* **8**, 217–36.

Rom, R.C. (1972). Herbicide effects on surface soil in orchards and rainfall infiltration. *Arkansas Farm Research* July–Aug., p. 5.

Steudle, E. (2000). Water uptake by roots: effects of water deficit. *Journal of Experimental Botany* **51**, 1531–42.

Stiles, W.C. (1982). Soil management practices which influence soil moisture and irrigation. In *Water Management and Irrigation of Tree Fruits. The Proceedings of the 1982 Pacific Northwest Tree Fruit Short Course*, ed. R.B. Tukey, pp. 153–69. Pullman: Washington State University.

Stoker, R. and Weatherley, P.E. (1971). The influence of the root system on the relationship between the rate of transpiration and depression of leaf water potential. *New Phytologist* **70**, 547–54.

Stoll, M., Loveys, B. and Dry, P. (2000). Hormonal changes induced by partial rootzone drying of irrigated grapevine. *Journal of Experimental Botany* **51**, 1627–34.

Stopar, M., Black, B.L. and Bukovac, M.J. (1997). The effect of NAA and BA on carbon dioxide assimilation by shoot leaves of spur-type 'Delicious' and 'Empire' apple trees. *Journal of the American Society for Horticultural Science* **122**, 837–40.

Taylor, G. and Davies, W.J. (1985). Water relations and leaf growth. *Acta Horticulturae* **171**, 131–8.

Unrath, C.R. (1972a). The evaporative cooling effects of overtree sprinkler irrigation on 'Red Delicious' apples. *Journal of the American Society for Horticultural Science* **97**, 55–8.

Unrath, C.R. (1972b). The quality of 'Red Delicious' apples as affected by overtree sprinkler irrigation. *Journal of the American Society for Horticultural Science* **97**, 58–61.

Wang, Z. and Stutte, G.W. (1992). The role of carbohydrates in active osmotic adjustment in apple under water stress. *Journal of the American Society for Horticultural Science* **117**, 816–23.

Warrit, B., Landsberg, J.J. and Thorpe, M.R. (1980). Responses of apple leaf stomata to environmental factors. *Plant, Cell and Environment* **3**, 13–22.

West, D.W. and Gaff, D.F. (1976). The effect of leaf water potential, leaf temperature and light intensity on leaf diffusion resistance and the transpiration of leaves of *Malus sylvestris*. *Physiologia Plantarum* **38**, 98–104.

West, D.W., Thompson, W.K. and Black, J.D.F. (1970). Polar and lateral transport of water in an apple tree. *Australian Journal of Biological Science* **23**, 231–4.

Wilson, E.M. (1990). *Engineering Hydrology*, 4th edn. London: Macmillan.

13

Diseases, pests, and resistance to these

Introduction

Apples and pears are subject to a large number of diseases and pests. Some are very obvious and may cause distinctive damage. Others are virtually symptomless other than leading to reduced growth and cropping.

The impact of the different diseases and pests may be reduced by confining susceptible cultivars to regions of climate unsuitable for the pathogen. It may also be reduced by the use of quarantine and 'plant health' procedures designed to ensure healthy planting material. Control of damage where conditions are such as to make this a serious threat may be achieved by chemical and biological control agents and by the deliberate breeding of resistant cultivars.

With the world-wide expansion of apple and pear growing, new pest and disease problems have arisen. Consumer fears of pesticide residues and concerns about ecological impacts have placed significant constraints on chemical control. Biological and 'integrated' control methods to reduce chemical inputs may require sophisticated localized monitoring of both pathogens and environmental factors. Genetic resistance may break down as new strains of disease organisms and pests evolve. Control of disease and pest incidence is thus very complex. Emphasis in this chapter is given to the pests and diseases of greatest importance in Europe, North America and Australasia and genetic resistance to these. Discussion of chemical and biological control and of specific pests and diseases prevalent in Asia, Africa and South America is more limited.

Virus and MLO diseases

Most commercial cultivars of *Malus* and *Pyrus* are highly tolerant of viruses which cause severe effects on non-commercial forms of these genera (Posnette, 1977). This presumably reflects a historic process of selection for tolerance; for example, a number of quince clones discarded by Hatton for incompatibility with pear were subsequently found to be sensitive to a virus almost universally present in pear cultivars (Cropley, 1967). In Europe, the main apple cultivars are tolerant of 'chat fruit' which has very severe effects on the cv. 'Lord Lambourne'. The 'Malling' and 'MM.' apple rootstocks are tolerant of the widely present stem-pitting virus whereas the cold-hardy rootstock 'Virginia Crab' is not. Virus-induced incompatibility has been discussed in Chapter 4 (pp. 133–4).

In common with most woody plants, apples and pears contain chemical inhibitors that inactivate viruses. The relative resistance of the trees may explain why virus spread is usually much slower in them than in herbaceous crops.

The degree of virus infection is, however, increased by the practice of vegetative propagation which perpetuates infection from one generation to the next. Heat treatment of shoot tips at 36 °C, producing shoots by tissue culture from meristems (which are generally free from virus), or a combination of these methods is now generally used to provide virus-free source material (Campbell, 1977).

Virus-free apple and pear trees grow more vigorously than those infected with some of the common European viruses (apple chlorotic leaf spot virus, Spy 227 decline, rubbery wood, stem pitting virus, stem grooving virus and scaly bark virus for apple, and vein yellows for pear). They give heavier crops but yields per unit tree size are similar to, or lower than, those of infected trees (Posnette, 1977; van Oosten *et al.*, 1982). The lower cropping efficiency of virus-free trees may be a consequence of their greater size (cf. the relationship between tree size and cropping efficiency shown in rootstock studies). Virus-free trees of 'Golden Delicious' apple have smoother fruit skin with less russeting (van Oosten *et al.*, 1982).

Apple chlorotic leaf spot virus (ACLSV)

This closterovirus was first described from apple but is now known to occur world-wide and to be common in all fruit tree and woody ornamental species of *Rosaceae*.

Most commercial apple cultivars are symptomlessly infected. Leaf mottle and ring mosaic of pears are associated with ACLSV. Many pear cultivars remain symptomless but 'Beurré Hardy', 'Beurré Bosc' and 'Williams'' are susceptible and there are significant reductions in growth and yield on mixed infection with pear vein yellows and quince sooty ringspot (Delbos and Dunez, 1988).

Apple chat fruit disease

The causal organism of this is not identified. It is restricted to apple. Symptoms are seen only on the cvs. 'Lord Lambourne' and 'Tydeman's Early Worcester' (Adams, 1988a). Fruits are small and discoloured.

Apple proliferation disease

The first symptoms of this disease are the growth of upright secondary shoots to give 'witches brooms' at the ends of branches through suppression of apical dominance. Average fruit weight can be reduced by 30–60%. Infected young trees grow poorly and although there may be some recovery in later years yields and fruit size usually remain depressed. In 'recovered' trees with normal fruit size crop losses can still be 20–40%. The disease is widespread in central and southern Europe under climatic conditions suitable for wine growing. 'Golden Delicious', 'Jonathan' and 'Cox's Orange Pippin' are more sensitive than other cultivars but damage is possible on all cultivars.

Trees with the symptoms of apple proliferation contain mycoplasma-like organisms (MLOs) in the sieve tubes. These are thought to be the causal organisms and leaf sucking insects are probably vectors (Kunze, 1988).

Apple rubbery wood disease

Affected apple trees are stunted. Shoots and branches up to 3 years old show incomplete lignification and are often bent to the ground. Strong shoots may develop from the base of the trunk. Sensitive cultivars include 'Lord Lambourne' (the standard indicator), 'Golden Delicious' and 'Gala'. Even cultivars like 'Cox's Orange Pippin' that do not show the characteristic symptoms have reduced vigour and cropping. The causal organism is not known with certainty. No vector has been identified and spread in the orchard is very slow or absent. The use of healthy planting material provides practical control (Adams, 1988b).

Pear decline

Trees infected with pear decline MLO may show quick decline, slow decline or leaf curl. Trees with quick decline wilt suddenly and die within a few days or weeks. This is most evident in hot dry weather and on *P. ussuriensis* and *P. pyrifolia* rootstocks. It may also occur on *P. communis* rootstocks but the incidence is less and the trees often show red-leaf symptoms before wilting. Slow decline is characterized by progressive and fluctuating weakening of the trees. Terminal growth and leaf production is restricted and the leaves, which are small, leathery and light green with uprolled margins, may redden and drop prematurely. Trees may die within a few years of infection or live for many years. Leaf-curl symptoms may be shown where the decline is less severe and on relatively tolerant rootstocks, e.g. *P. communis*, *P. calleryana* and *P. betulifolia* (Seemüller, 1988).

Although quince rootstocks are much less susceptible than *P. communis* rootstocks, to the extent that their use was recommended as a control measure, recent research has shown that the widespread disorder of 'Conference' pear on quince rootstocks known as Parry's Disease is associated with pear decline MLO (Davies *et al.*, 1992, 1995; Davies and Eyre, 1996). The symptoms are similar to the mild or slow forms of pear decline. The leaves of affected trees develop a premature red colour followed by early leaf fall. In the following spring leaves remain pale and small, there is little or no shoot growth and no fruit production. A necrotic line is often evident in the bark at the stock/scion union. The majority of trees are affected for only one growing season but the loss of early yield can be economically damaging. In some cases death or severe stunting may occur. The cvs. 'Comice' and 'Concorde' are relatively unaffected. Pear decline is spread by the pear psyllid *Cacopsylla pyricola* Foerster, the most common insect pest in pear orchards.

Bacterial disease

Fire blight

Fire blight, caused by *Erwinia amylovora* (Burrill) Winslow *et al.*, is the most important bacterial disease of apples and pears. It has been reported from 40 countries around the world (Bonn and van der Zwet, 2000) and is so lethal and fast-spreading under suitable conditions that it has become a major factor in determining where different cultivars of apple and pear can be grown.

During the growing season blighted blossoms and shoots wilt and turn dark brown or black. The scorched appearance of affected leaves, twigs and branches gives the disease its name. Young shoot tips bend to form a 'shepherd's crook' due to loss of turgor and cell death. In humid weather in spring and summer bacterial slime may ooze from affected shoots, blossom trusses, branches and fruits. The bacterium moves in the intercellular spaces of parenchymatous tissues, spreading into spurs and along branches at up to 25–50 mm per day in susceptible hosts, forming cankers and sometimes affecting and rotting fruits. The cortex of the outer bark of the affected branches is typically dark green and water-soaked at first but later shrinks and turns brown. The inner bark is typically moist and red where the disease is active and red-brown streaks are often found in the sapwood (van der Zwet and Beer, 1995). Leaves and fruits on affected branches characteristically remain attached after normal leaf fall in the autumn. In autumn or after prolonged drought the cankers cease extending and cracks may appear at the margins. In mild wet autumns cankers may remain unsealed and these are the ones most likely to be active in the following spring. In less vigorous pear trees and in apple trees the spread of bacteria within the tissues is slower and the damage proportionately less (Lelliot, 1988).

In the following spring active overwintered cankers extend and ooze is produced externally. This inoculum can be spread by rain or a variety of insects, not including bees, to open flowers. Pollinating insects, bees and flies, then spread inoculum rapidly between flowers on warm sunny days. The pathogen multiplies rapidly on stigma surfaces at temperatures above 18 °C and is transported in water from rain, fog or dew to the region of the nectaries where infection occurs. Blossom infection is particularly likely on warm (21–30 °C) sunny days when insect activity is high. In north-western Europe the coincidence of these conditions with open blossom is greatest with secondary blossoms, to which some pear cultivars are particularly prone, and with late-opening flowers on one-year-old wood on apple trees.

Storms greatly increase the chances of infection, especially when flowers or young shoots are damaged or leaves detached by hail or strong winds. Exposure of young vascular tissue aids infection but the pathogen must invade cortical tissues to produce typical symptoms.

The pathogen has only a very short life on the leaf surface. Persistence within host tissues, overwintering in inconspicuous cankers, and the existence of symptomless carriers are, however, important factors in its epidemiology. Spread can result from the movement of infected plants from nurseries and, at a more localized level, by transfer of bacteria on pruning tools.

Tissue susceptibility is generally much greater in young tissues and trees than in older ones. It is generally greater in pears than in apples but there are also large differences between cultivars. Tables 13.1 and 13.2 show relative tissue susceptibility, as measured by response to artificial inoculation of shoots in the field (Lespinasse and Aldwinkle, 2000).

The 'Delicious' group of apples are relatively resistant, 'Golden Delicious' and its progeny and 'Fuji' moderately susceptible, 'Granny Smith' and 'Braeburn' susceptible, and 'Rome Beauty', 'Cox' and 'Idared' very susceptible. Spur-types tend to be less susceptible than the corresponding non-spur cultivars. The dwarfing apple rootstocks in Table 13.1 are all susceptible or highly susceptible. 'MM.106' which appeared highly susceptible in inoculation studies (Table 13.1) has shown little natural infection in the USA (Lespinasse and Aldwinkle, 2000). A rootstock breeding programme at Geneva, New York has produced a number of dwarfing and semi-dwarfing apple rootstocks that are resistant to fire blight. This resistance is derived from 'Beauty Crab', *M.* × *floribunda* or 'Robusta 5' (Robinson *et al.*, 1997). Some are resistant to some other diseases and pests and have given promising initial results in terms of effects on cropping (Robinson *et al.*, 1999).

Some of the most important *P. communis* cultivars, 'Bartlett', 'Passe Crassane', 'Packham's Triumph' and 'Doyenné du Comice' are very susceptible. Most *P. pyrifolia* (Nashi) cultivars are susceptible, *P. ussuriensis* and *P.* × *bretschneideri*

Table 13.1 *Relative susceptibility of apple scion and rootstock cultivars to fire blight following artificial inoculation*

Incidence reflects frequency of infection (scale 1–5). Severity is also on a scale 1 (least severe) to 5 (most severe). Index reflects both these factors on a scale of 0 to 100 and Class reflects a broad band grouping from 1 (most resistant) to 5 (most susceptible).

Cultivar	Incidence	Severity	Index	Class
Apple cvs.				
'Delicious'	1	3	15	1
'Jonagold'	2	2	21	2
'Gala'[a]	3	3	24	2
'Golden Delicious'	3	3	33	2
'Elstar'	3	4	34	2
'Fuji'	5	2	36	2
'Granny Smith'	5	3	45	3
'Braeburn'	5	3	52	3
'Cox'	5	4	68	4
'Idared'	5	4	68	4
Apple rootstocks				
'MM.104'	1	1	10	1
'MM.111'	1	3	17	1
'M.7'	4	2	24	2
'M.25'	3	3	28	2
'Pajam 2' ('M.9')	5	3	51	3
'Pajam 1' ('M.9')	5	4	60	3
'Supporter 4' ('P.180')	5	4	69	4
'M.26'	5	4	73	4
'MM.106'	5	4	79	4

[a] Although 'Gala' is only slightly susceptible on shoots it is highly susceptible on flowers.

Adapted from Lespinasse and Aldwinkle (2000), with permission.

(Li) cultivars less so. Quince rootstocks are moderately susceptible. New pear rootstocks with 'Old Home' as a parent are relatively resistant. Pear cultivar breeding for resistance is concentrating on selecting for resistance within high quality susceptible cultivars and by selecting parents on the basis of phenotypic values in the absence of strong evidence for major gene control (Lespinasse and Aldwinkle, 2000). Transfer of resistance genes from non-pear sources is discussed in Chapter 14, on biotechnology.

Susceptibility to fire blight in the field also depends very largely on time of flowering in relation to weather conditions suited to infection. 'Passe Crassane', 'Packham's Triumph' and 'Bartlett' pear all produce secondary (late) blossom and 'Golden Delicious' and its progeny flower late on one- year-old wood. The correlation between flower and shoot susceptibility is sometimes weak. 'Gala'

Table 13.2 *Relative susceptibility of pear scion cultivars and of rootstocks for pear to fire blight following artificial inoculation.*

Incidence reflects frequency of infection (scale 1–5). Severity is also on a scale of 1 (least severe) to 5 (most severe). Index reflects both these factors on a scale of 0 to 100. Class gives a broad band grouping from 1 (most resistant) to 5 (most susceptible).

Cultivar	Incidence	Severity	Index	Class
Pear cvs.				
'Conference'	3	1	16	1
'Beurré Bosc'	2	3	18	1
'Blanquilla'	4	1	18	1
'Coscia'	2	2	18	1
'Jules Guyot'	4	2	21	2
'Beurré Hardy'	2	3	25	2
'Abbé Fetel'	3	4	43	3
'Bartlett'	4	4	57	3
'Passe Crassane'[a]	4	5	71	4
'Packham's Triumph'	5	4	72	4
'Doyenné du Comice'	5	5	90	5
Pear rootstocks				
INRA Pyriam (OH11)	1	2	11	1
Farold 87 Daytor	3	2	14	1
Quince BA29	5	3	53	3
Quince C (EM)	5	3	55	3
Quince 'Adams'	5	3	58	3
Quince 'Sydo'	5	3	58	3

[a] Very susceptible on secondary blossom.
Adapted from Lespinasse and Aldwinkle (2000), with permission.

flowers, for example, are very highly susceptible although shoot susceptibility is only moderate.

CONTROL OF FIRE BLIGHT

Approaches to control involve minimizing tree susceptibility, reducing inoculum levels and interfering with the infection process (Aldwinkle and Beer, 1979). There are, as yet, no curative treatments.

Breeding programmes for fire blight resistance, detailed earlier, show promise. However, the economic superiority of high-density planting systems using the susceptible 'M.9' rootstock, and the increasing market demand for newer apple cultivars such as 'Fuji' and 'Gala', which are more susceptible than the hitherto dominant 'Red Delicious', coupled with the shifting of the bulk of apple production to warmer areas than hitherto, have all increased rather than reduced fire blight problems. Reducing the succulence of growth by limitation of N fertilizer application and irrigation can reduce fire blight risk. These treatments may be difficult to reconcile with the need for rapid growth of young trees to attain high cropping levels soon after planting. Control of shoot vigour by the use of spur-type cultivars and the application of some plant growth regulators can reduce susceptibility, but the use of dwarfing rootstocks does not achieve this. Pruning and training techniques that minimize vigorous regrowth may be useful.

The first step to reduce inoculum level is by plant quarantine to prevent the importation and movement of infected material. The explosive spread of the disease indicates that quarantine is less effective than previously but at present some major apple and pear producing countries are still free of fire blight and prevention of spread by close supervision of nurseries is still very important. Removal of potential hold-over cankers in winter has a major effect in reducing subsequent disease development. Eradication of early-season infections by pruning out and burning reduces further spread of inoculum as well as checking damage to the trees. Disinfection of pruning tools to prevent spread by these is important also. Over-tree sprinkling may lead to increased inoculum production and spread so should be discontinued if fire blight is present. Early-season copper sprays may reduce inoculum production.

Interference with the infection process is achieved by treatments to inhibit bacterial multiplication on blossoms and to reduce populations of insects likely to spread shoot blight. Fire blight is a sporadic disease and control of infection needs to be based on risk assessment. This is because routine, rather than targeted, monitoring of other potential host plants as well as apples and pears is laborious and expensive and, especially, because the most effective control treatment at present is with streptomycin, where this is permitted. As well as being expensive, use of this when not needed can contribute to development of antibiotic resistance. Mills (1955) identified 18 °C, together with precipitation or high relative humidity, as the threshold temperature during or shortly after bloom for fire blight development in New York State. In California excellent control of fire blight was achieved when streptomycin was applied only after the mean of maximum and minimum temperatures exceeded a 'prediction line' drawn from 16.7 °C on 1 March to 14.4 °C on 1 May (Thomson *et al.*, 1977, 1982). This gave appreciable cost savings compared with routine sprays (Aldwinkle and Beer, 1979). Zoller and Sisevich (1979) found that the incidence of epiphytic bacteria on pear flowers was related to the accumulated number of degree-hours above 18.3 °C, and a total of about 150 accumulated degree-hours indicates the need for bactericide application just prior to the next rainfall (van der Zwet *et al.*, 1988). This dependence on temperature is largely explained by Billing's (1974) finding that *E. amylovora* increases rapidly from 6.5 °C to 18 °C, which is the approximate threshold for rapid growth. The minimum conditions for blossom infection defined by the Maryblyt^{TM}

computer model, which has proved reliable for apples in eastern USA (Lightner *et al.*, 1999), are: open intact flowers, a wetting event comprising 0.25 mm of rain or heavy fog or dew, an accumulation of 110 degree-hours above 18.3 °C prior to this and a mean daily temperature of 15.6 °C or above. Blossom blight is estimated to appear when 57 day-degrees above 12.7 °C have accumulated after infection (Billing, 2000). In southeast England, Billing (1996, 1999) found similar but not identical temperature dependence of the infection process, and a likely time between direct blossom infection and visible disease of 47 day-degrees above a mean temperature of 13 °C for apple and 17 such day-degrees for pear and hawthorn. She also emphasized the importance of orchard factors. Fire blight risk assessments based solely on regional weather and its expected effects on the population growth of the pathogen can result in underestimation of risk or, more commonly, overestimation and unnecessary spraying. This has led to the development of approaches using orchard weather data and categorized orchard-specific factors such as the local history of fire blight, cultivar susceptibility, orchard sanitation, etc. (Shtienberg *et al.*, 1999; Smith, 1999). Summer infections can occur on secondary blossom of pear, on storm-damaged young shoots and hail-damaged young fruits. Infection-control measures can be targeted accordingly. Infection can be reduced by controlling insects which contact and feed on the ooze and carry the pathogen so acquired to natural infection sites and which also, through creating wounds by feeding, initiate infection sites and infect vegetative shoots (van der Zwet and Beer, 1995).

Streptomycin and other medical antibiotics have been widely used to control bacterial multiplication at the time of, and just after, inoculation, but are not permitted to be used in many countries because of the potential risk to human health of developing antibiotic-resistant bacteria. They are only locally systemic and are not effective when applied to unopened blossoms. A streptomycin spray one day before or one day after inoculation with *E. amylovora* prevents infection (Gouk *et al.*, 1999). This short time-span of effectiveness may result in the need for multiple applications. However, *E. amylovora* has developed resistance to streptomycin in many fruit-growing areas in the USA (Breth *et al.*, 1999) and in Israel (Manulis *et al.*, 1999). Copper compounds can be effective but can cause excessive russeting (Momol *et al.*, 1999). Resistance-inducers, natural plant extracts, bacterial powders and mineral powders are also being evaluated for fire blight control (Momol and Saygili, 1999).

Fungal diseases

Fungal attack in the orchard requires frequent and expensive chemical spraying for its control, and many apple and pear breeding programmes give high priority to resistance to fungal pathogens.

Apple mildew

Apple mildew, caused by *Podosphaera leucotricha* (Ell. & Ev.) Salmon, is currently one of the most important diseases of apple in England and many other fruit-growing countries. Its conidia and hyphae overwinter in buds. Floral and foliar organs emerging in spring from infected buds emerge totally or partially coated with powder mildew. This is primary mildew (Butt, 1988). Infected fruit buds give dwarf leaves and the flower buds remain closed: these organs wither in late spring. Infected vegetative terminal buds produce totally mildewed shoots. Those not totally colonized in the previous year grow as 'white shoots' for several weeks then shed leaves. During spring and summer new conidial infections give colonies of secondary mildew, mainly on the undersurface of leaves. The shoot tip, newly formed buds and, in some cultivars, e.g. 'Jonathan', fruitlets, may also be infected.

The quantity of primary mildew depends on the intensity of bud infection in the previous year and on winter temperatures. A few hours at −20° to −25 °C, or even at −12° to −15 °C in late winter, can almost eradicate the overwintering fungus. Infection in spring and summer is favoured by high humidity, especially at night.

Mildew reduces leaf area, photosynthesis and the number of fruiting spurs. It also induces russet and reduces fruit size. Butt *et al.* (1983) showed that an increase from 2% to 19% of mildewed leaves in mid-summer resulted in a reduction of 32% in number of shoots per tree, 19% in number of leaves per shoot, 27% in fruit size, 15% in crop weight and 33% in crop value for 'Cox's Orange Pippin'.

There is considerable variation in resistance among commercial cultivars, 'Discovery' being highly resistant and 'Crispin' ('Mutsu') and 'Jonathan' highly susceptible. Breeding programmes have involved the use of partially resistant parents, e.g. 'Prima'. More closely targeted breeding involves the gene Pl_2 derived from *M.* × *zumi* open-pollinated (Knight and Alston, 1968) and Pl_1 from *M.* × *robusta* and Pl_w from a North American ornamental crab apple 'White Angel' (Laurens, 1999).

Control is by sprays of sulphur on cultivars which are not damaged by this, of dinitrophenyl esters, bupirimate, fenarimol, nitrothal-isopropyl, triadimefon, myclobutanil and penconazole. Sprays are generally applied, where needed, every 7, 10 or 14 days from before flowering until late summer when most extension growth has ceased. It is important that the sprays do not damage fruit skin, do not interfere with biological control programmes (some, e.g. binapacryl, are incompatible with the use of predaceous mites to control red spider mite) and do not check shoot growth excessively (e.g. triadimefon is an ergosterol biosynthesis inhibitor). Control is also achieved by hand pruning to remove infected shoots, fruits, etc. Use of a bud-penetrating surfactant, either alone or mixed with a fungicide, in winter can markedly reduce overwintering mildew.

Pear leaf blight and fruit spot

This disease is caused by *Fabraea maculata* Atk. It leads to defoliation of both nursery and orchard trees and to disfigured, cracked and misshapen fruits. 'Clapps Favourite' and 'Bartlett' have been found to be resistant in some studies but not in others. Highly resistant material of potential use in breeding programmes has been identified (Bell *et al.*, 1996).

Apple rust

The cedar-apple rust caused by *Gymnosporangium juniperi-virginianae* Schw. is the most important rust disease of apple and has the North American red cedar as its alternate host. A number of good quality cultivars, e.g. 'McIntosh', are resistant: others, e.g. 'Jonathan', fully susceptible (Brown, 1975).

Apple scab

Apple scab, caused by *Venturia inaequalis* (Cooke) Winter, is evident as scabbed areas on leaves and fruits and, under favourable conditions, on shoots, buds and flowers. The spots on leaves are initially light in colour and then become dark olive-green, showing a visible mat of fungus. The spots expand and run into each other, extending along the veins. Infected leaves tend to fall prematurely. The scab spots on fruit are almost black in colour. Scabby fruits cannot be sold in the high quality grades and the scab lesions are readily colonized by rotting organisms.

The fungus can infect unripened wood and may persist for up to 5 years in the bark, from which it produces asexual spores in spring and early summer. It also overwinters on fallen infected leaves on which it produces sexual spores (ascospores) in spring. The ascospores are released within 30 minutes of wetting, a maximum rate being reached after 1–2 hours. The spores are then carried by the wind to leaves on which they will germinate if the surface remains wet for long enough for a 'Mills period' of, for example, 25 h at 6 °C or 9 h at 16–24 °C. Infection pressure is very limited in areas with dry spring and summer conditions.

Initial infection pressure can be reduced by pruning, by soil cultivation which increases earthworm populations and consequently leaf burial, and by post-harvest, pre-leaf-fall, applications of urea to accelerate decay of fallen leaves. Control during the cropping season is based on routine sprays from budburst and repeated at 1–3-week intervals. Knowledge of infection periods can allow more efficient fungicide use. Captan and dithianon will stop germ tube growth before infection is established and systemic fungicides can be effective up to 96 hours after infection (Byrde, 1977; Richter, 1988a).

The main commercial cultivars are susceptible to scab. Genes for resistance to this have been identified, notably the dominant allele V_f from *Malus × floribunda* which was used to produce cultivars including 'Prima', 'Priscilla', 'Liberty', 'Freedom', 'Priam', 'Primicia' and 'Goldrush'. The *M. × floribunda* donor has totally unacceptable fruiting characteristics and, despite back-crossing programmes to produce these cultivars, none has yet gained widespread acceptability. 'Antonowka' is used as a polygenic source of field resistance to all races of apple scab and 'Laxton's Fortune', 'Dülmener Rosen' and 'Discovery' appear to have some breeding value (Janick *et al.*, 1996; White and Bus, 1999).

Pear scab

Venturia inaequalis may be found on pears but pear scab is generally caused by *V. pirina* Aderhold. This fungus attacks only pear where it causes scab of shoots, leaves, fruits and buds. Shoot infections can give swellings which later burst. Infected fruits are deformed and deeply cleft. Pear scab development is facilitated by surface wetness and high temperature (Richter, 1988b). *V. nashicola* is an important pathogen of Nashi pear, *P. pyrifolia* (Burm.) Nakai, in Japan (Kajiura, 1994).

Apple canker

This is caused by *Nectria galligena* Bres. The fungus invades injuries to the bark such as pruning wounds, scab lesions and, especially, leaf scars. It causes a sunken area of bark, often surrounded by a ridge of tissue only partly invaded. The cankers increase in size and may girdle the stem, killing the part above the girdle. A rot around the eye of the fruit can also occur. Control is by cutting out infected wood and restricting nitrogen supply. Fungicidal paints may be applied to the cankers and benomyl and related fungicides are effective when sprayed in spring and summer (Byrde, 1977). Canker is readily carried over on young trees at planting time and nursery trees should be treated, e.g. with copper compounds, to ensure clean orchard planting material.

Collar rot and fruit rot

These are caused by *Phytophthora cactorum* (Leb and Cohn) Schroet. and *P. syringae* Kleb. which attack the rootstock or scion cultivar around soil level.

Irregular-shaped lesions develop in the bark, which looks moist, soft and rather spongy. When the affected bark is cut the tissue is orange-brown with lighter and darker brown striped areas. The pathogen is soil-borne and its spores are dispersed by splashing.

The lesions can girdle and kill the tree if not checked by being cut out at an early stage and the wounds so made treated with a copper fungicide.

Rootstocks differ greatly in their resistance. 'B.9' is very resistant, 'M.9', 'MM.111' and 'Robusta 5' are resistant, but 'MM.106' is susceptible (Cummins and Aldwinkle, 1974; Janick *et al.*, 1996). Several new rootstock clones from Cornell and East Malling are also resistant (Robinson *et al.*, 1997; Webster *et al.*, 1997). Scion cultivars are generally much less tolerant than the resistant rootstocks, so high-working on such rootstocks, so that the scionwood is out of the soil and splash zone, is an important control measure. *P. syringae* also can cause fruit rotting.

Storage rots

A number of other fungi cause rots in store. These include *Pezicula* (*Gloeosporium*) spp., *Alternaria alternata*, *Penicillium* spp., *Botrytis cinerea*, *Monilia fructigena* and *Mucor piriformis*. Control focuses on several areas. Appropriate orchard management, including effective spray programmes, lowers the load of inoculum. Reduction in injuries to fruit through rough handling can greatly reduce post-harvest decay: Benic and Combrink (1996) reported 49% of the decay in apples was due to infections resulting from mechanical injuries. Maintenance of high sanitary standards in packing-houses is essential to avoid fungal inoculum from decaying fruit, etc. being carried over from one season to another and within a season. Rapid fruit cooling to below 5 °C greatly slows down fungal growth. Chemical treatment, e.g. benomyl just before picking or in post-harvest dips greatly reduces, for example, *Gloeosporium* damage (Burchill and Edney, 1972). High fruit Ca status increases resistance to *Gloeosporium*. 'Cox' and 'Ingrid Marie' apples are very susceptible, 'Jonathan' and 'Jonared' very resistant and 'Golden Delicious' resistant (Alston, 1967).

In Japan, *Alternaria alternata* is also an important pathogen as a result of causing black spot disease of Nashi pears. The cvs. 'Kosui' and 'Hosui' are resistant to this (Kajiura, 1994).

Apple bitter rot caused by *Glomerella cingulata* is a very important fungal disease in Brazil, being potentially very serious wherever high temperature and high humidity (24–28 °C and 80% RH) occur simultaneously. 'Fuji', 'Delicious', 'Golden Delicious', 'Granny Smith' and 'Gala' are susceptible and 'Jonathan' and 'Red Rome' fairly resistant. Some potential parental material including *M.* × *zumi*, *M. prunifolia* and *M.* × *purpurea* clones is highly resistant (Camilo *et al.*, 1988).

Pests

Insects and mites can have a major impact on profitability through effects on tree productivity, on disease transfer and on the marketability of the fruits.

Insecticidal sprays generally gave good insect control but led to the increasing incidence of some pests as a result of killing their natural predators. Use has also been influenced by concern about residues. 'Integrated' control including use of natural predators is increasingly favoured. There are considerable variations between different apple and pear species and cultivars in their resistance to different pests, and apple rootstock breeding for resistance to woolly aphid has been outstandingly successful.

Moths

The codling moth (*Cydia pomonella* L.) is a very widespread pest of apple. The damage is caused by the caterpillar, which burrows into the fruit. The winter is spent as a fully fed caterpillar in a cocoon under loose bark on the tree, etc. The pupal stage occurs in late spring. The first moths emerge in late May or early June. After mating they lay eggs on leaves or fruits which hatch after 10–14 days. The young caterpillars find suitable entry points into fruit, often at the eye (calyx) and burrow into the fruit. Sometimes there are two generations in a season (Anon., 1992a). Control is by insecticide spraying.

The caterpillars of the fruit tree tortrix moth (*Archips podana*) eat away small shallow areas of skin from maturing fruits. Winter moth (*Operophtera brumata)* caterpillars attack fruit buds, blossoms, leaves and developing fruitlets.

The major insect pests of Nashi pears in Japan are Oriental fruit moth (*Grapholitha molesta*), Oriental pear moth (*Monema flavescens*) and Peach fruit moth (*Carposina niponensis*).

Aphids

The woolly aphid (*Eriosoma lanigerum* Hausmann) produces the mass of white waxy strands that give the pest its common name. It causes galls on shoots and dense aphid infestations soil pickers' hands and clothes. The entire life cycle is spent on the host tree and in some countries breeding colonies are found on roots below ground level and provide a source for re-infestation. The aphids can be a particularly serious problem in nurseries. Above-ground infestations can be controlled by insecticidal sprays but rootstock resistance is very important. 'Northern Spy' is highly resistant and was used in the breeding of the 'MM'. series of resistant rootstocks (Knight *et al.*, 1962). The resistance is controlled by a single dominant *Er* gene and although there have been reports of infection of resistant rootstocks on occasion there has been no serious breakdown of resistance.

The rosy apple aphid (*Dysaphis plantaginea* Pass) causes leaf curling on young shoots. Resistance is available in material derived from *Malus robusta*.

The rosy leaf curling aphid (*Dysaphis devecta* Wlk) also causes leaf curling and galls. 'Fiesta', a resistant cultivar derived from 'Cox's Orange Pippin', carries the Sd_1 gene for resistance to biotypes 1 and 2 of the aphid (Roche *et al.*, 1999).

Suckers

The apple sucker (*Psylla mali* Schmidberger) is usually well controlled by routine sprays, but pear sucker (*Cacopsylla pyricola* Foerster) is a major pest. Infested leaves turn brown and often drop, and the fruits drop prematurely or are small and of poor quality. The feeding on foliage suppresses root growth and reduces tree vigour. The psylla also transmits the MLO which is the cause of pear decline. Pear psylla is controlled by chemical sprays, but appreciable resistance to many insecticides has developed (Anon., 1992b). Host resistance to psylla has been identified in several species of Asian pear as well as in *P. communis* but no genotype combines this resistance with resistance to Fabraea leaf spot (*Fabraea maculata* Atk) and high fruit quality (Bell and van der Zwet, 1999).

Mites

The fruit tree red spider mite (*Panonychus ulmi* Koch) became a serious orchard pest largely because of insecticidal spray programmes which killed its predators and because of its own ability to develop resistance (Anon., 1992c). The first symptom is a minute speckling of the leaves; then the foliage loses colour and becomes first dull green, then brownish green or bronze, and brittle. White skins cast by mites while moulting are conspicuous on the underside of the leaves.

Injury is caused by the mites sucking sap and damaging leaf cells. Infestation checks shoot growth before it affects net CO_2 uptake per unit leaf area (Avery, 1964; Avery and Briggs, 1968). Reduction in CO_2 assimilation is slight until 50% of the leaf is speckled and bronzing begins. The photosynthetic surface of the plant is decreased by up to 15% by infestations which cause reductions in leaf size and number, and premature defoliation also decreases the leaf surface area. The growth of the root systems is decreased before that of the shoots and to a greater extent.

The multiplication of fruit tree red spider mites is greater on some cultivars, e.g. 'Discovery' and 'Worcester Pearmain', than on others such as 'Cox'.

Biological control is practised by the encouragement or introduction of the predatory mite *Typhlodromus pyri* (Scheuten), strains of which are resistant to some common pesticides.

Two-spotted spider mites also reduce net photosynthesis of apple leaves (Ferree *et al.*, 1986). 'Golden Delicious' and 'Jonathan' show appreciable effects under conditions where the photosynthesis of 'Paula Red' and 'McIntosh' is not affected and only slightly affected, respectively.

Replant problems

One of the most serious problems in apple production is the poor growth of trees planted in land previously occupied by apple trees. The reduction in growth in old apple soil compared with fresh soil may be up to 90% (Hoestra, 1968). In pot experiments in which 'apple soil' and fresh soil were mixed in varying proportions, apple seedling growth was related to the proportion of apple soil in a non-linear way: 10% of 'apple soil' gave 25% growth reduction, 50% a 66% reduction and 90% a 78% reduction. Trees tend to show recovery starting 2 or 3 years after planting, but Oehl (1980) found that fumigation of land previously cropped with apples using a broad spectrum biocide (chloropicrin) led to a 35% increase in accumulated crop over the first 10 years from planting. The delayed cropping on 'replant' soils has a particularly adverse effect on the economics of high-density planting systems. Geldart (1994) and Peterson and Hinman (1994) demonstrated that in British Columbia, Canada, and Washington State, USA it was not economically viable to plant apple trees on land previously cropped with apples unless remedial measures were taken.

Treatments with a broad spectrum of action against all categories of soil organisms (nematodes, fungi, bacteria and actinomycetes) restore the growth potential of replant soils to that of 'fresh' soils, the response being much greater on the former than the latter (Hoestra, 1968). Such broad-spectrum treatments include chloropicrin, steaming and heating to 60 °C. The effect of chloropicrin fumigation of soil samples in pot tests of seedling growth is taken as an index of the potential problems in replanting and the need for pre-planting orchard treatments.

Studies on poor growth of newly planted apple trees in the absence of soil sterilization have indicated two distinct, though sometimes interacting, problems. The first is caused by nematodes. The second is found even when nematode numbers are low and cannot be overcome by the use of nematicides (Hoestra, 1968). This was initially referred to as specific apple replant disease (SARD) but is now generally referred to as apple replant disease, ARD (Utkhede, 1998) following evidence, discussed below, that it can be induced by crops other than apple.

Nematode effects

On light soils damage by the endoparasitic nematode *Pratylenchus penetrans* (Cobb) may cause patchy distribution of poor growth over a field, with nematodes being visible in the young roots. Root lesions develop on the unsuberized roots. When these coalesce rootlets, or even complete root systems, may be destroyed. Infected roots are often discoloured and stunted. Many plant species including grasses can be attacked by nematodes, so the problem is not restricted to old apple land. Use of nematicides, e.g. D-D and methyl-isothiocyanate, as a preplanting treatment can lead to major increases in yield, and marigold (*Tagetes patula*) can be used as a preplant cover crop to reduce nematode populations (Mai *et al.*, 1994). Nematodes can, however, be transferred from infected nurseries to orchards when the trees are planted there (Hoestra, 1968). Nematodes are not a major cause of replant problems of apple trees in the UK (Way and Pitcher, 1971), in most Dutch apple growing areas (Hoestra, 1968) or in Washington State (Willet *et al.*, 1994), although they are a significant problem in New York State (Pruyne *et al.*, 1994).

Replant disease

In general the most severe check to growth of young apple trees planted on the sites of old apple orchards is that caused by apple replant disease. This is not due to nematodes, residual toxins, or induced nutrient deficiencies (Hoestra, 1968; Savory, 1966, 1967). Pot tests of responses to soil sterilization showed 60% of soils in Dutch fruit-growing regions to have replant disease, half being seriously infected (Hoestra, 1968). In England 67% of orchard soils showed economically significant responses to fumigation following adjustment of their P content (Sewell *et al.*, 1988).

The main field symptoms are poor growth of both shoots and roots. Leaves tend to be small and internodes short. Seedlings planted into replant soil develop root lesions within a few days. Root hairs are reduced in size and number, especially when in close contact with the soil. The primary cortex and epidermis may rot away, leaving a thin, light-coloured stele covered by remnants of the cortical layer. No abnormalities are found in the stele until an advanced stage of decay is reached. If roots in 'replant' soil are allowed to grow into sterilized soil they develop into healthy root systems with side roots bearing many root hairs, i.e. the replant condition is relatively immobile in the soil. The response of orchard trees to pre-planting fumigation is a function of the soil volume fumigated (Jackson, 1973).

EFFECTS OF PREVIOUS CROPS

Although the poor growth of apples after apples had long been recognized, interest in the effect of the previous crop was enhanced following some observations of nursery tree growth by Thompson (1959). He found that when nursery trees of apples, quince, cherries and some plums were planted in rows cutting at right-angles across the sites of previous rows of these species the growth of apples, and to a lesser extent of quince, was severely depressed when they were growing on land previously occupied by apples. The growth of cherry rootstocks was not depressed when growing on old apple land but was less good where they followed stone fruits. Apple tree growth shows large responses to soil fumigation of old apple land whereas cherry tree growth on the same land does not (Pitcher *et al.*, 1966; Jackson, 1973). From such observations and experiments Savory (1966, 1967) concluded that cherry and apple were subject to separate specific replant diseases and the fungus *Thielaviopsis basicola* (Berk. & Br.) Ferraris, which does not affect apple tree growth, was subsequently identified as the cause of cherry specific replant disease (Sewell and Wilson, 1975). Further work on apple replant disease, however, led Sewell (1979) to conclude that this is not strictly specific. The growth of apple is considerably diminished when planted after a number of other crops and the growth of some other crops may be diminished when these are planted after apple. Soil sterilization by heating increased *Malus* seedling growth more on ex- *Robinia* than on ex-*Malus* soil and *Robinia* and *Tilia* growth in ex-*Malus* soil was greatly increased when this was fumigated (Sewell and Roberts, 1986).

A replant disease condition can be induced by growing apple trees in the soil for only one year (Oostenbrink and Hoestra, 1961) and can persist for at least five years after grubbing (Thompson, 1959). Responses to fumigation are a function of soil pH, being much greater on alkaline and neutral soils than on acid soils. Cropping responses can be very large, e.g. 212–414% (Hoestra, 1968). Under some conditions where soil phosphorus levels are low, both pre-planting pot tests and field responses to fumigation may be low or even negative. This is because of the effect of fumigation in killing mycorrhizal fungi. In such circumstances both fumigation and P fertilizer treatment may be needed: the need for fumigation being shown only by pot tests carried out with P-fertilized soil (Sewell *et al.*, 1988).

CAUSAL ORGANISMS

The etiology of apple replant disease involves the build-up of pathogenic organisms in the soil of apple orchards and the persistence of these for an extended time after grubbing, either because they form resting stages or they possess a strong saprophytic ability.

Actinomycetes

Circumstantial evidence, including the effect of soil pH and responses to different fumigants, suggests that *actinomycetes* are involved (Hoestra, 1968). Otto and Winkler (cited by Utkhede and Smith, 1994b) found that the severity of apple replant disease in Germany was correlated with *actinomycete* presence. Apple seedlings, and rowan (*Sorbus aucuparia*) and pear (*Pyrus communis*) rootstocks were strongly infected with *actinomycetes* when planted in soil from an apple orchard but rootstocks of cherry, plum and rose were not (Otto *et al.*, 1994). The *actinomycetes* penetrate the epidermal cells and damage the cortex in such a way that the cortex and rootlets die off. They enshroud root hairs and cause their collapse (Otto and Winkler, 1998; Szabo *et al.*, 1998).

Pythium

Several studies have shown an association between *Pythium sylvaticum* and replant disease (Utkhede and Smith, 1994a). Sewell (1981) found that all tested isolates of *P. sylvaticum* and some isolates of other *Pythium* species significantly reduced the growth of apple seedlings and that the isolates of most virulence to apple were of low virulence to cherry. He noted that the disease profile of apple replant disease is compatible with a causal rôle for *Pythium*.

Other fungi

Penicillium janthinellum, *Constantinella terrestris*, *Peniophore sacrata*, *Penicillium claviforme* and *Cylindrocarpon* spp. have all been reported to be associated with apple replant disease.

Bacteria

Bacillus subtilis and *Pseudomonas putida* have been found to stunt the growth of young apple trees.

RESISTANCE/TOLERANCE

Given the range of potential causal agents, resistance might seem unlikely. However, when sensitivity is assessed by response to soil fumigation, 'M.9' and 'MM.106' appeared to be particularly sensitive and more vigorous rootstocks especially 'M.25', 'M.2' and 'Crab C' less so. The effect is not solely associated with vigour; 'M.26' and 'M.7' appeared to be relatively tolerant (Jackson, 1979) and 'M.27' to have a high degree of field tolerance (Oehl and Jackson, 1979). Some rootstock clones produced by crossing 'M.27' with 'MM.106' (the AR series) also show appreciable field tolerance. This may imply a single dominant cause of ARD in the East Malling soils and it is notable that 'Northern Spy', which also showed some apparent tolerance there, gave much smaller responses to soil fumigation on an old apple orchard site in New Zealand than did 'MM.115', 'MM.106' and 'M.12' (Ryan, 1975).

CONTROL MEASURES

Even though other plants appear able to induce apple replant disease in the soil, the main problem is that of replanting on old apple orchard sites. This is because in most countries the best land for apple production has already grown that crop and because the disease is extremely persistent.

Use of broad spectrum fumigants such as chloropicrin (Jackson, 1979), methyl bromide and metam sodium (Smith, 1994) has proved to be very effective in increasing early growth and yield on old apple land. Formalin treatment may be equally effective, less expensive, easier to apply and effective at low temperatures (Sewell and White, 1979; Daemen, 1994). These methods are, however, being questioned on environmental grounds. Field steaming as a method of heat treatment has been shown to be promising (Moyls *et al.*, 1994). Fumigation or steaming are often best accompanied by P fertilization because of their negative effects on mycorrhizal fungi.

Use of more vigorous rootstocks at a spacing more relevant to a dwarfing rootstock is practised, but a vigorous rootstock checked by replant disease does not have the beneficial effects of 'M.9' on precocity and fruit size. Some of the AR rootstocks may both prove relatively tolerant and convey beneficial effects on growth and cropping.

There is some evidence that biological control of the replant problem can be achieved by application of strain EBW4 of *Bacillus subtilis* (Utkhede and Smith, 1994b) and that mixing some organic substrates in the planting hole may give very beneficial results (Szczygiel and Zepp, 1998). The latter effect may in part result from modifications of the soil microflora.

References

Adams, A.N. (1988a). Apple chat fruit disease. In *European Handbook of Plant Diseases*, ed. I.M. Smith, J. Dunez, D.H. Phillips, R.A. Lelliot and S.A. Archer, p. 118. Oxford: Blackwell Scientific Publications.

Adams, A.N. (1988b). Apple rubbery wood disease. In *European Handbook of Plant Diseases*, ed. I.M. Smith, J. Dunez, D.H. Phillips, R.A. Lelliot and S.A. Archer, pp. 120–1. Oxford: Blackwell Scientific Publications.

Aldwinkle, H.S. and Beer, S.V. (1979). Fire blight and its control. *Horticultural Reviews* **1**, 423–76.

Alston, F.H. (1967). Varietal response to *Gloeosporium perennans* in the apple. *Report of the East Malling Research Station for 1966*, 132–4.

Anon. (1992a). Codling moth. In *Crop Pests in the U.K.*, ed. M. Gratwick, pp. 108–11. London: Chapman and Hall.

Anon. (1992b). Apple and pear suckers. In *Crop Pests in the U.K.*, ed. M. Gratwick, pp. 11–15. London: Chapman and Hall.

Anon. (1992c). Fruit tree red spider mite. In *Crop Pests in the U.K.*, ed. M. Gratwick, pp. 341–6. London: Chapman and Hall.

Avery, D.J. (1964). Carbon dioxide exchange by plum and apple leaves damaged by fruit tree red spider mite. *Report of the East Malling Research Station for 1963*, 94–7.

Avery, D.J. and Briggs, J.B. (1968). The aetiology and development of damage in young fruit trees infested with fruit tree red spider mite, *Panonychus ulmi* (Koch). *Annals of Applied Biology* **61**, 277–88.

Bell, R.L. and van der Zwet, T. (1999). Breeding for host resistance to pear psylla: evaluation of parental germplasm. *Acta Horticulturae* **484**, 471–5.

Bell, R.L., Quamme, H.A., Layne, R.E.C. and Skirvin, R.M. (1996). Pears. In *Fruit Breeding*, Vol. 1. *Tree and Tropical Fruits*, ed. J. Janick and J.N. Moore, pp. 441–514. New York: John Wiley and Sons.

Benic, L.M. and Combrink, J.C. (1996). Control of post-harvest diseases. In *Integrated Management of Post-harvest Quality*, ed. J.C. Combrink, pp. 57–63. Stellenbosch: Infruitec.

Billing, E. (1974). The effect of temperature on the growth of the fire blight pathogen, *Erwinia amylovora*. *Journal of Applied Bacteriology* **37**, 643–8.

Billing, E. (1996). BIS95, An improved approach to fire blight risk assessment. *Acta Horticulturae* **411**, 121–6.

Billing, E. (1999). Fire blight risk assessment: Billing's Integrated System (BIS) and its evaluation. *Acta Horticulturae* **489**, 399–405.

Billing, E. (2000). Fire blight risk assessment systems and models. In *Fire Blight: the Disease and its Causative Agent, Erwinia amylovora*, ed. J.L. Vanneste, pp. 293–318. Wallingford, UK: CAB International.

Bonn, W.G. and van der Zwet, T. (2000). Distribution and economic importance of fire blight. In *Fire Blight: the Disease and its Causative Agent Erwinia amylovora*, ed. J.L. Vanneste, pp. 37–53. Wallingford, UK: CAB International.

Breth, D.I., Bentley, A., Momol, M.T. and Aldwinkle, H.S. (1999). Alternative control measures for blossom blight in commercial apple and pear orchards. *Acta Horticulturae* **489**, 559–64.

Brown, A.G. (1975). Apples. In *Advances in Fruit Breeding*, ed. J. Janick and J.N. Moore, pp. 3–37. West Lafeyette: Purdue University Press.

Burchill, R.T. and Edney, K.L. (1972). An assessment of some new treatments for the control of rotting of stored apples. *Annals of Applied Biology* **72**, 249–55.

Butt, D.J. (1988). *Podosphaera leucotricha* (Ell & Ev) Salmon. In *European Handbook of Plant Diseases*, ed. I.M. Smith, J. Dunez, D.H. Phillips, R.A. Lelliot and S.A. Archer, pp. 263–5. Oxford: Blackwell Scientific Publications.

Butt, D.J., Martin, K.J. and Swait, A.A.J. (1983). Apple powdery mildew: damage, loss and economic injury level. *Proceedings of the 10th International Congress of Plant Pathology*, Brighton, 1, 184.

Byrde, R.J.W. (1977). Fungal diseases of fruit trees. *Scientific Horticulture* **29**, 14–18.

Camilo, A.P., Lamb, R.C. and Aldwinkle, H.S. (1988). Genetic resistance to bitter rot incited by *Glomerella cingulata* (Stoneman) Spaulding & von Schrenk in apple (*Malus domestica* Borkh). *Acta Horticulturae* **232**, 37–45.

Campbell, A.I. (1977). Improved planting material. *Scientific Horticulture* **28**, 151–4.

Cropley, R. (1967). Decline and death of pear on quince rootstocks caused by virus infection. *Journal of Horticultural Science* **42**, 113–15.

Cummins, J.N. and Aldwinkle, H.S. (1974). Breeding apple rootstocks. *HortScience* **9**, 367–72.

Daeman, E. (1994). The use of formalin to control replant problems of fruit trees. *Acta Horticulturae* **363**, 191–3.

Davies, D.L. and Eyre, S. (1996). Detection of phytoplasmas associated with pear decline in pear psyllids by polymerase chain reaction. *BCPC Symposium Proceedings* No. 65, Diagnostics in Crop Production, 67–72.

Davies, D.L., Barbara, D.J. and Clark, M.F. (1995). The detection of MLOs associated with pear decline in pear trees and pear psyllids by polymerase chain reaction. *Acta Horticulturae* **386**, 484–8.

Davies, D.L., Guise, C.M., Clark, M.F. and Adams, A.N. (1992). Parry's disease of pears is similar to pear decline and is associated with mycoplasma-like organisms transmitted by *Cacopsylla pyricola*. *Plant Pathology* **41**, 195–203.

Delbos, R.P. and Dunez, J. (1988). Apple chlorotic leaf spot virus (ACLSV). In *European Handbook of Plant Diseases*, ed. I.M. Smith, J. Dunez, D.H. Phillips, R.A. Lelliot and S.A. Archer, pp. 5–7. Oxford: Blackwell Scientific Publications.

Ferree, D.C., Hall, F.R. and Ellis, M.A. (1986). Influence of mites and diseases on net photosynthesis of apple leaves. In *The Regulation of Photosynthesis in Fruit Trees*, ed. A.N. Lakso and F. Lenz. Symposium Proceedings Publication. New York State Agricultural Experiment Station, Geneva, N.Y.

Geldart, H.G. (1994). The impact of replant problems on the economics of high density apple plantings. *Acta Horticulturae* **363**, 11–18.

Gouk, S.C., Spink, M. and Laurenson, M.R. (1999). Firework, a windows-based computer program for prediction of fire blight on apples. *Acta Horticulturae* **489**, 407–12.

Hoestra, H. (1968). *Replant Diseases of Apple in The Netherlands*. Mededelingen Landbouwhogeschool, Wageningen, 68–13, 105 pp.

Jackson, J.E. (1973). Effects of soil fumigation on the growth of apple and cherry rootstocks on land previously cropped with apples. *Annals of Applied Biology* **74**, 99–104.

Jackson, J.E. (1979). Soil fumigation against replant disease of apple. In *Soil Disinfestation*, ed. D. Mulder, pp. 185–202. Developments in Agricultural and Managed-Forest Ecology, 6. Amsterdam: Elsevier.

Janick, J., Cummins, J.N., Brown, S.K. and Hemmatt, M. (1996). Apples. In *Fruit Breeding*, Vol. 1. *Tree and Tropical Fruits*, ed. J. Janick and J.N. Moore. New York: John Wiley and sons.

Kajiura, I. (1994). Nashi (Japanese pear). In *Horticulture in Japan*, ed. K. Konishi, S. Iwahori, H. Kitagawa and T. Yakuwa, pp. 40–7. Tokyo: Asakura Publishing Co. Ltd.

Knight, R.L. and Alston, F.H. (1968). Sources of field immunity to mildew (*Podosphaeria leucotricha*) in apple. *Canadian Journal of Genetics and Cytology* **10**, 294–8.

Knight, R.L., Briggs, J.B., Massee, A.M. and Tydeman, H.M. (1962). The inheritance of resistance to woolly aphid, *Eriosoma lanigerum* Hsmnn in the apple. *Journal of Horticultural Science* **37**, 207–18.

Kunze, L. (1988). Apple proliferation MLO. In *European Handbook of Plant Diseases*, ed. I.M. Smith, J. Dunez, D.H. Phillips, R.A. Lelliot and S.A. Archer, pp. 118–20. Oxford: Blackwell Scientific Publications.

Laurens, F. (1999). Review of the current apple breeding programmes in the world: objectives for scion cultivar improvement. *Acta Horticulturae* **484**, 163–9.

Lelliot, R.A. (1988). *Erwinia amylovora* (Burrill) Winslow *et al.* In *European Handbook of Plant Diseases*, ed. I.M. Smith, J. Dunez, D.H. Phillips, R.A. Lelliot and S.A. Archer, pp. 187–9. Oxford: Blackwell Scientific Publications.

Lespinasse, Y. and Aldwinkle, H.S. (2000). Breeding for resistance to fire blight. In *Fire Blight: the Disease and its Causative Agent* Erwinia amylovora, ed. J.L. Vanneste, pp. 253–73. Wallingford, UK: CAB International.

Lightner, G.W., van der Zwet, T. and Steiner, P.W. (1999). Fifteen year summary of the efficacy of the Maryblyt prediction system on apple in West Virginia (1984–1998). *Acta Horticulturae* **489**, 445–7.

Mai, W.F., Merwin, I.A. and Abawi, G.S. (1994). Diagnosis, etiology and management of replant disorders in New York cherry and apple orchards. *Acta Horticulturae* **363**, 33–41.

Manulis, S., Zutra, D., Kleitman, F., Dror, O., David, I. Zilberstaine, M. and Shabi, E. (1999). Streptomycin resistance of *Erwinia amylovora* in Israel and occurrence of fire blight in pear orchards in autumn. *Acta Horticulturae* **489**, 85–91.

Mills, W.D. (1955). Fire blight development on apple in western New York. *Plant Disease Reporter* **39**, 206–7.

Momol, M.T. and Saygili, H. (1999) Editors. *Proceedings of the Eighth International Workshop on Fire Blight. Acta Horticulturae* **489**, Leuven: ISHS.

Momol, M.T., Norelli, J.L. and Aldwinkle, H.S. (1999). Evaluation of biological control agents, systemic acquired resistance inducers and bactericides for the control of fire blight on apple blossom. *Acta Horticulturae* **489**, 553–7.

Moyls, A.L., Hocking, R.P., Neilsen, G.H. and Hogue, E.J. (1994). Apple tree growth response in greenhouse pot tests using heat-treated replant soil versus orchard replant trees with *in situ* heated soil.*Acta Horticulturae* **363**, 57–64.

Oehl, V.H. (1980). The long term effects of pre-planting soil fumigation on the growth and cropping of cvs 'Laxton's Fortune' and 'Cox's Orange Pippin' on 'M.2' rootstock, planted on land affected with a replant disease of apple. *Journal of Horticultural Science* **55**, 259–66.

Oehl, V.H. and Jackson, J.E. (1979). Evaluating rootstocks for resistance to apple replant disease. *Report of the East Malling Research Station for 1978*, 53.

Oostenbrink, M. and Hoestra, H. (1961). Nematode damage and specific sickness in *Rosa, Malus* and *Laburnum. Tijdschrift voor Plantenziekten* **63**, 264–72.

Otto, G. and Winkler, H. (1998). Influence of root pathogenic *Actinomycetes* on the trimming of the rootlets of some species of *Rosaceae* with root hairs. *Acta Horticulturae* **447**, 49–54.

Otto, G., Winkler, H. and Szabo, K. (1994). Proof of actinomycetes in rootlets of species of *Rosaceae* from a SARD soil – a contribution to the specificity of replant diseases. *Acta Horticulturae* **363**, 43–8.

Peterson, A.B. and Hinman, H. (1994). The economics of replanting apple orchards in Washington State. *Acta Horticulturae* **363**, 19–23.

Pitcher, R.S., Way, D.W. and Savory, B.M. (1966). Specific replant diseases of apple and cherry and their control by soil fumigation. *Journal of Horticultural Science* **41**, 379–96.

Posnette, A.F. (1977). Virus diseases of woody plants. *Scientific Horticulture* **29**, 7–13.

Pruyne, P.T., Merwin, I.A., Mullin, P.G. and Gibson, D.M. (1994). Diagnosis of apple replant problems in New York orchard soils and evaluation of nematode-suppressive cover crops. *Acta Horticulturae* **363**, 121–8.

Richter, J. (1988a). *Venturia inaequalis* (Cooke) Winter. In *European Handbook of Plant Diseases*, ed. I.M. Smith, J. Dunez, D.H. Phillips, R.A. Lelliot and S.A. Archer, pp. 385–6. Oxford: Blackwell Scientific Publications.

Richter, J. (1988b). *Venturia pirini* Alderhold. In *European Handbook of Plant Diseases*, ed. I.M. Smith, J. Dunez, D.H. Phillips, R.A. Lelliot and S.A. Archer, p. 386. Oxford: Blackwell Scientific Publications.

Robinson, T.L., Cummins, J.N., Hoying, S.A. and Smith, W.H. (1997). Commercial orchard evaluation of the New Cornell-Geneva Apple Rootstocks. *Acta Horticulturae* **451**, 113–19.

Robinson, T.L., Hoying, S.A., Cummins, J.N., Johnson, W.C., Norelli, J.L. and Aldwinkle, H.S. (1999). Orchard performance of fire blight-resistant Geneva apple rootstocks. *Acta Horticulturae* **489**, 287–94.

Roche, P., Brown, L.M., King, G.J., Alston, F.H., Evans, K.M, Maliepaard, C., van Heusden, S., Vrielink, R., Laurens, F., Dunemann, F., Markussen, T. and Tartarini, S. (1999). Identification and development of markers linked to aphid resistance in apple. *Acta Horticulturae* **484**, 519–22.

Ryan, C.L.J. (1975). Specific replant disease in Hawke's Bay. Part III. Apple rootstock evaluation for replant sites. *Orchardist of New Zealand* **48**, 191–3.

Savory, B.M. (1966). *Specific replant diseases, causing root necrosis and growth depression in perennial fruit and plantation crops.* Research Review. Commonwealth Bureau of Horticulture, East Malling, No. 1. 64 pp.

Savory, B.M. (1967). Specific replant diseases of apple and cherry. *Report of the East Malling Research Station for 1966*, 205–8.

Seemüller, E. (1988). Pear decline MLO. In *European Handbook of Plant Diseases*, ed. J.M. Smith, J. Dunez, D.H. Phillips, R.A. Lelliot and S.A. Archer, pp. 127–9. Oxford: Blackwell Scientific Publications.

Sewell, G.W.F. (1979). Reappraisal of the nature of the "specific replant disease" of apple. *Commonwealth Mycological Institute Review of Plant Pathology* **58** (6), 209–11.

Sewell, G.F.W. (1981). Effects of *Pythium* species on the growth of apple and their possible causal role in apple replant disease. *Annals of Applied Biology* **97**, 31–42.

Sewell, G.W.F. and Roberts, A.L. (1986). Reciprocal plantings of four species in their respective soils. *Report of the East Malling Research Station for 1985*, 119.

Sewell, G.W.F. and White, G.C. (1979). The effects of formalin and other soil treatments on the replant disease of apple. *Journal of Horticultural Science* **54**, 333–5.

Sewell, G.W.F. and Wilson, J.F. (1975). The role of *Thielaviopsis basicola* in the specific replant disorders of cherry and plum. *Annals of Applied Biology* **79**, 149–69.

Sewell, G.W.F., Preece, D.A. and Elsey, R.F. (1988). Apple replant disease: the influence of soil phosphorus and other factors on the growth responses of apple seedlings to soil fumigation with chloropicrin. *Annals of Applied Biology* **113**, 605–15.

Shtienberg, D., Kritzman, G., Herzog, Z., Openhaim, D., Zillberstein, M. and Blatchinsky, D. (1999). Development and evaluation of a decision support system for management of fire blight in Israel. *Acta Horticulturae* **489**, 385–92.

Smith, T.J. (1994). Successful management of orchard replant disease in Washington. *Acta Horticulturae* **363**, 161–7.

Smith, T.J. (1999). Report on the development and use of Cougarblight 98C – a situation-specific fire blight risk assessment model for apple and pear. *Acta Horticulturae* **489**, 429–36.

Szabo, K., Winkler, H., Petzold, H. and Marwitz, R. (1998). Evidence for the pathogenicity of *Actinomycetes* in rootlets of apple seedlings from soils conducive to specific apple replant disease. *Acta Horticulturae* **477**, 53–65.

Szczygiel, A. and Zepp, A.I. (1998). Results of pot experiments on control of apple replant disease. *Acta Horticulturae* **477**, 103–6.

Thompson, J.A. (1959). The occurrence of areas of poor growth in a fruit tree nursery. *Report of the East Malling Research Station for 1958*, 80–2.

Thomson, S.V., Schroth, M.N. and Moller, W.J. (1982). A forecasting model for fire blight of pear. *Plant Disease* **66**, 576–9.

Thomson, S.V., Schroth, M.N., Moller, W.J., Reil, W.O., Beutel, J.A. and Davis, C.S. (1977). Pesticide applications can be reduced by forecasting the occurrence of fireblight bacteria. *California Agriculture* **31**(10), 12–14.

Utkhede, R.S. (1998). Influence of cultural practices on the growth and yield of young apple trees planted in replant disease soil. *Acta Horticulturae* **477**, 27–38.

Utkhede, R.S. and Smith, E.M. (1994a). Biotic and abiotic causes of replant problems of fruit trees. *Acta Horticulturae* **363**, 25–32.

Utkhede, R.S. and Smith, E.M. (1994b). Development of biological control of apple replant disease. *Acta Horticulturae* **363**, 129–33.

van der Zwet, T. and Beer, S.V. (1995). Fire Blight – Its Nature, Prevention and Control: A Practical Guide to Integrated Disease Management. *US Department of Agriculture, Agriculture Information Bulletin* No. **631**, 97 pp.

van der Zwet, T., Zoller, B.G. and Thomson, S.V. (1988). Control of fire blight of pear and apple by accurate prediction of the blossom blight phase. *Plant Disease* **72**, 464–72.

van Oosten, H.J., Meijneke, C.A.R. and Peerbooms, H. (1982). Growth, yield and fruit quality of virus-infected and virus-free 'Golden Delicious' apple trees, 1968–1982. *Acta Horticulturae* **130**, 213–20.

Way, D.W. and Pitcher, R.S. (1971). Specific replant diseases of tree fruits and their control by soil fumigation. *Proceedings of the 6th British Insecticide and Fungicide Conference*, pp. 263–73.

Webster, A.D., Tobutt, K.R., James, D.A., Evans, K.M. and Alston, F.A. (1997). Rootstock breeding and orchard testing at Horticultural Research International East Malling. *Acta Horticulturae* **451**, 83–8.

White, A.G. and Bus, V.G. (1999). Breeding commercial apple cultivars in New Zealand with resistances to pests and diseases. *Acta Horticulturae* **484**, 157–61.

Willet, M., Smith, T.J., Peterson, A.B., Hinman, H., Stevens, R.G., Ley, T. and Tvergyak, P. (1994). A successful apple replant educational program in Washington State. *Acta Horticulturae* **363**, 153–9.

Zoller, B.G. and Sisevich, J. (1979). Blossom populations of *Erwinia amylovora* in pear orchards *vs* accumulated degree hours over 18.3 degrees Celsius, 1972–1976. *Phytopathology* **69**, 1050 (Abstr.)

14

Biotechnology of apples and pears

Propagation *in vitro*

Propagation *in vitro* provides a method of rapid propagation of clonal plant material. It has a number of specific uses in the production of apple and pear plants which extend and complement the traditional means by which these, both scions and rootstocks, are vegetatively propagated. It also provides the reliable and efficient regeneration systems from somatic tissues that are essential to the development of systems of transfer of individual genes in the process of genetic engineering.

Shoot culture involves the use of explants which may be nodal buds or shoot tips ranging from 0.3 mm to 1.0 cm. Explants are surface-sterilized, usually by washing in solutions of sodium or calcium hypochlorite. They are then cultured in a medium based on that of Murashige and Skoog (1962), containing mineral salts, sucrose, cytokinin and possibly some auxin and gibberellin, and solidified with agar. Sorbitol may be more effective than sucrose with some apple cultivars and phloridzin or its breakdown product phloroglucinol may increase shoot growth (Jones, 1993). Shoot cultures are maintained on the culture medium in illuminated growth rooms. Their axillary buds extend to give new shoots which are excised at approximately monthly intervals and transferred to a fresh medium where they in turn produce axillary shoots. Shoot culture lines may be multiplied indefinitely by sequential subculture.

Shoot cultures may become slow-growing with tightly-rolled translucent leaves. This condition is known as vitrification. It involves excess water uptake and inhibition of lignin and cellulose synthesis and is associated with low agar concentrations and a high ratio of ammonium to nitrate.

The production of plantlets is achieved by excising shoots and transferring them to a medium with IBA but no cytokinin in which adventitious roots are initiated. The plantlets are then transferred to pots of compost and grown on in a glasshouse. Alternatively, shoots from *in vitro* culture can be rooted directly in potting compost after being dipped in IBA powder.

Mass propagation of apple and pear rootstocks *in vitro* is, at present, relatively expensive compared with conventional vegetative propagation. It may play an important rôle in the rapid multiplication of new rootstock cultivars and their distribution in disease-free form, but under some circumstances *in-vitro*-produced rootstocks if used directly produce excessive numbers of burr-knots and suckers, possibly associated with enhanced juvenility.

Sequential shoot subculture may greatly increase the readiness with which the shoots root. This is apparently due to tissue rejuvenation (i.e. development of juvenile characteristics including ready-rooting, but also spinyness and slowness to bear fruit) and provides a means of increasing the rooting potential of normally difficult-to-root but otherwise desirable rootstocks.

The apparent rejuvenation that takes place *in vitro* persists following establishment of nursery hedges, and both winter and summer cuttings from such hedges of 'M.9' show improved rooting. Stoolbeds established from micropropagated (*in-vitro*-cultured) plants of 'M.27' produce between 1.5 and 5.0 times as many shoots as their conventional counterparts and these shoots are better rooted (Jones, 1993). Conventional cuttings from micropropagated plants of a very difficult-to-propagate *Pyrus communis* rootstock show greatly improved rooting (Jones and Webster, 1989). Scion cultivars which are difficult to root as cuttings can readily be produced on their own roots by *in vitro* culture. Some cultivars which are naturally compact and precocious in flowering appear to be satisfactory when self-rooted via *in vitro* culture, but in general self-rooted cultivars are too vigorous and are slower to come into crop than when grafted on precocity-inducing rootstocks (Webster *et al.*, 1985; Zimmerman and Steffens, 1996).

Leaf discs and leaf strips produce shoots *in vitro* when grown on an MS (Murashige & Skoog) basal medium supplemented with 2.0 mg BA l^{-1} and 0.5 mg NAA l^{-1} (James *et al.*, 1988). This is particularly important because leaf mesophyll cells are competent for transformation, i.e. the insertion of 'foreign' genes. An N6-based medium may also be used for leaf disc culture (Welander, 1988).

Regeneration systems for *Malus* have been developed from a wide range of other explants including embryo, nucellus, cotyledon, hypocotyl, immature seed, flower part, and anther tissues (Korban and Chen, 1992).

Somaclonal variation may express itself as a result of *in vitro* culture. Much of this variation reveals deleterious traits but some, e.g. in fireblight resistance (Chevreau *et al.*, 1998), is potentially useful.

Genetic transformation

Genetic transformation is a means of transferring a specific gene by biotechnology from an organism, which may or may not be of the same species, into a cultivar to achieve a specific improvement.

The basic processes involved are:

1 Identification of the desired gene.
2 Its isolation.
3 Creation of a modified genetic sequence by fusing the desired gene, a promoter sequence which controls the functioning of the gene and a marker gene which allows the gene's presence to be detected.
4 Multiplication of the recombinant sequence.
5 Insertion of copies of the desired gene into the cultivar.
6 Selection of the organisms which have taken up the gene as indicated by presence of the marker gene.
7 Multiplication of the modified plants.

Apples and pears show enormous genetic variation, with genes relating to tree vigour control, precocity of flowering, aspects of fruit quality and storage life, and resistance to many diseases and some pests already identified. Understanding of the mechanisms of these processes has also facilitated the selection of potentially useful genes from bacteria and insects. Some potentially useful gene transfers have also been developed semi-empirically.

James *et al.* (1989) and Mourgues *et al.* (1996) achieved transformation of apple and pear, respectively, using disarmed binary vectors with genes for kanomycin resistance transferred by *Agrobacterium tumefaciens*. Leaves from shoot cultures were inoculated either by cutting out discs or strips and placing these in the suspension of *A. tumefaciens* carrying the binary vector, or by wounding with a scalpel dipped in the bacterial suspension. Selection of transformed material was by growing on media containing kanomycin. Subsequent cloning was by the standard tissue culture techniques. Optimization of the different stages has been described by De Bondt *et al.* (1994, 1996). Most major apple cultivars, e.g. 'Braeburn', 'Delicious', 'Elstar', 'Fuji', 'Gala', 'Golden Delicious' and 'Jonagold' have been transformed with different bacterial constructs as have the pears 'Beurré Bosc', 'Comice', 'Conference' and 'Passe Crassane' (Norelli and Aldwinkle, 2000). Transformation using *Agrobacterium* spp. is the most widely used method of gene transfer to apples and pears. In the field situation *A. tumefaciens* causes crown gall disease and *A. rhizogenes* causes hairy root disease. They naturally infect the host plants and insert some of their own DNA, the transfer or T-DNA, into the chromosomal DNA of the infected plant cells. This makes them particularly useful for insertion of genes of interest.

As yet no commercial cultivars have been produced by transformation of apples or pears but transformed plants of interest have been produced along a number of different lines.

Rol gene transfer

Agrobacterium rhizogenes carries genes that alter the growth of infected plant cells. Some of the genes on the T-DNA within the *A. rhizogenes* R_i-plasmid control auxin and cytokinin levels, others alter sensitivity to these hormones (Schell *et al.*, 1995; Tamas, 1995). A group of these genes are known as the *rol* (root inducing locus) genes. A number of plant species regenerated from 'hairy root' cells following *A. rhizogenes* infection show shortened stem internodes, increased branching and reduced apical dominance. Plants transformed with single *rol* genes or combinations of these also exhibit altered growth characteristics. Transformed plants overexpressing the *rol*A gene show reduced growth and increased root development. Plants transformed with the *rol*B gene show reduced apical dominance, increased rooting and reduced leaf senescence, possibly because the *rol*B gene product alters auxin sensitivity. Plants overexpressing the *rol*C gene are more dwarfed, with reduced apical dominance, increased branching and increased root growth. The effects are most pronounced when plants are transformed with several *rol* genes in combination (Holefors, 1999). Some of the induced characteristics are potentially important with regard to tree size control for high-density orchards.

The problem of vigour control is particulary acute for pears when grown in conditions unsuitable for dwarfing quince rootstocks. Bell *et al.* (1999) established shoot proliferation cultures of 'Beurré Bosc' pear and transformed leaf explants from these using the disarmed *A. tumefaciens* strain EHAIOI containing pGA-GUSGF *rol*C, the *rol*C gene (ORF-12 in *A. rhizogenes*) being controlled by its native promoter. The inoculation of freshly harvested leaves was achieved by cutting them with sterile scalpel blades dipped in the bacterial suspension and the leaves co-cultivated prior to transfer to antibiotic selection plates (containing SIM, kanomycin and timentin), followed by growth on a shoot expression medium and subsequent shoot proliferation culture.

Transformation with the *rol*C gene was confirmed by DNA, RNA and protein blot analysis. The transformed pears, grown on in a greenhouse after budding on 'Bartlett' seedling rootstock, had fewer nodes, much shorter internodes and at 6 weeks were only about 20% of the height of the controls. Holefors *et al.* (1998) produced transformed clones of the apple rootstock 'M.26' by infection of *in-vitro*-produced leaves with *A. tumefaciens* strain GV3 101 containing a binary vector carrying the *npt*II gene and the *rol*A gene on the T-DNA. All transformed plants exhibited reduced stem growth compared with untransformed controls and two transformed clones showed reduced (*c*. 50%) dry weight, internode length and shoot and root dry weight.

Dwarfing gene transfer

The GA20 oxidase gene controls shoot elongation in apple stems. It has been cloned and inserted back into the apple cv. 'Greensleeves', producing trees identical with the parent in every way except that they are dwarfed (James, 2000).

Gene transfer to prolong storage life

Ethylene plays a key role in fruit ripening (see Chapter 10). Its precursor, methionine, is converted via *S*-adenosyl-L-methionine (Ado-met) to ACC by ACC synthase, and ACC is converted to ethylene by ACC oxidase. Ripening-associated ethylene production in fruits is regulated by both ACC synthase and ACC oxidase activities (Oetiker and Yang, 1995) and can be inhibited, in tomato, to varying degrees by creation of transgenic plants with down-regulation of either the oxidase or the synthase genes (Picton *et al.*, 1995).

In apple Castiglione *et al.* (1999) found that slow-ripening (e.g. 'Fuji') and rapid-ripening (e.g. 'Golden Delicious') cultivars had corresponding differences in ethylene production, and these have different allelic forms of the ACC oxidase gene. James (2000) found that adding an extra copy of the apple ACC synthase or oxidase genes to the cultivar 'Greensleeves' (a rapid-ripening progeny of 'Golden Delicious') resulted in the fruits, in many cases, producing much less ethylene than the controls.

Gene transfer for resistance to fungi

Plants which are infected by pathogens synthesize defence-related proteins such as PR proteins, and anti-microbial peptides including defensins and lipid transfer proteins. An anti-fungal defensin, Rs-AFP2, has been isolated from seeds of *Raphanus sativus* and an anti-microbial peptide, Ace-AMPI, from seeds of *Allium cepa*. The cDNAs encoding each of these have been cloned in a binary plant transformation vector and transferred to 'Jonagold' apple using *Agrobacterium*-mediated transformation. Protein extracts from transformed shoots showed an 8–32-fold increase in anti-fungal activity (Rs-AFP2 expressing shoots), or a four-fold increase in anti-fungal activity (Ace-AMPI expressing plants) relative to control plants (De Bondt *et al.*, 1999).

Trichoderma harzianum produces chitinolytic enzymes, endochitinase and chitobiosidase, which are active against a number of fungi: chitin being a major component of the cell walls of ascomycete and basidiomycete phytopathogens, though not of vertebrates or higher plants. Endochitinase and chitobiosidase inhibit spore germination and germ tube elongation of *Venturia inaequalis*, the apple scab pathogen. The cDNA of endochitinase has been cloned into vectors and the construct used to transform 'Royal Gala' apple using *Agrobacterium tumefaciens*. Some but not all of the transgenic 'Royal Gala' lines produced show increased scab resistance in greenhouse tests (Wong *et al.*, 1999).

Gene transfer for resistance to fire blight

Genes encoding proteins that lyse bacterial cells are found in the giant silk moth, *Hyalophora cecropia*. These are cecropin B and attacin E. Two 38-amino acid peptides, SB-37 and Shiva-1, are synthetic analogues of cecropin B. Hen egg white lysozyme is also potentially effective. The relevant genes have been transferred to 'M.7' rootstock and 'Royal Gala' apple scion using an *Agrobacterium tumefaciens* mediated leaf piece transformation system. Transformed shoots with the gene encoding attacin E have significantly increased resistance to fire blight caused by *Erwinia amylovora* (Norelli *et al.*, 1999). Transgenic clones of the pear cv. 'Passe Crassane' containing the attacin E gene show enhanced fire blight resistance (Reynoird *et al.*, 1999a,b). One transgenic line with the SB-37 cecropin transgene was significantly more resistant to fire blight than untransformed 'Royal Gala' (Norelli and Aldwinkle, 2000). The bacteriophage T4 lysozyme has been found to give increased fire blight resistance to transgenic 'Galaxy' apples.

Fire blight resistance in apple may be positively associated with sorbitol concentration, and transgenic apple trees with elevated levels of sorbitol synthesis have been produced. A key enzyme in the synthesis of sorbitol (S6PDH) has been cloned from apple (Norelli and Aldwinkle, 2000).

Gene transfer for resistance to insects

A gene encoding a cowpea trypsin inhibitor (*Cp*T1) that is anti-metabolic to a wide range of Lepidopteran and Coleopteran insects has been transferred to apple (James *et al.*, 1993). The *CryI*A(c) gene from *Bacillus thuringiensis*, which codes for endotoxins with insecticidal effects, has also been transferred to apple.

Use of transformed plants

No commercial cultivars have yet been developed using transformation. There is currently popular concern about the use of genetically modified food plants. Utilization of clones of major, well-accepted, commercial cultivars that have been genetically modified to overcome specific problems or incorporate novel benefits will therefore depend on consumer acceptance of the technology as well as on the comparative advantages of its products.

Molecular markers

The long juvenile period of apples and pears and their strong self-incompatibility have limited the speed with which understanding of apple genetics has been gained and genetic improvement achieved. Molecular markers are showing promise as tools for the detection of important genes without the need to maintain progenies for years in the field. Markers have been identified which are linked to genes relevant to important apple breeding targets.

Isoenzyme studies linked LAP-2 to a mildew resistance gene (Manganaris and Alston, 1992) and PGM-1 to V_f scab resistance (Manganaris *et al.*, 1994). PCR-based markers have been linked to the powdery mildew resistance gene Pl_1 (Markussen *et al.*, 1995), and SCAR and CAPS markers developed from RAPDs linked to V_f scab resistance (Gianfranceschi *et al.*, 1996). Roche *et al.* (1999) identified molecular markers linked tightly to the gene Sd_1 for resistance to rosy leaf curling aphid.

Ethylene production, which is involved in fruit ripening and can determine storage life in most Japanese pears, has been clarified at gene level. DNA markers in 1–3-month-old seedlings can be used to predict the maturation time and storage potential of their fruits (Itai *et al.*, 1999).

References

Bell, R.L., Scorza, R., Srinivasan, C. and Webb, K. (1999). Transformation of 'Beurré Bosc' pear with the *rolC* gene. *Journal of the American Society for Horticultural Science* **124**, 570–4.

Castiglione, B.P., Pirola, B., Sala, F., Ventura, M., Pancaldi, M. and Sansavini, S. (1999). Molecular studies on ACC synthase and ACC oxidase genes in apple. *Acta Horticulturae* **484**, 305–9.

Chevreau, E., Brisset, M.N., Paulin, J.P. and James, D.J. (1998). Fire blight resistance and genetic trueness-to-type of four somaclonal variants from the apple cultivar Greensleeves. *Euphytica* **104**, 199–205.

De Bondt, A., Eggermont, K., Druart, P., De Vil, M., Goderis, I., Vanderleyden, J. and Broekaert, W.F. (1994). *Agrobacterium*-mediated transformation of apple (*Malus* × *domestica* Borkh): an assessment of factors affecting gene transfer efficiency during early transformation steps. *Plant Cell Reports* **13**, 587–93.

De Bondt, A., Eggermont, K., Penninckx, I., Goderis, I. and Broekaert, W.F. (1996). *Agrobacterium*-mediated transformation of apple (*Malus* × *domestica* Borkh): an assessment of factors affecting regeneration of transgenic plants. *Plant Cell Reports* **15**, 549–54.

De Bondt, A., Zaman, S., Broekaert, W., Cammue, B. and Keulemans, J. (1999). Genetic transformation of apple (*Malus pumila* Mill) for increased fungal resistance: *in-vitro* antifungal activity in protein extracts of transgenic apple expressing *RS*-AFP2 or *ACE*-AMPI. *Acta Horticulturae* **484**, 565–70.

Gianfranceschi, L., Koller, B., Seglias, N., Kellerhals, M. and Gessler, C. (1996). Molecular selection in apple for resistance to scab caused by *Venturia inaequalis*. *Theoretical and Applied Genetics* **93**, 199–204.

Holefors, A. (1999). *Genetic Transformation of the Apple Rootstock 'M.26' with Genes Influencing Growth Properties.* Doctoral Thesis, Alnarp, Swedish University of Agricultural Sciences.

Holefors, A., Xue, Z.T. and Welander, M. (1998). Transformation of apple rootstock 'M.26' with the *rolA* gene and its influence on growth. *Plant Science* **136**, 69–78.

Itai, A., Kawata, T., Tanabe, K., Tamura, F., Uchiyama, M., Tomomitsu, M. and Shiraiwa, N. (1999). Identification of 1-aminocyclopropane-1-carboxylic acid synthase genes controlling the ethylene level of ripening fruit in Japanese pear (*Pyrus pyrifolia* Nakai). *Molecular & General Genetics* **261**, 42–9.

James, D.J. (2000). Genetic modification of temperate fruit crops–why we do it and why it's necessary. *British Association Festival of Science,* Imperial College, London, 8 September, 2000.

James, D.J., Passey, A.J., Barbara, D.J. and Bevan, M. (1989). Genetic transformation of apple (*Malus pumila* Mill) using a disarmed Ti-binary vector. *Plant Cell Reports* **7**, 658–61.

James, D.J., Passey, A.J. and Rugini, E. (1988). Factors affecting high frequency plant regeneration from apple leaf tissues cultured *in-vitro*. *Journal of Plant Physiology* **132**, 148–54.

James, D.J., Passey, A.J., Webster, A.D., Barbara, D.J., Viss, P., Dandekar, A.M. and Uratsu, S. (1993). Transgenic apples and strawberries: advances in transformation, introduction of genes for insect resistance and field studies of tissue cultured plants. *Acta Horticulturae* **336**, 179–84.

Jones, O.P. (1993). Propagation of apple *in-vitro*. In *Micropropagation of Woody Plants*, ed. M.R. Ahuja, pp. 169–86. Dordrecht: Kluwer.

Jones, O.P. and Webster, C.A. (1989). Improved rooting from conventional cuttings taken from micropropagated plants of *Pyrus communis* rootstocks. *Journal of Horticultural Science* **64**, 429–34.

Korban, S.S. and Chen, H. (1992). Apple. In *Biotechnology of Perennial Fruit Crops*, ed. F.A. Hammerschlag and R.E. Litz, pp. 203–27. Wallingford, UK: CAB International.

Manganaris, A.G. and Alston, F.H., (1992). Genetics of leucine aminopeptidase in apple. *Theoretical and Applied Genetics* **83**, 345–52.

Manganaris, A.G., Alston, F.H., Weeden, N.F., Aldwinkle, H.S., Gustafson, H.L. and Brown, S.K. (1994). Isoenzyme locus PGM-1 is tightly linked to a gene (V_f) for scab resistance in apple. *Journal of the American Society for Horticultural Science* **119**, 1286–8.

Markussen, T., Kruger, J., Schmidt, H. and Dunemann, F. (1995). Identification of PCR-based markers linked to the powdery-mildew-resistance gene Pl_1 from *Malus robusta* in cultivated apple. *Plant Breeding* **114**, 530–4.

Mourgues, F., Chevreau, E., Lambert, C. and De Bondt, A. (1996). Efficient *Agrobacterium*-mediated transformation and recovery of transgenic plants from pear (*Pyrus communis* L.). *Plant Cell Reports* **16**, 245–9.

Murashige, T. and Skoog, F. (1962). A revised medium for rapid growth and bioassays with tobacco tissue cultures. *Physiologia Plantarum* **15**, 473–97.

Norelli, J.L. and Aldwinkle, H.S. (2000). Transgenic varieties and rootstocks resistant to fire blight. In *Fire Blight: the Disease and its Causative Agent, Erwinia amylovora*, ed. J.L. Vanneste, pp. 275–92. Wallingford, UK: CAB International.

Norelli, J.L., Mills, J.Z., Jensen, L.A., Momol, M.T. and Aldwinkle, H.S. (1999). Genetic engineering of apple for increased resistance to fire blight. *Acta Horticulturae* **484**, 541–5.

Oetiker, J.H. and Yang, S.F. (1995). The role of ethylene in fruit ripening. *Acta Horticulturae* **398**, 167–78.

Picton, S., Gray, J.E. and Grierson, D. (1995). Ethylene genes and fruit ripening. In *Plant Hormones*, ed. P.J. Davies, pp. 372–94. Dordrecht: Kluwer.

Reynoird, J.P., Mourgues, F., Aldwinkle, H.S., Brisset, M.N. and Chevreau, E. (1999*a*). First evidence for differences in fire blight resistance among transgenic pear clones expressing Attacin E. gene. *Acta Horticulturae* **489**, 245–6.

Reynoird, J.P., Mourgues, F., Brisset, M.N. and Chevreau, E. (1999*b*). Expression of SB-37 gene in transgenic pears enhanced resistance to fire blight. *Acta Horticulturae* **489**, 243–4.

Roche, P., Brown, L.M., King, G.J., Alston, F.H., Evans, K.M., Maliepaard, C., van Heusden, S., Vrielink, R., Laurens, F., Dunemann, F., Markussen, T. and Tartarini, S. (1999). Identification and development of markers linked to aphid resistance in apple. *Acta Horticulturae* **484**, 519–22.

Schell, J., Palme, K. and Walden, R. (1995). Molecular approaches to the study of the mechanism of action of auxins. In *Plant Hormones*, ed. P.J. Davies, pp. 354–71. Dordrecht: Kluwer.

Tamas, I.A. (1995). Hormonal regulation of apical dominance. In *Plant Hormones*, ed. P.J. Davies, pp. 572–97. Dordrecht: Kluwer.

Webster, A.D., Oehl, V.H., Jackson, J.E. and Jones, O.P. (1985). The orchard establishment, growth and precocity of four micropropagated apple scion cultivars. *Journal of Horticultural Science* **60**, 169–80.

Welander, M. (1988). Plant regeneration from leaf and stem segments of shoots raised *in-vitro* from mature apple trees. *Journal of Plant Physiology* **132**, 738–44.

Wong, K.W., Harman, G.E., Norelli, J.L., Gustafson, H.L. and Aldwinkle, H.S. (1999). Chitinase-transgenic lines of 'Royal Gala' apple showing enhanced resistance to apple scab. *Acta Horticulturae* **484**, 595–9.

Zimmerman, R.H. and Steffens, G.L. (1996). Long-term evaluation of micropropagated apple trees: vegetative growth, cropping and photosynthesis. *Scientia Horticulturae* **66**, 69–76.

Cultivar Index

Cultivar differences in morphology and physiology are discussed throughout the book in the appropriate chapters. Only the descriptions and some of the most important features are listed here.

General Index

For EU product safety concerns, contact us at Calle de José Abascal, 56–1°, 28003 Madrid, Spain or eugpsr@cambridge.org.

www.ingramcontent.com/pod-product-compliance
Lightning Source LLC
LaVergne TN
LVHW090256210926
848480LV00001B/38